POTATO
Health Management

Second Edition

Edited by

Dennis A. Johnson
Department of Plant Pathology
Washington State University, Pullman

Steering Committee

Mary L. Powelson
Department of Botany and Plant Pathology
Oregon State University, Corvallis

Randall C. Rowe
Department of Plant Pathology
Ohio State University, Wooster

Walter R. Stevenson
Department of Plant Pathology
University of Wisconsin, Madison

Robert E. Thornton
Department of Horticulture and Landscape Architecture
Washington State University, Pullman

APS PRESS
The American Phytopathological Society
St. Paul, Minnesota, U.S.A.

Front cover photograph by William J. Schaefer, used by permission
Back cover photograph (left) by Dennis A. Johnson

Mention of various pesticides is made throughout this book. To the best knowledge of the editor, these pesticides were registered for use on potatoes in the United States or Canada at the time of this writing. Pesticide labels are constantly changing, and thus current labels must be in hand at the time of pesticide use, and applications must be made in strict accordance with these labels. Mention of specific pesticides in this book is to be construed not as a recommendation or endorsement of their use but rather as information for readers of this book to consider when devising their own potato health management plans.

Reference in this publication to a trademark, proprietary product, or company name by personnel of the U.S. Department of Agriculture or anyone else is intended for explicit description only and does not imply approval or recommendation to the exclusion of others that may be suitable.

Library of Congress Control Number: 2007935822
International Standard Book Number: 978-0-89054-353-5

First edition published 1993
Second edition published 2008

Printed in the United States of America on acid-free paper

The American Phytopathological Society
3340 Pilot Knob Road
St. Paul, Minnesota 55121, USA

Dedication

The Land-Grant Mission

The Morrill Land-Grant College Act of 1862, which established a state university or land-grant college in every state, was possible only because of foresight and persistence. It initiated a series of legislation that revolutionized higher education and extended research-based knowledge and technology to agriculture and industry nationwide. In the midst of the Civil War, which preoccupied the nation from 1861 to 1865, as the young republic fought for survival, the Morrill Act was passed by Congress and signed by Abraham Lincoln, setting a course for true change—extending higher education to the "common people" and enabling agricultural and industrial production to achieve levels never before envisioned.

The passage of the bill was the culmination of a long and often bitter fight. The land-grant idea was championed throughout the North by farm organizations, labor unions, educators, preachers, newspapers, and public-spirited reformers. One of the first supporters was Jonathan Baldwin Turner, of Illinois, who espoused the creation of "industrial universities" financed with land endowments to each state. Later, Horace Greeley's *New York Tribune* advocated a system of people's colleges. Hard-driving New York businessman Ezra Cornell threw himself into the educational reform movement, with the belief that higher education should be a democratic possession. Thomas Clemson, son-in-law of John C. Calhoun, added support from the South. Justin Morrill, of Vermont, authored the legislation (Nevins, 1962).

Land-grant institutions replaced the old-style classical curriculum with one that championed science and included agriculture and mechanical arts. Higher education, which had been accessible primarily to the sons of the leisure class, government leaders, and members of the professions, was opened to the common people. It has been said that no true university existed in the United States before the land-grant system was created (Nevins, 1962). The land-grant philosophy evolved further through legislation to fund scientific and agricultural research at state agricultural experiment stations (the Hatch Act of 1887) and to extend new knowledge to farmers and industry through the Cooperative Extension Service (the Smith-Lever Act of 1914). A second Morrill Act, in 1890, provided college access to blacks, and legislation in 1994 established Native American tribal colleges and funded universities serving Hispanic students.

Greater educational access and equality of opportunity were the primary goals of the land-grant movement. The principal motivation for change was the belief that every person should have the opportunity to complete an education as his or her tastes and abilities warranted. Education is not just for aristocrats and the wealthy; it is for all people. At the heart of the land-grant movement was the idea that liberty and equality could not survive unless all people had the opportunity to pursue all occupations at the highest practicable level (Nevins, 1962).

As a result of the land-grant college acts, there is at least one land-grant institution in every state and territory of the United States and the District of Columbia. Out of nearly 100 major state universities and colleges in the United States, about 70 are land-grant institutions. They enroll the majority of students and award the majority of all graduate degrees in the United States (Nevins, 1962). Agricultural experiment stations and extension programs in every state have played major roles in promoting scientific knowledge and increasing agricultural production. The land-grant system, along with favorable soils, water, and climate, a free-market system, and a hard-working people, has been a major factor in the development of America's tremendous agricultural strength.

Society and agriculture have changed greatly since the land-grant system was created. The United States is now mostly urban, with less than 2% of the population involved in farming. Educational and employment opportunities are vastly more diverse. Agriculture has become a major business, rather than a way of life. Research is focused on basic rather than applied topics. Diminishing proportions of federal and state money currently support the land-grant system. For example, less than 27% of budgeted funds at Washington State University in 2005 was provided by state appropriations. Students pay a relatively high percentage of the cost of education, and private sources and federal grants are required to fund research and to extend new technologies and information. Less financial support from federal and state sources is projected for the future. Land-grant universities are becoming privatized, in a reversion to the pre-land-grant era. The land-grant system may all be history in the near future.

Commitment is needed for the land-grant mission to continue and to build on its success—commitment from land-grant institutions and those who work there to continue to pursue research that yields economic and social returns while maintaining environmental quality, along with commitment from state and federal decision makers to provide adequate and consistent funding for university instruction, agricultural and natural resource research, and the extension of technology to agriculture and business.

This book is dedicated to the future and continued success of the land-grant mission and American agriculture.

Reference

Nevins, Allan. 1962. The Origins of the Land-Grant Colleges and State Universities: A Brief Account of the Morrill Act of 1862 and Its Results. Civil War Centennial Commission, Washington, DC.

Preface

Potatoes have been cultivated in the Andean highlands of South America for at least 8,000 years and were a staple crop of the native peoples of Peru when the Spanish explorers arrived in the mid-1500s. The agriculture of the Incas and their predecessors was sustained by planting a diversity of crop species and multiple potato cultivars; practicing fallowing, crop rotation, and multiple cropping; adding organic amendments to the soil; planting on raised beds; using sanitation practices and tillage; and altering the time of planting to avoid pest outbreaks (Thurston, 1994). These practices were discovered through trial and error and intentionally integrated over time to support a sustained food supply for human survival (Thurston, 1994). Today, successful agriculture requires more uniform products and energy-intensive inputs. The fundamental goals of both past and present production strategies, however, are similar: stability, resilience, productivity, and efficiency. Agricultural systems of the past are much different and cannot feed the millions of people that modern agriculture currently does, except we are now beginning to recognize the value of sound, sustainable practices, similar to those of the Andean people thousands of years ago.

The potato has hundreds of pests (weeds, deleterious insects, and rodents) and pathogens (disease-causing agents, including nematodes, fungi, oomycetes, bacteria, phytoplasmas, and viruses). Many pest and disease problems have followed potatoes to wherever they are grown, because the potato is vegetatively propagated from tubers, which easily carry some pathogens and pests. Some pathogens and pests were transported to new locations as the potato was moved from South America throughout the world.

The late blight pathogen migrated on seed tubers throughout the world on several different occasions. Evidence indicates that the causal organism originated in central Mexico, and the A1 mating type was carried to the eastern United States, eastern Canada, and northern Europe in the early 1840s. Late blight was unknown in Europe until it devastated the Irish potato crop in 1845, causing the infamous Irish Potato Famine. The moist, mild weather typical of Ireland favored the development of late blight, which spread like wildfire first in the latter part of the 1845 growing season and then in full force during much of the 1846 season. Nine-tenths of the potato crop in Ireland was destroyed, and the epidemic spread across Europe. Additional crop failures in following years further extended the devastation. One million Irish died, and another 1.5 million emigrated to North America as a result of the famine. More recently, migrations of the A2 mating type likely occurred in the late 1970s into Europe and in the 1990s into North America (Fry and Goodwin, 1997).

In contrast to the early agricultural system in the highlands of Peru, where several staple crops and multiple potato cultivars were grown, the Irish relied on a single, genetically uniform staple crop. Crop uniformity and a lack of understanding of the potato plant, the pathogen, and how both are affected by the environment intensified the Irish late blight epidemic.

Modern agriculture has made great strides since the Irish Potato Famine. However, late blight and many other diseases and pests are still not adequately managed. Many problems challenge modern agriculture. Today, not only are satisfactory yields, quality, and markets needed, but also human health and the environment must be protected. North American agriculture competes in a global market, with fierce competition from areas of the world with inexpensive labor. Energy inputs in agriculture are high and expensive. Markets are sometimes restricted because of political decisions. Environmental conditions and legal decisions dictate that production practices be modified to protect human health and the environment. All these factors are pushing modern agriculture toward more management-intensive systems.

Potato production is no exception. To thrive, production management strategies must be improved at all levels. A better understanding of both the potato plant and the underlying principles of modern pest and disease management is needed, along with an appreciation of how various factors interact to affect the overall health of the potato crop. A comprehensive, integrated strategy for potato health management must be implemented that results in a high-quality product, reasonable profit, and environmental health. This is **holistic health management**, as explained in Chapter 1.

A holistic approach to potato health management is emphasized throughout this book. Our goals are twofold: first, to provide the most current information on potato production practices, with an emphasis on pest and disease management, and, second, to provide a knowledge base so that these practices can be integrated into a comprehensive management scheme in the context of today's agriculture.

Some constraints encountered in culturing and storing potatoes are still not adequately addressed, because of insufficient knowledge. A lack of scientific knowledge contributed to the Irish late blight epidemic; likewise, it is a contributing factor for the crop constraints we must currently manage. Research funded through the land-grant system and other sources has greatly enhanced agricultural production, but additional research is warranted if a stable agriculture system is to meet today's needs. For example, cultivars resistant to many insect pests and diseases are needed. Relatively few resistant potato cultivars are available, in comparison to small grains, in which extensive breeding efforts have produced a wealth of resistant cultivars that are widely grown. The relatively small number of resistant potato cultivars is not due to a lack of genetic resources in potato but rather is due to past decisions to manage potato pests and diseases with what were then relatively inexpensive pesticides with broad-spectrum activity. Profits from potatoes, per unit of area, are greater than profits from small grains, and thus pesticides have generally been more affordable for potato production, allowing breeders to focus their efforts on quality factors. Today, pesticides are more expensive, and society has come to realize that there are environmental benefits from utilizing host resistance. As a result, research is ongoing to develop potato cultivars with durable resistance to disease and insect pests and to learn how to culture, process, and market these new cultivars. Research is also needed in many other areas, such as soil health and how to promote it. The challenge today is to keep potato production viable and strong through research and to apply the new knowledge in a holistic approach in managing potato health.

Dennis A. Johnson and Randall C. Rowe

References

Fry, W. E., and Goodwin, S. B. 1997. Re-emergence of potato and tomato late blight in the United States. Plant Disease 81:1349–1357.

Thurston, H. D. 1994. Andean potato culture: 5,000 years of experience with sustainable agriculture. Pages 6–13 in: Advances in Potato Pest Biology and Management. G. W. Zehnder, M. L. Powelson, R. K. Jansson, and K. V. Raman, eds. American Phytopathological Society, St. Paul, MN.

Acknowledgments

This book is the result of the dedicated efforts of many people. Planning began in the summer of 2004, when Rose Gergerich, Editor-in-Chief of APS Press, requested that the first edition of *Potato Health Management* be revised. The first edition was a tremendous success, with over 7,000 copies sold. However, new research had been conducted and several diseases and pests had emerged since the printing of the first edition, and it was time to include this information in a revision of the book. Randy Rowe edited the first edition, and it served as a solid foundation for this revision. I thank him for that publication.

The best decision I made as editor was in selecting the steering committee: Mary Powelson, Randy Rowe, Walt Stevenson, and Bob Thornton. These four are dedicated plant scientists with solid backgrounds in potato research and the potato industry, and each has keen editorial skills that helped me greatly. We initially met for a day-long meeting at the headquarters of the American Phytopathological Society (APS), in St. Paul, Minnesota, in October 2004. There we decided on the organization of the book, the contents, chapter topics, and the principal author of each chapter. Subsequently, we met various times in groups of two or three, and each member of the committee additionally met or corresponded with me on several occasions and offered extremely valuable counsel. Each member of the steering committee thoroughly reviewed and edited each chapter and then reread either portions of the book or the entire book as the manuscript was submitted for peer review. I greatly appreciate the advice and editorial suggestions given me by each of them.

For each chapter, a scientist knowledgeable about the topic was invited to be the principal author, and that person then selected co-authors when desired. I thank the authors for their contributions and most authors for meeting timelines. Tim Paulitz served as APS Press Senior Editor, reviewed the entire manuscript, and organized a peer review by two anonymous scientists. I appreciate their help and contributions. I thank Ryan Johnson and Brenda Schroeder for suggestions on specific chapters. I also thank Maegan Davis for drawing several of the illustrations and Erin and Frank Jones for clerical and technical help. The APS headquarters staff handled the production of the book, and I thank them for their dedicated service.

Dennis A. Johnson

Contents

POTATO HEALTH MANAGEMENT

CHAPTER 1

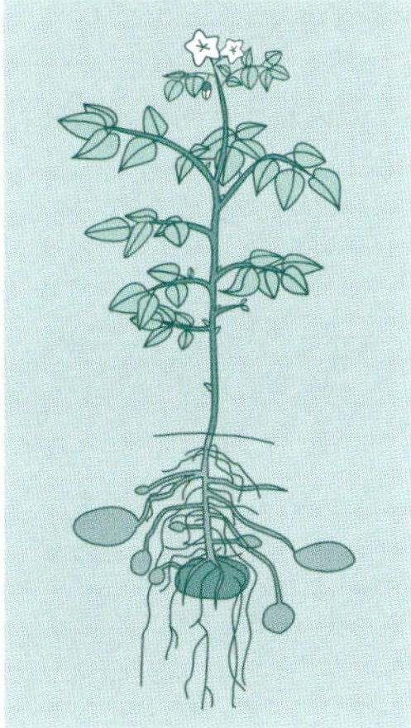

Randall C. Rowe
Department of Plant Pathology
Ohio State University, Wooster

Mary L. Powelson
Department of Botany and Plant Pathology
Oregon State University, Corvallis

Potato Health Management: A Holistic Approach

There is no single correct way to raise potatoes. Not only do local conditions and market requirements vary greatly, but biological constraints on production often differ considerably among fields and across geographic regions. Despite this variability, all potato producers face the challenge of developing and implementing management practices by which the health of the potato crop can be maximized in the face of threats from diseases, insects, and weeds.

This book is written for those involved in managing the health of potato crops—producers of commercial and seed potatoes, extension specialists, farm advisors, consultants, home gardeners, students, and anyone interested in growing healthy potatoes. The information presented here pertains directly to potato production in the United States and Canada, but much of it will be of value worldwide.

What Is Potato Health?

As a preliminary to a discussion of potato health management, an understanding of the concept of health is needed. In human terms, health means freedom from biological, environmental, and genetic factors that adversely affect bodily functions, daily activities, and personal well-being. If this concept is applied to potatoes, a healthy potato crop would be defined as one that is *free of biological constraints (insects, diseases, and weeds) and physical limitations (nutrition, water, and chemical and environmental stresses) that prevent the plants from achieving their full genetic potential.*

This definition is not adequate from a practical standpoint, since perfect potato health is seldom if ever achieved in the real world. Furthermore, although humans are valued as individuals, plants are usually valued only as a population (a crop) for the yield it produces. Since potatoes are grown to produce tubers, a healthy potato crop could be defined as one *free from any constraints that limit tuber yield or quality.*

This definition of potato health focuses on the final product, the tubers, but it is still unsatisfactory as a basis for management. A potato crop is grown for the economic benefit that is returned to the producer. Thus, a valid concept of potato health must include economic considerations. Many decisions are made when growing a crop, and each has a cost attached, which will affect the total profit from the crop. Total freedom from biological and physical constraints in potato production is not achievable. Nor is that really the issue, because there is some tolerance for these constraints before economic loss occurs. The foundation of effective potato health management is understanding how each constraint affects the economics of production and when it is appropriate to take action. Thus, from a practical standpoint, a healthy potato crop is one in which *biological and physical constraints are managed in such a way that economic gain can be maximized.*

Holistic Management of Potato Health

Holistic is a word that is not commonly used in discussing the management of potato crops. Implicit in a holistic viewpoint is that choices among management strategies cannot be made in isolation—the big picture must be considered. Each management practice must be evaluated not only for its direct effects on a particular biological or physical constraint but also for its potential indirect and unintended effects on the overall agricultural enterprise and on surrounding farms, communities, and natural resources. A holistic approach addresses the reality facing today's agriculture, where emphasis has shifted from maximizing total yield to maximizing total economic return with due regard for environmental protection and social responsibility.

Holistic management can be visualized as a pyramid in which the four sides represent the criteria that any production practice must satisfy: production efficiency, economic viability, environmental compatibility, and social responsibility (Fig. 1.1). Each practice must be effective and efficient in managing an identified constraint. Production practices must also be economically viable components of the total production system. Beyond being effective and affordable, each practice must not pose a threat to local or regional environments; must not jeopardize the health of workers, neighbors, and consumers; and must be acceptable to the local community. If all four criteria are satisfied, the pyramid will stand firm, whereas if even one is not met, the stability of the system is threatened.

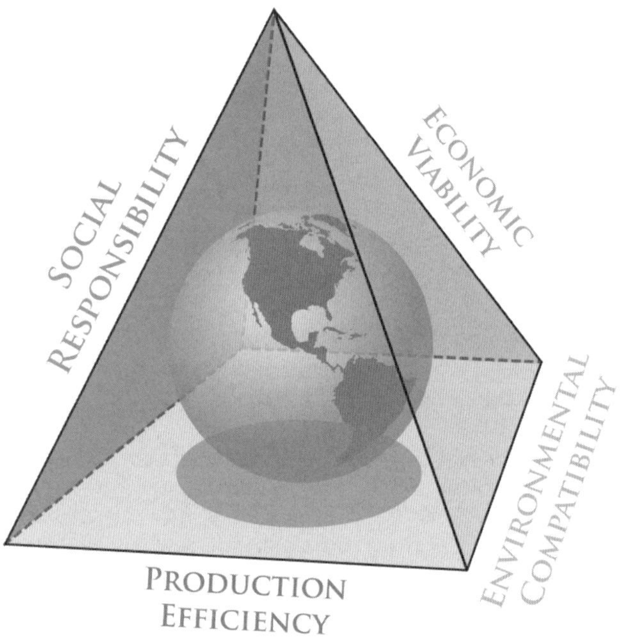

Fig. 1.1. Holistic potato health management visualized as a pyramid in which the four sides represent the criteria that any production practice must satisfy. When all four are satisfied, the pyramid will stand firm, whereas if even one is not met, the stability of the system will be threatened. (Adapted from a logo of the College of Food, Agricultural, and Environmental Sciences, Ohio State University)

Holistic potato health management is *the application of a totally integrated set of production practices that results in a quality crop produced in a manner that optimizes economic return and minimizes environmental impacts and undesirable societal effects.* A thorough understanding of the biology of the potato plant and the underlying principles of modern pest and disease management, together with an appreciation of how these factors interact to affect the overall health of the potato crop, is central to holistic management. With this knowledge, various production options can be evaluated and then assembled into an effective and appropriate holistic management plan.

Principles of Holistic Potato Health Management

Production of a healthy potato crop is a constantly moving target. Factors that affect potato health are continually changing, and effective management requires that production systems be continually adjusted to improve the management of existing constraints and to address emerging threats. Thus, practices must be selected that are effective in multiple ways and interlink appropriately, like pieces of a jigsaw puzzle (Fig. 1.2). Because of the biological complexity of a potato production system, changes in one management practice may cause **rebound effects** (Chapter 3) throughout the system and may even result in the emergence of a new and unanticipated problem. Recognizing that management practices are all interlinked is key to understanding holistic potato health management.

Effective management of potato health requires that production factors be manipulated to provide conditions

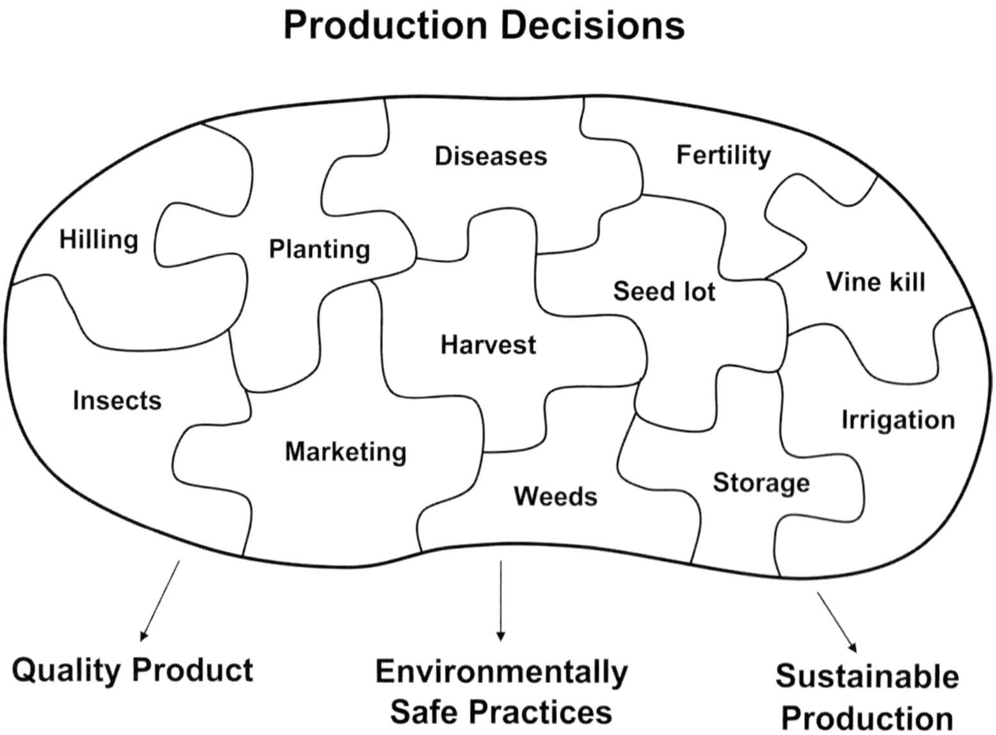

Fig. 1.2. Practices selected for a holistic potato health management plan must be effective in multiple ways and must interlink appropriately, like pieces of a jigsaw puzzle. Recognizing that management practices are all interlinked is key to understanding the concept of holistic health management. (Drawing by Maegan Davis)

that both optimize the crop environment and minimize the effects of pests and diseases. Once the requirements of the crop and the production limitations of the planting site are fully understood, management options can be selected and implemented to minimize stresses throughout the production cycle. A holistic health management plan includes decisions about crop rotation, site selection and land preparation, tillage, seeding practices, water and fertility management, and harvest and postharvest procedures. After these have been addressed, specific pest and disease management decisions are made and implemented before and as the crop develops.

Three concepts can be considered in pest and disease management decision making: the damage level, the economic loss level, and the action threshold.

The **damage level** is *the pest population or disease level that is of sufficient magnitude to limit crop yield or quality.* One Colorado potato beetle chewing on a potato leaf will not affect the yield of a crop, but a larger population can cause severe damage. The level at which a pest population becomes large enough to affect the yield or quality of the crop is the damage level for that pest.

Given that the most practical definition of potato health is one that is linked to maximizing economic gain from the crop, a second, more useful management concept is the **economic loss level** (often referred to as the **economic injury level** in insect pest management). This is *the pest population or disease level, usually somewhat beyond the damage level, at which the potential financial loss due to the pest or disease exceeds the cost of management.* Thus, to avoid a net financial loss, corrective action must be taken sometime *before* the economic loss level is reached. *The pest population or disease level at which action should be taken* is called the **action threshold** (often referred to as the **economic threshold** in insect pest management). Fig. 1.3 illustrates the relationship between the economic

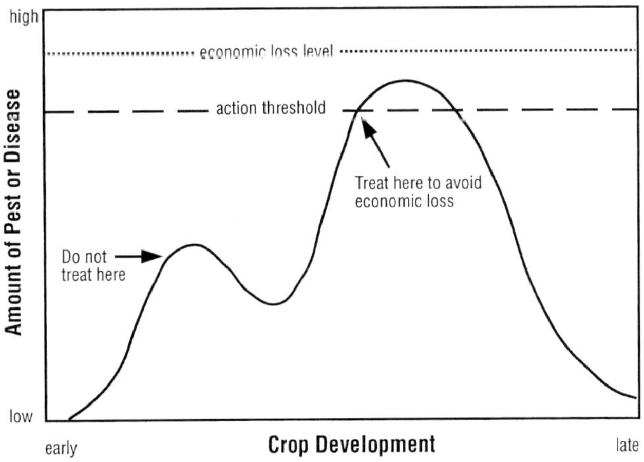

Fig. 1.3. Implementation of pest and disease management practices based on the concept of the economic loss level. The **economic loss level** is *the pest population or disease level at which the potential financial loss due to the pest or disease exceeds the cost of management.* The **action threshold** is *the pest population or disease level at which corrective action should be implemented to avoid reaching the economic loss level.* (Adapted from C. S. Hollingsworth, D. N. Ferro, and W. M. Coli, eds., 1986, Potato Production in the Northeast: A Guide to Integrated Pest Management, Publ. C-178, Massachusetts Cooperative Extension Service, Amherst)

loss level and the action threshold. As the crop develops, the population of a particular pest or the amount of a particular disease may go up or down. Beginning treatment too early results in an unnecessary production expense, whereas applying a treatment when the action threshold is reached prevents the potential loss from exceeding the cost of the treatment.

Building a Holistic Health Management Plan

An effective holistic health management plan will contain multiple integrated components selected on the basis of strategies that work together to protect the crop. Pest outbreaks and disease epidemics originate from small populations of organisms that increase into larger populations capable of causing economic loss. The population that begins the process may be introduced on seed tubers or equipment or by natural means, or it may already be present in a production field. This reservoir of pest and disease organisms is referred to as the **initial population** in the case of insects or weeds and as the **initial inoculum** (or **primary inoculum**) in the case of microbes that cause disease.

The first strategy for protecting potato health is to **exclude or evade** pests and disease organisms (Box 1.1). The goal is to avoid contact between the crop and organisms that might cause damage. Exclusionary practices are particularly effective in avoiding the introduction of pests and disease organisms that might be carried on or in seed tubers. Planting seed tubers produced under strict certification procedures is an important foundation of any health management plan. Other exclusionary practices include sanitary precautions to prevent the introduction of soilborne pests and disease organisms into uninfested lands (such as cleaning and sanitizing field equipment before moving it to a new production field), governmental regulations (such as quarantine and inspection of seed shipments), planting in uninfested production sites, and selection of appropriate planting and harvest times.

In many cases it is not possible to avoid contact with some organisms that may damage the crop. Thus, a second strategy for protecting potato health is to target the initial populations of pests or the initial inoculum of disease organisms. The goal is to **reduce initial populations** or **limit the effectiveness of initial populations** of pests and disease organisms that cannot be avoided (Box 1.1). Practices that reduce initial populations of weeds, insects, and disease organisms include crop rotation, soil fumigation, removal of long-term survival sites, prevention of weeds from going to seed, deep plowing of crop residues, and sanitation of equipment and facilities. Practices that limit the effectiveness of the initial inoculum of disease organisms include the selection of cultivars with some types of disease resistance, selection and preparation of planting sites to ensure satisfactory soil drainage, use of chemical seed piece treatments and other soil-applied fungicides, adjustment of soil pH, incorporation of soil amendments and green manure crops, early- and late-season irrigation management, planting at an appropriate seeding depth, adjustment of the planting date to avoid planting in cold soil, and harvest timing. Treatment with soil-applied insecticides, selection of cultivars resistant to insect feeding, and application of pheromones to disrupt mating can reduce the activity of initial populations of insects. Application of

Box 1.1

Health Protection Practices That Can Be Integrated into a Holistic Health Management System

Practices that EXCLUDE or EVADE pests and disease organisms

- Planting certified seed tubers
- Federal inspection of seed tubers
- Sanitation of field equipment between cropping sites
- Quarantines
- Site selection
- Timing of planting and harvest dates

Practices that REDUCE INITIAL POPULATIONS of pests and disease organisms

- Crop rotation
- Soil fumigation
- Removal of long-term survival sites
- Preventing weeds from going to seed
- Deep plowing of crop residues
- Sanitation of equipment and facilities

Practices that LIMIT THE EFFECTIVENESS OF INITIAL POPULATIONS of pests and disease organisms

- Selection and preparation of planting sites to ensure satisfactory soil drainage
- Use of chemical seed piece treatments
- Use of soil-applied pesticides
- Selection of cultivars resistant to certain diseases or insects
- Pre- and postemergence herbicide application
- Adjustment of soil acidity (pH)
- Incorporation of soil amendments or green manure crop residues
- Planting at an appropriate seeding depth
- Cultivation
- Early- and late-season irrigation management
- Adjustment of the planting date to avoid planting in cold soil
- Pheromone treatment to disrupt insect mating
- Timing of harvest

Practices that MINIMIZE THE RATE OF POPULATION GROWTH of pests and disease organisms

- Selection of cultivars resistant to certain diseases or insects
- Foliar pesticide treatment
- Row spacing and orientation
- Nutrition management
- Irrigation management
- Harvest management to minimize mechanical damage
- Storage environmental management

pre- and postemergence herbicides and cultivation can reduce competition from initial populations of weeds.

Once an insect or weed population is expanding or a disease epidemic has begun, management practices that **minimize the rate of population growth** of pests and disease organisms come into play (Box 1.1). Practices in this category include selection of cultivars with some types of resistance to insects or diseases, most foliar pesticide applications, cultural practices such as row spacing and orientation, management of fertility and irrigation, harvesting and handling practices that minimize mechanical damage, and management of the storage environment.

The challenge is to select an appropriate set of management practices and integrate them into a holistic health management plan that results in a quality crop and maximizes total economic return. Because all management options have associated costs, economic loss levels and action thresholds must be considered before implementation. The successful manager understands the genetic potential of the crop and the limitations of the production site and then makes a series of decisions that take into account the changing weather, pest populations, and disease levels. Economic, legal, and societal constraints also must be considered as they affect management options (Fig. 1.1).

Devising a holistic potato health management plan that will be effective in a specific production situation can best be approached in a systematic fashion (Fig. 1.4):

Step 1: Identify and prioritize the potato health problems at the production site. What problems have historically affected the crop at this site, and what new problems

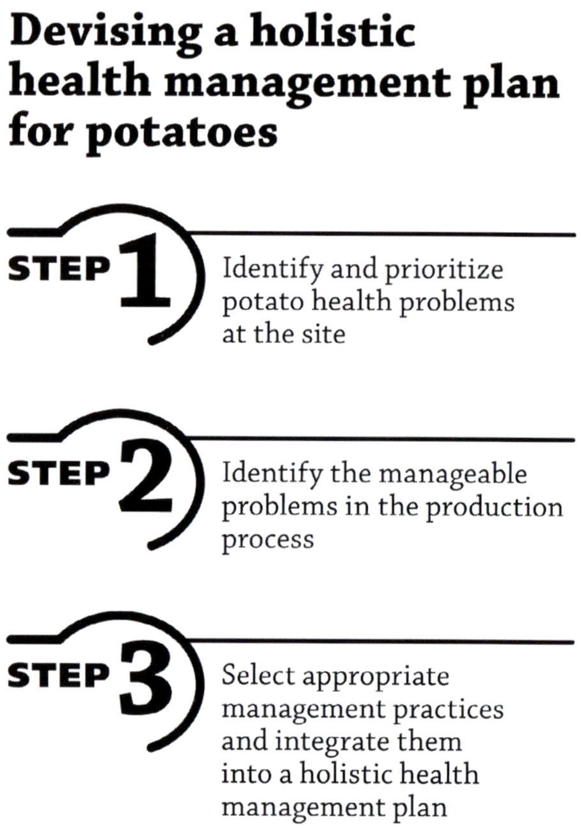

Fig. 1.4. A systematic approach for devising a holistic potato health management plan.

are likely to develop? Of these potential problems, which are most likely to limit the quantity and quality of potato tubers produced? Which ones are of major importance, and which are tolerable?

Step 2: Identify the problems that are manageable in the production process. Some problems may not be manageable in certain situations, because of physical, economic, environmental, or legal limitations.

Step 3: Select a set of practices capable of managing the high-priority problems and integrate them into a holistic health management plan.

Individual management practices must be compatible and must not be implemented without consideration of their effects on the entire agricultural enterprise. In most cases, various trade-offs must be evaluated. Certain options aimed at managing one problem may aggravate another, and the priorities established in Step 1 must be reevaluated. The economics of various management options must be considered in terms of input costs and effects on the profitability of the production enterprise. Legal constraints and concerns for environmental stewardship also affect choices among management options.

Throughout this book, a holistic approach to potato health management is emphasized. Chapter 2 contains a checklist of important health management practices. Chapters 3 through 14 discuss various aspects of potato production and present a generalized, comprehensive plan for producing a healthy potato crop, starting before planting and proceeding through production, harvest, and storage. Chapters 15 through 23 provide information on management strategies for specific problems affecting potato health—insect pests, diseases, weeds, and environmental stresses. A holistic health management plan can be devised for any potato production situation by selecting relevant management practices from Chapters 15 through 23 and then integrating them into the overall production plan presented in Chapters 1 through 14. A glossary and a list of relevant resource materials and Web sites are included at the end of the book.

CHAPTER 2

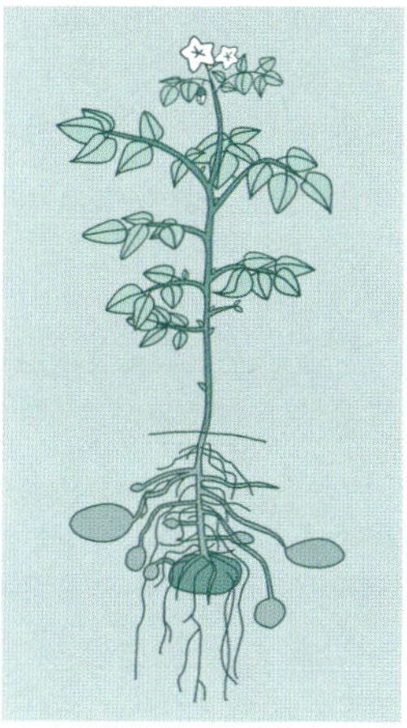

Jeffrey S. Miller
Aberdeen Research and Extension Center
University of Idaho, Aberdeen

Bryan G. Hopkins
Department of Plant and Animal Sciences
Brigham Young University, Provo, Utah

Checklist for a Holistic Potato Health Management Plan

A holistic plan for potato health management is required whenever potatoes are grown to combine management practices employed in the years prior to growing a potato crop, during the growing season, at harvest, and during tuber storage. After planting, the potato plant develops through five growth stages (Fig. 2.1): sprout development (Growth Stage I), vegetative growth (Growth Stage II), tuber initiation (Growth Stage III), tuber bulking (Growth Stage IV), and maturation (Growth Stage V). The stages of production and the growth stages of the plant provide a timeline for the following checklist of management practices that can be implemented in a holistic plan for potato health management.

Management in Years Prior to Growing Potatoes

☐ Establish an appropriate long-term crop rotation for each production unit, ideally with potatoes grown no more often than every fourth year, and less frequently if possible. Plant rotational crops that help suppress diseases, nematodes, weeds, and insect pests.

☐ In applying pesticides, alternate treatments with chemicals having different modes of action, or use tank mixes, to avoid the development of pesticide resistance (Chapter 14).

Fig. 2.1. Growth stages of the potato plant. (Drawings by Maegan Davis. Adapted from Western Regional IPM Project, 1986, Integrated Pest Management for Potatoes in the Western United States, Publ. 3316, University of California, Division of Agriculture and Natural Resources, Oakland)

☐ Follow an herbicide plan to manage weeds that harbor potato pests and weeds that are difficult to control in rotation crops (Chapter 22). Avoid herbicide carryover.

☐ Plant a green manure crop, if desired, to improve soil properties and control soilborne pests and pathogens (Chapters 5, 19, and 20).

☐ Establish permanent records of fertility, rotation, cultural management, and pesticide use for each field.

☐ Select fields with minimal rock, low to moderate slopes, and friable soil having adequate nutrient- and water-holding capacity, but not susceptible to clod formation.

☐ Enhance the physical, chemical, and microbiological properties of the soil. Avoid destruction of the soil structure, and prevent compaction and the development of poor drainage. Build the soil fertility by replenishing nutrients removed by the previous crop. Add as much organic matter to the soil as is reasonably possible. Incorporate appropriate soil amendments to adjust soil pH if needed and economical. Avoid accumulations of excess nutrients, salts, sodium, and other potential toxins. Apply pesticides or soil fumigants only if necessary and feasible (Chapters 3 and 5).

☐ In irrigated production, evaluate the availability of water in the field where potatoes will be planted. If a water shortage is likely, select a cultivar that is relatively drought tolerant. Design and maintain the irrigation system for maximum uniformity of water application.

☐ Select cultivars appropriate for the intended market, production conditions, and storage expectations. If possible, select a cultivar with resistance to diseases and nematodes expected to be present in the field.

☐ Identify reputable seed potato growers, and visit seed farms to examine production practices, seed lots, and certification records. Plant only certified seed tubers (Chapter 6).

Management Prior to Planting

☐ Collect soil samples in each unique zone for analysis to determine soil chemical properties and pest populations (e.g., nematodes, wireworms).

☐ Broadcast soil amendments and immobile nutrients (phosphorus, zinc, manganese, iron, and copper) prior to tillage, as determined by soil analyses. A banded application of immobile nutrients at row formation or planting often is also advantageous. Most of the other nutrients that are more mobile in soil can also be applied before planting, in all but very sandy soils. A portion of the nitrogen (25–40%) should be applied prior to tuber initiation (Chapter 8).

☐ Perform preplant tillage operations necessary to break up compacted layers, manage plant residues, and provide tilth for planting. Forming rows prior to winter buries pathogen inoculum, facilitates residue decomposition, allows clods to break up with freeze-thaw action, conserves moisture, avoids the need to work the ground when it is too wet in the spring, and minimizes erosion, compared with flat tillage. However, highly erodible soils should have a green manure or heavy crop residue until immediately prior to planting.

☐ In areas of low rainfall, irrigate prior to planting to establish sufficient soil moisture throughout the root zone to provide adequate water through emergence.

☐ Ensure that refuse tubers from a previous crop or from storage have been destroyed or rendered nonviable.

☐ Clean and sanitize storage facilities before seed potatoes are received. Clean and sanitize seed-cutting and planting equipment and all other equipment that will come into contact with seed tubers.

☐ Examine seed tubers for diseases and defects upon delivery. Test tubers for soft rot potential, as explained in Box 7.1 in the section entitled Soft Rot and Blackleg, to determine the potential for seed piece decay.

☐ Handle and store seed properly to maintain tuber health. Do not hold seed in a storage area that has been treated with a sprout inhibitor if the area may still be contaminated.

☐ Warm seed tubers to 50–55°F for 2–6 weeks prior to handling again or planting. Maintain conditions under which cut seed will suberize properly (55–60°F, adequate oxygen, and high relative humidity), or plant immediately after cutting, if soil conditions permit.

☐ Manage seed-cutting and healing operations carefully to avoid damage and produce uniform, properly sized seed pieces (Chapter 7). Sharpen knives and sanitize cutting equipment at least daily and before cutting each new seed lot. Apply seed piece fungicide and insecticide as needed.

Management During the Growing Season

PLANTING

☐ Delay planting until the soil temperature is above 45°F. Try to schedule planting to coincide with favorable soil and weather conditions. Ideally, seed and soil should be the same temperature at planting (Chapter 7). **Do not plant cold seed in warm soil.**

☐ Avoid allowing free moisture to collect on seed.

☐ Check the depth and spacing of seed pieces at the beginning of planting and throughout the planting operation. Plant at a depth that will ensure rapid emergence (about 6 inches, depending on the region).

☐ Apply soil insecticide and fungicide, instead of or in addition to seed treatments, depending on anticipated problems with particular pests or diseases (Chapters 7, 13, 15, and 19).

GROWTH STAGE I: SPROUT DEVELOPMENT

Sprouts develop from eyes on seed pieces and grow upward to emerge from the soil, and roots begin to form at the base of the emerging sprout. The seed piece is the sole source of energy for growth during this stage, because photosynthesis (the process by which the plant uses solar energy to produce carbohydrates) has not yet begun.

☐ Cultivate weeds. Where hilling is practiced, hill up 1–2 inches of soil on the plants at emergence or shortly after, but before the plants are 4–6 inches tall (Chapter 22). Reservoirs formed in furrows during cultivation (e.g., by the Dammer

Diker tillage system in fields with pivot irrigation) will aid water infiltration and uniformity. Avoid working the soil when it is too wet. Avoid injuring roots.

☐ Apply herbicide at cultivation. Apply insecticide, if it has not been applied earlier. Incorporate the chemicals into the soil by tillage or watering, according to the instructions on the product labels.

☐ Apply the remaining pretuberization fertilizer (especially nitrogen applied on sandy soils) as a side-dressing on the row at hilling (Chapter 9).

☐ Do not irrigate planted fields prior to emergence unless the soil becomes excessively dry at the depth of the seed pieces (Chapter 9).

GROWTH STAGE II: VEGETATIVE GROWTH

Sprouts emerge from the soil, and all vegetative parts of the plant are formed. After a sprout has emerged from the soil, leaves and stem branches grow from aboveground stem nodes, and roots and stolons grow from belowground stem nodes. While the seed piece still provides some nutrients during this stage, the leaves and stems begin to produce carbohydrates by photosynthesis. The formation of a plant canopy helps to reduce the soil temperature, slow weed growth, and increase the capture of sunlight by the plant.

☐ In irrigated fields, maintain more than 65–75% available soil moisture, but do not exceed the needs of the crop (Chapter 9). See Chapter 19 for descriptions of irrigation management practices that can reduce the severity of Verticillium wilt, powdery scab, and common scab. Monitor the irrigation system daily.

☐ Scout for weed escapes, and apply appropriate postemergence herbicides if needed (Chapter 22). Complete hilling and cultivation well before row closure, and avoid root pruning during tillage.

☐ Begin scouting for insects and diseases (Chapters 9, 15, 16, and 21). If an action threshold has been determined for a targeted pest or disease, the first insecticide or fungicide application should be made when the action threshold has been reached. If no action threshold has been determined, apply insecticide or fungicide before the targeted pest or disease has become established.

☐ Sample plants as soon as they have a fourth petiole. Analyze petioles to determine fertilizer requirements for optimum plant growth, and apply fertilizer as needed (Chapters 8 and 9).

GROWTH STAGE III: TUBER INITIATION

Tubers begin to form at the ends of stolons. The tuber initiation stage is very short (10–14 days) and occurs when the plant begins to transport excess carbohydrates (those not required for normal growth) downward, storing them in the tubers. This process is controlled by hormones in the potato plant, soil moisture and temperature, and nitrogen levels in the soil and plant. Tuber initiation usually occurs when the leaf area index (the ratio of the surface area of the leaves to the surface area of the ground) is 1.5 to 2.0. The formation of the first flowers is often associated with tuber initiation, even though these processes are unrelated.

☐ In irrigated fields, manage irrigation to avoid excessively wet or dry soil in order to minimize the development of common scab, powdery scab, and tuber physiological disorders and to avoid leaching of nutrients and pesticides (Chapters 19 and 23).

☐ Continue weekly petiole analysis and scouting for insects and diseases. Apply fertilizers and pesticides as needed (Chapters 8, 9, 15, 16, and 21).

GROWTH STAGE IV: TUBER BULKING

Tuber cells expand seven- to 18-fold with the accumulation of water, starch, and nutrients such as nitrogen, phosphorus, and potassium. Tuber growth is linear if all growing conditions are adequate. Aboveground growth slows significantly. Carbohydrates produced by the plant are transported to the tubers.

☐ Continue petiole analysis, and apply nitrogen after tuber initiation. Most of the nitrogen needed by the crop should be applied during the tuber bulking stage to avoid delaying tuber development. Late application of nitrogen delays tuber maturation and impedes vine desiccation. Apply other nutrients if needed.

☐ In irrigated fields, maintain uniform and adequate soil moisture. Avoid overwatering to minimize the development of tuber diseases and defects and minimize leaching (Chapter 9).

☐ Continue insect and disease scouting. Apply crop protection chemicals as needed.

☐ If desired, apply a foliar sprout inhibitor to plants according to the instructions on product label. In most cases, a sprout inhibitor is applied when the smallest tubers are at least 2 inches in diameter.

GROWTH STAGE V: MATURATION

The rate of photosynthesis decreases, leaves drop from stems, and vines slowly die. The dry matter content of the tubers reaches a maximum, and the tuber skin sets (thickens). Eyes on tubers of most cultivars are dormant.

☐ Reduce irrigation as plants begin to use less water to promote tuber skin set, minimize tuber diseases, and manage late-season weed growth (Chapter 9).

☐ Petiole nitrate concentration should drop to facilitate the gradual senescence of the canopy (ideally, for most cultivars, about 40% senescence at vine kill). Roll vines several days before vine kill, if necessary, to facilitate senescence. Rolling can also seal soil cracks and prevent damage to tubers from sunlight, disease, and insects.

☐ Schedule vine-killing operations to allow complete desiccation of foliage before harvest, if a green vine harvest is not appropriate (Chapters 10 and 21). Do not kill vines that are heat stressed or moisture stressed. Continue foliar applications of fungicides and insecticides, if appropriate, until foliage is completely dead.

☐ Inspect, repair, and sanitize storage facilities and harvest equipment. In dirt-floor storages, replace the top 1–2 inches of soil, and wet the floor to avoid tuber dehydration.

☐ Make necessary modifications to harvest equipment to minimize mechanical damage.

☐ Train personnel to conduct a safe, low-damage harvest.

☐ Conduct test digs to determine the storability of the crop. Plan to market tubers directly from the field if the incidence of late blight, pink rot, soft rot, or ring rot is greater than 5%. Severe stress-related problems (jelly end, sugar ends, and high sugar levels) may also necessitate direct marketing. Market tubers directly or plan for a short storage season if net necrosis caused by *Potato leafroll virus* is present.

Management During Harvest

☐ Conduct all harvest, transportation, and bin-loading operations with bruise management as a primary goal.

☐ Keep harvester chains well-filled with soil and tubers. Consider using a windrower to help keep the harvester conveyors filled.

☐ Ensure that tubers are not dropped from heights of more than 6 inches.

☐ Ensure that tubers are hydrated for at least 8 days prior to harvest (Chapter 10).

☐ Time the harvest to coincide with tuber skin maturity appropriate for the cultivar and the intended market (processing, fresh, or seed).

☐ If late blight is present, delay harvesting until the foliage is completely dead, but harvest early enough in the season to avoid frost damage.

☐ Coordinate harvest operations with current and expected weather conditions, so that tubers are dug when conditions are as close as possible to optimal (60–65% available soil moisture and tuber pulp temperatures of 50–65°F) (Chapter 10). In fields known to be at risk of pink rot, harvest when tuber pulp temperatures are 45–50°F (Chapter 19).

☐ Avoid harvesting in the early morning. Consider harvesting into the evening to avoid moisture from dew.

☐ Remove as much soil and debris as possible from tubers during harvest.

☐ Avoid harvesting in wet weather, if possible. If tubers are damp at harvest, dry them immediately. Do not allow a film of water to remain on the tubers.

☐ Tarp loaded trucks to protect harvested tubers from rain, direct sun, and adverse temperatures.

☐ Isolate damaged or diseased lots in separate bins for immediate grading and marketing. Sanitize all handling equipment prior to grading tubers in other bins.

☐ If tuber diseases are present, remove infected tubers as they are being piled into storage.

Management During Storage

☐ Begin storage ventilation as soon as the first few ducts are covered. Do not wait until the storage is filled to begin ventilation. Close the storage as soon as it is filled.

☐ Handle tubers carefully to avoid bruising. Keep pile heights less than 18 feet. Pile the tubers in a stair-step fashion to avoid rolling. The surface of the pile should be flat upon completion.

☐ If desired, apply postharvest disinfestants or fungicides according to the instructions on the product label to reduce disease development in storage.

☐ Rapidly cool tubers below 60°F within the first 2–3 days of storage. Maintain temperatures at 50–60°F and relative humidity at 95% during wound healing (the first 2–3 weeks of storage). After wound healing is complete, gradually lower storage temperatures to the level desired for the cultivar and the intended market use (Chapter 10). Ensure that the difference between the temperature at the top of the pile and the temperature at bottom of the pile is 0.5–2°F in all phases of storage.

☐ Take precautions to avoid condensation on walls and ceilings of storages and on tubers. Ventilate storages without humidity to remove excess moisture. Continually monitor the pile for signs of decay. If moderate to high levels of decay are observed, market the tubers as soon as possible.

☐ If appropriate, a certified custom applicator can apply a chemical sprout inhibitor after the curing process has been completed and prior to the end of tuber dormancy.

☐ Before removing tubers for marketing, warm the storage to raise the pulp temperature above 50°F. Manage bin-unloading operations to minimize bruising, following the same principles applied at harvest.

☐ When washing tubers in flume water, avoid holding them in deep water pits (several feet deep), since this increases the threat of bacterial soft rot (Chapter 18). Tubers should sit in water for the shortest possible time.

☐ Use fresh water to flume and wash tubers, if possible. Always use fresh water for the final rinse. Sanitizing agents can be added to wash water.

☐ Ensure that washed or flumed potatoes are thoroughly dried before packing. Use ventilated bags.

CHAPTER 3

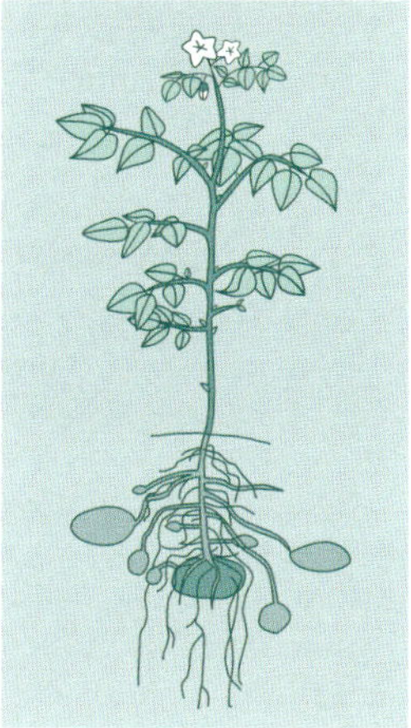

Linda Kinkel
Department of Plant Pathology
University of Minnesota, St. Paul

Soil Health: Managing the Soil Microflora to Enhance Potato Health

The terms *soil health* and *soil quality* have been used widely over the last decade to refer broadly to the potential for soils to sustain vigorous crop growth. More specific definitions for these terms have focused on a wide range of criteria, including soil tilth, drainage, chemical composition, resilience, and various characteristics of the soil microbial community. At present, there is no consensus on how soil health should be measured in agroecosystems. However, research over the past few decades has pointed strongly to the significance of soil microbes to plant health. The focus of this chapter will be on the management of soil microbial communities to minimize plant diseases in potato production.

Soils support a diverse collection of organisms, including fungi, bacteria, nematodes, and microarthropods. A single gram of soil may contain more than 10^{10} microorganisms. Soilborne microbes can have significant positive or negative impacts on plant health and crop yields through their roles as plant pathogens, plant growth promoters, pathogen antagonists, and saprophytes and by their effect on nitrogen fixation. Potato crop management tactics, including tillage, application of fertilizers and pesticides, irrigation, soil fumigation, and crop rotation, can all have significant influences on soil microbial communities and, consequently, on potato crop quality and yield. The potential for active management of soil microbial communities to enhance potato crop performance, reduce disease, and reduce the need for pesticides in potato production has been an area of significant research activity over the past decade.

Various microbes, including bacteria in the genera *Pseudomonas, Streptomyces, Bacillus, Paenibacillus,* and others, as well as fungi in the genera *Gliocladium, Coniothyrium,* and *Trichoderma,* contribute to the suppression of pathogens in soil. Suppressive activity against many pathogens causing severe diseases in potato has been noted. These pathogens include *Verticillium dahliae,* the fungus causing Verticillium wilt; *Streptomyces scabies,* the bacterium causing common scab; *Rhizoctonia solani,* the fungus causing black scurf; *Sclerotinia sclerotiorum,* the fungus causing white mold; *Pratylenchus penetrans,* the root-lesion nematode; *Meloidogyne* spp., the root-knot nematodes; *Helminthosporium solani,* the fungus causing silver scurf; and *Fusarium* spp., the fungi causing dry rot and seed piece decay. Mechanisms of pathogen suppression by indigenous soil microbes include competition for resources, production of antibiotics, hyperparasitism, and production of toxic substances or enzymes (e.g., nitrous oxide, chitinase). Pathogens are often relatively poor competitors in soil and are frequently sensitive to antibiotic inhibition, enhancing their susceptibility to suppression by microbial saprophytes in soil.

Some soils in long-term potato production have been reported to become strongly suppressive of common scab (Fig. 3.1). Even very susceptible potato cultivars remain largely scab-free when grown in these suppressive soils. Common scab suppression has been attributed to the enhancement of a microbial community that inhibits the population of *Streptomyces scabies* in soil (Fig. 3.2). This illustrates the potential for soil microbial communities to provide highly effective and long-lasting disease suppression. A better understanding of the dynamics of indigenous soilborne microbes and the impact of management practices on these communities has the

Fig. 3.1. Biological control of potato scab following spring inoculation of field soil with antibiotic-producing, nonpathogenic *Streptomyces* isolates. The tubers on the left are from a scab-suppressive soil. The tubers on the right are from a control plot. (Courtesy D. Liu)

potential to contribute significantly to the identification of tactics for pathogen and disease suppression in potato production systems.

How Can Soils Be Managed to Enhance Pathogen Suppression and Potato Yields?

Some management strategies presently used for controlling plant diseases may work against the long-term development of a pathogen-suppressive field soil community. Fumigation has been used to reduce pathogen and nematode population densities in soil and can be very effective in controlling many soilborne pathogens, but soil fumigants are generally nonspecific and toxic to a large proportion of soilborne microbes. Consequently, soil fumigation can lead to significant reductions in naturally occurring communities of antagonists of pathogens in soil. These reductions can lead to higher disease levels 3 or 4 years after fumigation, and they contribute to significant decreases in the diversity of the soil microbial community. Fumigants may also reduce or eliminate colonization of crop plants by beneficial *arbuscular mycorrhizal fungi,* which may further enhance plant susceptibility to diseases. In total, the nontarget effects of fumigation on the soil microbial community may contribute to the phenomenon of **fumigation rebound** (Fig. 3.3), in which fumigation treatments provide a short-term benefit in reduced disease intensity, but disease intensity subsequently becomes worse than it was prior to fumigation. Fumigation eliminates a large fraction of the soil microbiota and is generally followed by planting a disease-susceptible crop. The presence of the susceptible crop provides selective enrichment of soilborne plant pathogens. In the absence of active efforts to reestablish a vigorous and diverse community of saprophytes, symbionts, and pathogen antagonists and the natural disease suppression that they can provide, greater disease intensity seems inevitable. This can lead to a "**fumigation treadmill**," in which growers become locked into applying fumigants, often with increasing frequency of application over time. The fumigation treadmill may be unavoidable without active management to promote a diverse community of nonpathogenic soilborne microbes, especially pathogen antagonists.

Green manures have a long history in cropping systems and have been used for centuries to add organic matter to soil and increase soil fertility. They can also contribute to disease suppression by their impact on soil microbial communities. As early as the 1920s, scientists reported on the effectiveness of green manure crops in reducing potato scab intensity, specifically through their beneficial effect on antagonistic bacteria in soil. However, though green manures were a component of many cropping systems in the early part of the last century, the availability of nitrogen fertilizers and pesticides and economic pressure to produce annually on each unit of farmland reduced reliance on green manures for crop production. In recent years, researchers have returned to considering the potential for green manures to enhance plant vigor by their impact on soil and soil microbial communities.

Green manures can have diverse effects on soil microbial communities. As nonhost crops, green manures can contribute to a reduction in pathogen populations that rely on living host tissue. Perhaps more importantly, the added organic matter can provide resources for soil microflora, and the breakdown products from some green manures can inhibit pathogen populations in soil. For example, many mustards and other green manure species inhibit pathogen populations directly through the production of glucosinolates, saponins, ammonia, and other fungitoxic compounds; the use of green manures producing toxic compounds that inhibit pathogens

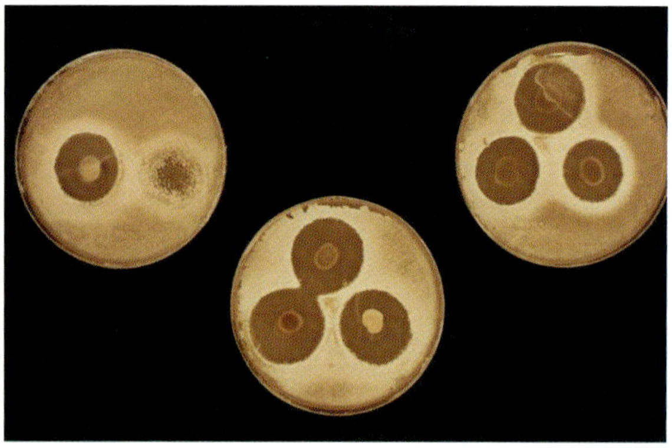

Fig. 3.2. Antibiotic inhibition of pathogenic *Streptomyces scabies* by nonpathogenic *Streptomyces* isolates from field soil. A pathogenic isolate was applied over the entire surface of each plate, and then two or three isolates of nonpathogenic *S. scabies* were plated on each plate. A zone of inhibition has formed around each of the nonpathogenic colonies.

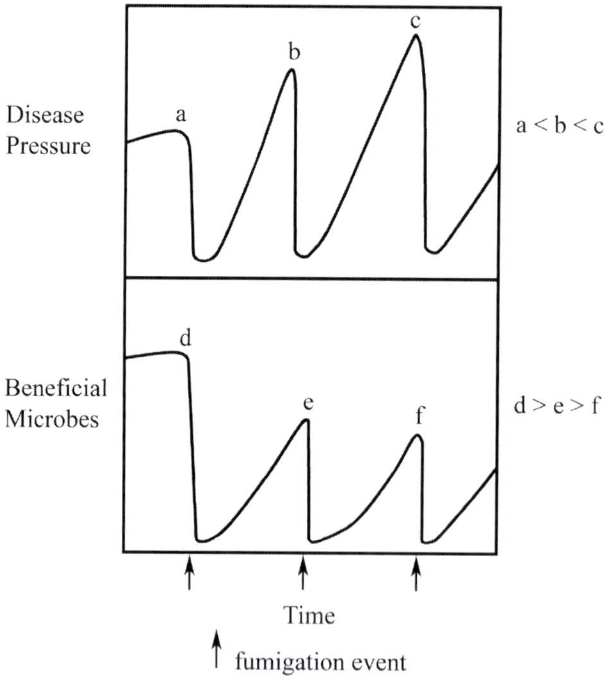

Fig. 3.3. The fumigation rebound effect can lead to a "fumigation treadmill." Research suggests that fumigation can provide a significant short-term benefit in reducing disease, but disease intensity a few years later can be even greater than before fumigation. Furthermore, fumigation eliminates a large fraction of soil microbiota, including many beneficial microbes, such as symbionts, saprophytes, and pathogen antagonists, whose populations may require some time to recover. These curves, though hypothetical, provide a general picture of the impact of soil fumigation on disease-causing and beneficial microbes in soil.

is known as **biofumigation**. In contrast, other green manure species appear to act indirectly, as a consequence of their influence on indigenous antagonist populations in soil. Pathogen antagonists, including hyperparasites, competitors, and antibiotic-producing microbes, are endemic in all soils and can sometimes be preferentially enhanced by green manure (Fig. 3.4). The effectiveness of some green manures in controlling plant disease has been shown to be directly related to their ability to influence specific nonpathogenic microbes in the soil.

Many crops, including canola, corn, mustard, sudangrass, sorghum-sudangrass hybrids, barley, oat, rye, and wheat, have been studied as green manures for the control of soilborne pathogens in potato (Fig. 3.5). Among these, sudangrass and sorghum-sudangrass have shown the most consistency and efficacy in the control of Verticillium wilt and common scab of potato. Research results suggest that these green manures can enhance the population density of beneficial microbes in soil and may also increase soil organic matter and nutrient levels. The benefits of sudangrass green manures in potato production appear to extend beyond disease-suppressive activity, with enhanced potato yield and quality observed following green manure treatment (Chapter 5).

Green manure treatment can require a full season out of crop production and income generation, with the green manure grown in the season immediately preceding potato. In some areas, sufficient plant biomass can be obtained by growing a green manure crop in late summer, immediately after potato harvest, and incorporating the green manure in late fall. Researchers have also considered the possibility of a precrop green manure. However, optimal timing of the green manure depends on the potato cultivar grown and the availability of soil moisture during late summer.

Unfortunately, despite their significant promise for enhancing the development of soil microbial communities that enhance potato vigor, green manures are not typically integrated in potato production systems. The lack of utilization of green manures is due partly to economic constraints on potato production systems, particularly where a full-season green manure crop is recommended. It is also due to the inadequate information on the factors that influence the success of green manures in enhancing potato yields and quality in integrated production systems. Active research is under way to provide the information necessary to increase the use of green manures in potato cropping systems with less loss of income.

Many other types of **organic amendments** have been studied for their effectiveness in suppressing plant diseases and enhancing potato yields, including cattle manure, poultry manure, swine manure, municipal wastes, pulp and paper wastes, blood meal and fish meal, and composts. Diverse amendments can have beneficial effects on soil microbial communities, often producing significant increases in total soil microbial populations and shifts in the numbers of targeted pathogen antagonists. Many amendments have also been shown to significantly reduce the incidence and severity of potato diseases, including Verticillium wilt and potato scab. Some soil amendments have also been shown to directly reduce the population density of nematodes and fungal pathogens in soil by producing toxic compounds (e.g., ammonia, nitrous oxide, volatile fatty acids) upon degradation in soil. Like green manures, organic amendments are not widely integrated into potato production systems. Ongoing research on the soil characteristics that are most conducive to the successful use of various organic amendments and on the varying effects of different amendments for controlling potato diseases is likely to be important to increasing the value of organic amendments to potato production.

Applications of **inorganic amendments** to soil can also significantly influence soil microbial communities. Active management of soil pH (e.g., by the application of sulfur) has been practiced for decades as a means of inhibiting the development of potato common scab, because of the preferential growth of the pathogen at high pH. Ammonium sulfate fertilizer is frequently recommended in preference to ammonium nitrate in soils that have a history of potato scab, because of its greater effectiveness in reducing soil pH. However, recent research suggests that ammonium sulfate may have a broader impact on soil microbial communities. The pathogen-inhibitory activity of microbial communities has been found to be greater (and common scab less severe) in soil treated with ammonium sulfate fertilizer than in soil

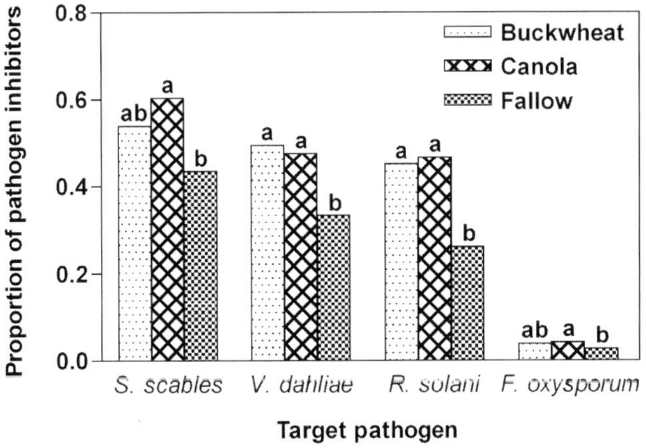

Fig. 3.4. Effects of green manures on the proportion of streptomycetes that inhibit the potato pathogens *Streptomyces scabies, Verticillium dahliae, Rhizoctonia solani,* and *Fusarium oxysporum.* For each pathogen, the lengths of bars with different letters are significantly different ($P < 0.05$, least significant difference; n = 11 or 12 replicates per treatment).

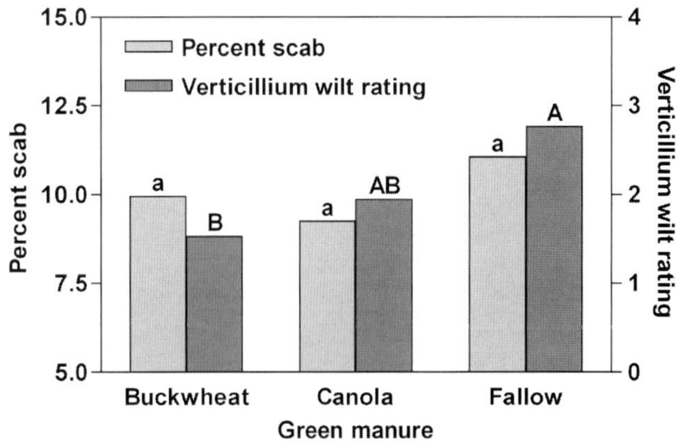

Fig. 3.5. Effects of green manures on the intensity of potato scab and Verticillium wilt. For each disease, treatments with different letters are significantly different ($P < 0.05$, least significant difference; n = 11 or 12 replicates for each green manure treatment).

treated with ammonium nitrate. Thus, in addition to shifts in soil pH, which may directly affect a pathogen population, pathogen-suppressive activity can vary significantly, depending upon fertilizer formulation.

Crop rotation, **crop sequence**, and tillage have all been reported to influence soil microbial communities and plant vigor (Fig. 3.6). Crop rotation has long been used as a means of reducing pathogen population densities in soil by reducing the availability of host tissue. Crop rotation, crop sequence, and tillage can also have significant effects on nonpathogenic microbial populations in soil. Reductions in the intensity of potato scab, pink rot, and Verticillium wilt have been reported after various crop sequence, rotation, or tillage treatments. Disease reductions are attributed to enhanced populations of microbial antagonists in the soil and within root tissues. Unfortunately, despite these findings, there is presently little in the way of a systematic database that can provide guidance on the specific crop sequences or tillage practices that are most likely to consistently support the development of a vigorous, pathogen-inhibitory microbial community in different soils. In general, crop plants not belonging to the plant family Solanaceae are the best choices for rotation. Where nematodes are a problem, nonhost crops should be selected.

Microbial inoculants offer what may appear to be a shortcut to the development of a vigorous, pathogen-inhibitory soil microbial community. *Rhizoctonia solani, Verticillium dahliae, Streptomyces scabies, Phytophthora infestans, Helminthosporium solani, Sclerotinia sclerotiorum,* and *Pratylenchus penetrans* have all been reported to be controlled following the inoculation of soil with microbial antagonists. While significant control can sometimes be achieved with microbial inoculants, they can lack the consistency and efficacy for reliable disease control desired by growers. However, promising field results for many plant pathogens suggest that further research may enhance the potential for microbial inoculants to contribute significantly to the development of pathogen-inhibitory microbial communities in soil.

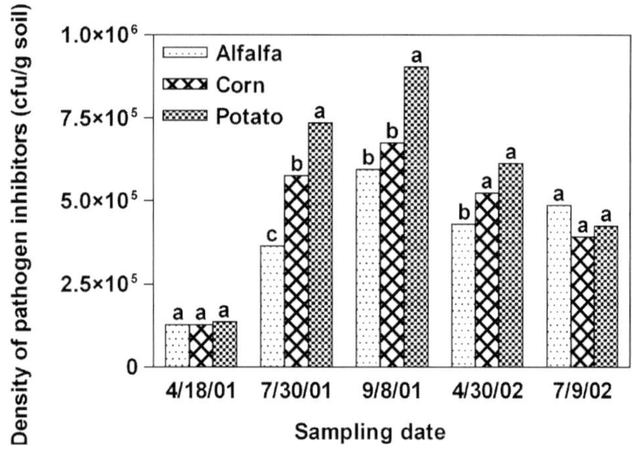

Fig. 3.6. Effects of primary production crop (alfalfa, corn, or potato) on the population density of pathogen-inhibitory microbes in field soil. The data are means of 12 replicates per treatment. For each sampling date, treatments with different letters are significantly different ($P < 0.05$, least significant difference).

Outlook for Management of Soil Microbial Communities

Virtually all crop management strategies, from basic tillage operations to fertilization to crop rotation, can have substantial impacts on indigenous soil microbial communities. These impacts can be positive, enhancing the capacity of the soil microbial community to inhibit plant pathogens and support the production of healthy potatoes, or they can be negative. Potato growers should think holistically when selecting crop management practices, to consider not only their impact on the potato plant but also their impact on the soil microbial community. While no single crop management strategy will be adequate for sustaining a vigorous, pathogen-suppressive microbial community in soil, research suggests that identifying appropriate combinations of management strategies to support pathogen-suppressive microbial communities in soil will provide significant economic benefits in potato production. Further research identifying management strategies that contribute most consistently to the development and maintenance of soil microbial communities in sustaining potato vigor, along with research on the factors that contribute to the success of those strategies in sustaining pathogen-inhibitory microbial communities in soil, is likely to contribute significantly to the enhancement of sustainable potato production systems in the coming decade.

Key Principles

- Soils support a diverse microbial community, including many organisms that can significantly influence potato health and yields.
- Indigenous soil microbes can play a major role in protecting potato plants against plant pathogens.
- Crop management strategies can have substantial positive or negative impacts on indigenous soil microbial communities, influencing their ability to support the production of a healthy potato crop.
- Management strategies that are conducive to the development of a vigorous, pathogen-suppressive soil microbial community will become increasingly important for sustainable potato production.

CHAPTER 4

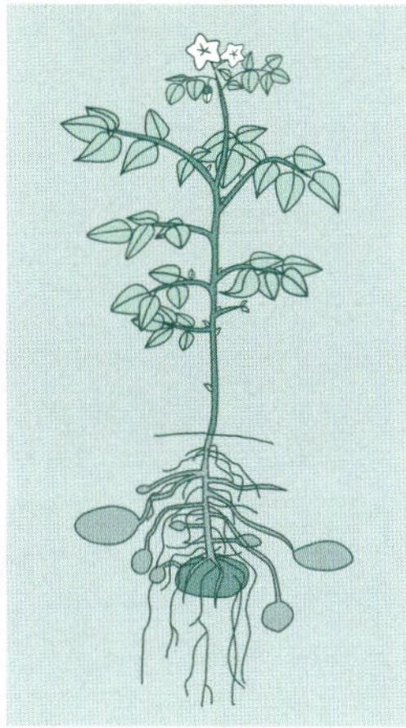

Joseph F. Guenthner
Department of Agricultural Economics and Rural Sociology
University of Idaho, Moscow

R. Thomas Schotzko
School of Economic Sciences
Washington State University, Pullman

Economics of Potato Plant Health

Holistic potato health management principles are presented in Chapter 1. Economics will now be added to provide a logical framework to determine whether benefits exceed costs for any potato health management decision. As simple as that might sound, when all four sides of the production pyramid described in Chapter 1 are brought into the decision model, the complexity increases. The intent of this chapter is to provide an introduction to economic decision making and an understanding of the information needed to analyze plant health decisions.

Economics is a logical system of thinking that enhances, but does not replace, production information. In what follows, an attempt is made to describe a way to view potato health management in terms of profitability. With the linking of revenues and costs to production, management decision rules can be developed. In the broadest sense, potato growers should choose the practice that most benefits them or has the least negative effect. On the surface that seems obvious, because that's what decision makers do.

There are three broad categories of potato health management decisions. One group includes decisions affecting costs only, and the obvious incentive is to minimize cost. Another group includes decisions affecting revenue only, and the incentive is to maximize revenue. A third group includes decisions affecting both costs and revenue, and these decisions require more complex analysis.

Several examples follow, beginning with relatively simple plant health decisions and then moving to more complicated issues. The selection of examples follows a list of production factors affecting potato yield and quality identified by R. Kunkel and R. E. Thornton at Washington State University. They identified 10 factors that growers can control:

1. cultivar
2. seed quality
3. seed piece size
4. number of plants
5. moisture supply
6. mineral nutrients
7. soil compaction
8. pests and diseases
9. days grown
10. timeliness of operations

First, the discussion of production is presented and how economics links to production. That reasoning is then illustrated by examples to describe the economics associated with analyzing the controllable factors in terms of profitability.

Total Yield Curve

A graph that shows how a production input—nitrogen (N) fertilizer, for example—affects potato yield forms a curve like that in Fig. 4.1. The shape of this curve demonstrates some basic economic concepts. Every point on the curve links a quantity of nitrogen fertilizer with some response level of potato yield. Research might show that an application of 100 units of N per acre results in a yield of 200 cwt of potatoes. This is represented by the point on the curve where 100 is on the horizontal axis and 200 is on the vertical axis. The research might also show that an application of 150 units of N produces 300 cwt of potatoes, and this would be represented by a higher point on the curve.

The curve shows how various levels of nitrogen impact potato yields. Total product or yield starts out slowly, increasing in response to small but increasing levels of N. Then yield increases more rapidly in response to larger N applications. Eventually the yield response begins to slow and at some level actually becomes negative. That is the point where excess nitrogen begins to be harmful to potato plants. The characteristics of this curve can be used to define reasonable and unreasonable levels of application of nitrogen fertilizer. Other concepts that are useful for making decisions will be shown using such graphs.

Yield Terminology

The total yield curve can be used to illustrate five different definitions of output: (1) absolute yield, (2) attainable yield, (3) economic optimum yield, (4) actual yield, and (5) average

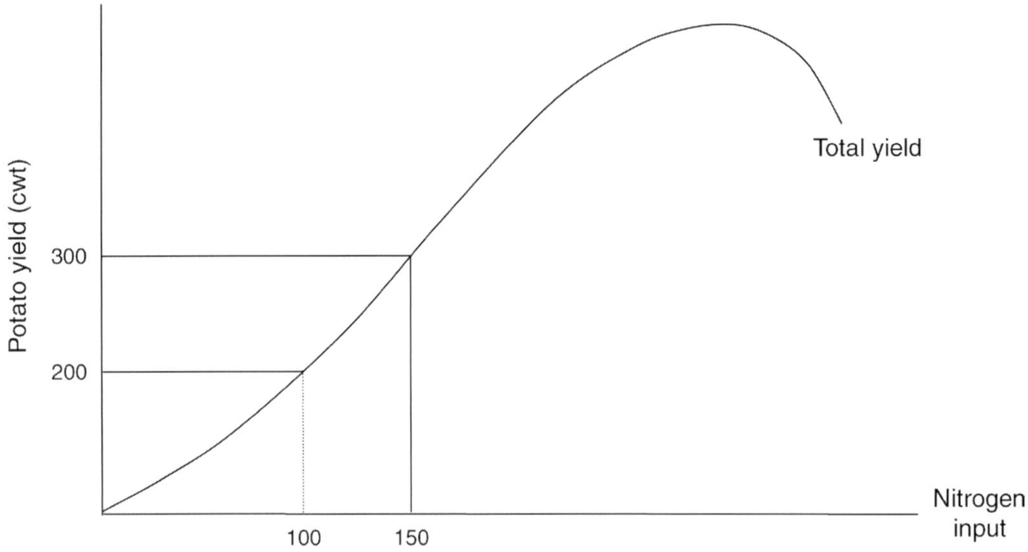

Fig. 4.1. Total yield curve.

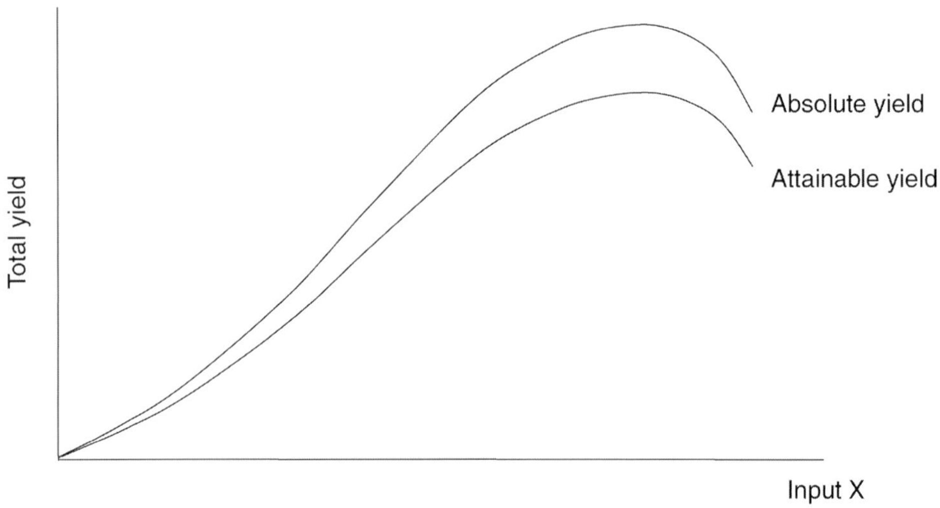

Fig. 4.2. Absolute yield and attainable yield.

yield (Box 4.1). These terms help to differentiate among various perspectives and outcomes.

Absolute yield is the amount that can be produced based on genetic potential when all factors of production (inputs) are applied in optimal quantities and with optimal timing (Fig. 4.2). These inputs include the nutrients required by the crop, management practices that eliminate all disease and pest problems and enhance yield, and the water, heat units, and sunlight that optimize production.

Attainable yield is the level that has been achieved under research conditions. In Fig. 4.2, the curve representing attainable yield is lower than the curve representing absolute yield, because in the field some inputs would not be utilized at their theoretical optimum, resulting in somewhat lower production. The absolute level of production implies a sophistication that is suitable only for theoretical discussions.

Economic optimum yield and **actual yield** are more appropriate concepts for the real world that growers face. **Economic optimum yield** becomes the goal associated with profitability. The **actual yield** is exactly that—the yield that is harvested at the end of the growing season.

The fifth category is the **average yield**, calculated from the yields obtained by individual growers. The average yield is most commonly used in policy discussions.

The term **average yield** expresses a piece of information different from the other yield terms as described here. While the other terms (absolute, attainable, etc.) are described with respect to changing some production input (N, for example) or group of inputs (the difference between absolute and attainable), the average yield is simply a measure of the actual yields obtained by growers on a given area of land (a farm, a county, a state, etc.). This is a descriptive term and offers no information on the use of any or all inputs employed in growing a crop of potatoes.

The goal here is to enhance appreciation of the use of economic reasoning in the evaluation of production management decisions. While the focus is plant health, the same process can be used any time there are management decisions that

Box 4.1

Definitions: Absolute, Attainable, Economic Optimum, and Actual Yield

Absolute yield is the yield possible with no limiting factors except the genetic potential of the crop. It is the theoretical maximum yield of potato and is at least as great as the world record yield.

Attainable yield is the yield possible in any given environment, year, and area. It is limited by the weather, depth of soil, and other factors that cannot be controlled by growers. These factors are given and cannot be changed. The attainable yield of potato is limited by available water, growing degree-days, temperature extremes, intensity of sunshine, and day length. It is at least as great as the best yield obtained in any given area, season, and management system on a given soil.

Economic optimum yield is limited by economics and is determined by the potential value of the actual yield offset by the price paid to achieve that yield. The price includes both the short-term direct costs to the farmer and the long-term costs to the industry and eventually to society.

Actual yield is the yield harvested in any given field and is the outcome of the ability of the crop to respond to growing conditions counterbalanced by weeds, insects, diseases, nematodes, fertilizers, compact soil, frost, or other production hazards.

The goal in potato research—and an important message of this book—is not that every potato crop produce a world record yield, but rather that the actual yield approach or equal the attainable yield by methods that are affordable and sustainable.

Adapted from R. J. Cook and R. J. Veseth, 1991, Wheat Health Management, American Phytopathological Society, St. Paul, MN

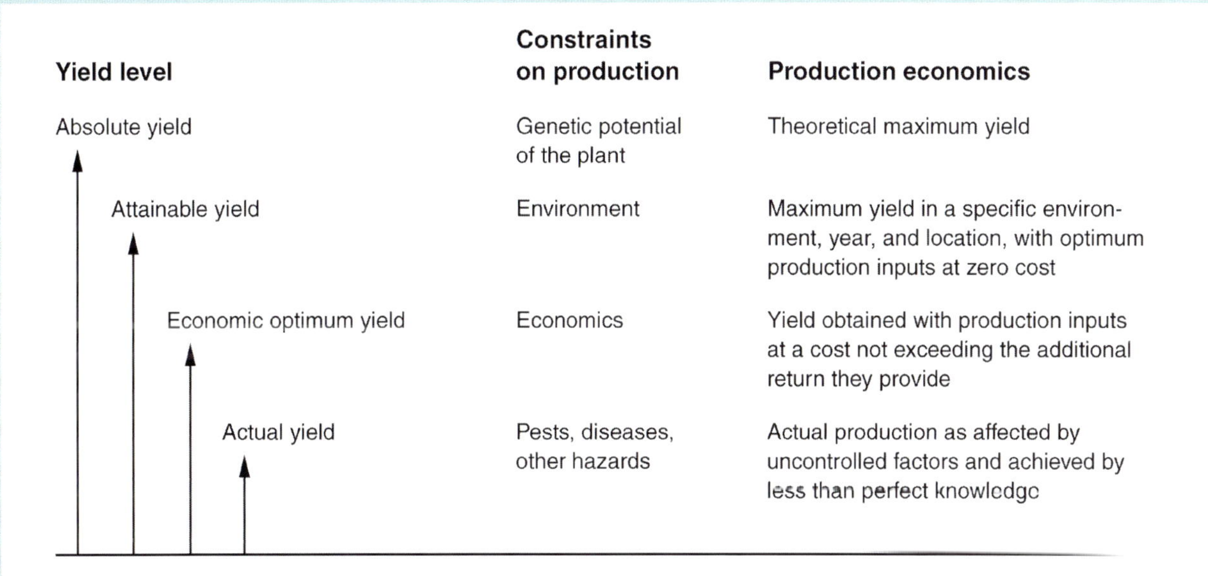

impact costs or receipts. In this terminology, what is being described is a process to identify **economic optimum** actions.

Some Production Economics Basics

Attainable, affordable, and actual yields can be demonstrated on a single total yield curve (Fig. 4.3). **Attainable yield** can be viewed as the yield that can be achieved when all production inputs are free or have zero cost. Attainable yield under those conditions would be at the highest point of the total yield curve (point A in Fig. 4.3) if the inputs have been applied in optimum quantities.

Economic optimum yield is something less than the attainable yield and is the result of the fact that inputs have costs. Input use eventually reaches a point where the cost of an additional unit of input is greater than the additional revenue from the yield increase it would produce. **Actual yield** comes about as a result of less than perfect knowledge about the impact of inputs, lack of precision in the application of inputs, and lack of control over other limiting factors, such as weather.

Since inputs are not free, the economic optimum level of production, or **economic optimum yield**, will be located somewhere to the left of point A in Fig. 4.3. The location of **actual yield** in the graph is more complex, because of uncontrolled factors such as sunlight and heat units. Because of the influence of the uncontrolled factors, the actual outcome may be represented by a point to either the right or the left of the **economic optimum yield** point. This is because the amount of input (nitrogen, for example) is based on unknown condi-

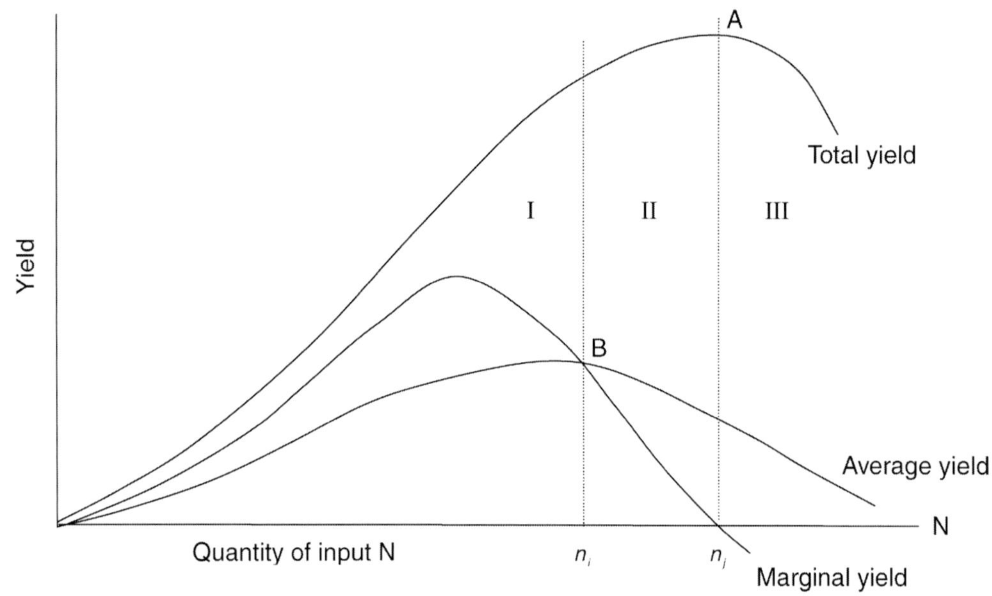

Fig. 4.3. Total yield, average yield, and marginal yield associated with the quantity of nitrogen (N) applied. Point A represents **attainable yield**. **Economic optimum yield** is located between n_i and n_j. Marginal yield equals average yield at point B. Three stages (I–III) are shown. Stage II, in which marginal yield is positive but less than average yield, is where the most profit is made.

tions during the growing season, whose ultimate impact is not known at the time of application.

Two other curves important to this discussion are included in Fig. 4.3. The **average yield** curve represents the average output per unit of N. As long as **total yield** is greater than zero, average yield will be greater than zero. Note that this average yield is not the same as the average yield mentioned above. Here the average is associated with a specific input. In Fig. 4.3 the average yield is the average per unit of N. Each point on the average yield curve represents the yield associated with the application of a particular amount of N, with the amount of N increasing as you move from left to right on the horizontal axis.

In terms of grower decisions, the **marginal yield** is more important. The marginal yield curve represents the change in output associated with a one-unit change in input. For example, if the potato yield obtained with 200 pounds of N per acre is 5 pounds (0.05 cwt) greater than the yield obtained with 199 pounds of N per acre, then the marginal yield associated with the 200th pound of N is 5 pounds.

There is another indicator that defines the economic optimal range of production. In Fig. 4.3, the **marginal yield** curve goes to zero (crosses the horizontal axis) when the **total yield** curve reaches the peak. A one-unit change in N beyond point A causes total yield to decline, so that the marginal yield is less than zero. Producing beyond the point where the marginal yield becomes zero is irrational, because additional inputs actually decrease the overall yield. That represents the upper limit of the range of potentially optimal production decisions. The lower end of that same range is identified by the point where the marginal yield curve crosses the average yield curve (point B in Fig. 4.3). At this point the marginal yield equals the average yield.

The total yield curve is segmented into three stages. Stage I is where marginal yield is greater than average yield (to the left of point n_i). In stage I, it is rational to continue to expand production by using more inputs, because the response to additional inputs is increasing. For example, if the 200th pound of N results in a yield increase of 5 pounds, then the 201st pound of N will increase output by more than 5 pounds.

Stage II is where marginal yield is positive but less than average yield (between points n_i and n_j). This is where growers make the most profit. The economic optimum or profit-maximizing input choice is found in this range. A specific optimal point could be calculated if it were possible to know the exact relationship between yield and inputs, the price at which the crop can be sold, and the cost of inputs.

Stage III is defined as the segment of the total yield curve that is downward-sloping. In this range, adding another unit of input reduces output. For nitrogen input, that would occur when excess applications would actually start to become toxic and decrease potato yield.

Economists use these relationships to analyze crop production decisions. For growers, the main point is that there is a range of profitable yields, but the maximum possible yield (absolute and attainable) is outside that range.

Converting production to revenue is as simple as multiplying the potato yield by potato price. That converts the total yield curve into a **total revenue** curve. Similarly, the average yield and marginal yield curves become average revenue and marginal revenue curves when multiplied by the price.

The process of determining whether a given activity should be undertaken is based, in part, on the marginal revenue that will be generated by that activity (or the revenue loss that will be "saved" by reducing or eliminating a given constraint). The other part of the decision process is determining the additional cost of the activity.

Fig. 4.4 contains curves representing total and fixed costs. In short-run economic analyses (a growing season, for example), costs are divided into two categories: fixed and variable. **Fixed costs** are costs that are incurred regardless of production. Property taxes and depreciation are two examples of

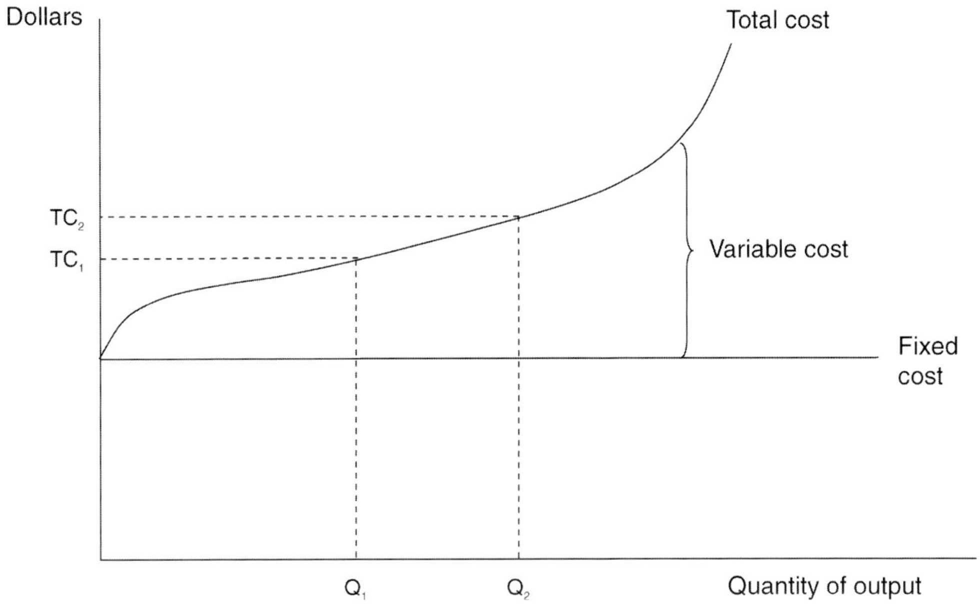

Fig. 4.4. Fixed cost, variable cost, and total cost (TC) associated with the production of different levels of output (Q).

fixed costs. They do not change when the production level changes.

In Fig. 4.4, with the horizontal axis representing total output, the curve representing fixed costs is a horizontal line. As total output increases (moving from left to right on the horizontal axis), fixed costs do not change (for example, property taxes are the same, regardless of whether production is 0 cwt or 500 cwt). Hence, the line representing fixed costs is horizontal (flat).

The other category, **variable cost,** is the difference between the fixed cost and the total cost. Variable costs are all the expenses of producing a crop. Since plant health decisions are production related, fixed costs can usually be ignored when such decisions are made.

Changes in total cost come from changes in variable costs (for example, the cost of fuel, fertilizer, control materials, and labor). In Fig. 4.4, as production increases from Q_1 to Q_2, total cost (TC) increases from TC_1 to TC_2. Since fixed costs have not changed, one needs to consider only the costs that have changed as the result of any action aimed at maintaining or improving plant health.

In the real world, changes in production do not come one unit at a time; nor are inputs often changed one unit at a time. There is also the issue of what units of measurement to use. Hence, the expressions *cost-benefit analysis, partial budgeting,* and *marginal revenue vs. marginal cost* are all used to describe the process of evaluating issues to determine the appropriate course of action. In some cases, such as the onset of late blight or an explosion in the population of a harmful insect, the decision may be easy. In other situations, the decision may not be as obvious.

Working to maintain plant health most often results in interactions among a number of factors. For example, improving seed piece size and spacing may result in a denser planting of healthy plants, which changes the physical environment of the field and may increase the likelihood of some diseases. The outcome is a whole series of cost changes that may occur (increased seed costs and the application of more plant protection materials). If revenue is affected only by yield, the calculation is relatively easy. However, if tuber grade and size are also affected by the changes in seed size and spacing, then the value per unit of production also changes, and another factor is added to the decision analysis.

Example: Fertilizer Response

The economics of mineral nutrients (production factor 6 in the list of factors identified by Kunkel and Thornton) is illustrated in research by E. Panique and co-workers (Table 4.1). In this study, a mathematical relationship was developed using research data to associate potato yields with potassium fertilizer in silt loam soil near Antigo, Wisconsin. The relationship between expected potato yield and application of potassium fertilizer at seven different rates, ranging from 50 to 350 pounds per acre, is shown in Fig. 4.5.

Potato yields increase rapidly with applications of 50, 100, and 150 pounds of potassium. The yield response slows at higher application rates, reaching a peak at 250 pounds of potassium. Going beyond that to 300 and 350 pounds of potassium actually decreases potato yield. Growers who want to produce maximum yield would apply 250 pounds of potassium and expect to harvest 366 cwt per acre.

Growers who want to maximize profit would apply less. They would make more money if they applied 200 pounds of potassium. Table 4.1 shows the yield response data with some economic calculations. The marginal revenue for the first 50 pounds of potassium is \$77.20 per acre. Since the cost of 50 pounds of potassium is \$11.00 per acre, it obviously is profitable to apply at least 50 pounds of potassium.

The last column in Table 4.1, marginal revenue minus marginal cost (MR – MC), shows the profitability of potassium application in 50-pound increments. Although maximum yield is obtained with the application of 250 pounds, growers

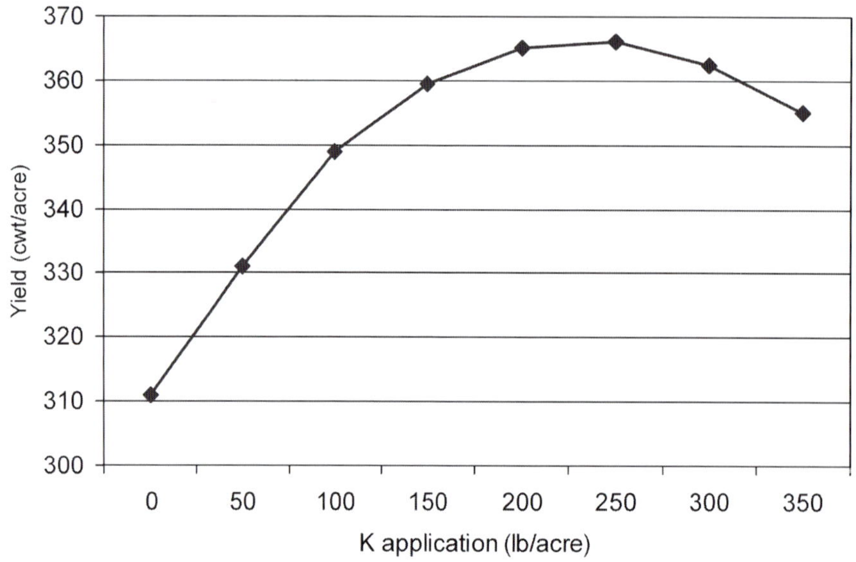

Fig. 4.5. Potato yield response to potassium.

Table 4.1. Economics of potassium fertilizer application[a,b,c]

Potassium application (lb/acre)	Potato yield (cwt/acre)	Marginal product (cwt/acre)	Marginal revenue (per acre)	Marginal cost (per acre)	MR – MC (per acre)
0	312				
50	332	20.6	$77.20	$11.00	$66.20
100	348	15.8	$59.09	$11.00	$48.09
150	359	10.9	$40.99	$11.00	$29.99
200	365	6.1	$22.89	$11.00	$11.89
250	366	1.3	$4.78	$11.00	–$6.22
300	363	–3.6	–$13.32	$11.00	–$24.32
350	354	–8.4	–$31.43	$11.00	–$42.43

[a] Source: E. Panique, K. A. Kelling, E. E. Schulte, D. E. Hero, W. R. Stevenson, and R. V. James, 1997, Potassium rate and source effects on potato yield, quality and disease interaction, American Potato Journal 74:379–398. Used by permission from The Potato Association of America.

[b] Potato price = $3.75/cwt. Potassium price = $0.22/lb.

[c] In English units the formula associated with the data in the table is $Y = 311.7409 + 0.186235X - 0.000158X^2$, where Y is tuber yield (in hundredweight per acre) and X is the amount of potassium applied (in pounds per acre). In metric units the equation is $Y = 35 + 0.046X - 0.000086X^2$, where Y is tuber yield (in metric tons) and X is the amount of potassium applied (in kilograms).

would lose $6.22 per acre by applying potassium at that rate instead of the lower rate of 200 pounds. Application of 300 or 350 pounds per acre would be even more unprofitable.

Any increase in expenditures for plant health enhancement should return at least as much in revenue or in other costs saved.

This example provides a framework of analysis that can be applied to more complex situations. Profitable decisions about potato health can be made by examining the marginal revenue minus marginal cost. Any increase in expenditures for plant health enhancement should return at least as much in revenue or in other costs saved. The profit-maximizing input choice is found where the marginal (additional) revenue from the last unit of the input equals its marginal (additional) cost. This is the economic optimum.

Potato health decisions can impact revenue through more than just yield. Tuber size and specific gravity and the incidence of hollow heart in relation to fertilizer application rates were also analyzed in the study by Panique et al. When potato quality is taken into consideration, with its impact on potato price, it might be reasonable to apply potassium at a lower rate than in the simple example above.

The cost analysis can also be more complex. There can be a cost of application if extra time is spent applying the material (production factor 10, timeliness of operations). Harvest, handling, and hauling costs might vary with different yields. Also, a decision to use one type of plant health enhancement method could reduce or eliminate the cost of another method that would no longer be needed.

In summary, the net impact of increased potassium application can be calculated by deducting the increased costs of material, application, and harvest from the increase in the value of production. Revenue estimates should include both yield and price changes.

Example: Analyzing Multiple Inputs

In addition to input quantity, another consideration is how efficiently an input is used. Research on missing and irregularly spaced potato plants conducted by M. Pavek and R. E. Thornton (American Journal of Potato Research 83:55–66) is one example. This research is related to production factors 2 (seed quality), 3 (seed piece size), and 4 (number of plants).

The agronomic and economic aspects of seed piece placement were analyzed in the study. In a survey of 70 commercial potato fields in Washington State, the average potato stand had 7% missing plants—6% due to planter skips and 1% due to nonviable seed pieces. Nearly 5% of the seed piece placements were doubles or multiples.

To estimate economic loss due to inefficient use of seed potatoes, research plots were established that imitated spacing errors in commercial fields. Yield, quality, price, and cost differences were determined between optimal spacing and simulated planting errors for two cultivars, Russet Burbank and Russet Norkotah.

Skips caused most of the loss of profit from both cultivars. In-row plants adjacent to skips partially compensated for the lost value of the missing plants, but plants in adjacent rows did not. Clumped seed pieces reduced tuber size and increased seed potato costs.

In an analysis applying processor contract pricing specifications to the Russet Burbank plot data, it was estimated that the Washington commercial fields in the survey lost an average of $81 per acre as a result of improper seed placement. In an analysis based on fresh-market prices, the Russet Norkotah fields lost $101 per acre as a result of inefficient use of seed.

These results suggest that it would be profitable for Washington potato growers to spend as much as $80 to $100 per acre to improve seed piece placement efficiency, depending on cultivar. This amount of money could be spent for improvements in planter technology, management expertise, and equipment operator training. Note that in this case the costs of solving the problem are short-term (improved equipment operation, for example), and the income stream may well be long-term, so that the return on investment could be substantial.

Example: Tuber Health

As plant health management becomes more sophisticated and the materials more selective, management decisions become much more complex. Research by T. L. Marsh at Washington State University provides an example of that complexity regarding production factor 8 (pests).

Marsh built a model to determine optimal levels of aphicide for the control of the green peach aphid, which is a vector of *Potato leafroll virus* (PLRV). In Washington, peach orchards provide overwintering habitat for the aphid. Modeling the system requires information on the life cycle of the aphid, the timing of infection, whether the aphids are carrying the virus, and the growth and development of the potato crop.

Modeling of the necrosis in the tubers is also necessary, as different levels of control may be possible if the tubers can be used before symptoms develop. Using an aphicide instead of a broad-spectrum insecticide also means that the natural predators of the green peach aphid are considered in the model.

The model needs two aphid life cycles, as it is during the second cycle that plant infection is most efficient. A model of potato plant growth is needed, because the effects of infection are greater in young plants than in older plants. Not only is the growth and development of the potato plant important, but so is the development of net necrosis over time.

Since not all plants will become infected, a probability function is needed to represent that characteristic. If an aphicide is to be substituted for a prophylactic material, then a predator-prey relationship model is also needed.

In addition, all of the economic factors mentioned above (base price, size incentive, net necrosis disincentive, possible storage payment, costs of materials and application) will be part of the analysis. Finally, a grower's aversion to risk can be included in the analysis, in effect, recognizing the importance of the reject clause in the typical contract (assuming the crop is to be processed into frozen products).

Estimates of the most economical method of control and estimates of the optimal timing of control activities can be made with the model. As long as the underlying models and assumptions reasonably reflect reality, this model offers usable information. As the insects adapt or cultivars change, the model must be recalculated (recalibrated) to ensure control activities are optimal.

In the future, growers will have increasing access to research-based models that will help them make profitable decisions regarding complex plant health management issues such as net necrosis. Regardless of the complexity of the plant health issue, the key to profitable decision making is to analyze changes in costs and revenue.

Summary

Plant health management decisions have economic ramifications. Some are so intuitively obvious that the decision is automatic—early control of late blight, for example. Others are not so obvious, especially when more than two alternatives are available.

This chapter is intended to help growers understand the need to **analyze costs and benefits** associated with potato health management decisions. Sometimes all that is needed is the back of an envelope. Other decisions can be handled with spreadsheet or accounting software. Some decisions can be enhanced with the use of sophisticated biological models developed by experts with specialized knowledge.

As time goes on, two things are likely to happen. First, the value of information will continue to increase. Continued grower success will be enhanced by the ability to correctly apply that information or knowledge in a timely fashion. Valuable information will come in three categories: (1) changes in potato value, (2) increases in potato yield, and (3) changes in input cost.

While the increase in value per unit is often recognized, sometimes there is a failure to consider that today's farming operations are much larger than they were 30 years ago. With both more acres per grower and more tons per acre, relatively small net benefits per ton or per acre can add up to significant amounts.

A second change is at least as likely as the first. Growers will change production practices, including the application of new technologies and the use of new control materials. Both of these should be accompanied by reevaluation of the decision rules that have been established for old technologies and materials. As additions to the knowledge base continue to occur more rapidly, it is necessary to consider more frequent updates to the decision-making tools to accommodate that new information.

As production processes become more complex and sophisticated, finding answers to problems arising in those systems becomes more expensive and time-consuming. Enhanced grower decision-making ability actually increases the rate at which answers can be generated by researchers for the complex problems by reducing the pressure to address less complex issues. This, in turn, allows for an improvement in potato health management decision making.

CHAPTER 5

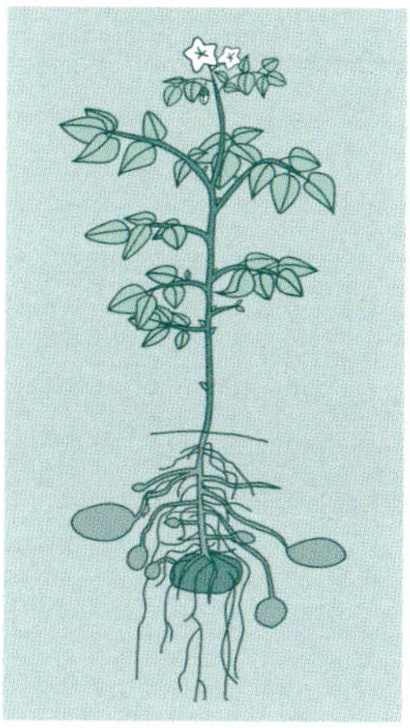

Mike Thornton
Parma Research and Extension Center
University of Idaho, Parma

Jeff Stark
Idaho Falls Research and Extension Center
University of Idaho, Idaho Falls

Bryan G. Hopkins
Department of Plant and Animal Sciences
Brigham Young University, Provo, Utah

Robert E. Thornton
Department of Horticulture and Landscape Architecture
Washington State University, Pullman

Selecting and Preparing the Planting Site

The field site, cropping sequence, and preplant tillage operations each affect the economic efficiency and environmental impact of the potato production system. Field selection largely depends on characteristics such as cropping history, topography, soil properties, and compaction. After a site is identified, an appropriate cropping sequence that favors the production of a healthy potato crop must be determined. When a rotation is established for commercial potato production, the maximum advantage from a particular rotational sequence can be gained only by implementing a long-term rotational management strategy. This strategy may differ from a more short-term one aimed only at maximizing the cash return from the potato crop. In rotational management, short-term economic returns from each crop or sequence of crops must be weighed against potential long-term benefits to the total agricultural enterprise, such as improvements in soil structure and fertility and reductions in populations of disease-causing organisms and pests. Various crops and green manures should be considered for inclusion in a rotation, as each may have economic, agronomic, and environmental value. The selection of rotation crops preceding potatoes and requirements for applying fumigants determines the tillage practices required to prepare the site for potato production. Since each field has its own unique set of characteristics, management procedures must be tailored to each field, with the goal of optimizing production efficiency of all crops grown over the long term.

Field Selection

The suitability of a field for potato production depends on topography, soil properties, cropping history, previous disease and pest problems, and ease of pesticide applications. It is unlikely that any site will be perfect in every respect, so the pros and cons for raising potatoes at any given location have to be determined. If potatoes have been raised previously on the site, the answer will come largely from the cropping history. If a new site is being evaluated, however, careful consideration of the characteristics discussed in this section is very important for identifying and avoiding potential problems. In addition to a thorough site inspection by an experienced agronomist, two key tools that aid the field selection process are historical crop production records and soil analysis reports.

Topography

Topography refers to the percent grade and aspect (direction) of slopes within a field. It also relates to the presence of features such as power lines, trees, homes, and other obstructions that may make some management practices difficult to perform. Selection of an appropriate site and the production practices to be used should take into account the potential for soil erosion, nutrient and pesticide leaching or runoff, and interference from nearby commercial or residential activities. Because of the possibility of non-point-source pollution, drainage patterns must be evaluated from the standpoint of possible off-site movement of surface water and groundwater carrying nutrients or pest control products.

Potato production is not well suited to highly **sloped fields**, because of tillage requirements and the risk of erosion. In fields with slopes greater than 10%, equipment such as planters, cultivators, and harvesters tends to slip sideways and therefore cannot be kept aligned properly. This creates a hazard for equipment operators and makes accurate row spacing difficult. Uneven row spacing often results in excessive damage to plants during cultivation and to tubers during harvest. The slope also affects air drainage from the field by concentrating cool air in low areas, where there is greater risk of frost damage. Low areas also accumulate excess moisture, resulting in greater risk of diseases such as late blight (Fig. 5.1).

The slope of a field also affects irrigation. Surface irrigation of slopes greater than 5% is difficult, if not impossible, without causing excessive erosion and loss of nutrients. Sprinkler irrigation can be used successfully on slopes in excess of 5%, although erosion may be severe in wheel tracks of center-pivot or linear-move systems. Regardless of the type of irrigation system, it is difficult to achieve uniform soil moisture in sloping fields. Low-lying areas tend to be too wet and slopes too dry, resulting in reduced tuber yield and quality. Even with sprinkler irrigation, successful potato production on slopes greater then 5% requires modification of tillage and cultivation practices. One common modification is reservoir tillage, in which specialized equipment, such as a Dammer Diker tillage system, forms a series of small depressions and ridges across the bottom of each furrow as the last tillage operation in the field. The small reservoirs formed down the rows hold water against erosion-causing runoff and increase the rate of infiltration of water into the soil.

In **irregularly shaped fields**, it may be difficult to effectively apply insect and disease control products, especially if the fields are bordered by trees or power lines. It may also be difficult to safely apply these products on fields located close to residential areas or other public facilities, because of potential problems with aerial drift. Epidemics of foliar diseases such as late blight and damaging insect infestations often start in areas where it is difficult to uniformly apply control materials (Chapter 13).

The **distance from fields cropped to potatoes the previous year** should also be taken into account, especially if pest problems were identified in those fields. Some insect pests, such as the Colorado potato beetle, overwinter near field edges and can move short distances in the spring to infest nearby fields. Where winter temperatures are not low enough

Fig. 5.1. A, Water runoff and ponding of irrigation water where potatoes are growing on a steep slope. **B,** Two large, coalescing late blight foci, appearing as chlorotic and blighted foliage in a low area of a potato field with center-pivot irrigation. A small late blight focus is in the middle foreground. (Courtesy D. A. Johnson)

to kill potato tubers left in the field, volunteer plants may serve as a source of plant-pathogenic fungi, bacteria, and viruses.

Soil Properties

Potatoes are adapted to a wide range of soil conditions, but not all soils consistently produce economic yields of high-quality tubers. Ideally, field sites should have low to moderate slopes and soils that are uniform, deep, well drained, fertile, and friable, with moderate to good water-holding capacity.

The biological, chemical, and physical properties of the soil influence potato productivity. Before a decision about suitability for potato production is made, every field should be extensively sampled to determine soil properties, nutrient concentrations, variations in nutrient concentrations, and nematode populations throughout the field. Laboratory analyses can be used to determine most soil properties. Other important properties can be determined through site inspection and bare soil imagery. Potatoes can be produced on sites with less than ideal uniformity, but successful production on such sites requires more careful management and increased inputs.

Biological properties are determined by the quantity and diversity of living organisms in the soil and are sometimes broadly characterized as soil health (Chapter 3). Although many soil organisms are detrimental to potatoes, most are beneficial. Larger soil organisms, such as earthworms, help build soil aggregates, aerate the soil, and ultimately make the soil more friable and conducive to root and tuber health. Many soil organisms are predators of detrimental insects, nematodes, and soilborne pathogens. They also serve as a major source of stored soil nutrients, which cycle to available forms throughout the growing season. Microbes are known to enhance the availability of certain mineral nutrients in soil. Unfortunately, testing soil for biological health is difficult at best, and quickly modifying soil health is even more difficult. The problem is that whatever stimulates beneficial organisms may also stimulate deleterious organisms. The best general recommendation is to evaluate field history for signs of problems with various soilborne pathogens and pests and to select a soil with chemical and physical properties that tend to promote a diverse and substantial microbial population.

The **chemical properties** of soil include pH, cation exchange capacity (CEC), organic matter content, and concentrations of nutrients and of various salts that are potentially toxic to plants. Soil pH affects each of the other chemical properties and greatly impacts plant growth. Potatoes grow well under a fairly wide range of soil pH levels. Soils with near neutral pH (pH 6.5–7.5) generally provide maximum nutrient availability. However, potatoes are frequently grown in acid soil (pH 5.0–6.0) and alkaline soil (pH 7.5–8.5). A soil test should be conducted to determine soil pH when the suitability of a field for potato production is being evaluated.

In alkaline soil (pH > 7.5), the availability of phosphorus, iron, zinc, copper, and manganese to plant roots is reduced. Much of the potato production in the United States is on alkaline soils in semiarid regions; hence, relatively high rates of fertilizer (especially phosphorus, manganese, and zinc) are required. Alkaline soils also are often calcareous (contain excess free lime), further exacerbating nutrient availability problems and requiring even higher rates of phosphorus and possibly other affected nutrients. Under these conditions, there is an increased need to apply a portion of fertilizers in a band and to make in-season applications to improve nutrient uptake efficiency. It is not generally cost-effective to lower the soil pH.

In acid soil, nutrient availability is also reduced, especially that of phosphorus. Acid soils also are often low in exchangeable bases and, as a result, are more likely to be deficient in calcium, magnesium, and potassium. Acid soils frequently require applications of lime to raise the soil pH to the optimal range for nutrient uptake and plant growth.

CEC is an important property, strongly influencing the ability of the soil to store water and retain positively charged ions (cations), such as potassium, calcium, and magnesium. The CEC also influences the leaching potential of negatively charged ions (anions), such as nitrate, sulfate, chloride, and borate. The primary determinants of CEC are clay and organic matter fractions in soil that have negatively charged surfaces, which attract cations and repel anions. Coarse-textured, sandy soils have relatively low clay contents and as a result have a lower ability to retain nutrients than fine-textured soils with high clay contents (Fig. 5.2). Sandy soils also tend to be warmer, because of their lower water content, and as a result contain less organic matter, because of the rapid decomposition of plant residues, further lowering the CEC. Soils with relatively high CEC (>28) tend to require less irrigation and have greater nutrient retention ability, but they also tend to be poorly drained and are more subject to problems with excess water. Soils with relatively low CEC (<10) have to be irrigated more often and usually require in-season nutrient applications.

Saline soils (having a high salt content) reduce potato growth and yield by decreasing the ability of the plant to absorb water from the soil. Salinity increases the osmotic potential of the soil water solution, making it more difficult for plant roots to extract water from the soil. Salinity can also interfere with plant physiological processes involved in dry matter production and distribution within the plant. Any soluble salt, including nutrients such as nitrogen and potassium, can contribute to this effect. Although excessive salt is toxic, some salts are necessary for plant growth and are found naturally in soil, fertilizer, manure, and water.

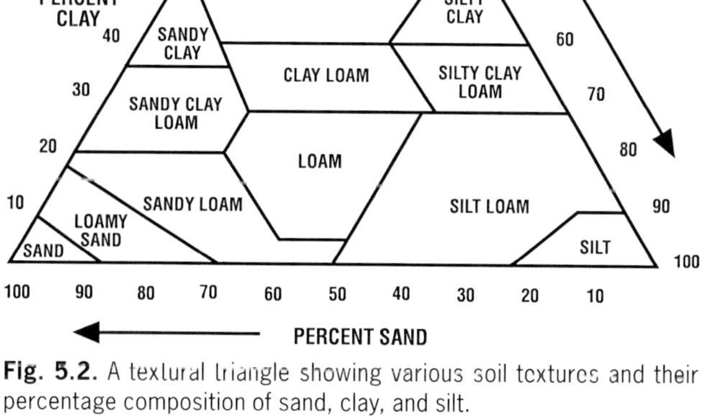

Fig. 5.2. A textural triangle showing various soil textures and their percentage composition of sand, clay, and silt.

Saline soil can be reclaimed if it has adequate infiltration and drainage and if sufficient amounts of low-salt water are available to leach salts out of the root zone. The amount of water needed to leach salts out of a soil will depend on the depth, texture, and permeability of the soil and the proportion of the initial salinity that must be removed. The salt concentration is measured by the electrical conductivity (EC) of the soil solution, expressed in decisiemens per meter (dS/m). For potatoes, the EC of a saturated extract of the soil should not exceed 1.7 dS/m to avoid yield loss, while the EC of the irrigation water should not exceed 1.1 dS/m.

One mineral element that causes problems for plant growth is sodium. In sodic soils (having a high sodium content), the high sodium concentration disrupts the soil structure, resulting in a substantial reduction in water infiltration and permeability. Sodic soils tend to become waterlogged, interfering with root growth. A high concentration of sodium in soil can be toxic to plants, causing disruption of physiological processes and tissue damage. The combined effects of a high sodium concentration on the physical characteristics of the soil and on plant physiological processes can result in severe stunting and even death of potato plants.

The sodium hazard can be identified by soil and water testing. The exchangeable sodium percentage of soil provides an indication of sodium hazard. No problem should be expected for soils with less than 5% sodium on the exchange complex. The risk of salt problems increases as the percentage increases above 5%, with almost certain problems above 15%. The sodium adsorption ratio (SAR) is the ratio of the sodium concentration in irrigation water to the concentrations of calcium and magnesium. If the SAR is below 6, there should be no significant reduction in soil permeability. When the SAR increases to between 6 and 9, there is an increasing hazard to soil permeability. Severe permeability problems are likely when the SAR exceeds 9.

To reclaim sodic soils, sodium has to be displaced from cation exchange sites by calcium. This can be accomplished by adding a soluble source of calcium, such as gypsum, or by adding acidifying agents, such as sulfuric acid or elemental sulfur, to dissolve existing limestone in the soil. Reclamation of sodic soils typically requires large amounts of amendments, often on the order of several tons per acre.

Physical properties of soils include texture (the proportions of clay, silt, sand, and gravel), structure (the arrangement of individual soil particles into definable aggregates), and bulk density (mass per unit of volume). Potatoes can be produced in soils with a wide range of textures. Soil texture determines the water infiltration rate and the ability of the soil to store water available for plant growth. The availability of soil water is a major factor determining potato yield and quality and can have a direct effect on root- and tuber-infecting pathogens. Inadequate water at critical times during the growing season increases the severity of diseases such as common scab (Chapter 19) and certain tuber disorders (Chapter 23). Excess water also reduces tuber yield and quality and accentuates disease problems associated with wet soils, both in the field and in storage. Excess water also leaches nutrients out of the root zone, thus affecting the growth of a shallow-rooted crop such as potato (Chapter 9).

The water infiltration rate of a soil is important in determining the method, application rate, and duration of irrigation. The relatively low infiltration rates characteristic of fine-textured soils make them more suitable for set-move sprinkler irrigation systems, such as wheel lines and hand lines, or solid-set systems. Surface irrigation is also more effective on fine-textured soils. Sandy soils, on the other hand, typically have high infiltration rates and low water-holding capacities, making them more suitable for center-pivot and linear-move irrigation systems, which have the ability to effectively apply less volume more frequently than other types of irrigation.

Soil water-holding capacity, permeability, and tilth tend to improve as soil organic matter content increases. However, in many soils in potato-producing areas, the organic matter content is less than 1%. As a result, consistent efforts must be made to increase soil organic matter levels by returning crop residues to the soil and minimizing residue burning and unnecessary tillage.

Rocky or gravelly soils often present tillage problems and may require specialized harvesting equipment. Other problems associated with such soils include unacceptable amounts of foreign matter in the harvested product and increased bruising of tubers during harvesting and handling.

Soil structure and bulk density are related physical properties that greatly impact potato production. A granular soil structure is best for passage of water and roots throughout the soil. Blocky structure is also acceptable. Soils with platy structure have reduced water movement and restricted root penetration. The worst soils have no structure at all. In soils where structure is absent, the particles tend to fit more tightly together, thus inhibiting penetration by water and roots. Lack of soil structure is usually indicated by high bulk density.

Compaction

Compaction is a consequence of any process that destroys soil structure and results in a denser soil, having less pore space to hold water and air. Compaction can be assessed formally with a penetrometer or more commonly with a shovel. The more energy that is required to push a penetrometer or shovel through the soil, the more difficult it will be for plants to grow a rooting system and high-quality tubers. The extent of compaction is determined by the size and weight of equipment, the moisture level of the soil, and the type of soil.

Most compaction is associated with tire compression and tillage operations. A force as low as 4 pounds per square inch (psi) is enough to cause significant soil compaction, and pressure on the soil surface from the edge of a tractor tire can exceed 25 psi. One way to reduce compaction is to use lighter equipment or spread the weight of the equipment out over dual tires or more axles. The number of trips over the field should also be reduced when possible. Plow pan or subsoil compaction occurs in a narrow zone just below the depth of normal tillage operations. It is most common in fields tilled at the same depth year after year.

Wet soils have weaker structure and are more susceptible to compaction than dry soils. When possible, allow fields to dry before conducting tillage operations. Even 1 or 2 days of drying may be enough to prevent serious soil compaction in some cases (Fig. 5.3). Generally, soils high in clay or low in organic matter compact more readily than coarse sandy soils and soils high in organic matter. However, hardpans in sands have been shown to severely limit potato root penetration. In addition, compaction of clay and loam soils can lead to clod formation, resulting in soil separation and tuber damage at harvest. Sandy loam soils are common in many potato production regions, and compaction may be one of the most unrecognized factors that limit potato yield and quality in these soils.

Compaction can occur at different depths in the soil profile. Soils compacted in the surface layer can usually be loosened by tillage operations, freeze-thaw cycles, and the growth of some deep-rooted crops, such as alfalfa. Subsoil compaction is difficult to treat, because it is below the zone of normal tillage. Compaction within the shallow zone where roots initially form is generally considered to be more damaging to crop growth than deep compaction.

Soil compaction affects potato plant growth in a number of complex ways. Air and water movement are restricted, so that their availability to plants is limited. Roots do not develop or penetrate the soil as well, resulting in a much restricted and shallow root system. The plant also has to expend more energy on root and tuber growth, reducing the energy available for other growth processes. The shallow root system limits the ability of plants to obtain nutrients and to withstand periods of moisture stress, so that plants are less vigorous and tuber yield and quality are lowered.

Potatoes have a very fine, branching root system that is especially susceptible to damage due to compaction. Compaction greatly restricts both the total root mass and the maximum depth of rooting. The result is a delay in emergence, slower canopy development, and reduced yields. Marketable yield losses of 20–30% have been attributed to compaction, with tuber size and quality reduced to a greater extent than total yield.

The negative impacts of compaction on potato growth cannot be overcome by adding additional water, fertilizer, or amendments to the soil. Compaction is partially remedied by tillage practices such as chiseling, which breaks up com-

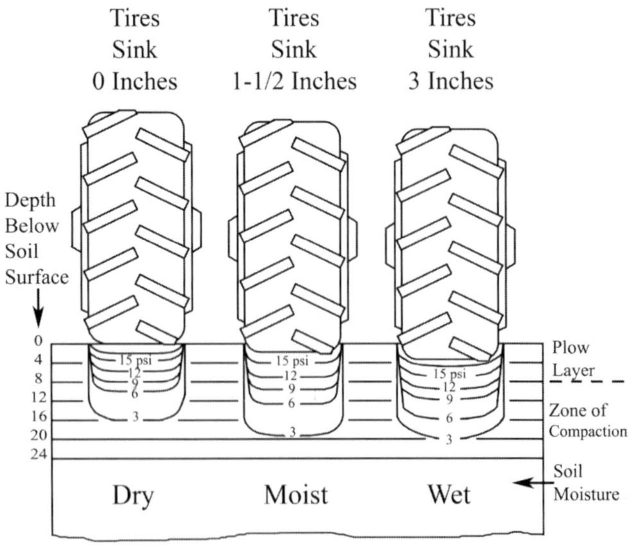

Fig. 5.3. Relationship between soil moisture, tire pressure, and compaction. (Redrawn from D. R. Daum and R. F. Shipp, 1999, Agricultural soil compaction—Causes, effects and cures, Cooperative Extension Fact Sheet B-79, Pennsylvania State University)

pacted layers, but tillage cannot substitute for good soil structure. Fields with significant compaction should not be used for potato production (Box 5.1).

Crop Rotation

The primary consideration in cropping history is the effect of previous crops on populations of disease organisms, nematodes, weeds, and insects in the soil. Pest control chemicals applied to previous crops should also be considered, because of potential carryover damage by herbicides and label restrictions on plant-back intervals for some pesticide products.

The inclusion of nonhost crops in a rotation can reduce populations of pathogens and pests that attack potato, particularly when the number of years between potato crops is increased. Increasing the length of the rotation also helps reduce the severity of several diseases in which soilborne inoculum plays a significant role, including white mold, pink rot, Pythium leak, common scab, early blight, and *Rhizoctonia* diseases (Chapter 19). Planting nonhost crops and establishing long rotations are also primary management tools for reducing nematode populations in soil.

Knowledge of **past weed problems** is an important factor in deciding whether to plant potatoes in a specific field. Perennial weeds can be a serious problem. Most perennial weed infestations should be controlled by using a combination of herbicides and tillage during the production of the crops preceding potatoes in the rotation (Chapter 22). In some fields with severe perennial weed problems, it may be advisable to delay potato production 1 or 2 years and plant other crops in which appropriate weed management programs can be implemented. Fields heavily infested with Canada thistle, bindweed, or Russian knapweed should be avoided completely, if possible.

Crops that can become weeds later in a rotation should be spaced two or more growing seasons ahead of the potato crop. **Volunteer potatoes** can also become weeds in subsequent crops, in places where the soil does not freeze deep enough to kill tubers left in the ground after harvest. In some climates, volunteer potatoes may reproduce every year and carry over several years into the next potato crop, resulting in cultivar mixtures and providing a reservoir for pathogens.

An appropriate crop rotation scheme should allow opportunities in the production cycle for the control of volunteer plants and weeds when the field is not planted with potatoes. A rotational management strategy can include such practices as well-timed applications of broad-spectrum herbicides, the use of herbicides that cannot be used during potato production but can be applied when other crops are in production, and irrigation timed to stimulate the germination of annual weeds or volunteer grain after harvest, followed by timely tillage to remove them.

Herbicide carryover from previous crops can be a problem for potatoes, especially after a dry growing season and a dry winter. Some herbicides applied to cereal crops at labeled rates, particularly sulfonylurea and imidazolinone products, can remain in the potato root profile the following year at concentrations high enough to adversely affect potato growth. The potential for herbicide carryover can be minimized by following label directions, particularly for specifications on rates and the time interval between application and planting a sensitive crop.

Box 5.1

Soil Compaction

The best way to deal with compaction is to treat the causes, not the symptoms. The following management practices will help reduce soil compaction.

- Minimize the number of trips across the field.
- Eliminate unnecessary field operations. Conduct minimum tillage or no-till throughout the rotation, where practical.
- Avoid field operations when soils are wet. The drier the soil, the less compaction (Fig. 5.3).
- Reduce axle loads by using lighter equipment or distributing the weight over more axles.
- Vary the depth of tillage operations to avoid creating plow pans, which restrict root growth and water infiltration.
- Increase soil organic matter. Incorporate green manure crops, manure, compost, and crop residues into the soil to build organic matter and improve soil structure and drainage.
- Increase the number of years between potatoes and other row crops in the rotation. Row crops return little crop residue to the soil, and the heavy equipment traffic associated with tillage and harvesting operations in these crops results in soil compaction. In contrast, deep-rooted crops, such as alfalfa, corn, sunflower, and small grains, help break up compacted layers and return lots of residue to the soil after harvest.
- Improve surface and subsurface drainage. Soils will dry and warm up faster, allowing more flexibility in crop management operations.

Information on all pesticides applied should be entered in a permanent record for each field, including at least the following: materials applied; rates, times, and method of application; soil texture (including organic matter content); and soil pH. The cropping history prior to potatoes and the climatic conditions during the previous growing season should also be recorded. Important climatic information includes soil temperature (winter and summer), the amount of soil moisture, and the movement of water through the soil profile. This information is needed to determine the degradation rate of previously applied pest control products. If accurate records are not available, soil can be sampled and tested to detect residues of potentially damaging herbicides. Potatoes should be planted on the site only if it can be determined that no potentially damaging residues are present and that no pest control products have been applied whose label instructions prohibit planting potatoes the following year.

If a field history is not available, **laboratory tests** can identify and quantify nematodes, some soilborne fungal pathogens, and some herbicides. The validity of laboratory test results depends on adequate, thorough sampling and reliable laboratory procedures. Consult local agricultural advisors for instructions on sampling, handling samples, and identifying a reliable testing facility. If lab test results indicate a potentially severe problem with nematodes, insects, or diseases for which current control methods are inadequate, the correct decision may be to not use the field for potato production.

If field history or lab tests indicate a potentially severe problem with nematodes, insects, or diseases for which current control methods are inadequate, the correct decision may be to not use the field for potato production.

An appropriate crop rotation can also help reduce or prevent **soil erosion** during potato production and throughout the entire rotation. Residues from rotation crops should be managed so as to maintain or improve the organic matter content, the aeration of the soil, and water infiltration and retention. Cereal crops, including small grains and corn, are commonly planted for this purpose. Perennial legumes, such as alfalfa, as rotation crops provide both nitrogen and organic matter, and their roots can penetrate and break up deeper soil layers, thus helping to reduce the effects of compaction. Annual legumes, such as beans, also provide nitrogen, although in lower quantities. Avoid planting potatoes after high-traffic crops with low levels of crop residue, such as sugar beets and onions, because of the negative effects of erosion and soil compaction on potato yield and quality.

A primary benefit of a well-designed crop rotation is improved soil quality. The addition of large amounts of **crop residue** to the soil over time can improve soil tilth and nutrient availability, increase water-holding capacity and infiltration rate of the soil, and improve drainage. These improved soil conditions also tend to reduce compaction, water runoff, and erosion.

Residues from previous crops in the rotation can affect the availability of plant nutrients, especially nitrogen. Large amounts of crop residues with a high ratio of carbon to nitrogen, such as wheat straw and cornstalks, may immobilize nitrogen early in the season, so that it is not available to the potato crop until the residues completely decompose. In contrast, residues with a lower ratio of carbon to nitrogen, such as bean straw, alfalfa hay, and residues of green manure crops, may release large amounts of nitrogen into the soil for use by the potato crop during the season. Both situations can complicate nitrogen management in a subsequent potato crop. The positive aspects of crop residues usually override difficulties they may cause in managing nitrogen. Appropriate nutrient management can be achieved by incorporating residues well ahead of planting, banding fertilizer below and to the side of the seed pieces at planting, and avoiding the placement of excessive amounts of residues in the vicinity of the seed pieces.

Green Manures

Green manures are crops grown specifically for incorporation into the soil while plant tissue is still green and growing. A green manure crop can be used in a crop rotation to improve soil physical conditions, increase organic matter, and reduce populations of some soilborne pathogens and pests. In some cases, incorporation of a green manure crop has provided enough disease suppression to replace fumigants. In other cases, green manures suppress pathogen populations to levels where they are more easily controlled by other management tactics.

The type of crop selected, the total amount of plant material produced, and the timing and method of incorporation determine the benefits derived from green manures. Legumes (e.g., Austrian pea, clover, vetch), cruciferous crops (e.g., mustard, oilseed radish, rapeseed), and sudangrass are commonly grown as green manures in potato cropping systems. Legumes have the potential to provide significant nutrient benefits and suppress some diseases, and cruciferous crops and sudangrass have been shown to be very effective in suppressing certain diseases, weeds, and nematodes (Chapter 20). Some of these plants produce substances belonging to a class of compounds called glucosinolates, which break down in the soil after incorporation to form biologically active compounds that are toxic to some soilborne pests.

To achieve the best results, green manure crops have to be managed to produce high amounts of biomass for incorporation. Crop biomass production in the range of 10–20 tons (fresh weight) per acre has been shown to be more effective in suppressing nematodes and pathogens than lower amounts. Important management practices include adequate nitrogen fertilization to promote plant growth and speed residue decomposition, seeding at the recommended density, and providing adequate soil moisture during crop establishment and growth. Weed control is a key factor, as a high weed density can reduce crop growth and negate the benefits of soilborne pest suppression by providing host plants for nematodes and pathogen reproduction.

Any factor that reduces the amount of green leaf material incorporated (frost, leaf senescence) may reduce the benefits of a green manure crop. A green manure crop should also be incorporated before it produces seeds, to prevent it from surviving as a weed in a subsequent crop in the rotation. Green manure crops should generally be flailed or mowed and then incorporated immediately by disking or plowing, to thoroughly mix the plant residue in the soil and speed decomposition. Direct plowing of green manure into the soil is the least recommended method of incorporation, because it does not adequately mix the residues into the soil. Thorough incorporation of green tissue promotes a rapid increase in bioactivity and release of nutrients and chemicals that suppress soilborne pathogens and pests.

A secondary benefit of green manure crops is that they increase the microbial diversity of soil as they decompose (Chapter 3). Microbially complex soils promote vigorous plant growth and suppress disease development by providing competition for pathogens, including parasitic nematodes. Green manures also tend to capture and recycle nutrients that are subject to leaching and volatilization loss (Box 5.2).

Field Preparation

Tillage

Selection of tillage operations will depend on soil type and structure, the susceptibility of the soil to erosion, and energy costs, as well as the residue management requirements of the previous crop. Tillage operations should be designed to effectively incorporate crop residues and fertilizers, break up compacted soil layers, improve water infiltration, and prepare the soil for planting, while minimizing the potential for soil erosion. A properly designed tillage program can also help improve the control of annual and perennial weeds and volunteer potatoes. The timing of tillage operations to achieve maximum destruction of weeds is critical for peak effectiveness.

In some areas, land preparation is done entirely in the fall. This includes plowing, disking, subsoiling, and forming beds. Fall bedding facilitates earlier planting in the spring and minimizes clod development by eliminating traffic on the beds, thus taking full advantage of the effects of winter freezing and thawing. It may also eliminate overwintering sites for some pests, such as the European corn borer in corn stubble. The main disadvantage of fall bedding is exposure of land to wind and water erosion during the winter.

In regions where fall bedding is not commonly practiced, preplant bedding, or "marking out," is done in the spring prior to planting. This is especially desirable when large, multirow planting equipment is used, to ensure accurate row spacing. Tractors with steering systems guided by a global positioning system (GPS) have recently been introduced. These systems help eliminate errors during the row-marking operation, which prevents plant and tuber damage during subsequent hilling and harvesting operations.

Sandy soils may need to be partially prepared in the fall to reduce wind erosion. On these coarse-textured soils and in areas where water erosion is a problem, some form of minimum tillage, or **conservation tillage**, may be appropriate. Conservation tillage leaves crop residues on the surface or incorporates them only slightly into the soil. The residues may be from a previous crop, such as cereal stubble, or a cover crop, such as rye, intentionally seeded to anchor the soil in place. Several concerns about minimum tillage often arise. Some growers are reluctant to accept trashy-looking fields or purchase special equipment to conduct minimum tillage operations. Minimum tillage also makes it more difficult to apply preplant, broadcast-incorporated fertilizers. Incorporation of preplant herbicides is also difficult, but there are alternative herbicides that can be used after planting.

Factors to be considered in selecting a method of preplant tillage include the amount of crop residue to be incorporated, the depth of tillage desired, and the condition of the soil surface required. **Excessive crop residue** should not be left in

Box 5.2

Advantages and Disadvantages of Green Manures

Potential Advantages of Growing a Green Manure Crop in a Potato Cropping System

- Increased supply of nutrients for uptake by subsequent crops
- Improved potato yield and quality
- Improved soil tilth and organic matter content
- Improved water infiltration and water-holding capacity
- Suppression of some soilborne pathogens, nematodes, and weeds
- Reduced soil erosion

Potential Disadvantages of Growing a Green Manure Crop in a Potato Cropping System

- Reduced income if the green manure crop replaces a rotation crop that would be harvested and sold
- Additional seed, tillage, and irrigation costs
- Additional labor and machinery costs
- Creation of a potential weed problem if the green manure crop produces seed
- Increased pest populations if an inappropriate green manure crop is chosen

Factors to Consider in Selecting a Green Manure Crop

- Ability of the crop to quickly produce substantial foliar growth
- Ability of the crop to fix nitrogen or take up nitrogen and other nutrients that would otherwise be lost as a result of leaching
- Compatibility with management requirements of other crops in the rotation
- Compatibility with the growing environment during the period when the green manure crop will be grown (for example, frost tolerance and the ability to grow in cool weather are important in many northern production regions)
- Availability of products registered for the management of weeds and other pests
- Availability of seed and lack of planting restrictions due to interference with seed production of closely related crops
- Tendency to produce and shed seeds that develop into volunteer plants causing problems in a subsequent crop

the vicinity of potato seed pieces. A large amount of undecomposed organic matter, especially a recently incorporated green manure crop or animal manure, may favor the development of tuber diseases such as common scab, Pythium seed piece decay, and bacterial soft rot of seed pieces. It may also increase the incidence of the seedcorn maggot damage to seed pieces after planting.

Managing **soil compaction** is another key factor determining the type and timing of preplant tillage operations. Compacted soil layers can be disrupted by subsoiling below the plow layer with deep rippers or chisel plows when the soil is relatively dry. Subsoiling when the soil is too wet will usually not adequately fracture the compacted layer and may in some cases make the compaction problem worse. For subsoiling to be effective, the tillage depth must exceed the compaction depth, so that hardpans are adequately disrupted. Ideally, shanks should be set at an operating depth that will lift and shatter the compacted soil layer. Shank spacing should ensure that the entire surface layer of soil is disrupted by the tillage operation. An integrated approach to reducing compaction should also include increasing the organic matter content of the soil through the use of soil-building crop rotations and appropriate residue management practices.

Soil moisture should be considered when preplant tillage operations are being determined. The optimal amount of moisture in a seedbed at planting time is 75–80% of available soil water. Subsoil moisture should also be considered, as it is an important source of water during the hottest time of the growing season, when evapotranspiration may exceed the capacity of the irrigation system. In unirrigated areas, if rainfall is not adequate to provide good seedbed moisture, planting may need to be delayed if sufficient rainfall is likely to occur in time to allow planting by a reasonable date. If spring rainfall is usually inadequate, tillage practices that conserve winter moisture, such as fall bedding, should be considered. In irrigated production areas, preplant irrigation in the fall or spring can be performed to increase soil moisture. Enough water should be applied before planting that further irrigation is not needed until after the plants have emerged. Irrigation after planting but prior to emergence should be avoided, as it promotes seed piece decay. In areas with fine-textured soils and abundant winter and spring rainfall, soil moisture at planting is often excessive. Fields in these areas should be equipped with drainage tiles to help remove excess water.

One of the best management strategies is to **prevent the introduction of soil pests** into clean fields. Tillage equipment brought in from other fields should be thoroughly cleaned of all soil before use, to prevent the introduction of nematodes, weed seeds, and disease-causing organisms.

Fumigation

In some production areas, parasitic nematodes, soilborne insects, and pathogens are limiting factors affecting tuber yield or quality. Preplant soil fumigation is commonly practiced in these situations, especially in areas with coarse-textured soils and relatively long growing seasons. Soil fumigation is most often used for suppressing nematodes and the fungi causing Verticillium wilt. In some areas, it is also used to reduce tuber damage from wireworms. Some fumigants reduce annual weed populations by lowering seed viability and reduce perennial weed populations by weakening or destroying rhizomes or stolons.

Soil fumigation has significant advantages and disadvantages. Properly done, it can effectively help manage soilborne pathogens, nematodes, and insect pests. However, it is an expensive procedure requiring special equipment and well-trained applicators, because most fumigants are highly toxic (Chapter 13). Furthermore, fumigation completely alters the community of living organisms in the soil and thus may have unintended side effects, such as a change in the response to applied fertilizers or higher levels of residual herbicides. These effects occur because fumigation reduces populations of soil microorganisms that play a role in nutrient cycling and herbicide breakdown.

Fumigation is usually done after tillage, because good soil preparation, along with proper application procedures, is important in achieving the desired results. Before a soil-injected fumigant is applied, the soil should be in good seedbed condition. It is important that clods and compaction layers are broken up and crop residues are finely chopped and thoroughly incorporated. If this is not done, target organisms in soil clods and large pieces of plant debris may not be exposed to the fumigant and thus may survive the treatment. Plant debris that is not well incorporated also allows the fumigant to escape from the soil before the target pests are exposed for the duration needed to achieve the desired results.

Soil fumigation is **not a cure-all**. It can reduce populations of soilborne pests and pathogens, but these organisms are not completely eradicated. Following fumigation, populations of nematodes, insects, and pathogens are likely to increase again to damaging levels, possibly after only one season of a susceptible crop (Chapter 3). Because of these limitations, fumigation should always be done in conjunction with appropriate soil sampling and analysis for populations of target organisms.

CHAPTER 6

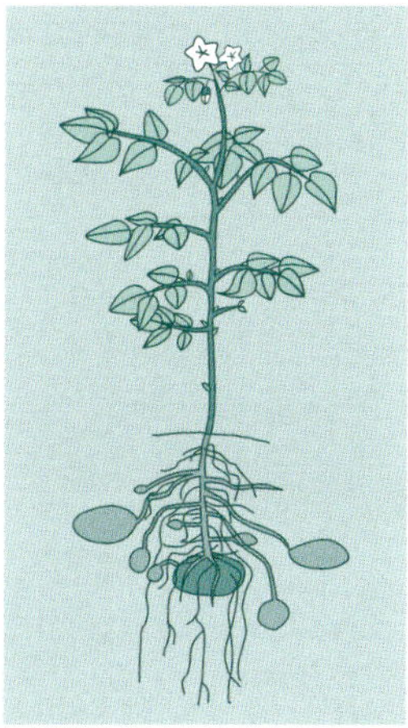

Jonathan L. Whitworth
U.S. Department of Agriculture, Agricultural Research Service
Aberdeen, Idaho

Robert D. Davidson
Department of Horticulture and Landscape Architecture
Colorado State University, Center

Quality Seed: Seed Improvement, Cultivar and Seed Lot Selection, and Certification

An essential tool for the holistic health management of a potato crop is quality seed stock. Healthy, productive seed provides the essential base to obtain optimum tuber yield and quality that result in obtaining maximum profits. Choosing seed stock based only on price and availability of seed may compound inherent difficulties in growing a potato crop. Thus, seed stock selection should be based on specific needs within the operation and a clearly identified market. Key components in choosing quality seed includes selection of the appropriate cultivar, understanding the certification process, obtaining complete information about potential certified seed lots, and the careful analysis of the market prospect.

The primary goals of seed potato improvement are to maximize productivity of marketable tubers by eliminating seedborne pathogens and pests and by selecting superior clones within a cultivar. The certification process is where most of these efforts take place. Selected cultivars must meet market needs. Once a cultivar is selected, it is moved into the certification system, which in turn sets in motion a series of steps that insures that seed lots meet the above criteria.

Cultivar Selection Based on Market Requirements

Both the processing market and the fresh market have specific requirements for tubers, including such characteristics as specific gravity, sugar level, skin and flesh color, and appearance. In addition to meeting these market requirements, a cultivar must be the capable of producing a high marketable yield.

Processing cultivars can be divided into three groups: fries, chips, and dehydrated. Tubers of cultivars acceptable for processing into **fries and chips** must have relatively high specific gravity and low levels of reducing sugars. More finished product can be obtained from tubers with high specific gravity than the same amount of tubers with low specific gravity. Specific gravity is a means of estimating the dry matter content of tubers. Dry matter, also referred to as solids, is primarily made up of starch. In the frying process, the amount of oil uptake is lower when the specific gravity is high. Low sugar content after storage (especially low-temperature storage) is also desirable, as sugars contribute to the darkening of fries and chips. Potatoes for processing are traditionally stored at 44–48°F or, if stored at a lower temperature, reconditioned at higher temperatures to convert sugars to starch. If a cultivar can be stored at a lower temperature and still fry a light color, then it can be held longer in storage with less artificial sprout control. For example, Snowden was the first commercial cultivar that had low sugar content even after storage at a low temperature. Additional characteristics apply to fries: the cultivar generally needs to be oblong and russet or white-skinned and needs to have a size profile with a high number of large, blocky tubers. For chips, round tubers with white flesh are desired. Other tuber shapes are used for wedge-type fries, including short, blocky types. New chip-cutting machines or processes may utilize tubers other than round types.

Potato **dehydration** was originally developed to produce instant reconstituted mashed potatoes but has evolved over the years to include reconstituted chip products, which have gained wide market acceptance. High specific gravity is a cultivar characteristic that helps maximize product output during dehydration. Typically, tubers used for dehydration are off-grade or culls, but the acceptance of cultivars that produce a high proportion of in-grade tubers reduces the amount of tubers available for dehydrating. In response, breeding programs have started to produce cultivars with high levels of dry matter for dehydrating.

Cultivars used for table stock, or **fresh-market tubers**, must have good tuber appearance and boil, mash, or bake well. Culinary characteristics of tubers are listed in Table 6.1. The categories utilized for fresh markets include long russet, long white, round white, red, and yellow-fleshed cultivars (such as Yukon Gold).

Decisions about the cultivar to grow are often made between the producer and the end user. For example, potatoes used for processing are usually grown under contract with the processing company specifying the cultivar. Many fresh-market cultivars are grown on the open market. If a dual-purpose cultivar is chosen that could be either fresh

Table 6.1. Physical characteristics of potato tubers related to culinary use

Specific gravity	Total solids (%)	Texture	Best use
<1.060	<16.2	Very soggy	Pan frying, salads, canning
1.060–1.069	16.2–18.1	Soggy	Pan frying, salads, boiling, canning
1.070–1.079	18.2–20.2	Waxy	Boiling, mashing; fair to good for chip processing and canning
1.080–1.089	20.3–22.3	Mealy, dry	Baking, chip processing, frozen french fry processing; some cultivars tend to slough when boiled
>1.089	>22.3	Very mealy or dry	Baking, frozen french fry processing; tendency to produce brittle chips and to slough when boiled

packed or processed, then more market opportunities may be available.

Disease-Resistant Cultivars

Knowledge of a cultivar's reactions to major diseases, such as blackleg, Verticillium wilt, common scab, powdery scab, early blight, late blight, and virus infections, is valuable when selecting a cultivar. Information on the strengths and weaknesses of 15 leading cultivars is given in Table 6.2. The foliage and tubers of cultivars may differ in reaction to a particular pathogen. For example, tubers of Umatilla are moderately resistant to late blight, but the foliage is susceptible. Jacqueline Lee has foliar resistance to late blight. Defender has late blight resistance in both the foliage and the tubers.

Field history is also an important consideration in selecting cultivars when soilborne pathogens may be important. For example, light-skinned cultivars are very susceptible to common scab and should not be grown in fields with a recent history of this disease. A russet-skinned cultivar with resistance would be a better choice.

Efforts to develop resistance to *Potato virus Y* (PVY) have resulted in a small number of cultivars with varying amounts of resistance, including Bannock Russet, Umatilla Russet, Defender, and Ranger Russet. All of these are less susceptible to PVY than Russet Burbank. Ranger Russet has a whole-plant hypersensitive response to PVY. For example, Ranger Russet seed pieces infected with PVY strain O (PVY^{O}) produce very stunted plants, which die about halfway through the growing season.

Cultural Requirements and Seed Condition

Potato cultivars have specific cultural requirements for optimal growth, and such requirements need to be considered when selecting cultivars. For example, if Ranger Russet receives the same rate of nitrogen fertilizer as Russet Burbank, the plants will not fully mature before harvest, so that skin set is poor and the storability of the crop is decreased. When a new cultivar is developed, the first few years in commercial production are critical in determining and utilizing the best practices to produce the cultivar. Growth requirements that may be cultivar-specific include plant nutrition, optimal seed piece size and spacing, water needs, length of growing season, and storage conditions.

History of Seed Potato Certification

Prior to 1900, the propagation of potatoes by tubers was believed to be responsible for the degeneration of cultivars and decreased yields, which were invariably observed over a period of years. It was assumed that the initial vitality of a cultivar that had become less productive, or "run out," could be restored only by propagation from true seed. Now it is known that the use of true seed eliminated many tuber-borne pathogens, especially viruses that had accumulated in tubers of seed stocks of a cultivar over several years of field propagation. However, plants produced from true seed were not true to type of the original cultivar clones.

By 1900, Dutch and German agriculturists found that certain symptoms of degeneration, such as leaf curling, rolling, crinkling, and blotching could be transmitted from plant to plant and persisted from one crop to another. Careful observation taught them that removing plants with these symptoms from seed fields was helpful in reducing these problems. A few U.S. scientists traveled to Europe in the early 1900s to study potato diseases and observe the systems for inspection and production of potato seed stocks used in Germany. It became evident to them that pests and disease organisms were being distributed with seed stocks, which were being shipped freely throughout the world. This situation was officially recognized by the U.S. government with the passage of the National Plant Quarantine Act of 1912. This law prohibited the introduction of potatoes with black wart, a disease caused by the fungus *Synchytrium endobioticum*. Two years later, a similar embargo was placed on potatoes with powdery scab, caused by the fungus *Spongospora subterranea*.

These events set the stage for the establishment of seed potato certification programs. In 1914, scientists from the U.S. Department of Agriculture presented a proposal for certification at the first annual meeting of the Potato Association of America and at the first official Potato Seed Certification Conference. Seed potato certification programs were initiated between 1913 and 1915 in New Brunswick, Prince Edward Island, Idaho, Maine, Maryland, Vermont, and Wisconsin. By 1922 programs were in place in 22 provinces and states, covering about 24,500 acres of seed potatoes.

Over the years, U.S. certification programs came to the realization that to be truly effective, uniformity among systems was necessary. To this end, the Certification Section of the Potato Association of America was formed in 1964. Membership included the managers of each certification program in the United States and Canada. This collaboration prompted several areas of uniformity, including developing the U.S. No. 1 Seed Potato Grade, uniform tolerances for

Table 6.2. Acreage, use, and reported disease characteristics of 15 leading cultivars grown in the United States and Canada in 2004

Cultivar and acreage[a]	Tuber type	Primary uses	Verticillium wilt[b]	Common scab[b]	*Potato virus* Y[b]	Leafroll net necrosis[b]	Late blight[b,c] Foliar	Late blight[b,c] Tuber	Storage rots[b]	Strengths	Weaknesses
Atlantic (7,102)	Round white	Fresh market, chipping	MS	S	MS	R	S	MS	S	Uniform tuber size and shape, high specific gravity	Susceptibility to internal heat necrosis and hollow heart
Chieftain (4,787)	Round red	Fresh market	MS	MS	S	MR	S	S	MS	Wide adaptation, attractive appearance, high yield	Susceptibility to skinning and internal heat necrosis
Dakota Pearl (2,127)	Round white	Fresh market, chipping	R	MR	S	MR	S	S	S	Resistance to blackspot bruise, good chipping quality from low-temperature storage	Rarely, internal necrosis
Goldrush (5,257)	Long russet	Fresh market, processing	MR	R	S	MR	S	S	S	Low external and internal defects, culinary quality	Medium dormancy, twin and glove-shaped tubers
Kennebec (3,593)	Oblong white	Fresh market, processing	VS	S	R	R	MR	S	MR	Wide adaptation, high yield, multipurpose, some resistance to late blight	Susceptibility to external defects and bruise
Norland (10,228)	Round red	Fresh	S	MR	MS	MR	VS	S	MS	Culinary quality, early maturity, attractive appearance	Short dormancy, high susceptibility to silver scurf
Ranger Russet (40,310)	Long russet	Frozen processing, fresh market, dehydrated products	MR	MS	MR	MS	S	VS	MR	Low internal defects, high yield	Susceptibility to blackspot bruise and common scab
Red LaSoda (3,348)	Oblong light red	Fresh market	MS	MS	S	MS	S	VS	MS	Culinary quality, high yield	Deep eyes, susceptibility to hollow heart
Russet Burbank (51,367)	Long russet	Frozen processing, fresh market, dehydrated products	MS	MR	S	S	S	S	MS	Wide market use, storability, culinary quality	Stress defects, uniform soil moisture required
Russet Norkotah (22,502)	Long russet	Fresh market	VS	MR	VS	MR	VS	VS	MR	Attractive appearance, high proportion of U.S. No. 1	Susceptibility to Verticillium wilt and *Potato virus Y*, low culinary quality
Shepody (11,521)	Long white	Frozen processing	MS	S	VS	S	VS	VS	S	Good fry processor, early maturity	Susceptibility to common scab and pink eye, inconsistent yield
Snowden (2,502)	Round white	Fresh market, chipping	MS	MR	S	MR	S	MR	MR	Good chipping quality from storage, high specific gravity	Mild leafroll symptoms, small tuber size, moderately deep eyes and netted skin
Superior (5,943)	Round white	Fresh market, processing	S	MR	S	R	S	S	S	Early maturity, scab resistance, low internal defects, good quality for processing	Inconsistent type and skin texture, susceptibility to Verticillium wilt and stress
Umatilla (2,581)	Long russet	Frozen processing, fresh market	MR	R	MS	VR	S	MR	S	Low internal defects, high yield	Susceptibility to shatter bruise and dry rot
Yukon Gold (7,216)	Oval yellow	Fresh market	S	MS	MS	MR	S	MS	S	Culinary quality, attractive appearance	Short dormancy, susceptibility to common scab and powdery scab

[a] Total acreage approved for certification in 2004 in the United States and Canada based upon a survey conducted by the Certification Section of the Potato Association of America.
[b] S = susceptible; MS = moderately susceptible; VS = very susceptible; R = resistant; MR = moderately resistant; VR = very resistant.
[c] Strains US11 and US8 of the late blight pathogen were used to test Ranger Russet, Russet Burbank, Russet Norkotah, Shepody, Umatilla Russet, and Yukon Gold at one or more of the following locations: Corvallis, Oregon; Mount Vernon, Washington; and Rosemount, Minnesota.

early-generation seed stocks, the U.S. Export Seed Potato Standards, and normalizing relations between U.S. and Canadian certification programs so that certified seed could move between countries with relative ease.

Today, seed potato production in North America is concentrated in the northern United States; in the central states, often at high altitudes; and in Canada. Fourteen states and 10 provinces annually certify nearly 200,000 acres of seed potatoes. Over 240 cultivars are certified, but over 85% of the acreage is used to produce only 16 cultivars (Table 6.3).

Production of High-Quality Seed: Requirements of a Seed Farm

Certified seed farms must meet certain standards to adhere to the rules and regulations for certified seed production and to maintain the framework for producing high-quality seed stocks with low disease levels. Seed certification standards do not specify where seed can be grown, but isolation of seed fields from commercial potato fields helps limit the amount of virus that may be moved into the crop. Most of the certified seed from North America is produced within a limited generation system. In addition, comprehensive **pathogen testing programs** are routinely utilized to verify that levels of specific pathogens are within prescribed tolerances. These testing programs are extremely important when working with latent virus infections, such as *Potato virus X* (PVX). Most certification systems in North America test for viruses (*Potato leafroll virus*, PVX, PVY, *Potato virus S*, *Potato virus A*, *Potato virus M*, and *Potato mop-top virus*), spindle tuber viroid, and the ring rot and blackleg bacteria at some point in the production cycle of the seed farm. The enzyme-linked immunosorbent assay (ELISA) (Fig. 6.1) has dramatically changed the industry, and it has become the workhorse of certification testing. This sensitive serological technique is well suited to mass indexing of seed lots. While many efforts are used to produce disease-free seed, a more accurate term to be used is **disease-tested** (Box 6.1). Simply put, a zero or a negative result on a lab test only means that the methods used did not detect the pathogen.

Table 6.3. Acreage tier structure for cultivars produced as certified seed in the United States in 2004

Acreage in production of certified seed	Number of cultivars in production	Percentage of total acreage[a]
1000+	16	85.3
100–1000	40	11.8
10–100	70	2.4
0.1–10	120	0.5

[a] A total of 122,116 acres was approved for certified seed production; 246 cultivars and clones were certified.

The best certified seed farms follow certain standard practices that make it easier to produce high-quality seed. This seed then will have the full potential to produce a quality commercial crop. These practices can be grouped into field, equipment, and storage operations. The following covers what a seed buyer should look for when visiting a seed farm.

Field Operations

Many seed lots may be grown on a single farming operation. These may consist of several lots of a single cultivar or many different cultivars. Identification of lots within the field and a clear separation of multiple lots grown within the same field or farm are essential. Absence of weeds not only increases yields but also reduces the risk of certain diseases, where weeds may act as infection reservoirs, and it improves roguing and certification inspections. Isolation of early-generation seed lots from later-generation seed lots helps reduce the transmission of viruses between lots.

A combination of cultural practices can be effective in controlling PVY in seed crops (Chapter 17). First, the use of **crop borders** planted around small seed plots extends the area where aphids would land first. Infected aphids probe the border of the field looking for the preferred host. With each probe, the aphid sheds virus particles and essentially becomes a noninfective vector again. These borders can be planted with soybeans or winter wheat in the spring so that it stays green during the entire growing season. Second, **crop oils** applied at regular intervals to protect new growth provide an aphid stylet barrier to prevent probing. Finally, regular **roguing** of diseased plants reduces the inoculum levels (Fig. 6.2). This

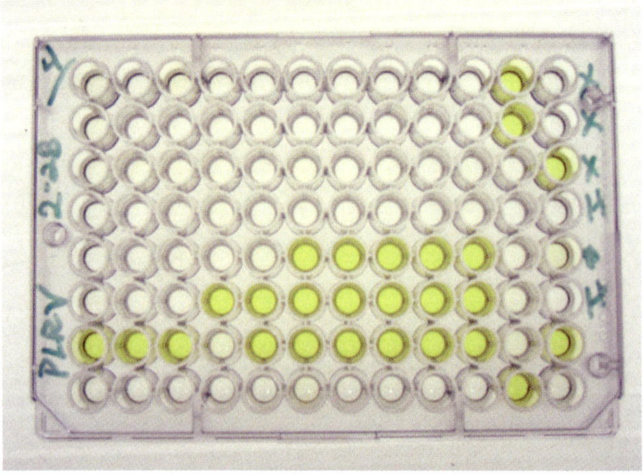

Fig. 6.1. ELISA plate with a typical positive reaction in the yellow cells. The color develops in response to the presence of a virus in tissue samples prepared from leaves or tubers.

Fig. 6.2. Mosaic symptoms on leaflets of cultivar Russet Burbank, caused by *Potato virus Y*. Virus-infected plants should be rogued from seed production fields. (Courtesy D. A. Johnson)

must be done not only in virus-susceptible cultivars but in all cultivars that are nearby. Early blight and late blight need to be managed. **Properly timed pesticide applications**, using pesticides with appropriate chemistries, including herbicides, fungicides, and insecticides, are essential to maintaining seed health. A visit to the seed farm during the season can give a seed buyer much information about practices on the farm as well as a look at the condition of the fields.

Several additional techniques can be utilized to keep diseases and pests within acceptable limits in seed lots. **Tuber unit production**, in which a single tuber is cut into two to six seed pieces and planted as a unit, makes it easier to identify and rogue plants with disease symptoms or off-types, since they will be grouped together. Strict **sanitation** at all phases of production is critical. This is especially crucial during the cutting of tubers, as this is one of the practices where pathogens can be easily introduced and spread. Novel cutting systems (such as cutting blades that are heated to high temperatures prior to cutting the tuber), the use of appropriate disinfectants on standard cutting blade surfaces during cutting, and properly sizing tubers so that fewer cuts are necessary, all help to eliminate or reduce the spread of pathogens between tubers. Growers sometimes withdraw seed lots from certification when there is evidence that they may contain excess levels of disease that cannot be properly removed or controlled. These lots move into the commercial market and are no longer used as seed. Early vine killing is also effective in limiting late-season infections by viruses transmitted by aphids. Finally, cultivar development programs in both the United States and Canada have made major strides in developing **disease-resistant cultivars**. Growers are routinely using the current certification system to move new cultivars into the industry in a rapid, appropriate manner.

Equipment Operation

All equipment used on a seed farm should be dedicated to the production of certified seed and should not used in any

Box 6.1

Definitions of Terms Used in Seed Certification

certification factor Any of the factors identified in the rules and regulations for seed potato certification as affecting the production of low-disease, high-quality seed stocks. Such factors include disease level, separation of equipment used in certified production, and isolation of certified production areas from other production areas.

common seed Seed that has not been certified by an official certifying entity. It is not subject to disease testing, and isolation from other production areas, and other safeguards required for certified seed. Common seed is often referred to as *year-out seed,* since it is usually derived from seed that was certified the previous year.

current-season infection An infection initiated during the growing season. Symptoms of current-season infection by certain viruses are often not visible during the growing season.

disease-tested Subjected to testing to detect the presence of specific pathogens in samples of leaves, stems, or tubers. Different tests have different levels of sensitivity, and results are best interpreted with regard for the limitations of the tests. A sample with a negative result may still harbor the pathogen at a level too low to be detected. A negative result should not be construed to mean that the seed lot or field from which the sample was collected from is also negative; only the sample itself is negative. Good management decisions can be made only if the sample is representative of the seed lot or field.

generation An official designation for certified seed lots, relative to the number of years of production in a field setting. In most certification programs, production from a single seed lot is limited to seven field generations.

plant health certificate An official certificate prepared by a certification agency detailing the history, results of disease testing, cultivar, etc., of a seed lot produced in a seed certification program. A plant health certificate is usually provided to the buyer either prior to purchase or at the time of shipment of certified seed.

postharvest test (winter test) A test conducted during the off-season, either as a field grow-out or by laboratory analysis, to detect pathogens or identify other potential problems that would affect the crop in the following growing cycle.

recertification Certification of a subsequent generation of seed tubers produced from certified seed. Seed lots that qualify for certification in one year but are no longer eligible for certification the following year because of excess disease or inability to meet certification rules and regulations are described as *not eligible for recertification.*

seed class An official designation for certified seed lots, classified according to the number of years of production and disease levels.

seed law A law requiring the use of certified seed in the production of potato crops. Seed laws of different states, provinces, or countries may vary slightly, but this requirement is the basis of any seed law.

shipping-point inspection, receiving-point inspection Inspection of certified seed potatoes at the time of shipping or upon delivery, conducted by a qualified inspector from a government agency, detailing the grade, size, cultivar, etc. Such inspections protect both the seed seller and the seed buyer in meeting contract stipulations.

other operation. This includes tractors, trucks, seed cutters, planters, harvesters, cultivators, and storages. Most certification agencies have regulations that prohibit equipment from being shared between different seed farming operations. Equipment should be properly padded and adjusted so that tubers are handled at harvest to minimize bruising. Bruises provide an important entry point for many pathogens, reduce seed piece vigor, increase the physiological age of seed, and may injure the eyes of seed tubers. Trucks and other conveyances used to transport seed over long distances should be properly cleaned and disinfected before seed potatoes are loaded.

Box 6.2

Checklist for Obtaining Healthy, Productive Seed Lots

The first five steps should be accomplished during the summer and up to harvest. Most buyers have booked certified seed prior to or immediately following harvest of the seed crop. It is crucial that buyers of certified seed potatoes have a basic understanding of how seed certification systems function in their state or province. This will go far in helping with decision making, reducing miscommunication, and avoiding mistakes.

- Select a cultivar on the basis of market conditions, sales potential, or contract obligations.
- Utilize information available for the cultivar and its specific production needs. Obtain information on best management practices specifically for the cultivar.
- Decide on the seed class and disease levels appropriate for intended operation. Check on seed availability. Choose a certified seed producer, considering the reputation of the grower, recommendations from other buyers, and the cost of the seed. Be proactive—do not accept anything!
- Contact the seed grower or the certification program manager in the state or province where the seed is grown. Visit the seed farm during the season and after potatoes are stored.
- Request a North American Certified Seed Potato Health Certificate prior to finalizing a contract. Ensure that the contract to buy a seed lot is pending satisfactory postharvest test results. The contract should include the price, size and grade, cutting preference, and shipping details, and it should contain provisions for handling potential problems.
- Set the shipping schedule and take delivery of the seed. Verify that the load information is correct and that the load has been properly sealed.
- Plant and evaluate seed as the season progresses. Expect a follow-up visit from the seed grower. Evaluate the performance of the certification system and the seed grower; the availability of information and ease of transfer of information; how problems were handled; and how the product looks in the field.

All equipment used on a seed farm should be dedicated to the production of certified seed and should not used in any other operation.

Storage Operations

Facilities that can maintain the correct temperature and humidity and appropriate airflow are essential for holding seed in good condition until planting. Storages and bins should be in good working order, cleaned, and disinfected prior to receiving seed. Adequate separation of seed lots, clear records of seed lot location and amount of each lot being stored are necessary. Barriers between seed lots can consist of wood, metal, straw, or other material that can reasonably keep seed lots from becoming mixed. A method for disinfecting boots or footwear should be available. This may be as simple as a dip pan filled with appropriate disinfectant for dipping footwear. Another option is to put on plastic disposable boots, worn over regular footwear, prior to entering the facility. Access to storages with certified seed should be limited to properly equipped personnel. These measures reduce the chance of contamination (Box 6.2).

Limited-Generation Programs

Certification of seed potatoes is a system in which individual producers voluntarily agree to accept rules and regulations promulgated by individual states and provinces. In the United States, the authorized certification agency in each state develops and publishes its own regulations. In Canada, a set of Federal regulations govern seed certification. These standards cover a wide range of factors which affect seed production including eligibility of lots to be certified, seed lot identity, disease tolerances, cultivar purity, specific production requirements, grade and shipping standards, and the establishment of the fees. In the United States, official legal status and program authority for each certification agency comes from the state department of agriculture, a land-grant university system, or a crop improvement association with authority from the state. While certification is a voluntary contract between the seed grower and the certification agency, the official status of each program is well understood, and legality has been verified through court rulings. Indeed, only certification programs and certified seed growers may use the term *certified seed potatoes,* to refer to potatoes that have met the legally constituted requirements of an official seed potato certification agency.

One of the most significant advances in seed potato certification has been the introduction and use of **tissue culture**

stocks at the start of the production cycle. These micropropagated, fully pathogen-tested seed stocks are the origin (pre-nuclear or nuclear) of all field-grown certified seed potatoes. Managing diseases, especially bacterial diseases, through the use of seed derived from tissue culture has been extremely effective. State and private tissue culture labs and greenhouses have been established with appropriate quality control measures to produce tissue culture stock. In three states and two provinces, seed farms have been developed to provide seed growers with base seed stocks.

The tissue culture program starts with selection of the best material (most often individual plants) from a field seed lot which is most representative of the cultivar being propagated. This process is followed for new selections of existing cultivars or lines from breeder seed. These lines are generally selected in early-generation material and are fully tested for pathogens prior to being moved into the lab to be introduced into an *in vitro* system. Each line is carefully evaluated to be sure that it meets the specific standards for the cultivar being grown. This is especially critical, since early selections will be the basis for the next 5 to 7 years of field production. A mistake at this juncture is not only costly but may jeopardize the acceptance of a new cultivar into the marketplace.

Tissue culture plants are grown *in vitro* (Fig. 6.3), in a clean, pathogen-free environment, to remain disease-free. Once the needed number of plants is obtained, they are used to produce *in vitro* microtubers or greenhouse minitubers, or they are transplanted directly in the field. **Minitubers** produced in a greenhouse are most commonly used for planting the first year for the field stocks (Fig. 6.4). Utilizing this nuclear material represents a **zero point** in terms of pathogens and years in the field and ensures that seed field production begins with clean stocks.

All-certified, limited-generation seed production also helps maintain low levels of disease in successive generations of certified seed. A seed farm should have only potatoes that are eligible and entered for certification. This removes the potential for introducing untested stocks that might contain diseases or introduce other problems into the grower's system. A limited-generation, flush-through system (Table 6.4) is intended to move the seed through a series of field generations until it is eventually phased out of certification, regardless of the disease content of the stocks. Growers start with the originating nuclear seed, and after each year of production,

Fig. 6.3. *In vitro* tissue culture of potato plantlets.

Fig. 6.4. Minitubers grown in 3½-inch pots in the greenhouse.

Table 6.4. Seed class terminology

U.S. seed class[a]	U.S. seed export class	Canadian seed grade	Field generation	Description
Pre-Nuclear	Nuclear	Pre-Nuclear	0	Stocks are maintained in tissue culture. Accession records are kept, including pathogen testing history. This class also includes *in vitro* microtubers and minitubers produced in a protected environment, such as a greenhouse.
Nuclear	Generation 1	Pre-Elite	1	Tubers from first field year
Generation 1	Generation 2	Elite I	2	Tubers from second field year
Generation 2	Generation 3	Elite II	3	Tubers from third field year
Generation 3	Generation 4	Elite III	4	Tubers from fourth field year
Generation 4	Generation 5	Elite IV	5	Tubers from fifth field year
Generation 5	Generation 6	Foundation	6	Tubers from sixth field year
Generation 6	Generation 7	Certified	7	Tubers from seventh and final field year

[a] Certain generations may be designated Foundation or Certified class in some states, depending upon disease incidence, field generation, and specific rules and regulations governing production systems. Some states follow the nomenclature of the U.S. Seed Export standards, in which Generation 1 is the first field generation.

the seed is automatically increased another generation. Most seed certification programs in North America limit the total number of field generations to seven or fewer, after which seed is no longer eligible for certification. Limitation of the number of generations and strict disease tolerances help to ensure that diseases and pests are within acceptable limits in certified seed.

Limitation of the number of generations and strict disease tolerances help to ensure that diseases and pests are within acceptable limits in certified seed.

Most states and provinces have enacted certified seed laws which effectively control the type of seed potatoes that can be planted. These laws restrict the planting of all potatoes within the region to sources derived from official certification programs. In some cases, seed stocks one year removed from the certification system may be eligible for planting if they are screened and in compliance with pertinent certification tolerances. Seed laws have effectively reduced overall disease inoculum in areas which share both certified seed and commercial industries.

Certification System

The certification process begins with an application submitted by a grower, listing seed source generation and lot history, area planted, and cultivar. This is the first step in creating a paper trail for each individual seed lot to be certified. It is at this crucial step that certification officials are given the opportunity to fully research the lots entered, verify that the information submitted is correct, and ensure that the seed lot meets the rules and regulations of the state or province. A unique lot number is assigned to identify the seed lot until it is ultimately sold and planted the following season. Acreage fees for certification are collected from the seed grower, including either all costs necessary for the certification agency or only the cost of field inspections. Additional fees may be collected for laboratory testing performed during the season, postharvest test costs, tags or bulk certificates, and shipping-point inspections.

Tolerances exist for each generation or class of seed, with the lowest tolerances for the early generations. Tolerances are also set within a given state or region to accommodate different production problems and to provide an acceptable product for their given markets. However, tolerances are similar among programs. In most programs in the United States, maximum tolerances are 1.0% for leafroll, 2.0% for mosaic, and 0.5% for varietal mixture. Only diseases and pests that can be seedborne and can be reliably detected are given tolerances. Certification systems are flexible. As new diseases or pests are discovered, and as new techniques are developed to detect seedborne problems, tolerances can be added or adjusted as appropriate. Tolerances at first inspection are higher than at second inspection, to give the seed producers the opportunity to rogue diseased plants from the field. Roguing is typically practiced to remove plants infected with viruses (e.g., *Potato leafroll virus* and PVY) or phytoplasmas, to eliminate cultivar mixtures, or to manage problems such as giant hill, calico (*Alfalfa mosaic virus*), and weeds.

Field Inspection

Inspections are conducted by trained certification personnel, who walk through the planted lots in a prearranged pattern, counting plants and visually examining them for disease, cultivar mixture, and other problems. Disease percentages are calculated by dividing the number of infected plants by the total number counted. Lots that exceed specific tolerances are downgraded in class to the next appropriate level or are rejected from the certification program. Certification agencies may also refuse certification because of other factors unrelated to meeting the prescribed tolerances. These factors include higher levels of other disease conditions not under tolerance (e.g., wilts or blackleg), poor stand, lack of isolation, equipment issues, conditions which interfere with inspection (e.g., chemical phytotoxicity or weeds), and any other condition which may lead to seedborne problems. Seed lots that pass inspection are assigned to a seed class (in the United States) or a seed grade (in Canada) (Table 6.4). Following each inspection, the certification agency will issue a report detailing the findings. These reports are given to the grower and become a permanent part of the certification records maintained by the certification agency.

Certain diseases, because of their seriousness within the crop, are considered to be at a *zero tolerance* within certified seed. **Zero tolerance** means that if at any time during the certification process a single plant or tuber is found to be infected with a zero-tolerance pathogen or pest, the lot is immediately rejected from certification. The farming operation where the infection was found must comply with a specified series of remediation-type steps. Zero tolerance means none found during visual inspection, thus, certification of the lot is not a guarantee that the lot inspected is free from the pathogen or pest. An example of a zero-tolerance pathogen is the bacterium that causes bacterial ring rot. Since its discovery in the certified seed systems in the 1930s, programs have developed various mechanisms to handle ring rot. Exclusion through extensive testing and removal of potentially infected stocks before introduction into the field is the most successful strategy. Improvement in testing systems, especially tests for specific antibodies (ELISA and immunofluorescent staining) and molecular techniques such as PCR (polymerase chain reaction), allowing the identification of genomic material of pathogens, have made significant inroads in the levels of bacterial ring rot found in certified seed. If bacterial ring rot is found at a seed farm, typically none of the seed produced there is eligible for recertification, so that the seed farm is forced to start with all new seed stocks. This is often necessary to break the cycle of ring rot infection on a farm, because the ring rot bacteria may be present at undetectable levels in other lots on the farm, as a result of the use of common equipment or storages, or as a result of mixing of seed lots.

Harvest and Storage

Certified seed lots must be harvested with equipment that has been sanitized and is designated for use in seed fields only.

Lots that pass all field inspection tolerances and meet the rules and regulations are harvested and stored for eventual sale or use as seed. All seed must be stored in clean, disinfested storages that contain only qualified certified seed stocks.

After harvest, seed storages are inspected by certification personnel, to verify production maps and lot identity and to check that storage conditions (relative humidity, airflow, and temperature) are appropriate for seed and that the bins are properly labeled. If seed fully passes the certification process to this point, it is eligible for listing in an official crop directory for sale. Crop directories contain the grower's name and address, lots that have been approved with acreage of each lot, and often information regarding disease levels found in the lot during summer field inspection. These directories are widely distributed among potential seed buyers and are one of the primary sources of information regarding seed certification and availability.

Postharvest Testing

Samples from seed lots eligible for recertification the following year are subject to postharvest testing, either in field trials (in areas where the winter climate is conducive to potato growth) or in laboratory tests on tuber sprouts. Certification programs use field plots in Florida, California, and Hawaii or greenhouses in their own state or province. Postharvest testing is conducted to detect late-season virus infections; mixing of cultivars, which can occur at harvest; chemical damage that carries over in the seed; and other factors that may adversely affect growth the following season. In some programs, the postharvest test also determines whether the seed is eligible to be tagged for sale. The postharvest test is the completion of the certification summer inspections. Some diseases do not express diagnostic visual symptoms until tissue grow-out from affected tubers. Without the postharvest test, there is no way to tell how much disease may have accumulated in a seed lot during the late season.

Each lot submitted for a postharvest test must meet the prescribed tolerances. Tolerances vary greatly among regions, depending upon the ability of the given system and its growers to manage the disease the following season. Generally, tolerances for recertification range from 0.5 to 3.0% total virus (e.g., leafroll and mosaic). Each lot submitted for recertification the following season must have had a postharvest test, at some level. All seed must meet the specific tolerances in order to be sold as certified.

Typically, tubers in the 2- to 3-ounce range are submitted for postharvest testing. A minimum sample is 400 tubers, but may be as high as 1,200 tubers, depending on the acreage grown and the seed class, with a higher percentage of tubers being tested in earlier generations. Samples representing the entire seed lot are required. Samples are often collected during harvest by selecting a certain number of appropriately sized tubers from each truck as it is unloaded at the storage.

In the postharvest grow-out, field plots are inspected in a manner similar to the summer field inspections. ELISA testing is used to supplement visual inspection in cultivars with mild or latent PVY symptoms. Some programs routinely lab test all lots to reinforce visual readings. It should be noted that occasionally postharvest test results differ from information obtained on the seed lot when field inspections are conducted the following season.

In postharvest testing, the sample size is many times smaller than that used in the summer field inspections, so there is some room for deviation built into the system. The detection of relatively few virus-infected plants can statistically distort the true level of infection, because of the small sample size. For example, if a lot has a 1% disease infection level, then the probability of detecting it with a 100-tuber sample is 63%, with a 400-tuber sample it is 98%, and with a 1,200-tuber sample it is 99.9%. If the disease level is lower than 1%, then all of these detection percentages also decrease. In general, postharvest results have been found to be quite accurate, on the basis of many years of testing by different certification agencies. Deviations are more often the result of poor sampling technique than the result of inherent bias or problem in the field plots or with laboratory testing. Finally, postharvest test results should be considered along with the summer inspection results. Summer results most accurately represent virus levels that are from the seed source used to plant the lot, while the postharvest test represents virus spread during the growing season. The postharvest test detects virus transmitted late in the season. The subsequent virus movement in the plant rarely results in visible symptoms during the summer inspections.

Sales, Tagging, and Shipping-Point Inspection

The decision to buy certified seed is based on many factors, including the reputation of the seed grower, the performance of seed purchased in the past, the quality of the current seed crop, and field visits. A **seed contract** specifying such things as cultivar, seed lot, price, grade, size, and time of delivery should be utilized (see Chapter 7). Once the contract is official and signed by both parties, shipping can take place. Commonly, a North American Certified Seed Potato Health Certificate (NACSPHC) (Fig. 6.5) will be provided to the seed buyer either prior to purchase or at the time of seed shipment. The NACSPHC contains all of the pertinent information about a seed lot: the state or province of origin, the buyer's and seller's information, cultivar, quantity of seed purchased, lot identification, lot history, summer and postharvest test inspection readings, results of laboratory tests, and any other pertinent information related to the lot or the regulations of the certification program. The NACSPHC was developed in the mid-1980s to assist buyers in making more informed decisions about seed purchases while allowing seed certification programs to standardize the information available to the market. It has become widely accepted as one of the best tools in use today for the informed buyer.

At the time of shipping, most certification programs will conduct a **shipping-point inspection**. The inspection is conducted by qualified inspectors either employed by the individual state or federal inspection or certification agency or conducted under a recognized quality assurance program monitored by federal officials. Inspectors conduct their evaluations based upon the examination of a set number of samples of a designated number of tubers pulled throughout the load as it is being loaded or packaged for shipment. These evaluations note size, grade, and levels of diseases. Most often, the seed must meet U.S. No. 1 Seed Grade, the Canadian tuber standards, or tolerances specified by the state or by contractual arrangement. Shipping-point inspections protect both the seed seller and the buyer. Inspectors may find

NORTH AMERICAN CERTIFIED SEED POTATO HEALTH CERTIFICATE - CROP YEAR 2004

	Grower		**Importer**
Name			
City, State/Prov.			
Variety		**Acres**	**Quantity Shipped**
			Size

Lot Certification

Certification #		**Lot origination from tissue culture**	No	Yes
Seed Class/Gen.		Year micropropagated for planting		
Certifying State/Prov.		by		

Production environment pedigree: Fill 1 column per production year, use different initials in Greenhouse and Field boxes for different farms (e.g. JSF for John Smith Farms); indicate a tuber-united lot with a '+" after farm initials; describe other footnotes in notes below.

1997	*1998*	*1999*	*2000*	*2001*	*2002*	*2003*	*2004*	Year of Production
								Greenhouse (insect excluding) & sterile soil
								Field (note special measures below)
								Certification No. Number of years produced
								Certifying State in field soil

Summer Field Readings

Field inspections

1st	2nd	3rd	Final		FINAL
				%LEAF ROLL	
				%MOSAIC	
				%VARIETAL MIXTURE	
		Less Than		%BLACKLEG	
		Less Than		%VERT + %FUSARIUM + %EARLY BLIGHT	

Post harvest readings

Location

Sample No.

Plant Count

ELISA test results for latent viruses

%PVY %PVX

Other Diseases	Not known to occur in growers area	No. of years since last found on this grower's farm, or NONE ON RECORD if free > 10 years	Not found this year during normal certification field inspections
Bacterial Ring Rot			
Late Blight			

Notes:

The above information is accurate to the best of our knowledge:

Program official / title

Date

Agency

Telephone

FAX

Approved for use by the Certification Section of the Potato Association of America

Fig. 6.5. North American Certified Seed Potato Health Certificate, used by seed certification agencies for verification of seed testing and status. The certificate is also used in the exchange of information between buyer and seller.

zero-tolerance diseases that result in rejection of a seed lot from certification and stops any future shipment of seed tubers from that lot.

A **seed tag** accompanies each load after the shipping-point inspection. In Canada, the seed tag color represents different seed classes, such as Elite, Foundation, and Certified, determined by the Canadian Standards for Classes of Seed Potatoes. In the United States, the seed tag color represents size and grade. A blue tag indicates that the load has met the criteria for U.S. No. 1 Seed Grade. U.S. certification programs are uniform in this approach, with the exception that individual programs may have standards that are stricter than absolutely necessary to meet the U.S. grade. However, no program has standards that are less strict than the U.S. grade. A yellow tag indicates variance from U.S. No. 1 Seed Grade, including size variance and increased disease tolerances or defects that do not materially affect seed performance. Some programs have additional tag colors to represent experimental seed stocks. Seed may be identified by either tags or bulk certificates. Information commonly found on these official documents includes the seed seller's name and address, cultivar, seed lot identification number, crop year, field generation, tag color, class, recertification eligibility, and specific grade information, if necessary. It is only after shipping-point inspection has been conducted and tags or bulk certificates have been issued that the seed is considered certified for all practical and legal purposes.

Shipping

For transportation, certified seed is loaded into new 100-pound sacks, new mini-bulk bags capable of holding up to 22 cwt of potatoes per unit, or in bulk into trucks. Sanitation of all equipment and conveyances is essential. Proper handling of seed as it is moved into the shipping containers includes sorting to remove any rotted or damaged tubers, minimizing drops to prevent excess bruising, and verification that the correct lot and cultivar is being shipped. Once seed is loaded, the container is sealed with a uniquely numbered metal seal. This seal must remain unbroken until the load is delivered to the buyer to ensure that there has been no tampering with the load. Upon receipt, the buyer removes the seal and verifies that the documentation accompanying the load, the bulk certificate or tag, invoice, and bill of lading, are correct. Once the seal is broken, all identity of the lot is lost if it is a bulk load. If there is any concern with the load at delivery, the buyer should immediately notify the official inspection service representative in the region to provide additional inspection of the seed (i.e., receiving-point inspection). A seed lot will occasionally be out of grade because of excess shipping time, improper temperature control in the truck, or tuber degradation from the seed storage unit. If this occurs and is verified, the buyer will often make new arrangements with the seed seller in terms of price or will refuse to accept and pay for the load. Only through satisfying both the seed seller's needs and the seed buyer's concerns can a shipment be counted a success.

Limitations of Seed Certification

Seed potatoes are constantly being improved due to research and implementation of research results by the seed growing community. Because of past successes, seed quality expectations are high in the industry. Certified seed programs have seen fewer rejected acres for many years, and seed buyers have had increased quality and yields from their purchased seed product. Even with these successes, there are limitations to the certification process. These limitations are twofold. First, there is an inherent weakness in the inspection process, since **not every plant can be visually examined**. Second, many diseases are often latent at some point in the season and difficult to detect visually. The first limitation is countered by (1) representative sampling, (2) conducting more than one inspection, (3) training and updating inspectors intensively, (4) using statistical models to determine appropriate plant counts, predicting with 95% accuracy when a zero-tolerance disease or pest might occur, (5) understanding how each cultivar will react to diseases of concern so that inspections can be properly timed, and, (6) relying on the seed grower's historical performance. The second limitation, **latent disease expression**, can be overcome by the utilization of comprehensive laboratory testing programs. These programs again rely on statistical models to determine how many samples to collect. These models coupled with increasingly sensitive lab techniques do an excellent job in predicting potential problems.

Overall, seed potato quality has continued to improve. Even with the acknowledged limitations, the certified potato seed industry has consistently produced high-quality seed stocks. The market has found that the use of certified seed is one of the critical components in a successful potato operation. In fact, this is so apparent that major processing companies require the use of certified seed in their contracts. Additionally, international export markets require certified seed or its first year progeny to meet phytosanitary concerns. As potato cultivars improve with added disease resistances, it is expected that the certification system will continue to demonstrate its effectiveness. In the future, certification will continue to be an essential component of potato production, especially as the demand for food and value-added food production increases. Thus, even as the certification industry changes to meet the demands of the future, the continued steady supply of high-quality certified seed must continue.

CHAPTER 7

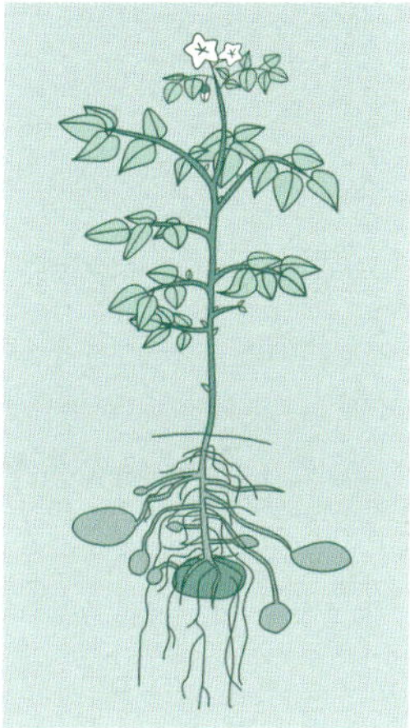

Gary A. Secor
Department of Plant Pathology
North Dakota State University, Fargo

Steven B. Johnson
Department of Plant, Soil and Environmental Sciences
University of Maine, Presque Isle

Seed Tuber Health Before and During Planting

The methods by which seed tubers are selected, handled, prepared, treated, and planted are as important as the quality of the seed itself. Once a high-quality seed lot has been identified and purchased, management practices determine seed performance to a great degree. Proper handling and planting practices promote optimum stand establishment by minimizing exposure to disease-causing organisms, reducing mechanical and physiological stress damage to the seed, and providing favorable conditions for rapid plant development. Establishing a good stand of healthy, vigorous plants is a critical step toward the dual goal of harvesting high yields of tubers of high quality. The consequences of improper seed handling and planting are seed decay, poor stand, weak plants, difficulty controlling weeds, season-long stress, more disease, early maturity, low yields, and poor quality of harvested tubers.

Planning and Preparation Before Seed Delivery

Clear communications and a good working relationship between the seed supplier and the seed buyer benefit both parties. Visits to the seed-growing operation by a prospective buyer during the growing season can provide insight into what to expect of the coming seed crop: a visitor can view the cleanliness of the operation and examine the fields and storage to assess management practices and conditions that may affect seed tuber quality, such as diseases and herbicide damage. Herbicide damage, either from carryover of compounds in the soil or from drift from an outside source during the growing season, can be manifested as abnormal or nonsprouting seed tubers the following year.

Whether a buyer is working through a seed broker or directly with a grower, the exact details of delivery, such as the cultivar, seed generation, seed tuber size, date of delivery, and price, must be specified. The details of the agreement should be confirmed with a written contract to prevent any misunderstanding of the transaction. Operations that do not have enclosed bins for accepting and holding seed should arrange delivery so the seed can be promptly planted. Seed quality will be better maintained in a seed storage building than on a concrete slab open to the elements. If seed must be loaded or delivered during extremely cold periods, several inches of sawdust on the truck bed will help insulate the tubers and reduce bruising when they are removed. In extreme conditions, a layer of paper between the truck body wall and the tubers can help prevent cold damage.

Preparing the Seed Storage

After a seed lot has been selected and a price and delivery time negotiated, storage facilities should be prepared to accept the seed. Seed tubers should never be placed in a storage that has been treated with the **sprout inhibitor** CIPC (chlorpropham) unless the entire ventilation system, including plenums, flumes, fans, and ducts, has been thoroughly cleaned with a high-pressure washer using soap and water. CIPC is routinely applied to tubers in storage to inhibit sprouting. It is applied once, and sometimes twice, as a very fine aerosol or smoke, which is injected into the ventilation system and carried throughout the storage by air movement. CIPC does not degrade quickly, and even a small amount of residue remaining in a storage facility can retard or prevent the sprouting of seed tubers. In contrast to storage-applied sprout inhibitors, field-applied sprout inhibitors, such as maleic hydrazide, do not contaminate storages.

Cleaning and **sanitizing storages** and seed-handling equipment each year to eliminate carryover of disease-causing organisms on surfaces or in potato debris are essential steps in seed potato health management, even if there was no disease problem in previous crops. The two pathogens of major concern are the ring rot bacterium and the silver scurf fungus. Ring rot bacteria can persist for 2–5 years on surfaces that come into contact with infected tubers. The silver scurf fungus produces large numbers of spores in storage that can survive from one season to the next and cause infection of tubers the following year (Chapters 18 and 19).

The cleaning process consists of two steps: disposing of old potato debris and sanitizing all surfaces and equipment that come in contact with seed tubers. All trash left from previous seasons, including tubers, foliage, soil, broken boxes, and old bags, should be removed from the storage and destroyed or discarded. Leftover tubers, which could be a source of disease organisms, should not be left near the storage or in a cull pile, but should be destroyed by burying, composting, or shredding. In northern areas during the winter, refuse tubers can be spread sparingly on the surface of fields not intended for potato culture the following spring, where they can undergo several cycles of freezing and thawing.

After cleanup, storage bins, walls, and floors should be thoroughly washed with a high-pressure washer using hot, soapy water and rinsed well. A **disinfectant** should then be applied according to the manufacturer's instructions. To kill bacteria, disinfectants generally must be in place for a minimum of 10 minutes. Porous surfaces retain larger populations of bacteria than smooth surfaces and require extra attention. Adding a foaming agent to the disinfectant helps retain the chemical on vertical surfaces for the required time.

All containers, tools, and equipment, such as seed trucks, planters, seed cutters, and conveyors, that will contact seed lots should be washed with a high-pressure sprayer, sanitized with a disinfectant, and then rinsed well with clean water. The ring rot bacterium is transmitted most efficiently by cutting, and therefore special attention should be given to the sanitation of equipment and tools used during seed cutting. Cutting equipment should always be sanitized before a new seed lot is cut. Open-cell foam rubber rollers on the cutter may carry ring rot bacteria deep in the foam from one season to another and cannot be properly sanitized. To avoid this problem, replace them with closed-cell foam rubber rollers.

Important characteristics of a good disinfectant are high germicidal activity, rapid action, effectiveness in the presence of organic material and in hard water, long shelf life, low human toxicity, low corrosiveness, and ease of use. Several types of disinfectants are available and marketed under many brand names. Most registered and commercially available disinfectants are effective sanitizers if properly used. Each type of disinfectant has certain characteristics that affect use. In general, all disinfectants are inactivated by organic matter, and prewashing to remove organic material from equipment and storage facilities is necessary in order for these compounds to work effectively (Table 7.1). Iodine, quaternary ammonium, and phenolic disinfectants are good choices for footbaths. Ethanol, isopropanol, and methanol are particularly suitable for small jobs such as sanitizing tissue culture instruments and hand tools. They should be used at a concentration of 70% for soaking instruments and tools. Most disinfectants also inactivate viruses, but hypochlorite bleach and general biocides such as chlorine dioxide and trisodium phosphate are the most effective.

Do not confuse condensed water vapor (fog) with steam, which is colorless and therefore invisible.

Live steam also can be used to kill ring rot bacteria, but it must be used properly to be effective. The surface to be sanitized must be exposed to live steam at temperatures exceeding 180°F for 15–30 seconds. Caution: Do not confuse condensed water vapor (fog) with steam, which is colorless and therefore invisible. The temperature of condensed water vapor may be lower than that required to kill bacteria. Steam is more useful for sanitizing equipment than for entire storages, because of the need for high temperatures, the time involved, and the small surface area that can be covered at one time with a steam appliance.

Table 7.1. Disinfectants used for sanitizing potato-handling equipment and storage facilities

Disinfectant	Inactivation by organic material	Safety	Corrosiveness to metal	Shelf life	Comments
Quaternary ammonium compounds	Some	Caution	Slight	1–2 years	Numerous brands; widely used; stainless. Always follow label directions.
Hypochlorite 5.25% bleach	Yes	Irritant; caustic to skin and clothing	Yes	5.25% bleach is stable for 6 months at room temperature	Quick-acting; inexpensive. Use 1:50 mix with water. More effective at pH 7.0 than at normal pH 11.0. For maximum efficacy, mix 1 part 5.25% bleach, 200 parts water, and 0.6 part white vinegar, giving a final concentration of 256 ppm.
Iodine compounds	Some	Relatively safe; caution	Yes	1–2 years	Loses activity when yellow brown color disappears. Tamed iodine works best.
Phenolic compounds	Some	Oral poison; caution	No	1–2 years	Has residual activity; usually fragrant; the word *phenol* is used on the label.
Chlorine dioxide	Some; not as much as bleach	Prepare with adequate ventilation; fumes caustic; safe at dilution prepared for use	Yes, especially soft metals, such as aluminum and brass	Activated product is stable for 2 weeks but light-sensitive; concentrate is stable for 1–2 years	Widely used general biocide with broad-spectrum activity against fungi, bacteria, viruses, and algae. Does not produce trihalomethane. Decomposes into salt and water.

Nonchemical alternatives are appropriate in some instances. Painting wooden surfaces can seal in disease organisms and prevent them from spreading. For sanitizing earth cellars, it may be necessary to cover earthen walls with plastic sheeting to prevent direct contact with tubers. If an earthen floor is contaminated with disease organisms, removing the top few inches of soil or adding a few inches of new soil may effectively remove or seal in pathogens.

Accepting Delivery and Handling Seed

A separate facility should be available and prepared to accept incoming seed tubers prior to delivery. Keep individual seed lots separate to maintain identity and to prevent contamination with pathogens and varietal mixtures. The seed pedigree, which is a record of the identity, generation number, seed grower, and certification record, should be maintained for every seed lot throughout storage, cutting, planting, and field operations. In addition, each lot of seed should have its own North American Certified Seed Potato Health Certificate (Chapter 6). Accurate records of seed pedigree allow a seed lot to be evaluated for performance and to be traced back to its origin if problems develop later.

Certain procedural and legal details should be taken care of before delivery. A signed confirmation or contract for the purchase should be in hand. Assurances should be obtained that the seed will be delivered in clean, sanitized containers or trucks. The seed should be inspected at the shipping point at the time of loading; if not, it is advisable to request a federal or state inspection prior to unloading. Bulk handling and shipping are common, because of reduced costs. Each bulk load should have its own tag and seal. Seed tubers should not be held in tightly sealed, unrefrigerated, unventilated containers during shipping for more than a few days. These conditions tend to create a low-oxygen environment and condensation on the tubers, which increases the rate of physiological aging of seed and favors the development of bacterial soft rot.

> **The most important consideration is to provide conditions favorable for the healing of wounds incurred during loading, transport, and unloading.**

Seed deliveries should be received by an experienced person capable of determining that the load meets contract specifications. When the seed arrives, this person should inspect the load to be certain that the seals are intact and examine the shipping-point inspection report. The shipment should be inspected before unloading to be sure that it complies with the order; e.g., it contains the correct cultivar, seed lot, and tags, and it has not been damaged or frozen in transit. Both external and internal inspections of tubers are necessary to detect diseases and disorders. If the seed is out of compliance, the supplier and shipping-point inspector should be contacted before unloading. Reinspection of the load by a federal or state inspector may be advisable. Because it is important to maintain a good working relationship between buyer and seller, the seed grower should be informed of any problems before the seed is unloaded. Not all sales are directly between the seed producer and the buyer; seed brokers often arrange sales, in which the seed producer does not have contact with the buyer. Copies of tags and all paperwork associated with each load should be retained for future reference. Once the seed is unloaded, it is the property of the buyer.

If seed tubers need to be stored for a short period before planting, they should be stored in a cool (45–50°F), well-ventilated building at 90–95% relative humidity. The most important consideration is to provide conditions favorable for the healing of wounds incurred during loading, transport, and unloading. Steps should be taken to minimize bruising during unloading and handling. Bruising seed potatoes before planting significantly increases the risk of seed piece decay, often resulting in a poor stand. Bruising also increases physiological aging of the seed. If possible, the tuber pulp temperature, i.e., the temperature of internal tissues measured with a pulp thermometer, should be above 45°F before unloading, as tubers sustain increased bruise damage when handled at temperatures below 45°F. Conveyors should be adjusted so that tubers drop no more than 6 inches. Delivered seed should not be piled higher than 8–10 feet, and the piles should be arranged to allow for good ventilation. As seed is put into storage, it is a good policy to examine it for disease. There are several tuber-borne diseases to watch for, which can be indicators of poor health (Box 7.1).

> **Bruising seed potatoes before planting significantly increases the occurrence and severity of seed piece decay, often resulting in a poor stand.**

Physiological Age of Seed

The physiological age of seed tubers is an important consideration in seed tuber selection and cutting. Factors affecting physiological age of tubers include growing-season stress, storage temperature, and time. Higher storage temperatures will speed the aging of tubers. Since the physiological age of seed can affect productivity (Chapter 10), an understanding of the stages of aging is necessary. The physiological age of seed tubers can be determined by gathering a sample of tubers, placing them indoors at 60–75°F in the dark, and allowing them to sprout. The physiological age of the seed can be estimated by observing the sprouts. There are distinct stages of aging through which potato seed passes (Box 7.2).

Managing the Seed-Cutting Operation

Careful management of seed cutting is an important part of a holistic health management program, because of the health hazards associated with this process. Tuber pulp temperature should be monitored, and seed tubers should be warmed to 50–55°F before handling. Beginning 7–10 days prior to cutting, storage temperatures should be raised slowly, about 0.5–1.0°F per day. Warming seed tubers before cutting

Box 7.1

Seed Tuber Diseases to Watch For

Bacterial Ring Rot

Characteristic sunken cracks in the skin can be seen when seed tubers are examined. As a precaution, cut about ½ inch of tissue from the stem end of at least 100 tubers selected at random, and examine the cut areas for discoloration of the vascular ring. This is visible as a slight to dark discoloration appearing as a ring or partial ring about ¼–½ inch from the skin. Squeeze suspected tubers and check for bacterial ooze emerging from the vascular ring at the cut surface (see the figure below). Any tuber with suspected symptoms should be sent out for laboratory testing, since other tuber problems can be mistaken for this disease. The presence of bacterial ring rot should be verified by an experienced plant pathologist acceptable to all parties involved, including the seed grower. If any tubers suspected of having the disease are found in a seed lot, delay planting that lot until laboratory confirmation is received. No seed lot should ever be planted with a confirmed diagnosis of bacterial ring rot. If bacterial ring rot is suspected, pending diagnosis or confirmation in a seed lot, all equipment and materials coming in contact with the seed lot must be thoroughly decontaminated before cutting the next seed lot, to prevent further contamination (Chapter 18).

Black Dot

Tuber symptoms of black dot resemble those of silver scurf, and it is difficult to distinguish between the metallic lesions caused by both diseases. Small, black sclerotia develop in the lesions; they can be seen when viewed with a hand lens or low-power microscope. The importance of seedborne inoculum in the disease cycle is not understood, and no threshold has been established for this disease (Chapter 19).

Common Scab

Seed tubers should be free or nearly free of common scab, to prevent the introduction of the pathogen into the field. Excessively scabby seed should not be planted (Chapter 19).

Early Blight

Tuber lesions are circular, brown to black sunken spots with well-defined margins and a gunmetal blue halo (see the figure below). The tissue under the lesion can take on the appearance of a corky dry rot in the periderm (Chapter 21).

Fusarium Dry Rot

Dry rot is characterized by a dry, crumbly decay. Ideally, no more than 2% of tubers in a seed lot should have obvious symptoms of Fusarium dry rot, and 2% incidence of dry rot is allowed under most certification standards. Seed lots in which more than 2% of tubers have dry rot symptoms can be used if the diseased tubers are removed before cutting and seed treatments are used on the remainder of the lot. Seed tubers with dry rot have the potential to develop Fusarium seed piece decay and should be treated with an appropriate fungicide after cutting (Chapter 19).

Late Blight

When a tuber infected with the late blight fungus is cut, a brick red or coppery-colored, granular-appearing discoloration of tissues can be seen just under the skin, penetrating as deep as ½ inch (Chapter 21). Infected seed tubers can serve as an important source of the pathogen and may lead to an epidemic later in the season. A 1% tolerance for late blight in seed lots is permitted by most certification agencies, but even this amount can lead to an early-season epidemic when cool, wet conditions persist during emergence and early growth. Select seed lots free of late blight.

Advanced decay of the vascular ring due to ring rot, with cheesy rot exuding where the tuber is being squeezed with the thumb. (Courtesy V. D. Peterson and G. A. Secor)

Sunken, circular lesions, an early blight symptom in tubers. (Courtesy D. A. Johnson)

Nematodes

Seed tubers known to be infected with plant-parasitic nematodes should not be planted. Certification agencies inspect seed potatoes for symptoms that indicate the presence of nematodes. Some species of root-knot nematodes cause small bumps on the skin of tubers or pinpoint discolorations just under the skin. Careful inspection for evidence of nematodes is especially critical if the seed tubers originated in a production area where the Columbia root-knot nematode or the potato rot nematode is present (Chapter 20).

Powdery Scab

Symptoms of powdery scab can resemble those of common scab, and the two diseases can be confused. Powdery scab lesions are open pustules with papery remains of periderm at the borders and a powdery mass of brown spores. Infected seed can infest an uninfested field. The only control is to plant clean seed in clean soil. Previously, certification agencies established zero tolerance for powdery scab, but it is now permitted in some certified seed lots (Chapter 19).

Black Scurf

Ideally, seed should be free of visible sclerotia of *Rhizoctonia solani,* the black scurf pathogen, but as much as 2–5% coverage of the surface can be tolerated. Greater coverage than this may result in yield and quality losses (Chapter 19).

Silver Scurf

Silver scurf is characterized by metallic lesions restricted to the periderm of tubers. The presence of the fungus causing silver scurf can be verified by incubating suspect tubers in a moist chamber for 3 weeks at room temperature and examining the tubers for conidia and conidiophores. The pathogen is primarily seedborne, and it is difficult to find seed lots in which the disease is not present. Seed lots with the least amount of silver scurf are the most desirable. A seed treatment can be used to minimize infection of progeny tubers. In general, there is a positive correlation between the amount of silver scurf in seed and the amount in progeny tubers grown from the seed lot, but a disease threshold for silver scurf in seed has not been established (Chapter 19).

Soft Rot and Blackleg

Soft rot is characterized by a cream-colored, mushy, wet decay with the consistency of soft butter, often with a black border (Fig. 18.6). Tuber blackleg is characterized by a soft decay of the interior of the tuber, usually beginning at the stem end (see the figure below). If more than 1% of the tubers in a seed lot show visible symptoms of soft rot, excessive bacterial seed piece decay or blackleg (or both) may develop. The bacteria that cause these diseases are likely to be present in symptomless tubers. The percentage of tubers infected with these bacteria is an indicator of the potential for the development of bacterial seed piece decay and blackleg after the seed is planted. The potential for soft rot development can be estimated by wrapping 100 randomly selected tubers in wet paper towels and then in clear plastic wrap and incubating them for 5–7 days at 50–60°F; alternatively, tubers can be incubated in a chamber at room temperature with intermittent misting to keep them wet for 4 days. The number of tubers that decay is an estimate of the incidence of bacterial seed piece decay and blackleg that could develop if conditions following planting favor these diseases. These are reasonable tests but are poor predictors of field performance. Extra care, including avoiding free moisture, warm temperature, and seed bruising, should be given to seed lots in which more than 50% of the tubers develop soft rot. If they are handled carefully, seed lots with high levels of symptomless contamination with soft rot bacteria can produce acceptable stands and yields unless conditions after planting (primarily, excess soil moisture) are highly favorable for disease development (Chapter 18).

Verticillium Wilt

Tubers infected with *Verticillium* often have a tan discoloration of the vascular ring, which is visible when they are cut at the stem end. Infected seed tubers are not as important in current-season disease as is inoculum in the soil. However, tuber-borne inoculum is an important source of the pathogen in previously uninfested fields. Infected seed pieces can produce diseased plants, introducing inoculum that continues to increase in the soil with each subsequent potato crop (Chapter 19).

Virus Diseases

Certification records, inspections, and results of winter grow-out tests indicate the amount of virus infection in a particular seed lot. Symptoms of most viruses are generally not visible in tubers, except for current-season infection by *Potato leafroll virus,* which may cause net necrosis in tubers of certain cultivars, notably Russet Burbank (Chapter 16). *Tobacco rattle virus, Potato mop-top virus,* and necrotic strains of *Potato virus Y* cause internal and external necrotic arcs and rings, best seen by cutting tubers and observing the flesh for symptoms.

Soft rot of internal tuber tissues. (Courtesy V. D. Peterson and G. A. Secor)

Box 7.2

Stages of Physiological Age of Seed Tubers

N. Richard Knowles
Department of Horticulture and Landscape Architecture
Washington State University, Pullman

The degree of apical dominance is indicative of the physiological age of seed tubers. Apical dominance is the correlative inhibition of sprout growth in lateral eyes by sprouts developing from the apical end (bud end) of the tuber. As age advances, apical dominance declines, resulting in multiple sprouts. The figure illustrates early sprout development at room temperature in the dark after various intervals of cold storage over a prolonged period (1–3 years). The actual physiological age depends on time and temperature, and the rate of aging is cultivar-dependent.

Dormant Seed

Most tubers undergo a dormant or resting period immediately following harvest. The length of the resting period varies with the cultivar and the storage temperature. If seed tubers do not sprout, they may still be in a period of dormancy. The duration of dormancy can be reduced by chemical and nonchemical means, including storage at higher temperatures (Chapter 10) and treatment with gibberellic acid or Rindite, but such treatments are not generally used for commercial seed lots.

Young Seed

Young seed is characterized by a high degree of apical dominance. One or only a few sprouts emerge from eyes on the apical end (or bud end) of the tuber. As these sprouts begin to grow, they inhibit sprouting from lateral eyes in the middle and at the stem end of the tuber. Eyes at the stem end may never sprout, depending on tuber age. Young seed produces plants with few (e.g., one to three) stems and consequently low tuber set and large tubers (Fig. 10.10).

Middle-Aged Seed

Middle-aged seed will produce multiple sprouts (e.g., three to six), because of reduced apical dominance between eyes. The number of sprouts depends on the physiological age; however, the number of sprouts from middle-aged seed will usually not exceed one per eye. The number of tubers set per plant increases with stem number, resulting in reduced tuber size (Fig. 10.10). Managing seed tuber age to control the degree of apical dominance and thus the number of main stems per seed tuber can effectively add value to a commercial crop by optimizing tuber size distribution in relation to the requirements of the market (seed, fresh, processing) (Chapter 10).

Old Seed

Physiologically old seed will have lost apical dominance between eyes as well as within eyes and thus produces multiple sprouts from all eyes. These sprouts are weak and do not produce a vigorous plant. Plant stand will be poor, and the plants that do emerge from old seed will typically set many tubers very early in development, before adequate leaf area has established to bulk tubers to a desirable size. Seed can be so old that small tubers form on sprouts soon after initiation or buds within eyes tuberize directly. These secondary tubers are often formed from seed that has been stored (at 38°F) for a year or longer. Seed tubers old enough to have formed such secondary tubers should never be used for production.

Drawing by G. N. M. Kumar

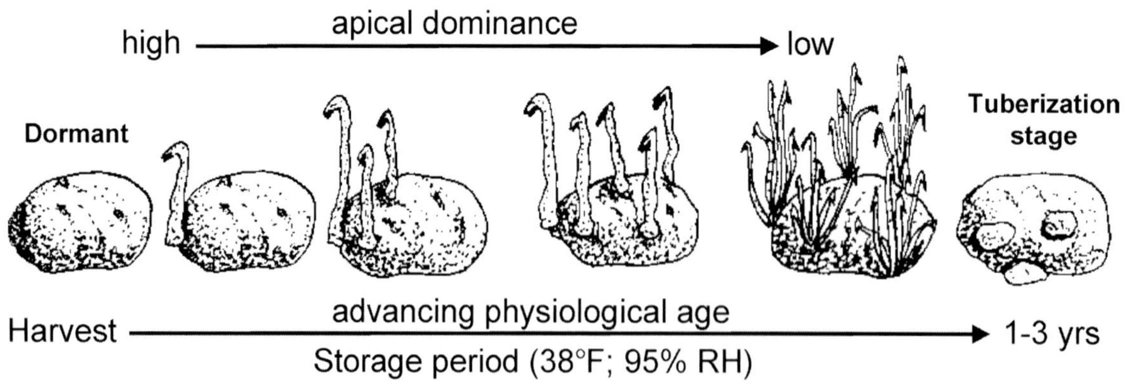

serves three purposes: it reduces bruising during handling, promotes rapid healing of cut surfaces, and initiates sprouting before planting. Bruising seed potatoes before planting significantly increases the subsequent occurrence and severity of seed piece decay, often resulting in a poor stand. Healing of cut surfaces, which is accelerated at temperatures above 50°F, retards infection by disease organisms. However, warming seed tubers for periods longer than 2 weeks or at temperatures higher than 55°F can result in excessive sprouting and physiological aging, leading to lower yields, decreased tuber set, and smaller tubers.

Controlling the **size of seed pieces** is an important management objective. Most seed tubers are cut with mechanical cutters that produce seed pieces of various sizes. The preferred size is 1.5–3 ounces. Small seed pieces may result in weak plants, while seed pieces above 3 ounces produce plants with an excessive number of stems and may fall off planting cups. A wide variation in seed piece size results in skips and doubles because of inconsistent feeding through the planter. Seed piece size is controlled by adjustments on the cutting machine and by the size distribution of the tubers being cut. If the machine is set to cut tubers in the 6- to 8-ounce range, those below 5 ounces or above 9 ounces may yield numerous seed pieces below the 1.5-ounce minimum. Large tubers of some cultivars with uneven distribution of eyes also yield a significant number of seed pieces without eyes. For these reasons, it is best to avoid seed lots that have an excessive number of very large or very small tubers.

Differences among cultivar characteristics such as the number and placement of eyes, dormancy, and desired seed piece size may dictate differences in handling and cutting procedures. For example, some cultivars have many eyes spirally distributed on the tuber; seed pieces of these cultivars can be cut smaller, with each seed piece still having an eye. Some cultivars have a long dormancy period and may require a longer warming period than others. Consult local advisors for recommendations on seed cutting for particular cultivars.

Seed tubers should be cut in a clean, relatively humid area, preferably indoors. Many types of seed cutters with various characteristics are commercially available. Regardless of the type used, **it is essential that cutters be cleaned and sanitized** at least daily while cutting, and always before cutting a new seed lot. Cutters with water-impermeable (closed-cell) sponge rollers are recommended, since open-cell rollers may harbor ring rot bacteria.

> **Seed cutters should be cleaned and sanitized at least daily while cutting, and always before cutting a new seed lot.**

Workers should be provided with disinfectants and hand-washing facilities to minimize bringing disease organisms, especially bacteria, into the seed-cutting area. A change of plastic disposable booties and new gloves at least daily is advisable. Disinfectant dip pans for boots are most effective if the boots are free of soil and soaked in them for about 10 minutes; a quick dip is ineffective. A good technique is to keep a pair of rubber boots soaking in fresh disinfectant and change boots when entering the warehouse.

The flow of seed tubers into the cutters should be adjusted so that they are not more than one tuber deep. The blades should be sharp and kept adjusted to deliver seed pieces in the appropriate size range. Periodic examination of the cut seed is advisable. A 12-pound tuber sample should contain about 100 seed pieces. Sorting the sample into size categories is also recommended, to determine the distribution of seed pieces that are too small (less than 1.5 ounces), within the preferred range (1.5–3 ounces), and too large (more than 3 ounces). Adjustments to the cutters can then be made, based on the distribution pattern for each seed lot.

Watch for diseased tubers during the cutting operation, especially tubers with ring rot. If the presence of ring rot is suspected, stop all cutting operations immediately. Remove anything and everything that has come in contact with the suspected seed lot, and thoroughly clean and sanitize all cutting equipment and facilities before continuing. If the presence of ring rot is confirmed by laboratory analysis, the seed lot should not be planted.

> **If the presence of ring rot is suspected, stop all cutting operations immediately. Remove anything and everything that has come in contact with the suspected seed lot, and thoroughly clean and sanitize all cutting equipment and facilities before continuing.**

As soon as a tuber is cut, a natural process of **wound healing** begins, which diminishes both moisture loss and infection of the cut surfaces by bacteria and fungi. Wound healing is favored by abundant oxygen (fresh air), high relative humidity (95–99%), and temperatures of 55–60°F. High relative humidity is critical, because low humidity may lead to death of cells on the cut surfaces, which prevents proper healing. However, the formation of free water (condensation) on tuber surfaces must be avoided, because a water film acts as an oxygen barrier to cells on the cut surface. Free water on seed pieces slows wound healing and creates an environment that allows soft rot bacteria to become established. The cut surface of a tuber heals by a twofold process. The first phase is **suberization**: a waxy, fatty compound called suberin is produced by cells just below the cut surface. This seals the wound, prevents water loss, and provides a barrier to some pathogens. Suberization begins in 1–3 days and is usually complete in 4–7 days. The second phase of healing is the formation of the **wound periderm**, which is a permanent, protective layer of cells replacing the epidermis (skin) that has been destroyed. The new periderm is a corky layer, which serves as the final covering to prevent infection and desiccation. Periderm formation begins shortly after suberization and is complete in 1–2 weeks (Chapter 10).

After cutting, seed pieces can be either planted immediately or held in storage under conditions that favor wound healing, a practice known as precutting. If seed is planted immediately, wound healing takes place in the soil. Wound healing occurs most rapidly when both the seed pieces and the planting soil are around 55–60°F. Planting seed pieces that have not been properly healed in cold, wet soil or in hot, dry soil usually results in a poor stand. Properly healed seed pieces seldom decay, and the best practice is to plant when soil conditions encourage wound healing and rapid emergence.

Wound healing is favored by fresh air, 95–99% relative humidity, and temperatures of 55–60°F.

When seed pieces are to be **planted immediately** after cutting, it is best to cut no more seed than can be planted the same day. Conveyors used to load seed pieces onto trucks and planters should be operated so that the seed pieces drop no more than 6 inches. Cut seed should be hauled in covered trucks, but not stored there for long periods, because of potential overheating as a result of poor aeration. Some trucks are constructed to facilitate air movement through the length of the truck and up through the cut seed. Growers with this kind of seed transport truck routinely cut directly into the truck, force humidified air to circulate through the seed to promote wound healing for 2 days, and then plant.

Another option to consider is whether to cut seed or not. Because potatoes are vegetatively propagated, many pathogens are maintained between generations and spread during the cutting operation. Bacterial and some viral and fungal pathogens need a wound to infect, and the cutting operation provides such an entry site. Most seed tubers produced in North America are too large to plant as whole tubers, but small, single-drop seed potatoes are routinely planted in the rest of the world. Using **whole seed** prevents pathogen spread during cutting (especially bacterial ring rot) and reduces seed decay, and whole tubers are less messy and easier to handle than seed pieces. However, it is difficult to produce small tubers for seed, and seed growers must generally ask a higher price for them, to compensate for the lower yield realized. Whole seed is most often used in North America in early seed generation production as minitubers, which are not cut, nor are B-size tubers (less than 1¾ inches), which are usually graded out at shipping time.

Precutting Seed

Once a healthy seed lot is selected, a decision must be made to determine if precutting is appropriate. In precutting, seed tubers are warmed, cut to size, and then cooled back down to a holding temperature that enables wound healing to take place.

Not all seed tubers should be precut. Tubers with excessive bruising or evidence of dry rot, soft rot, or other tuber-rotting pathogens are poor candidates for precutting. If late blight is present, it is usually preferable to plant fresh-cut seed. Precutting seed has been shown to encourage the spread of the late blight pathogen and subsequent infection of the cut surfaces by the pathogen, so that the amount of initial inoculum is increased.

The condition of the seed determines whether precutting can be performed. Only seed that is physiologically young or middle-aged should be precut, since the age of the seed is advanced in the precutting process. Seed that is physiologically young can be precut up to one month before planting. If the seed has previously sprouted, it should not be cut less than 2 weeks before planting. Middle-aged seed that has not sprouted can be precut up to 2 weeks prior to planting. Middle-aged seed that has sprouted and has been desprouted should be considered old seed. Physiologically old seed should not be cut more than a few days before planting. Cutting any earlier may result in additional aging and advance the seed tubers into producing secondary tubers instead of normal sprouts. Seed tuber age determines the temperature to which seed tubers are to be warmed before cutting and the temperature at which seed is to be held after cutting. Young seed tubers require a higher temperature than middle-aged seed tubers. Young seed can be cut and held at about 50°F, whereas older seed should not be warmed or held above 45°F. Since sprouting increases the physiological age of tubers, the temperatures should be lower for seed that has previously sprouted.

Precutting offers several advantages to help potato seed overcome adverse soil conditions at planting and to avoid dormancy factors in some cultivars. Precutting advances the physiological age of the tubers and provides a better opportunity to heal under controlled conditions in storage. Also, cutting can begin before planting starts, when labor is less in demand. Properly healed cut seed, if held for 3 or 4 weeks, will overcome dormancy factors and provide more uniform sprout development. Precut seed will provide earlier emergence, vigorous early growth, and higher plant and stem populations. Tubers of cultivars in which wound healing is slow and tubers that become wet during cutting are good candidates for precutting, because these characteristics make them particularly prone to seed piece decay, which leads to poor stands.

Successful use of precutting depends on careful management of the operation. It is essential to avoid condensation on freshly cut seed pieces, as this retards wound healing and favors decay-causing organisms. The extent of this problem may vary with different cultivars. To avoid condensation, whole seed tubers should be warmed to approximately 55°F for several days before precutting. Seed tubers should not be cut outdoors on rainy days or in extremely humid areas. The best plan is to set up the seed-cutting operation in the same area where the cut seed pieces will be stored.

After cutting, it is critical that precut seed be stored under conditions that favor rapid wound healing: temperatures of 55–60°F, relative humidity of 95–99% (but no free water!), and air movement to provide adequate oxygen and prevent the accumulation of carbon dioxide. Some growers use pallet bins or boxes to hold precut seed, while others place it in piles no more than 4–6 feet deep. Blowing air through ventilation ducts, as in normal storage management, may provide too much air for freshly cut seed and result in excess dehydration. If so, over-the-pile fans, lower fan speed, or using every other duct are suggested alternatives. It may be best not to humidify the air in the storage during the first 24–48 hours after cutting, to ensure the removal of any condensation and surface moisture that may develop, but this may vary with the cultivar and local storage conditions. In any case, humidity should be monitored carefully, and humidification should be provided as needed to keep the relative humidity of the air being delivered to the seed above 95% but managed to prevent condensation.

Suberization of cut surfaces should occur in a few days under proper conditions. A layer of dry, dead cells on the cut surface can be mistaken for a suberized layer. This nonprotective layer can be differentiated from a suberized layer by holding a seed piece in the hand and applying gentle pressure across the cut surface with the thumb. A layer of dead cells is usually pushed off the cut surface, while a suberized layer remains in place.

Precutting seed tubers can be a useful management practice if done properly. However, improper storage of precut seed, especially at temperatures above 60°F, can result in severe problems with seed piece decay, particularly from Fusarium dry rot. *If appropriate curing and storage conditions cannot be provided before planting, precutting is an extremely risky procedure and should not be attempted.*

Using Seed Piece Treatments

Seed treatments are products that can be applied to whole seed tubers or cut seed pieces to manage disease, promote wound healing, and facilitate easy handling. Compounds labeled for potato seed treatments are primarily fungicides formulated to prevent infection by seedborne pathogens, which can be spread during the handling and cutting operations, and by pathogens in the soil in close proximity to the seed piece. Seed treatments can reduce seed piece decay caused by *Fusarium* and *Pythium* species; reduce the spread of late blight during handling and cutting; reduce stem and stolon cankers caused by *Rhizoctonia solani;* and reduce infection of daughter tubers by *Helminthosporium solani,* the silver scurf pathogen. Seed treatments may also be of some use in managing other diseases caused by seedborne pathogens, including pink rot and common scab. Seed treatment fungicides are not systemic; they do not affect fungi already inside seed pieces, but affect only those on the surface of the seed piece and in the soil close to the seed piece. Seed piece treatments do not control viruses and provide little direct control of bacteria. None of the treatments affect lenticel-borne soft rot bacteria. However, because *Fusarium*-infected seed is more susceptible to blackleg, managing *Fusarium* indirectly helps reduce blackleg. Seed treatments do not prevent the spread and infection of ring rot bacteria. The main point is that seed treatments do not eradicate existing infections but only prevent new infections. They cannot improve seed quality!

Seed treatments, in the form of a dust or liquid, can be applied to whole seed, seed that is cut and immediately planted, and precut seed. Several products are available as seed piece treatments, containing the fungicides fludioxonil, thiophanate-methyl, mancozeb, flutolanil, or cymoxanil, alone or in combination. Fludioxonil, thiophanate-methyl, and mancozeb have broad-spectrum activity against *Fusarium* species, *Rhizoctonia solani,* and *Helminthosporium solani.* Flutolanil is active only against *Rhizoctonia solani.* Many seed treatment formulations contain mancozeb to control the spread of the late blight fungus during cutting. Cymoxanil is also used in combination with a protectant fungicide for this purpose. If late blight has been present in the seed production area, treatment with cymoxanil should be strongly considered. Liquid formulations should be applied at low application volumes to avoid excess water, which favors soft rot bacteria and subsequent seed decay. Application of seed treatments, especially fludioxonil, to successive generations of potato seed in sequential years increases the risk that fungicide resistance will develop. For resistance management, products should always be applied with companion and contrasting chemistry, such as mancozeb. *Selected trade names of fungicides are given in Table 14.5.*

Seed treatments are also formulated with a fungicide and a systemic insecticide to provide both disease control and early-season insect control. A liquid formulation of fludioxonil plus thiamethoxam and a dust formulation of thiophanate-methyl plus imidacloprid are currently registered. Seed treatments for control of nematodes are not available. Finally, many seed treatment formulations containing organic compounds or microbes purported to have activity in preventing disease have been developed and marketed. In general, most of these biological compounds have exhibited limited benefit as seed treatments in university and industry field trials. The final selection of a treatment should be based on product label information, the target organism, results of local professional evaluations, and personal experience.

Seed treatment products should not interfere with normal suberization and periderm formation as long as environmental conditions are appropriate. The inert ingredients of seed treatments are usually finely ground minerals (e.g., talc, gypsum, diatomaceous earth, mineral clays) or finely ground bark of trees such as fir or alder. Mineral carriers may absorb moisture under wet conditions and adhere as a wet plaster on the cut surface. This reduces the availability of oxygen to the cut surface and may promote bacterial seed piece decay. Bark carriers tend to make the seed flow more easily through planting equipment, and they wick away moisture from cut surfaces. Fir bark has been reported to enhance wound healing. Because of the advantages of bark, some growers use bark treatments that do not contain any fungicide. The disadvantages of bark-only treatments are that they do not kill fungal pathogens on the surfaces of seed pieces, and they may encourage the growth of nonpathogenic molds on cut surfaces, especially if treated seed pieces are held for more than a few days before planting.

The fungicides used in seed piece treatments are not highly toxic, but like all pesticides they must be handled and applied with care to ensure safety. Seed treatments are usually dusted onto cut seed as it flows by on a conveyor directly from the cutter. The seed is often then moved through a rotating drum apparatus to distribute the treatment more evenly. The applicator should be checked often to be certain that the treatment is being applied at the labeled rate. Because of the proximity of the cutting and treatment operations, dust control can be a problem, and personnel should wear appropriate dust masks or other protective devices. Improved applicators that contain and recirculate dusts are being developed and will greatly facilitate seed treatment. Liquid formulations of fludioxonil are available and minimize the exposure of workers to dust. As with all pesticides, only materials registered for use as potato seed piece treatments should be applied, and they should be applied according to the label directions.

Planting the Crop

The three principles of planting are to select high-quality seed, handle it carefully, and encourage quick emergence. Proper seedbed condition, especially soil temperature and moisture, is one of the most important factors in ensuring a healthy stand of potatoes. Ideally, the seed temperature and the soil temperature should be the same at planting. The optimum is 50–55°F for seed that has at least partially undergone wound healing. This temperature encourages quick emergence without promoting the growth of seed piece decay organisms. The seedbed should be neither too wet nor too dry, e.g., 75–85% available soil water. Planting in a wet soil, either hot or cold, favors bacterial seed piece decay, which may result in a poor stand. Pathogenic bacteria can enter through cut surfaces and wounds, but more importantly they may already be present in the tuber lenticels (Chapter 18). Under the low-oxygen conditions that develop in waterlogged soils, these bacteria become active and can rapidly decay seed pieces. Fungal seed piece decay is usually caused by several species of the fungus *Fusarium* (Chapter 19). These pathogens require wounds in order to enter seed pieces, and decay is favored by a wide range of temperature and moisture conditions. Seed tubers with Fusarium dry rot in storage continue to decay when planted. Serious problems with seed piece decay are often related to adverse weather during the planting season. If high-quality seed stock is used and properly healed after cutting, seed piece decay is usually minimal if soil conditions at planting and in the first few weeks following planting favor rapid emergence. Planting seed pieces immediately after cutting in a dry, hot soil (especially sandy soil) results in poor wound healing and increases the potential for seed piece decay and poor stands. Cold seed planted in warm soil often sweats after planting, and the moisture film around the seed piece increases the potential for bacterial decay. Some physiological disorders may also result from planting in excessively wet or dry soil (Chapter 23). In extremely sandy soil, preplant irrigation may be necessary to provide appropriate soil moisture at planting and to prevent wind erosion during and after planting. This is preferable to preemergence irrigation after planting, which may favor seed piece decay.

The three principles of planting are to select high-quality seed, handle it carefully, and encourage quick emergence.

When to plant is an important management decision that can have significant implications for crop health. Try to plant when soil conditions favor **rapid emergence**. In unirrigated areas where spring rains are often plentiful, planting is usually a waiting game controlled by the weather. Because of problems with planting in cold, wet soil, it is usually a good management decision to wait for warmer, drier conditions. Other factors affecting the choice of a planting date include the maturity class of the cultivar and the timing of harvests needed for certain markets. Planting should be delayed at least 2 weeks after plow-down of a green manure crop, since the undecomposed residue may heat and favor Pythium seed piece decay.

Potato planters are of the following general designs: pick, cup, assist-feed, and tuber-unit planters. By far the most common is the pick planter. It operates by forcing sharp metal picks into seed pieces. The impaled seed is carried out of the hopper and dropped into the planting shoe. The picks must be of appropriate length, straight, sharp, and spaced correctly to obtain the desired seed spacing. A major liability of this type of planter is that disease organisms from infected tubers can be carried on the picks and inserted into healthy seed. This can be significant in the spread of the ring rot and blackleg bacteria and some viruses.

Cup planters use a series of cups instead of picks to convey the seed to the soil, thereby lessening the hazard of disease spread. Seed planted with cup planters must be quite uniform in size, to avoid skips and doubles and to obtain even spacing of the seed pieces. This type of planter is well adapted for use on flat terrain and with whole seed, as long as the planter has been adjusted to accommodate the seed size. Air planters use cups attached to hollow arms with a vacuum device to pick up seed pieces. Seed pieces are discharged into the planting shoe when the vacuum is released.

Operator-assisted or assist-feed planters are sometimes used in seed-growing operations or in small-acreage production. Seed pieces are mechanically fed onto a rotating spacing disk before being dropped into the furrow. With this type of planter, additional labor is usually needed to place seed in skipped seed pockets and remove excess seed to prevent doubles.

The tuber-unit planter combines the cutting and planting operations. It is used for early-generation seed potato production. Whole tubers are placed in the seed box, and individual tubers are transported to a cutting knife, where they are cut into four seed pieces. Each seed piece is then planted separately but sequentially, with one empty space between groups of four pieces, to facilitate roguing of diseased tubers and cultivar mixtures. Equipment may be added to sanitize cutting knives and planter parts between tubers. The machine has a small capacity, and proper sizing of tubers is essential.

Appropriate **spacing of rows and seed pieces** within rows depends on several factors, including the cultivar, fertility program, availability of irrigation, soil type, and expected market. Row widths generally range from 30 to 36 inches, with 36 inches being most common. Spacing requirements within the row vary with the cultivar and the growing area, and within-row spacing may affect the development of some physiological disorders. For example, in highly susceptible cultivars, hollow heart may be minimized by closer spacing (Chapter 23). For cultivars with a tendency to develop oversized tubers or to set few tubers, seed pieces should be spaced closer within the row. Close within-row spacing is also appropriate where irrigation is used or where moisture is not expected to be a limiting factor. For cultivars that generally produce a high number of tubers per stem, seed pieces should be spaced farther apart in the row. For some cultivars grown for specific markets, bed planting is often used in conjunction with drip irrigation.

Market considerations also affect spacing. Frozen french fry processors prefer large tubers, within limits, and often offer an incentive for tubers over 10 ounces. Production of large tubers is enhanced by wide within-row spacing. The

Table 7.2. Weight of potato seed pieces of different sizes and number of seed pieces required to plant an acre with various between-row and within-row spacings

Between-row spacing (inches)	Within-row spacing (inches)	Weight (in hundredweight) of seed pieces per acre, with seed piece size of: 1.75 ounces	2.00 ounces	2.25 ounces	Number of seed pieces per acre
34	7	28.8	32.9	37.1	26,356
	8	25.2	28.8	32.4	23,061
	9	22.4	25.6	28.8	20,499
	10	20.2	23.1	17.3	18,449
	11	18.3	21.0	26.0	16,772
	12	16.8	19.2	21.6	15,374
36	7	27.2	31.1	35.0	24,891
	8	23.8	27.2	30.6	21,780
	9	21.2	24.2	27.6	19,360
	10	19.1	21.8	24.5	17,424
	11	17.3	19.8	21.8	15,480
	12	15.9	18.2	20.4	14,520

seed market, however, prefers tubers less than 8 ounces and less than 3¼ inches in diameter, and tubers less than 2 inches in diameter often command a premium price. Production of tubers in this range is best accomplished with close within-row spacing. Table 7.2 lists seed requirements for various spacings and seed piece sizes.

Appropriate **planting depth** and soil cover vary in different production areas, because of variations in soil and moisture conditions and established practices. A general practice in northern regions to encourage rapid emergence is to plant shallow to minimize seed piece decay and the development of Rhizoctonia stem canker. In the West and in any area where moisture may be limiting, seed pieces are usually planted deeper, to ensure that they are covered at all times with moist soil. In some cases, they are planted 6 inches deep in preformed beds. In areas with fine-textured (clay-type) soils, some growers install rippers behind the planter wheels, to allow early-season rains to drain below the planted seed pieces.

Fertilizer may be applied before, during, or after planting or at all of these times, depending on the production area and cultural system used (Chapter 8). Some starter fertilizer is usually broadcast before planting or banded at planting (or both). Proper fertilizer placement is important, to avoid burning or stressing young sprouts. Fertilizer should not come in direct contact with planted seed pieces.

Fungicides may be also applied in-furrow at planting for disease control instead of or in addition to seed treatment. For example, a combination of a seed treatment plus an in-furrow fungicide application provides better control of silver scurf than either treatment alone, but this treatment is more expensive. Mefenoxam can be applied in a single application in-furrow at planting or in a split application in-furrow and at row closure for the suppression of pink rot and leak. Both flutolanil and azoxystrobin can be banded in-furrow at planting for control of seed- and soilborne *Rhizoctonia* (Chapter 19).

The **systemic insecticides** imidacloprid, thiamethoxam, and phorate can be applied at planting to provide early-season control of certain insects. These insecticides target mainly aphids, leafhoppers, Colorado potato beetles, and psyllids during the first 30–60 days of growth (Chapters 15 and 16). At planting, application of oxamyl and ethoprop can provide suppression of certain insects and nematodes (Chapter 20). The need for early-season insect or nematode management and the potential for groundwater contamination should be taken into account if this practice is under consideration. *Selected trade names of insecticides are given in Table 14.4.*

Herbicides are usually applied either preplant or postplant but not at planting (Chapter 22). Where a preemergence herbicide is used, minimizing field operations between planting and emergence prevents excess soil mixing and thus maximizes the effect of the herbicide.

Summary

Adequate planning and preparation prior to receiving seed for planting will pay for the time required for the effort. Proper sanitation of all potato handling facilities will help prevent ring rot outbreaks. Seed bruising contributes to poor stand establishment, and thus careful handling of seed to avoid bruising is critically important. Managing physiological age, sanitation, promoting wound healing, and proper use of seed treatment fungicides all help the seed to perform at the highest level. The three principles of planting are to select high-quality seed, handle it carefully, and encourage quick emergence. Planting recommendations include good seedbed preparation; planting seed that is at the same temperature as the soil; handling seed gently; avoiding planting in wet, hot, or cold soil; using a seed treatment when necessary; and following good cultural practices.

CHAPTER 8

Jeff Stark
Idaho Falls Research and Extension Center
University of Idaho, Idaho Falls

Dale Westermann
U.S. Department of Agriculture, Agricultural Research Service
Kimberly, Idaho

Managing Potato Fertility

Efficient potato fertility management ensures that all essential plant nutrients are available in appropriate amounts throughout the growing season and provides for optimal plant health and growth. Essential nutrients are those that are required for normal growth and development and for which no other nutrient can be substituted. Sixteen elements are known to be essential for plants. These elements function as necessary constituents in metabolic processes in which they are converted into cellular material or used in energy transfer. The three most abundant nutrients in the environment—carbon, hydrogen, and oxygen—are acquired from air and water, while the remaining nutrients are absorbed from the soil solution by roots or from the aerial environment through leaves.

Potato Nutrient Requirements

Potato growth is commonly divided into five stages (Fig. 2.1). Cumulative uptake curves for most major nutrients generally follow a pattern similar to tuber growth in which uptake rates increase during tuber bulking before leveling off during tuber maturation late in the growing season (Fig. 8.1). Large portions of phloem-mobile nutrients, including nitrogen (N), potassium (K), phosphorus (P), magnesium (Mg), and chlorine (Cl), are eventually transported to the tubers and have accumulation and distribution patterns similar to that of dry matter accumulation. Nutrients that are not mobile in phloem, including calcium (Ca), sulfur (S), zinc (Zn), boron (B), manganese (Mn), and iron (Fe), tend to accumulate in the foliage during tuber growth, and their accumulation and distribution patterns are less affected by tuber nutrient demands.

Potential Tuber Yield

The maturity class of the cultivar and length of the growing season both affect the optimal rate and timing of nutrient applications. Early-maturing cultivars, such as Russet Norkotah, have a shorter development period and typically have higher nutrient uptake rates during vegetative growth and tuber initiation. As a result, it is important to have an adequate preplant fertilization program for early-maturing cultivars. In late-maturing cultivars, such as Russet Burbank, nutrient application rates during tuber bulking must be slightly higher than those required for tuber development. This allows for the production of the new vegetative growth needed to maintain effective photosynthesizing leaf area as the older leaves senesce. The potential yield of late-maturing cultivars increases as the length of the growing season increases, as long as the plants remain healthy and supplies of water and nutrients are adequate. As a result, crop nutrient requirements increase proportionally, to be able to sustain the higher potential yield.

Soil Characteristics

Soil physical and chemical characteristics can have a substantial effect on nutrient availability to plants, which in turn affects nutrient use efficiency (Chapter 5). Restrictions in root growth caused by hardpans, soil compaction, or shallow soils reduce the volume of soil from which nutrients can be extracted, thereby decreasing the available nutrient supply. Sandy soils have relatively low water-holding capacities, high infiltration rates, and a greater potential for leaching of nutrients, carrying them beyond the root zone. They also have a relatively low cation exchange capacity, which limits their ability to hold and exchange positively charged nutrients, such as potassium, calcium, and magnesium, compared to soils with higher clay contents.

Soil pH has a major effect on nutrient solubility and availability, with optimal availability of most plant nutrients at pH 6.5–7.0. At pH 5.0 and below, Ca, Mg, P, and molybdenum (Mo) deficiencies and ammonium, manganese, and aluminum toxicities may develop. At pH 7.5 and above, B, copper (Cu), Fe, Mn, and Zn deficiencies may occur. In calcareous soils, P availability may also decrease, as P is precipitated as calcium phosphates.

Soil **organic matter** functions as a storehouse for many essential elements, releasing a variety of compounds during decomposition to provide a relatively stable source of plant nutrients. Plants grown in soils that receive regular additions of organic residues, such as animal manure, usually have few

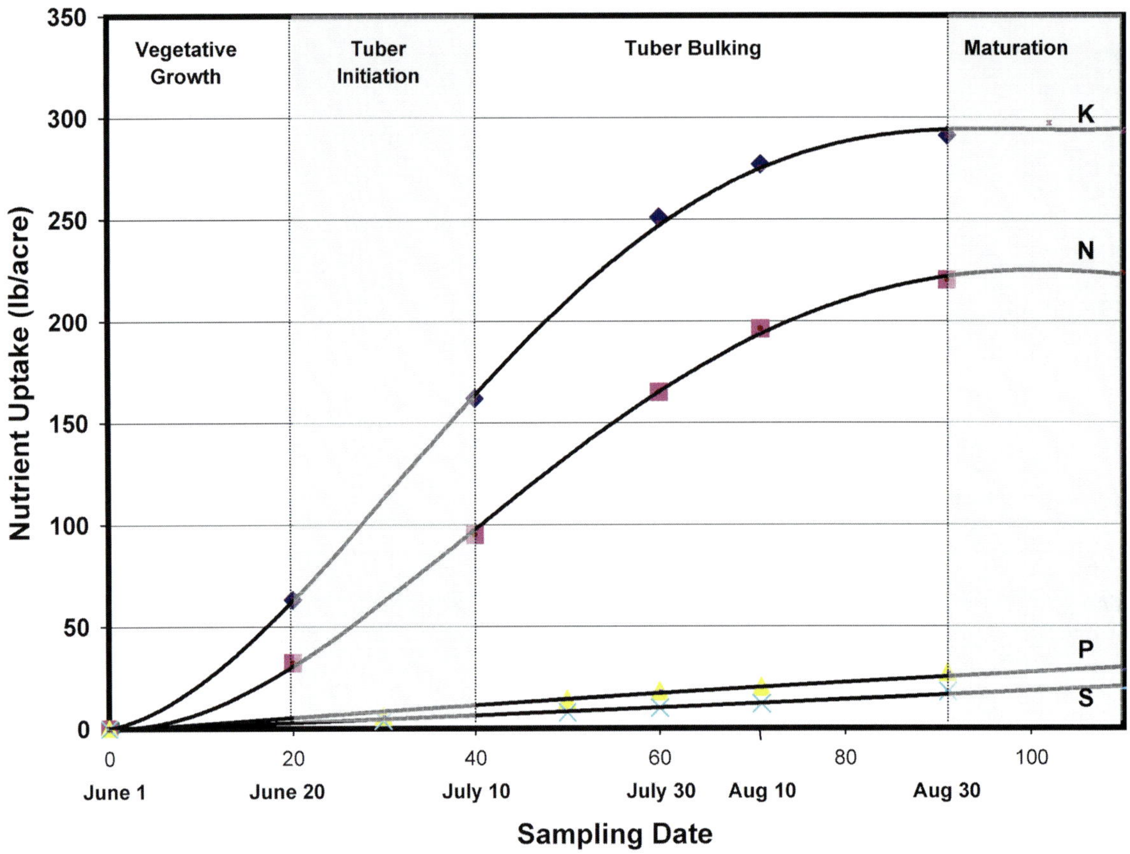

Fig. 8.1. Cumulative uptake of nitrogen (N), phosphorus (P), potassium (K), and sulfur (S) by Russet Burbank potato plants.

micronutrient deficiency problems. Soil organic matter also influences nutrient availability through the process of chelation, in which metallic ions form bonds with organic molecules. The solubility of metallic chelates is usually greater than that of the inorganic forms, but the activity of the ion in solution is usually lower. As a result, chelation can either increase or decrease nutrient availability. The availability of Cu, Mn, and Zn usually decreases as soil organic matter content increases to high levels.

Environmental Factors

Most micronutrients exist in several oxidation states but usually are more available in the reduced form. Excessive irrigation or rainfall and poor soil drainage can lead to waterlogged conditions, which in turn lead to reducing conditions resulting from the decreased availability of oxygen. Consequently, in wet areas, micronutrient availability can increase to levels that cause toxicity. Conversely, waterlogged soil conditions can reduce plant nutrient availability and uptake by reducing root zone oxygen supply and root activity or by increasing nutrient solubility, resulting in the leaching of nutrients below the root zone.

Low soil temperatures early in the growing season reduce the physiological activity and growth of roots and thus lead to reduced nutrient uptake. This effect is particularly significant for P, which moves very slowly in soil and requires active root uptake. Banding fertilizers or applying higher rates can partially compensate for the effects of low temperature. The goal of early-season fertilizer adjustments should be to provide sufficient nutrients for early canopy development and optimal photosynthetic efficiency.

Cool soil conditions also reduce the rate of N release or mineralization from soil organic matter and slow the conversion of ammonium to nitrate. High soil temperatures accelerate emergence and early-season growth, creating a corresponding increase in nutrient demand. Nitrogen mineralization rates also increase in warm soils; in abnormally hot weather, soil N availability can increase enough to cause excessive uptake of N by plants. If soil temperatures get too high (e.g., above 100°F), mineralization decreases considerably. Excessive N uptake resulting from high N mineralization rates favors vegetative growth at the expense of tuber growth.

Nutrient Interactions

The plant's response to the application of a specific nutrient often depends on the relative availability of other nutrients. Correcting deficiencies of the other limiting nutrients usually improves nutrient use efficiency overall and increases the likelihood of achieving the potential yield for that location. Conversely, the uptake of certain nutrients can be affected by excessive amounts of other nutrients. Excessive applications of phosphorus, for example, can induce zinc deficiency when zinc availability is marginal.

Nutritional Disorders

Nutrient deficiencies and toxicities are broadly classified as nutrient disorders and are influenced by a wide variety of

Box 8.1

Foliar Symptoms of Nutrient Deficiency in Potatoes

Nitrogen. Initially, the entire plant begins to turn light green. Younger leaves remain green, while older leaves change to yellow and then light brown and eventually senesce and dry up.

Phosphorus. Deficient plants are often stunted and are darker green than adequately fertilized plants. Severe deficiency can cause leaf blades to curl upward and a purplish color to appear. Leaf rolling usually increases as the deficiency becomes more severe.

Potassium. Young leaves initially develop a glossy appearance, with surface crinkling and a slightly black pigmentation. Severe deficiencies can cause scorching of the leaf margins of older leaves, which eventually become necrotic.

Boron. Plants appear bushy and have shortened internodes. Leaf blades thicken and roll upward, and their edges turn light brown. Eventually, growing buds die, and leaf tissue darkens and collapses.

Calcium. The youngest leaves, near the growing point, roll upward and become chlorotic with brown spotting. Severe deficiency can cause the growing point and growth buds to die. In tubers, a brown discoloration develops within the vascular ring before symptoms appear on the foliage.

Chloride. Young leaves are initially light green and then turn purplish bronze and may also curl upward and appear pebbled.

Copper. Young leaves remain green and are of normal size but exhibit a pronounced rolling and cupping. Leaf tips eventually wilt and die.

Iron. Young leaves become yellow to nearly white, usually without necrosis. Leaf veins appear greener than the interveinal tissue. Leaf tips and edges usually remain green longer than the rest of the leaf.

Magnesium. Initially, a slight interveinal chlorosis develops on young, fully expanded leaves, with some brown spotting, which develops into interveinal leaf scorch. Leaves near the growing point remain green.

Manganese. Younger leaves initially exhibit interveinal chlorosis, a slight upward cupping, and gray and black spotting. The density of the spots gradually increases along the larger leaf veins, where spots eventually develop into necrotic patches.

Molybdenum. Leaf blades become uniformly yellow, similar to those of plants with nitrogen or sulfur deficiency.

Sulfur. Symptoms are similar to those of nitrogen deficiency, except that in sulfur deficiency the yellowing appears first in younger leaves, which turn uniformly light green to yellow.

Zinc. The youngest leaves become narrow, erect, and chlorotic and develop tipburn and leaf cupping. Blotching and spotting of interveinal tissues develops in severe deficiency, and lower leaves turn brown and die.

soil characteristics and environmental and cultural factors. Deficiencies can be caused by inadequate fertilization, soil conditions that reduce nutrient availability, antagonism between nutrients, or reduced root growth and activity resulting from disease or from physical or chemical characteristics of the soil. Potato nutrient deficiencies can sometimes be identified from visible symptoms (Box 8.1), but visible diagnosis can often be misleading, particularly with multiple deficiencies.

Nutrient deficiency symptoms are related to the function of each specific nutrient in the plant. Nutrients typically have multiple functions, which sometimes make it difficult to associate visual symptoms resulting from inhibition of a specific physiological process with a deficiency of a single nutrient. For example, N deficiency typically causes plant leaves to become pale green or light yellow, as a result of reduced chlorophyll production and increased visibility of the yellow pigments carotene and xanthophyll. However, a number of other nutrient deficiencies, herbicide toxicities, and diseases also cause leaves to be pale green or yellow. As a result, the visual examination may need to be extended to include the location of the symptoms on the plant and the distribution of symptomatic plants in the field.

Integrated Nutrient Management Systems

Potato nutrient management in North American production areas is highly intensive, relying heavily on chemical fertilizers. This trend has resulted in a reduced emphasis on crop rotation and plant and animal residue management, leading to an overall reduction in soil carbon and plant nutrient concentrations. By comparison, cropping systems in which large amounts of organic residues from green manures, animal wastes, or composts are applied to the soil often have improved soil nutrient availability, tilth, water-holding capacity, aeration, and overall quality.

A fully integrated nutrient management system involves a set of practices designed to efficiently utilize and conserve available resources and maintain or enhance the long-term productivity of the crop ecosystem. Nutrient management systems should take into account the interactive effects of the various cash crops, cover crops, green manures, animal manures, and soil amendments used in the rotation with respect to nutrient cycling, soil quality, and effects on diseases and other plant pests.

Legumes in Crop Rotation

Legumes have received considerable attention as an important component of potato cropping systems because of their significant effect on the N cycle. In addition to contributing biologically fixed N to the cropping system, legumes can also improve potato yield and quality and improve physical properties of the soil. Total N contents of legume crops grown in potato rotations can be as high as 214 pounds per acre, most of which is released during the first year after incorporation. It is important to match N release patterns from legume residues with potato N uptake patterns to optimize N use efficiency. Indeterminate potato cultivars are particularly sensitive to the seasonal pattern of N availability. Nitrogen released too early from rotation crops or green manures can delay tuber bulking and promote excessive vine growth, resulting in reduced yield and an increased proportion of immature tubers. Nitrogen released too late in the growing season can reduce N use efficiency and increase the potential for nitrate leaching.

Methods for determining legume N contributions vary widely but are usually based on some form of a fertilizer N replacement value, generally described as the amount of fertilizer N required to produce a given yield following a nonleguminous crop that is identical to that produced following a legume when no N fertilizer is applied. This method is based in part on the assumptions that the increased yield of the nonleguminous crop is completely due to legume N contributions and that fertilizer N and crop residue N are equally available. Both of these assumptions are often invalid. However, the fertilizer replacement value does provide a good estimate of the total effect of the legume crop on yield at low N fertilizer rates, but it does not separate N effects of the rotation crop from non-N effects.

Another approach to estimating legume N credits is based on the difference in fertilizer N rates needed to maximize crop yields following legumes and nonleguminous crops. This approach accounts for the fact that the N fertilizer response curve following legumes is often quite different from the response curve following a nonleguminous crop. It also allows for a determination of optimal N rates regardless of cropping sequence or the amount of residual soil N in the root zone. In most cases, the economically optimum N rate can be derived from these data by factoring in the cost of N fertilizer.

Integrating Legume and Fertilizer Nitrogen

One approach for addressing the uncertainty associated with estimating legume N mineralization rates is the use of split N fertilizer management. With this approach, early-season N requirements are determined from a combination of inorganic N results from spring soil testing, estimated legume N credits, and estimated total crop N requirements. For indeterminate cultivars grown in irrigated production systems, one-third to two-thirds of the total N requirement should be top-dressed or applied with irrigation water during tuber growth, based in part on the concentration of nitrate nitrogen (NO_3-N) in petioles. The petiole test allows the producer to adjust for seasonal variation in N mineralization rates by changing the timing and amount of top-dressed N applications. Although a split N management approach is not effective in all areas, it does provide management flexibility in many irrigated production regions, utilizing conservative N fertilizer rates at planting with subsequent applications based on in-season measurements of plant N status.

Animal Manures and Composts

Animal manures and composted organic materials often play an important role in supplying nutrients to potatoes. Manure contains large quantities of nutrients, but the specific amounts of these nutrients can vary appreciably, depending on the type of animal and its age, the composition of the feed ration, the type and amount of bedding, climatic conditions, storage and handling systems, and application methods.

To effectively utilize animal manure as a nutrient source without increasing the potential for nutrient leaching and runoff, both the amount and availability of the nutrients contained in the manure must be known. Laboratory analysis of manures or composts can be utilized to make estimates of the potential nutrient contributions that these organic materials can provide. However, the availability of nutrients and the timing of nutrient release can be difficult to predict, because of complex interactions with the soil and biological and environmental variables affecting the amendments.

Despite these uncertainties, animal manure and compost amendments can make significant contributions to potato nutrient management. The application of animal manure can rapidly increase the concentrations of labile and stable soil organic matter while also increasing soil N concentration, cation exchange capacity, water-holding capacity, and soil fertility levels. It can also reduce soil bulk density and increase soil aggregation, and it can increase the activity and diversity of soil microbial populations, which can stimulate root activity and nutrient uptake. Compost and manure applications have also been shown to increase potato root length.

In addition to the difficulty in predicting the timing and amount of nutrients released by manures and composts, the transportation and application of these amendments is often costly. In the short term, manure and compost amendments may be less cost-effective per unit of nutrient applied than chemical fertilizers, even when they enhance crop yields. However, this extra cost may be justified in the long term, because of the positive effects on the soil and other biological components of the cropping system.

Determining Fertilizer Requirements

Nutritional Management Tools

Because of the many factors that affect plant nutrient requirements and availability, appropriate nutrient management strategies need to be developed for each cropping system that take into account the specific characteristics of each field. Soil and plant analysis are two primary tools that can be used to develop an efficient and environmentally responsible nutrient management program.

Soil Analysis

The relative availability of soil nutrients at a specific field site can be determined by using appropriate soil testing procedures that are locally calibrated with crop yield responses. Laboratories that use procedures developed and calibrated for the local growing region generally give the best recommendations. Laboratory test results should then be used in formulating fertilizer recommendations and selecting appropriate nutrient sources and methods of application.

The accuracy of a soil test result can be affected by both the laboratory analysis and quality of the soil sample. A soil sample that is not representative of the corresponding field area will likely produce misleading data resulting in inappropriate fertilizer application rates. Therefore, it is essential that each field be sampled using procedures that represent the majority of the soils in that field. Areas with significant differences in soil texture, color, topography, cropping history, or fertilization history should be sampled and fertilized separately.

A **composite soil sample** for a field zone should consist of about 20 soil cores from the 0- to 12-inch soil depth collected with a soil probe that provides a uniform sample throughout the entire sampling depth. Each composite sample should represent no more than 20 acres, even if the soil appears uniform. Typically, the 20 cores are taken in a selectively random pattern, such as a zigzag pattern throughout the sampling area. The samples should be thoroughly mixed together, and a subsample should be collected and immediately air-dried before being sent to the lab. Keeping soil samples warm and moist over a period of a few days or more will allow N mineralization to continue, leading to erroneously high soil inorganic N concentrations.

Site-specific soil sampling for variable-rate fertilization can be accomplished by either systematic grid sampling or some form of directed sampling based on differences in soil color, crop growth patterns, or landscape characteristics. Grid samples are typically collected from 1.5- to 2.5-acre grids, with irregular sampling positions such as those used with a systematic, unaligned grid pattern. Data are then analyzed by various statistical procedures that provide an estimate of nutrient patterns within the field.

Directed sampling, also known as zone sampling, typically involves collecting soil samples from three to six management zones per field. Various sources of information are used in delineating the zones, including aerial photographs, satellite imagery, soil survey maps, topography maps, yield maps, and producer experience. Six to 10 soil cores from each zone are mixed thoroughly to produce composite subsamples that represent each zone. The subsamples are analyzed separately, and fertilizer recommendations are developed for each zone.

Plant Tissue Analysis

Plant tissue analysis is widely used as a diagnostic tool for monitoring the nutrient status of potato plants. It is based on established relationships between nutrient concentrations in a standard plant part and plant growth rate or yield. In potatoes, the petiole of the fourth leaf from the top of the plant (Fig. 8.2) is generally used to determine plant nutrient status. It is important to consistently sample the fourth petiole, since analysis of samples taken from leaves in higher or lower positions will produce different values that are not calibrated with the crop's nutritional status. Petiole sampling usually begins at tuber initiation and is continued weekly throughout most of the tuber growth period.

For laboratory analysis, usually 40–50 petioles are collected from representative areas of the field, in a manner consistent with the soil-sampling procedure. Alternatively, separate samples can be collected from selected areas or management zones differentiated by factors such as soil type, crop history, and topography. All leaflets should be stripped off the petiole immediately after sampling. The petioles should then be placed in a clean container or paper bag and quickly dried at 150°F or kept cool (below 40°F) until submitted to the lab. This procedure should minimize changes in nutrient concentrations that can occur in warm, moist tissue samples.

Petiole nutrient concentrations are divided into low, marginal, and sufficient ranges on the basis of their relationship to growth and yield (Table 8.1). These ranges can vary according to the growth stage and the cultivar; the ranges for nitrate nitrogen concentration, in particular, can vary widely over the course of the growing season. Sufficient concentrations are generally adequate for foliar and tuber growth at the time of sampling. However, trends observed from weekly petiole samples are more reliable in anticipating and detecting nutrient deficiencies.

Nutrient Uptake Ratios

The N, P, and K nutritional status of plants during tuber growth can also be evaluated by comparing the ratio between the total plant uptake rate and the uptake rate required for tuber growth. When this ratio is greater than 1.0, more of the nutrient is being taken up than is required by the tubers. However, when the ratio is less than 1.0, some nutrients will be translocated out of the vegetative portions of the plant to the tubers. These relationships can be tracked by comparing measured ratios at different points in the growing season to standard ratios developed previously from well-fertilized plants, which are referenced to days after emergence, heat units, or radiation absorbed. Alternatively, if the concentration of the

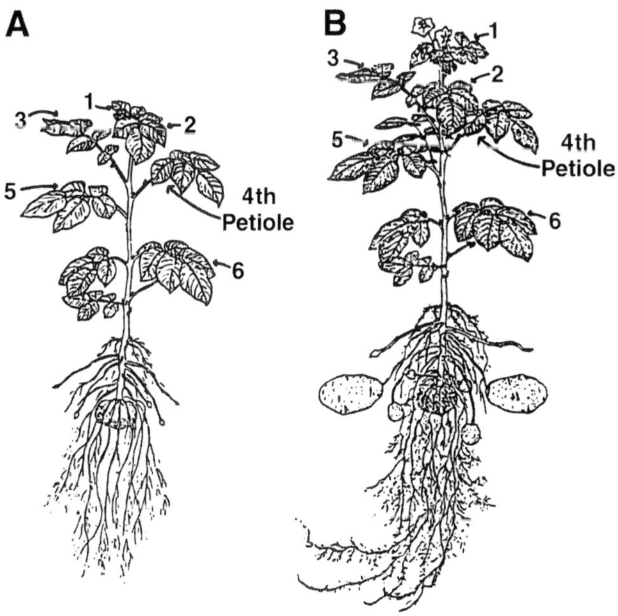

Fig. 8.2. Vegetative shoot (**A**) and shoot with a floral spike (**B**). The petiole of the fourth leaf from the top of the plant is used in tissue analysis to determine the nutrient status of the plant during growth.

Table 8.1. Suggested ranges of nutrient concentrations in the fourth leaf from the top of the potato plant during tuber bulking (Growth Stage IV)

	Low	Marginal	Sufficient
Petiole without leaflets[a]			
Nitrate nitrogen, ppm	<10,000	10,000–15,000	15,000–20,000
Phosphorus, %	<0.17	0.17–0.22	>0.22
Potassium, %	<7.0	7.0–8.0	>8.0
Calcium, %	<0.4	0.4–0.6	>0.6
Magnesium, %	<0.15	0.15–0.3	>0.3
Sulfur, %	<0.15	0.15–0.2	>0.2
Zinc, ppm	<10	10–20	>20
Boron, ppm	<10	10–20	>20
Manganese, ppm	<20	20–40	>40
Iron, ppm	<20	20–50	>50
Copper, ppm	<2	2–4	>4
Entire leaf (petiole + leaflets)[a]			
Nitrogen, %	<2.5	2.5–3.5	>3.5
Phosphorus, %	<0.15	0.15–0.25	>0.25
Potassium, %	<2.25	2.25–3.5	>3.5
Calcium, %	<0.30	0.3–0.6	>0.6
Magnesium, %	<0.15	0.15–1.25	>0.25
Sulfur, %	<0.12	0.12–0.20	>0.2
Zinc, ppm	<15	15–20	>20
Boron, ppm	<10	10–20	>20
Manganese, ppm	<10	10–20	>20
Iron, ppm	<11	11–30	>30
Copper, ppm	<2	2–5	>5

[a] Petiole concentrations are for Russet Burbank. Concentrations for the entire leaf are suitable for many cultivars.

nutrient in plant tissue is related to these ratios, plant tissue tests can also be used to monitor crop nutrient ratios and determine the need for additional fertilizer.

Petiole Sap Analysis for Nitrate Nitrogen

Petiole sap testing for NO_3-N has been reported to provide a quick, inexpensive, and reasonably accurate assessment of potato N status. Sap extracted from potato petioles can be analyzed directly by the use of an ion-specific electrode, or it can be subjected to colorimetric analysis, in which test strips are immersed in diluted sap extract, and then the color of the strips is read with a hand-held reflectometer. Highly significant linear relationships have been found between sap NO_3-N test results and petiole NO_3-N concentrations determined from laboratory analysis. As with conventional laboratory analysis, reliable N recommendations from sap testing will require appropriate calibration to establish sufficiency ranges at different growth stages.

Visual Symptoms

Nutrient deficiencies can sometimes be identified from visible symptoms (Box 8.1). However, by the time the symptoms develop it is often too late to make corrective fertilizer applications that will prevent a loss of yield or quality. Diagnosis can also be complicated by symptoms due to multiple nutrient deficiencies, diseases, and environmental stress. More accurate and timely identification of developing nutrient deficiencies can be obtained by chemical analysis of plant tissue.

Fertilizer Application Strategies

Fertilization practices should be consistent with characteristics of the cropping system; efficiently utilize available equipment, fertilizer materials, and other resources; and provide flexibility in responding to changing nutrient requirements. Fertilizer application methods commonly used for potatoes include

- Preplant broadcasting followed by incorporation
- Banding at mark-out or planting
- Side-dressing after planting
- Applying foliar nutrient sprays
- Applying liquid fertilizer through the irrigation system
- Top-dressing

Broadcast fertilizer applications before planting can be effective as long as soil fertility levels are relatively high and limited soil fixation of nutrients is expected. Broadcast fertilizer applications made in the fall or spring should be incorporated into the surface 6–12 inches of soil to provide ready access for the potato roots. Usually less than half of the seasonal N supply should be applied prior to planting in irrigated systems, but the bulk of all other fertilizers can be broadcast at this time. Preplant fertilizer applications can greatly reduce the need for additional applications at planting, allowing more time for management of the planting operation.

Fertilizers can be **banded** at mark-out or at planting. The bands should be placed close enough to the seed piece to provide early root access. However, direct contact with the seed piece should be avoided, particularly with fertilizers applied at high rates or having a high salt index. For mark-out applications, an effective band placement is 3–4 inches to the side of the seed piece and 1–2 inches below seed piece depth. Fertilizer bands applied at planting are best placed 3–4 inches to the side and 1–2 inches above seed piece depth, to provide greater early-season availability to the developing plant. Liquid formulations of nutrients, including N, P, and various micronutrients, are commonly applied in bands to enhance early-season availability.

Side-dressed N fertilizer is often applied during vegetative growth prior to tuber initiation. This method is particularly effective in sandy soils that are prone to nitrate leaching. Later applications risk damage to the crop from root pruning, which can occur during the side-dressing operation. Side-dressing of other nutrients is usually not as effective as broadcast or band application.

Foliar nutrient sprays can be an effective treatment for existing or developing nutrient deficiencies, particularly deficiencies of B, Cu, Fe, Mn, and Zn, and usually produce a quicker response than soil applications. Tank-mixing surfactants usually improves nutrient absorption through the leaf surfaces. However, the amount of any nutrient that can be safely applied is limited, because concentrated sprays can result in leaf damage.

Water-soluble fertilizers are commonly applied to potatoes through **sprinkler irrigation systems** (fertigation) in many potato-producing regions. Nutrients such as N, P, K, and S can be applied according to crop needs and partially incorporated into the root zone with the irrigation water. Careful irrigation management is essential when fertigation is practiced, because of the high potential for nutrient leaching and runoff when the irrigation rate is excessive. The uniformity of irrigation has a corresponding effect on the uniformity of fertilizer application, which in turn affects fertilizer use efficiency. The compatibility of the fertilizer materials with the irrigation water should be checked before application, since precipitation of nutrients can reduce availability and clog nozzles. In-season application of both liquid and dry water-soluble fertilizers can also be made with the use of **aircraft** or **ground application** equipment. Incorporation must be accomplished by irrigation or rain following application.

Nitrogen Management

Efficient management of nitrogen (N) is an essential part of any potato fertilization program. Nitrogen management can affect vine and tuber biomass production and the size, grade, specific gravity, and internal and external quality of tubers. The potential for nitrate leaching is also closely related to the efficiency of the N management program. Crops can acquire N from a number of sources, including (1) inorganic soil nitrate (NO_3) and ammonium (NH_4), (2) N mineralized from soil organic matter, crop residues, and animal wastes, (3) N present in irrigation water, and (4) fertilizers. To determine the amount of N fertilizer to apply to a potato crop, an assessment needs to be made of the crop N requirements, which are proportional to yield and the availability of N from all sources. Nitrogen applications should be based on locally calibrated recommendations adjusted according to soil N supply and established yield potential at the site.

For **determinate cultivars**, dry matter in the foliage and roots usually peaks at or before the midpoint of tuber growth and then levels off or gradually declines. The dry matter distribution pattern for determinate cultivars is fairly independent of nutrient supply. For indeterminate cultivars, however, high nitrogen availability can cause foliage and roots to become the dominant sink for carbohydrates and mobile nutrients at the expense of tubers. As a result, tuber growth is slowed, while the plant diverts greater amounts of dry matter to foliage and roots. Applying the entire seasonal N requirement preplant can delay the start of the linear tuber growth phase in some indeterminate cultivars by up to 2 weeks. The same N management strategy does not delay linear tuber growth in determinate cultivars but rather hastens the development of optimum photosynthetic capacity in the plant canopy, thereby facilitating early tuber growth. In N-deficient situations, vine growth in indeterminate cultivars is reduced, and carbohydrates that would normally be used for leaf growth are shifted to tuber growth.

About 50–60% of the seasonal N requirement is taken up by early tuber bulking. Consequently, adequate N must be available to the crop early in the season to allow for sufficient canopy development. Excessive early-season N supply can reduce overall plant N recovery and increase nitrate leaching potential. It also increases susceptibility to brown center and hollow heart in some cultivars. Excessive late-season N application can also increase nitrate leaching potential and usually reduces tuber specific gravity and skin set.

Once the linear tuber growth phase begins, tubers become the dominant sink for carbohydrates and mobile nutrients. If dry matter production is not sufficient for both vines and tubers because of environmental or cultural factors, carbohydrates and nutrients will be remobilized from the vines and transported to the tubers. This normally occurs in determinate cultivars later in the season. Vine and root growth can be maintained in indeterminate cultivars as long as nutrient supply and growing conditions are optimal. However, early-season nutrient deficiencies can reduce canopy leaf area and photosynthetic capacity to suboptimal levels, which in turn can reduce the plant's ability to maintain high tuber bulking rates for the entire growing season. Nutrient deficiencies can also result in premature senescence and increase the plant's susceptibility to diseases such as early dying and early blight.

In **indeterminate cultivars**, excessive N supply during tuber bulking can promote vegetative growth at the expense of tuber growth. This often occurs when large amounts of N are applied during the latter part of tuber bulking in an attempt to increase low petiole nitrate concentrations or delay senescence or disease development. A primary goal of potato nutrient management is to maintain adequate but not excessive nutrient supplies throughout the growing season, resulting in an optimal balance between vine and tuber growth.

In-Season Nitrogen Applications in Irrigated Systems

The most efficient potato N management strategies in many irrigated systems utilize split N management with one-half to three-fourths of the total seasonal N supply side-dressed after emergence and/or applied through the irrigation system or by aircraft in several small applications. This is particularly true for late-season indeterminate cultivars. Sprinkler application of N should not begin until the end of tuberization. Prior to that time, the plant root system is not sufficiently developed to capture most of the N applied by this method. With side-dressed N management, it usually is best to decrease the amount of N in the starter with indeterminate cultivars and delay application until emergence and hilling. Posthilling N applications can also increase yields and reduce N leaching where rainfall is high early in the season.

> **Sprinkler application of nitrogen should not begin until the end of tuberization. Prior to that time, the plant root system is not sufficiently developed to capture most of the nitrogen applied by this method.**

Once tuber bulking begins, weekly crop N requirements can be estimated from the relationships between tuber growth rate and plant N uptake. Daily N uptake rates for different potato cultivars range from about 1.8 to about 5.5 pounds per acre, depending on the tuber bulking rate. An average Russet Burbank crop requires about 3 pounds of N per day to prevent the loss of both N and dry matter from the tops and roots to the tubers during tuber bulking. Assuming 75% plant uptake efficiency for water-run N fertilizer, about 28 pounds of N per acre would satisfy the crop N requirements for a week, and 40 pounds per acre would be adequate for 10 days, without a contribution from N mineralization. Adjustments to these projected rates to account for contributions from N mineralization should be based on weekly petiole nitrate tests.

In intensively managed sprinkler-irrigated fields, N application frequencies range from almost continuous injection through center-pivot and linear-move systems on sandy soils to 10- to 14-day intervals with set-move systems on loam soils. A common approach is to apply 20–40 pounds of N per acre every 7–14 days during tuber bulking, which is usually adequate to maintain petiole nitrate concentrations in the optimal range.

Several liquid N fertilizers, such as **urea**, **ammonium nitrate**, and **ammonium polyphosphate**, can be applied through sprinklers. Nitrogen is also a component of many other liquid fertilizers containing phosphorus, potassium, and sulfur that can be injected through sprinklers. Dry N fertilizers may be successfully top-dressed by aircraft or ground application equipment during crop development if the application is followed by sufficient irrigation or rainfall to incorporate the fertilizer. Urea should be incorporated soon after application, to avoid losses due to the volatilization of ammonia. Anhydrous ammonia and aqua ammonia should not be applied by sprinkler irrigation, because of potentially high N losses from volatilization. Volatilization losses of these materials may be even greater if the pH of the water is high.

The NO_3-N concentration in the top 18 inches of soil can also be used to monitor N availability in irrigated systems, particularly in coarse-textured soils. The combination of petiole and soil NO_3-N concentrations provides a good indication of the effectiveness of the N management program and potential causes of decreasing petiole NO_3-N concentrations. For example, the maintenance of an adequate soil NO_3-N concentration when the petiole NO_3-N concentration is falling indicates that other factors are limiting NO_3-N uptake or translocation.

Nitrogen Application in Rain-Fed Systems

Split N management often works well in irrigated cropping systems with late-maturing cultivars, but there are potential drawbacks to this practice in many areas, such as the rain-fed production regions of North America. For example, preplant soil nitrate tests are often not a good indicator of future N availability in these areas. In addition, potato N application guidelines based on soil N concentrations have not been completely developed and validated for all areas. Any diagnostic tools used must reliably detect deficiencies early enough in the season to allow corrective applications before yield or quality loss occurs. As a result, locally developed or adapted fertilizer management recommendations should be followed to maximize N use efficiency.

Some additional precautions should be considered in planning N applications in unirrigated areas. In rain-fed potato production systems, timing and incorporation of split N application may be problematic, and the entire seasonal N requirement may be more effectively applied at or before planting. Delayed N applications are also usually less effective with highly variable moisture availability and will probably be less useful for early-maturing, determinate cultivars than for later-maturing cultivars.

Under conditions where water supply is limiting plant growth, the response to N fertilizer will be greatly reduced. As water deficits increase during tuber bulking, the maximum attainable yield is reduced, and the N requirement decreases accordingly. Consequently, adequate amounts of water need to be available, or N rates must be modified, to maximize N fertilizer efficiency and crop yield.

However, excessive amounts of water and N can also be detrimental to potato yield and quality and can result in an increased potential for nitrate leaching. Efficient N management is challenging in rain-fed potato cropping systems, since adequate, but not excessive, supplies of both water and N must be maintained in the root zone throughout the tuber growth period for optimal efficiency.

Nitrate Leaching

Nitrate leaching losses are minimized by maximizing the amount of time nitrate remains in the root zone. However, this goal must be balanced against the need to frequently replenish the water supply in the root zone, to avoid plant water stress. Excessive irrigation can dramatically accelerate nitrate-leaching losses, which can produce crop N deficits unless additional N fertilizer is applied. In reality, even excessive N applications cannot completely offset yield losses resulting from overirrigation. Reduced root activity and increased root and vascular diseases resulting from excessive irrigation produce yield and quality losses that cannot be compensated for by changes in other cultural practices.

Nitrogen management practices that reduce nitrate leaching and improve N use efficiency include splitting N applications to closely match N uptake patterns and using controlled-release N fertilizers. These approaches distribute the application or release of N through the root zone over longer periods during the growing season, which in turn reduces the exposure of root zone nitrate to leaching, compared to preplant applications with conventional N sources.

Winter Cover Crops

Appreciable amounts of NO_3-N can accumulate in the soil at the end of the growing season. This is attributable to reduced late-season plant N uptake efficiency, continued mineralization of soil organic matter and plant residues, and N applied in excess of that required by the crop. A cover crop such as winter wheat or rye following potato harvest can capture some of the residual root zone N and retain it for use during the following growing season. Nitrogen absorbed by the cover crop root system is stored over the winter in the plant biomass, until the cover crop is tilled into the soil in the spring and begins to release the stored N. Conserving soil N in this manner prevents it from leaching during the winter and reduces the potential for groundwater contamination. Winter cover crops also provide ground cover for reducing wind erosion and can increase soil organic matter content and tilth.

Phosphorus Management

Adequate availability of phosphorus (P) in the soil is important for early crop development and tuber initiation and to enhance tuber maturity. Phosphorus deficiencies, on the other hand, significantly reduce tuber yield and tuber size and specific gravity. Phosphorus moves very slowly in soil and therefore needs to be adequately incorporated into the soil to facilitate plant uptake. Phosphorus is not readily leached but can be lost in runoff from areas prone to soil erosion. Phosphorus availability is strongly influenced by soil pH, reaching a maximum in most soils at pH 6.5 and decreasing substantially at low and high pH.

Soil solution P concentrations are usually very low (0.01–0.3 parts per million) and consequently need to be constantly replenished from available soil P sources during the growing season. Daily potato P uptake requirements for different cultivars are only 0.3–0.8 pounds per acre during tuber bulking, but severe deficiency can develop if the available soil P concentration is inadequate. It is important to use locally calibrated procedures for soil analysis to provide a good representation of the available soil P fraction that can be related to crop yield responses.

In **alkaline soils**, the primary factors used in determining potato P recommendations are the soil test P concentration, amount of free or excess lime, and yield goal. Excess lime (calcium carbonate) in the soil increases P adsorption on $CaCO_3$ surfaces and also increases the precipitation of soil solution P in the form of Ca-P minerals. The combined effect of these processes is an overall reduction in P availability to plants. As a result, the relative reduction in soil P availability in many alkaline soils is directly proportional to the amount of free lime in the soil.

In **neutral to acidic soils**, the soil test P concentration and the yield goal are typically used to develop P recommendations. Phosphorus availability decreases with low pH, as a result of reaction with Fe and aluminum (Al) oxides and precipitation as $AlPO_4$ and $FePO_4$. If the soil pH is less than 5.2, higher than normal P fertilizer rates may be required to compensate for low P availability.

Phosphorus fertilizer may be **broadcast** in either fall or spring or may be **banded** at mark-out or during planting. Banding generally improves early-season P availability by concentrating the fertilizer in a narrow zone near the seed piece. Banding fertilizers containing ammonium may increase P availability in alkaline soils by reducing soil pH in the band. Banding has also been shown to be beneficial in low pH soils by concentrating P near the early developing root system. For full-season P availability, it is important to raise the soil test P concentration in the root zone to adequate levels by broadcasting P to provide the entire root system with ready access to available soil P. Banding applications at planting are often adequate if soil test P concentrations are relatively high. On medium- to low-testing soils, however, broadcast P applications are often needed in addition to banding to provide adequate P availability to the crop later in the growing season. Liquid fertilizers (such as ammonium polyphosphate) and dry fertilizers (such as monoammonium or diammonium phosphate) are equally effective P sources for potatoes as long as they are managed properly.

In-Season Phosphorus Applications

Petiole P concentrations can provide a good indication of potato P status during the growing season. Studies have shown that the P concentration of the plant top during tuber growth is significantly related to the P concentration of the petiole and the active leaves. Therefore, the total phosphorus concentration in the fourth petiole from the growing point should be maintained above the sufficiency level to provide adequate P to satisfy both vegetative and tuber growth requirements. Petiole sampling for P should begin at or shortly after tuber initiation and continue at weekly intervals through most of the tuber-bulking period.

The total phosphorus concentration in the fourth petiole from the growing point should be maintained above the sufficiency level to provide adequate phosphorus to satisfy both vegetative and tuber growth requirements.

Sprinkler application of soluble P fertilizer is effective for correcting in-season potato P deficiencies in irrigated systems as long as there are active roots near the soil surface and the fertilizer materials applied are compatible with the irrigation water. Potato root density increases near the soil surface once the canopy starts to shade the hill. This is important because P applied by sprinklers moves only about 2 inches into the soil. Maximum penetration occurs when the soil surface is wet before the P fertilizer is applied. Petiole P concentrations usually respond within about 10–14 days after application. Because of this delay, P fertilizer should be applied through the sprinkler system before P deficiencies develop as indicated by plant tissue analysis. Studies have shown that applying 20–40 pounds of P per acre through the irrigation system can increase total P uptake by 4–5 pounds per acre and maintain adequate petiole P concentrations throughout tuber bulking.

Liquid P fertilizer sources applied with irrigation water must remain soluble to be effective. Compatibility of the fertilizer materials with the irrigation water should be checked beforehand to determine solubility (Box 8.2). Precipitated P materials can coat or plug sprinkler nozzles, preventing them from operating properly.

Foliar sprays are also effective and are usually the best option if the fertilizer materials are not compatible with the irrigation water or if surface irrigation is being used. Repeated sprays may be necessary, since the uptake of foliar-applied P by potato leaves is usually less than 50%.

Potassium Management

Potatoes use substantial amounts of potassium (K). Seasonal K uptake rates range from 200 to 400 pounds per acre and are directly proportional to tuber yields. Potassium availability influences tuber yield, tuber size, and a number of tuber quality factors, including specific gravity, blackspot bruise susceptibility, chip and fry color, and storage quality. Potassium deficiencies decrease photosynthesis, resulting in a reduction in dry matter and starch formation. When K uptake is excessive, surplus K is translocated to the tubers, causing increased tuber water content and decreased specific gravity.

The optimum tuber K concentration for maximum dry matter production is 1.8%. At this concentration, about 0.5 pound of K_2O is required to produce 1 cwt of potatoes. Daily K uptake ranges from 3 to 6 pounds per acre during tuber bulking, depending on the cultivar and rate of tuber growth. A potato crop bulking at 7 cwt per acre per day requires about 3.0–3.5 pounds of K per acre per day to maintain optimum dry matter production, and higher bulking rates require proportionately higher K uptake rates. Therefore, it is important that potassium fertilization programs should be designed to provide sufficient K to maintain optimum plant concentrations throughout the tuber-bulking period.

Potassium fertilizer requirements vary considerably among soils, particularly in relation to texture and soil K-fixing capacity. The probability of obtaining a positive yield response to K fertilization generally increases as the sand content of the soil increases (i.e., loamy sands > sandy loams > loams > silt loams). In general, the K source has relatively little effect on

Box 8.2

Compatibility of Phosphorus Fertilizer with Irrigation Water

Insoluble calcium phosphate deposits can form when liquid phosphate fertilizer is injected through a sprinkler system with irrigation water containing an appreciable concentration of Ca (above 50 parts per million). These deposits can plug nozzles and reduce the effectiveness of sprinkler heads. As the Ca concentration in the irrigation water increases, the potential for sprinkler problems increases proportionately. The liquid P source can also influence the potential for calcium phosphate precipitation.

Compatibility can be tested by adding a proportional amount of the specific liquid P fertilizer to a gallon of fresh irrigation water.

In center-pivot systems, the amount of fertilizer solution that can be safely added to the irrigation water can be estimated from the following relationship:

$$X = \frac{0.0283 \times F}{D \times P}$$

where

X = teaspoons of fertilizer solution per gallon of irrigation water
F = fertilizer application rate (in pounds of P_2O_5 per acre)
D = depth of water application (in inches per acre)
P = P_2O_5 concentration in the fertilizer solution (in pounds per gallon)

If a fine, white precipitate forms after the calculated amount of fertilizer solution is thoroughly mixed with the irrigation water, the P fertilizer application rate should be reduced. The test should be repeated until no visible precipitate forms.

For wheel lines, hand lines, and solid-set irrigation systems, the relationship is

$$X = \frac{0.0283 \times F}{W \times T \times P}$$

where

X = teaspoons of fertilizer solution per gallon of irrigation water
F = fertilizer application rate (in pounds of P_2O_5 per acre)
W = water application rate (in inches per hour)
T = injection time (in hours)
P = P_2O_5 concentration in the fertilizer solution (in pounds per gallon)

If a precipitate forms after mixing, the P application rate should be reduced or the length of the injection period should be increased, and the test should be repeated.

Irrigation waters with high Ca and Mg concentrations or high pH can be partially acidified to improve compatibility with P fertilizers. Phosphoric acid or urea phosphoric acid can be injected through the sprinkler system in appropriate amounts. Alternatively, sulfuric acid or urea sulfuric acid can be injected along with ammonium polyphosphate (10-34-0) to minimize calcium phosphate precipitation.

Maintaining irrigation water pH between 5.0 and 7.0 can substantially reduce the potential for nozzle plugging. The final pH should be kept above 5 to prevent corrosive damage to the irrigation system and nozzles.

total yield, although K_2SO_4 tends to produce slightly higher percentages of large U.S. No. 1 tubers and higher tuber specific gravity than KCl, particularly when applied at high rates shortly before planting. The differences between K_2SO_4 and KCl with respect to tuber quality usually decrease as the interval between application and planting is increased.

Banding generally increases K efficiency in soils where significant K fixation occurs. However, high rates of banded K should be avoided, because of potential salt toxicity, particularly if the fertilizer is placed near the seed piece. Banded K fertilizer is generally not as effective as broadcast K when the bulk of the seasonal K requirement is applied at or before planting. High K fertilizer rates should also be split between fall and spring applications, to avoid yield losses. Fall applications can be effective but may be prone to leaching in sandy soils in areas with high winter precipitation.

In-Season Potassium Applications

Applying all K preplant is usually more effective than applying most or all of the seasonal K supply in irrigation water. Application of over 50% of the seasonal K requirement during tuber bulking reduces tuber yield and specific gravity, compared with preplant application. Therefore, in-season K applications should not be made unless K deficiency is likely to develop by the end of tuber bulking.

There is no consistent difference in effectiveness among the K sources KCl, K_2SO_4, and KTS (potassium thiosulfate) when applied through the irrigation system. However, in-season application through irrigation water supplying more than 50 pounds of K_2O per acre should be avoided, because of the increased probability of tuber quality reductions. In irrigated production systems, the K content of the irrigation water should be taken into consideration when calculating K fertilizer requirements.

Petiole K concentrations generally decrease with time following tuber initiation. The rate of decrease depends on soil K availability and foliage and tuber growth rates. Petiole K concentrations will respond to in-season K applications when plant K levels are deficient. However, there usually is a 2- to 3-week lag period following the application of K fertilizer with irrigation water before petiole K concentrations change appreciably. In some cases, petiole K concentrations do not increase after K application but level off instead of continuing downward. As a general guideline, K fertilizer should be applied 15–20 days before the date when petiole K concentrations are estimated to drop below the sufficiency level, and it should not be applied less than 30 days before vine kill. Future petiole K concentrations may sometimes be estimated by plotting the previous test results against time and fitting a trend line to the data points. This procedure should be done separately for each field, since the relationship between petiole K concentration and time is highly variable.

Sulfur, Calcium, and Magnesium Management

Sulfur (S) fertilizer may be needed for potatoes when the concentration of sulfate sulfur (SO_4-S) in the soil is low or where it has not accumulated from previous applications. Irrigation water low in SO_4-S is another indicator that S fertilizer may be needed. Other plant-available forms of S include S mineralized from soil organic matter and crop residues and S stored as gypsum ($CaSO_4$) in the crop root zone.

Sulfur is applied as a sulfate source or as elemental S. Sulfate sulfur is readily available for plant uptake but is susceptible to leaching. Elemental S, in contrast, needs to be oxidized to SO_4 before being taken up by plant roots. When elemental S is applied, there often is a significant time lag in the conversion to SO_4, due to the initially low activity of S-oxidizing bacteria. The time lag is longer in cold, wet soil, since the oxidation process is slowed under these conditions.

Preplant application of SO_4-S as ammonium sulfate, potassium sulfate, or sulfuric acid should satisfy the S requirement of the crop. However, potatoes also respond well to soluble S applied through the sprinkler system during tuber growth. A total S concentration in petioles below the sufficiency range indicates a potential need for supplemental S applications. As with other nutrient concentrations, petiole S concentrations need to be monitored weekly to provide a reliable estimate of plant S status and trends.

Both calcium (Ca) and magnesium (Mg) are usually present in adequate amounts in calcareous, alkaline soils and associated irrigation waters. Some deficiencies may develop in very sandy soils or in acid soils. Low exchangeable soil Ca or Mg concentrations indicate a need for supplemental Ca and Mg applications, which should be made before or at planting. High rates of K fertilization coupled with low levels of soil Mg can increase the potential for Mg deficiency. Broadcast or band applications of calcium sulfate should satisfy crop requirements for Ca, and broadcast or band applications of magnesium sulfate or potassium–magnesium sulfate should satisfy the requirements for Mg. Calcium and magnesium can be applied as dolomitic lime when an increase in soil pH is desired. Calcium nitrate can be applied if both N and Ca are needed.

Some tuber quality disorders, such as internal brown spot and brown center, are associated with Ca deficiency. To improve tuber Ca uptake, Ca fertilizer should be placed in the zone of tuber formation to facilitate uptake by stolon roots. Fertilizers such as calcium nitrate, calcium–ammonium nitrate, and calcium sulfate can be used to supply Ca without significantly increasing soil pH.

Calcium applied to foliage is not translocated to tubers but may help satisfy some of the Ca requirement of leaves when deficiencies develop. Immobility of Ca in soil and plants also limits the effectiveness of sprinkler-applied Ca. Foliar sprays of magnesium sulfate can be applied to correct Mg deficiencies when petiole concentrations are low.

Micronutrient Management

Micronutrients are largely supplied to plants from soil mineral and organic sources. Zinc, manganese, iron, and copper are positively charged ions that are exchanged with the surfaces of soil clay particles and are released as soil minerals weather and organic matter decomposes. Boron is also contained in soil minerals and organic matter and can be present in significant amounts in irrigation water. The availability of each of these micronutrients decreases substantially as the soil pH increases above 7.0.

Zinc

Zinc (Zn) fertilization may be required on calcareous soils, particularly in areas where topsoil has been removed by erosion or land leveling. For soils with low Zn concentrations, fertilizer should be broadcast and incorporated prior to planting. Compared to broadcast rates, banded Zn rates can be reduced by up to 50%, because of higher fertilizer uptake efficiency. Zinc fertilizer sources with a high percentage of water-soluble zinc, such as $ZnSO_4$, ZnDTPA (zinc diethylenetriaminepentaacetate), and zinc lignosulfate, should be used to maximize uptake efficiency. Zinc will move through phloem tissue from leaves to tubers, and foliar sprays of Zn sulfate or chelates can correct Zn deficiency.

Manganese

On neutral to acid soils, manganese (Mn) can be broadcast and incorporated prior to planting to correct or prevent Mn deficiency. On calcareous soils, Mn should be banded or applied as a foliar spray, to minimize soil fixation. Applying Mn in a band near the seed piece has been effective in providing adequate Mn to potatoes when soil concentrations are low. Manganese will also move from leaves to tubers when applied as a foliar spray. Manganese chelates and $MnSO_4$ are effective for foliar applications.

Boron

Boron (B) fertilizer may be needed where soil and irrigation water have naturally low B concentrations. Broadcast applications should be made prior to planting when soil test B concentrations are low. Band applications should be avoided, because of the increased likelihood of B toxicity. Sodium borates, Solubor, and boric acid can be applied as foliar sprays, but B is not translocated from leaves to tubers in appreciable amounts.

Iron

Band applications of chelated iron (Fe) on calcareous soils may be beneficial when soil Fe concentrations are low. Iron is relatively immobile in soils and plant tissue. Consequently, multiple applications of Fe sulfate or chelates to the foliage may be required to correct Fe deficiency.

Copper

Copper (Cu) concentrations in mineral soils are usually sufficient for adequate plant growth. Copper has intermediate mobility in plant tissue, and foliar sprays of Cu sulfate or chelates applied during tuber bulking increase petiole Cu concentrations. Copper is also a common component of certain potato fungicides, which can provide significant amounts of Cu to plants.

Chloride

There is often enough chloride (Cl^-) present in irrigation water or potassium fertilizer (KCl) to provide adequate chloride for a potato crop. Chloride deficiency may occur when irrigation water contains little chloride and when the applied K is from a source not containing chloride, such as K_2SO_4 or KTS.

CHAPTER 9

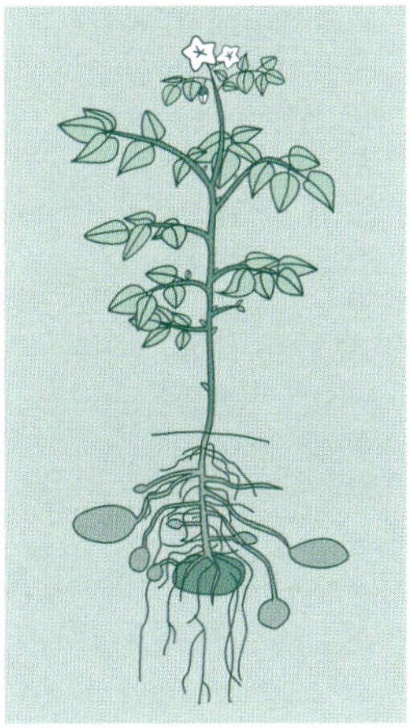

Neil C. Gudmestad
Department of Plant Pathology
North Dakota State University, Fargo

Potato Health from Sprouting to Harvest

A number of practices that influence plant health should have already been implemented by the time the potato crop is planted and seed tubers are beginning to sprout (Chapters 2 and 5–8). While the potato crop grows, there are many environmental changes that influence potato health. The crop nutritional status and management of water during the growing season have large effects on the crop and on disease development. An understanding of how management strategies affect the health of the crop is important so that they can be more effectively executed to positively influence potato health.

The disease triangle illustrates that plant disease results from the interaction of a susceptible host plant, a virulent pathogen, and an environment favorable for disease development. In reality, the classic disease triangle is a complex array of factors that interact at various levels (Fig. 9.1). Production practices modify the environment and impact disease development. For example, the environment under which the

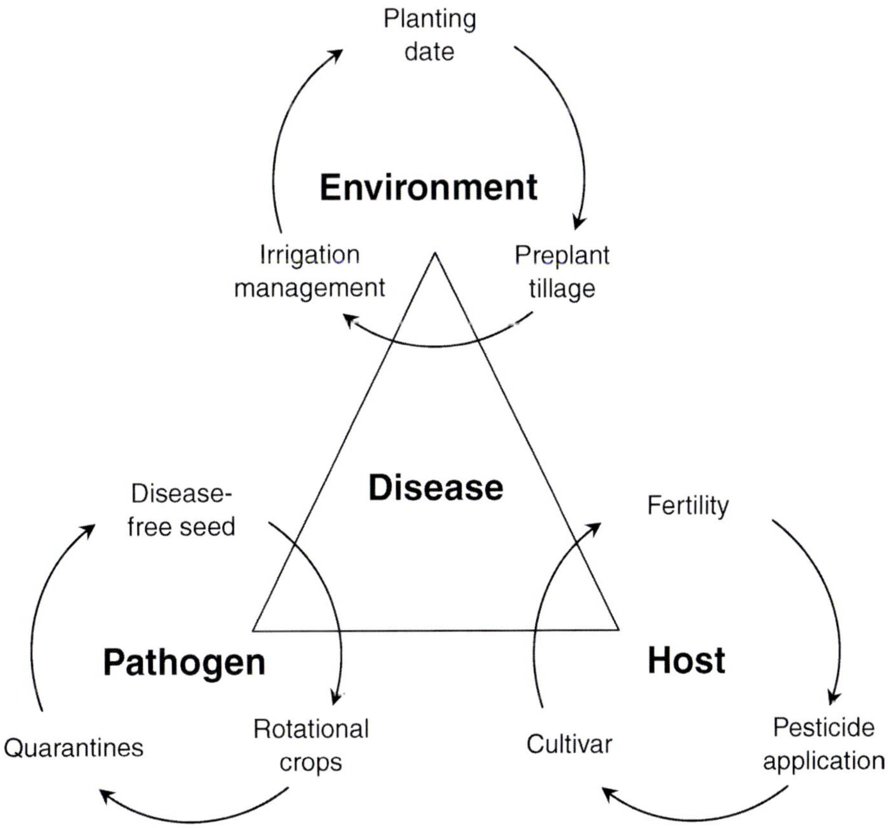

Fig. 9.1. Interaction of production practices with the host, the pathogen, and the environment in the disease triangle.

potatoes are grown will be affected by preplant tillage operations, the date of planting, and irrigation management (Fig. 9.1). Irrigation management is important in crops grown under sprinkler irrigation, which provides moisture and high relative humidity favoring several foliar diseases. Preplant tillage operations that influence soil compaction also affect water infiltration, and a number of root and tuber diseases are exacerbated in soils that are waterlogged because of compaction. Soil temperatures at planting have a dramatic effect on the rate of sprout elongation, which in turn affects the development of diseases such as Rhizoctonia canker and seed piece decay. The choice of cultivar, and therefore the susceptibility of the host, will influence which diseases will be problematic, in addition to affecting fertility levels and the application of pesticides.

Excessive nitrogen stimulates overly large plant canopies, which are difficult to penetrate with fungicides used to control late blight. The potato pathogens that will require management are also influenced by the choice of seed and the crop rotation used on the farm. Short rotations between potato crops can increase populations of a number of plant pathogens, and the choice of rotational crops influences the survival of soilborne pathogens. A buildup of soilborne pathogens that infect roots will ultimately result in reduced water and nutrient uptake efficiency and therefore detrimentally affects crop health. These examples illustrate that the overall health of the potato crop is directly influenced by management decisions and that the complex interactions of all variables leading to potato crop health requires a holistic approach for successful crop production. The goal of this chapter is to develop a fundamental understanding of the effects of management decisions such as plant nutrition, water management, and other cultural practices on potato crop health.

Fertility and Water Management

Fertility and soil moisture should be managed to maintain uniform plant growth throughout the growing season. Fertility and water management are inexorably tied, especially when potatoes are produced under irrigation, and the nutritional and water status of the crop will affect the diseases that develop and determine the degree to which they can be managed.

Proper nutrient management should be directed at maintaining nutrient concentrations throughout the growth cycle (Table 8.1). The bulk of the nutritional requirements (major and minor elements) of potatoes grown without irrigation is applied as either a preplant broadcast application, followed by incorporation, or a band application during mark-out (row marking) or at planting. Potatoes grown under irrigation receive some of the required nitrogen (N) and phosphorus (P) as a sidedress after planting, during cultivation, or as liquid fertilizer injected through the irrigation system. Adjustments in minor nutrients to correct deficiencies can also be made with chelated foliar sprays. In irrigated potato crops, a major proportion of the N applied in the fertility program is generally applied through irrigation. An important balance exists between too little and too much nitrogen, both of which can have detrimental effects on potato health. Failure to make adequate N available during the early stages of crop development (Growth Stages I–III) can increase the susceptibility of potato to diseases such as black dot, early blight, and early dying. Excessive postemergence applications of N can force the development of an overly lush canopy, which is more susceptible to late blight, white mold, and bacterial stem rot. Once again, balanced fertility and irrigation programs are essential for effective potato health management.

Pathogens that infect roots, such as *Pythium ultimum, Rhizoctonia solani, Colletotrichum coccodes,* and nematodes, and those that infect the vascular tissue, such as *Verticillium dahliae* and *Fusarium oxysporum,* can significantly reduce nutrient uptake and transport within the potato plant. Therefore, it is important to implement the necessary disease management tactics prior to planting. If foliar diseases, such as early blight and late blight, are not effectively managed, the plant canopy will be reduced, and thus photosynthetic efficiency will be decreased, ultimately reducing the amount of energy available for water and nutrient uptake by the plant. Water management, nutrient management, and disease management tactics must be integrated to produce a healthy and profitable crop.

Water is the most important input in a potato crop. Water makes up 90–95% of the green tissues and 75–85% of the tuber tissue. Under optimal growing conditions, the rate of transpiration of a well-watered potato plant will require replacement of the entire water content of the plant approximately four times each day. This means that if water is withheld, a potato plant would wilt in about 12 minutes. Obviously, water is extremely important to the potato plant, but it also favors many potato diseases.

Although excess water can be a problem in poorly drained soils or in rain-fed potato production areas, the most common water management problem is a plant water deficit resulting from too little water. A water deficit occurs when soil water is depleted below a critical level. Soil water storage is frequently expressed as a percentage of **available soil moisture** (also referred to as available soil water or available moisture-holding capacity). In potato, soil moisture reaches a critical stage when the available soil moisture drops below 60–65% (Fig. 9.2). In other words, when 35–40% of the available soil moisture has been used by the crop, a water deficit develops in potato plants in most soils.

Water management is greatly influenced by soil type. Soils vary in ability to accept and retain soil moisture (Table 9.1 and Fig. 9.3), and these differences must be taken into account. For example, fine to coarse sand-based soils will hold less than 1 inch of water per foot of soil, while loam-based

Table 9.1. Available soil water and approximate infiltration rates in soils of various textures[a]

Soil texture	Available soil water (in./ft)	Infiltration rate (in./hr)
Coarse sand	0.4–0.7	0.5–1.0
Fine sand	0.7–0.9	0.5–1.0
Loamy sand	0.9–1.3	0.5–1.0
Sandy loam	1.2–1.9	0.5–1.0
Loam	1.8–2.6	0.3–0.5
Silt loam	2.0–3.0	0.3–0.5
Clay loam	2.0–2.6	0.1–0.3
Clay	1.8–2.4	0.1–0.3

[a] Source: D. Curwen, 1993, Water management, pages 67–75 in: Potato Health Management, 1st ed., R. C. Rowe, ed., American Phytopathological Society, St. Paul, MN.

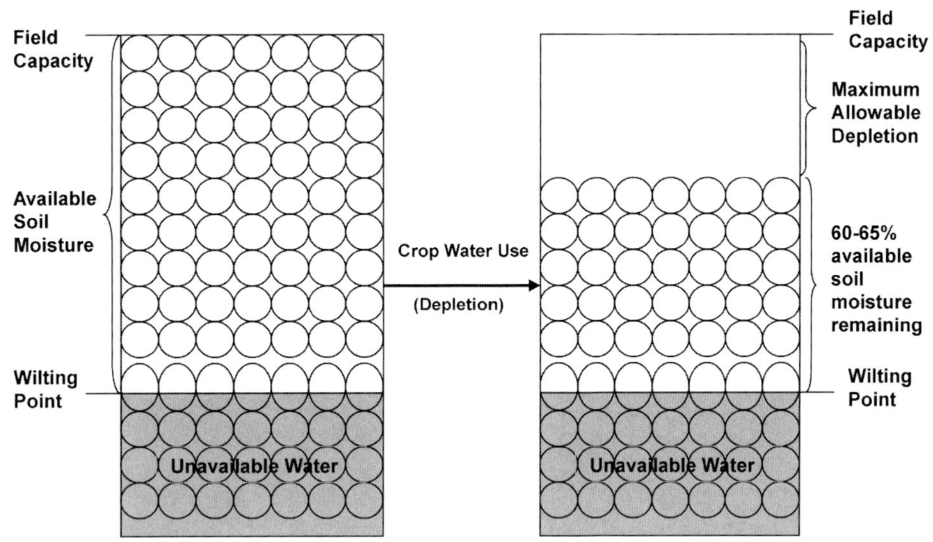

Fig. 9.2. Available soil moisture is the difference between the field capacity of the soil (the amount of water retained in the total pore space of a saturated soil after the soil has drained) and the permanent wilting point (the point at which plants can no longer obtain water from the soil). Allowable depletion is the point at which available soil moisture cannot be depleted further without inducing plant stress. For potatoes, the soil must always be maintained above 60–65% available soil moisture. (Adapted from Curwen, 1993)

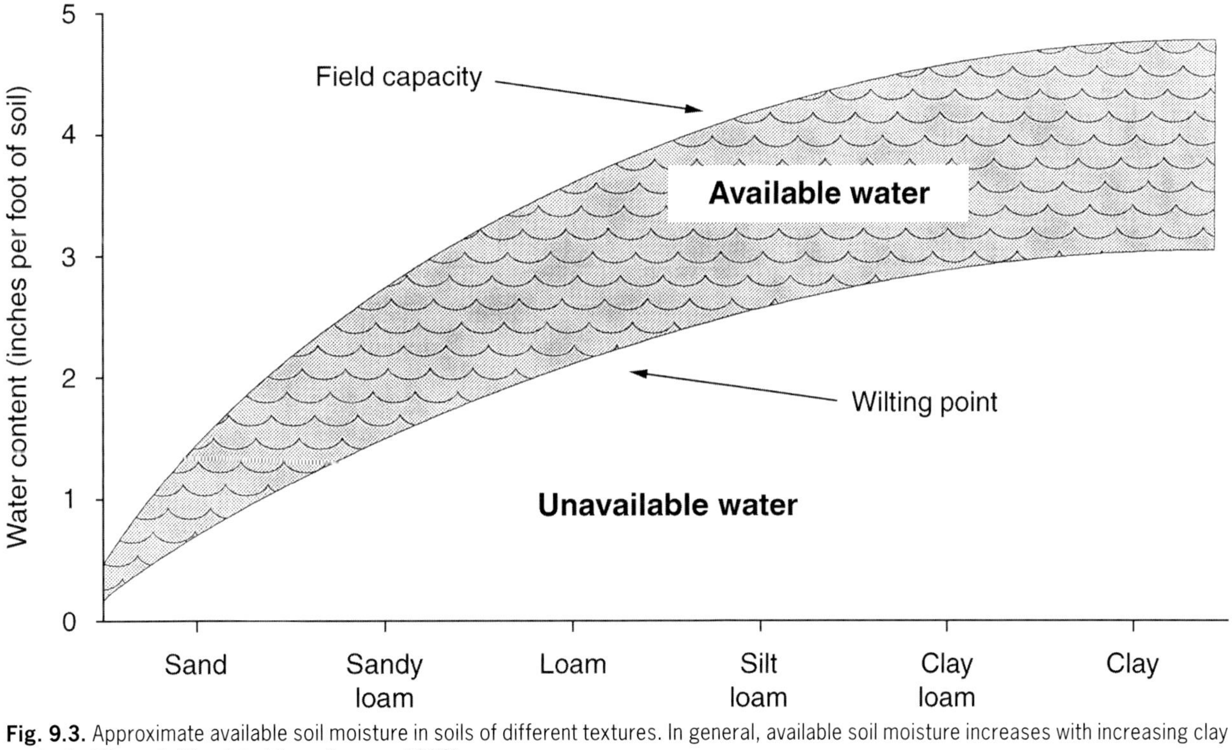

Fig. 9.3. Approximate available soil moisture in soils of different textures. In general, available soil moisture increases with increasing clay content of the soil. (Reprinted from Curwen, 1993)

soils will hold more than 2 inches of water per foot of soil. The frequency of irrigation will be much greater in fields with sand-based soil than in fields with loam-based soil. However, the threat of nitrate leaching is much greater in a sand-based soil than in a loam-based soil. The amount of water that can be applied and the rate of application will also differ greatly between these two broad categories of soil textures (Table 9.1). While it is relatively easy to apply ¾ inch of water on a sand-based soil in a single application with little or no runoff, the same amount of water could not be applied on a loam-based soil in a single application without detrimental effects. Therefore, full awareness of the soil types present on a farm is important for the proper application irrigation water. Fields with large variation in soil types should be avoided if possible, since it will be extremely difficult, if not impossible, to irrigate them properly (Chapter 5).

Soil moisture should be managed to avoid excessively wet or dry conditions, as these conditions can lead to diseases and disorders that affect yield and tuber quality (Box 9.1). For example, allowing soil moisture to drop below 65% available soil moisture during tuber initiation increases the risk of common scab and promotes several physiological disorders of tubers. Wetness of foliage favors the development of several foliar diseases, such as late blight and white mold, and soil moisture levels close to saturation late in the season contribute to tuber diseases, such as leak, pink rot, and bacterial soft rot (Chapters 18 and 19). In addition, excessive soil moisture can cause nitrogen to be leached out of the root zone, thereby affecting overall crop health. Therefore, in irrigated production areas, a precise water management plan should be implemented, to meet but not exceed the moisture demands of the potato crop. Although moisture and nutrient stress cannot be avoided entirely, the more closely this goal is achieved, the better the yield and quality of the potato crop.

Water in excess of the holding capacity of the soil will leach N from the root zone and thus leads to N deficiency, compromising crop health and limiting productivity.

Fertility and water management are interrelated. The productive potential of any nutrient management program can be fully and efficiently realized only if water is not limiting. Nutrient uptake occurs through the water and transpiration stream. The uptake of nitrogen, in particular, is directly related to the availability of water. Potato plants stressed by a water deficit will demonstrate little response to N fertility. Water in excess of the holding capacity of the soil will leach N from the root zone and thus leads to N deficiency, compromising crop health and limiting productivity.

Box 9.1

Potato Diseases and Disorders Associated with Deficient or Excess Soil Moisture at Various Growth Stages

	Deficient soil moisture	**Excess soil moisture**
Growth Stage I: planting to emergence	Prevents proper wound healing of seed pieces, thus favoring seed piece decay Inhibits sprout and root development	Increases clod formation during planting Promotes seed piece decay
Growth Stage II: vegetative growth	Inhibits plant development and response to fertilizer May lead to foliar tipburn and reduction in total foliage and inhibit full root development	Increases soil compaction and clod formation during cultivation May favor increased infection by *Verticillium dahliae*
Growth Stage III: tuber initiation	Favors infection by the common scab pathogen Increases potential for tuber disorders such as translucent end and off-type shape	Favors infection by the powdery scab fungus Favors development of some tuber disorders, such as brown center and stem-end hollow heart
Growth Stage IV: tuber bulking	Favors development of common scab May initiate premature senescence, leading to increased early blight and early dying Intermittent dry periods favor some tuber disorders, such as internal brown spot, growth cracks, and off-type shape	Promotes nitrate leaching, which can increase early blight Promotes lush canopy, which favors late blight, Sclerotinia stalk rot, and bacterial stem rot Increases potential for development of rots such as leak and pink rot May favor development of bud-end hollow heart
Growth Stage V: tuber maturation	Dehydrates tubers, making them more susceptible to blackspot bruise at harvest May cause stem-end vascular discoloration	Leads to enlargement of lenticels, allowing entry of soft rot bacteria and the pink rot pathogen Favors infection by leak and pink rot fungi May delay tuber skin set
Harvest	Increases tuber susceptibility to blackspot bruise Increases formation of clods, leading to increased bruise at harvest	Increases tuber susceptibility to shatter bruise and thumbnail cracking Makes separation of soil and tubers at harvest very difficult Favors development of tuber rot in storage

Adapted from Curwen, 1993.

Field Scouting

The potato is one of the more complex crops to grow, requiring intensive management to ensure success. Much of this is due to ever-changing growing conditions and changes in the potato from growth stage to growth stage (Chapter 2). The pests and diseases that afflict the potato crop also vary with environmental conditions and the growth stage. The best way to effectively manage crop health is to be in the field regularly to routinely determine the status of the crop and check for the presence of pests and diseases. For success, potato fields must be scouted continually to make effective decisions to maintain crop health.

Scouting activities to provide information on crop and pest status can be done by the grower or by a knowledgeable professional scouting service. Field scouting must be done routinely to accurately monitor weed status, plant growth, insect populations, and disease development. In addition, frequent petiole sampling, to determine nutrient status, and soil probing, to determine the amount of available soil moisture in various parts of the field, should also be conducted. Timely application of a pesticide or nutrient must be based on the information gathered during field scouting.

Throughout each growth stage, representative areas of each field should be scouted thoroughly. Individual plants and groups of plants should be dug and examined to evaluate the health of all plant parts, from foliage to roots and tubers. At the same time, a knowledgeable scout can determine the soil moisture status both within and below the tuber zone, information that will be very helpful in scheduling future irrigations. In addition to digging plants, effective scouting involves examination of leaves and stems for insects and disease symptoms and sweeping the foliage to determine insect populations that cannot easily be determined by direct examination of individual leaves.

Management programs for key insects and diseases have been developed in most potato production areas. Once the **action threshold** for an insect pest or disease has been exceeded, specific management options should be implemented, which usually involves the application of a pesticide according to the label. Other management options may include adjustments in irrigation scheduling or fertility. While it may be true that more plant pests are effectively managed by nonchemical means than through the use of agrochemicals, pesticide treatment is frequently the only effective tactic as an immediate response to an outbreak of a plant disease or insect pest. Care must be taken to avoid nontarget effects when pesticides are used. Routine pesticide applications should be avoided whenever possible and should be made only in response to action thresholds determined by scouting or a disease forecast model.

Details of managing specific insect pests and diseases are described in other chapters of this publication. The following discussion will focus on the integration of management practices during the five growth stages of potato (Fig. 2.1). Although nutrient, insect, and disease management strategies vary among regions and according to local differences in cultural practices and the requirements of irrigated versus unirrigated production, the goal of the discussion that follows is to develop a logical progression of health management. The goal is to provide a template to be used in devising a holistic management program that can be fine-tuned for specific production systems, production areas, and market classes of potato.

Growth Stage I: Sprout Development

Sprouts and roots develop from seed pieces at a rate that is greatly influenced by the physiological status of the seed, soil moisture content, and soil temperature during Growth Stage I. Sprouts will elongate much more rapidly in soil that is closer to field capacity than in drier soil, a phenomenon that is enhanced in warm soil. Sprout development and elongation occur much more rapidly at 64–68°F than at 45–50°F. The physiological status of the seed tuber or seed piece, whether it has been taken directly out of cold storage without bud development or whether considerable sprout development has occurred, will also influence the rate at which further sprout elongation will occur. Seed tubers that are warmed prior to cutting and planting to stimulate bud development will sprout and emerge much more rapidly and uniformly than seed that has not been warmed. Prior to emergence, seed tubers and seed pieces are the sole source of energy for the developing plant. Therefore, managing the seed to encourage **rapid emergence** is essential to establishing a healthy crop from the onset (Chapter 7). All these factors will influence the length of time it takes for sprouts to emerge from the soil, develop leaves, and become plants independent of the seed piece. During this period, developing sprouts are particularly susceptible to Pythium damping-off and Rhizoctonia stem canker. The pathogens that cause these diseases are very aggressive on juvenile plant tissues, given the proper environmental conditions, but are much less able to attack mature potato plants, in which the physiological process of lignification has formed materials that strengthen cell walls. Soft rot bacteria can also be problematic, particularly if soils become saturated from excessive rainfall or irrigation. Cold seed planted in warm soil or in a light, sandy soil that warms quickly is highly susceptible to soft rot seed piece decay. When a cold seed piece is surrounded by warm, moist soil, water will condense on the seed piece, creating an anaerobic environment that favors soft rot bacteria.

Water management is critical prior to emergence. Most potato crop management recommendations generally advise against postplant, preemergence irrigation, to avoid soft rot seed piece decay and Pythium damping-off. However, in many production areas with light, sandy soils, preemergence application of water is necessary to minimize wind erosion and replenish soil moisture to ensure uniform sprout development and emergence, and failure to provide such irrigation can dramatically impede root development and sprout elongation. The seed piece should be planted in moist soil with at least 75% available soil moisture. Once sprouts are in excess of 1 inch, the soil moisture profile above the seed piece should also be in this moisture regime to guarantee uniform root and sprout development. In unirrigated fields, the lack of control over soil moisture prior to emergence means that careful consideration and attention must be given to soil preparation, preplant seed handling, and soil temperature and moisture status at planting to encourage rapid sprout development.

Root development progresses more rapidly than sprout development during this growth period. For example, if the seed pieces are approximately 6 inches below the soil surface at the time of emergence, rooting depth can be 1.5 to 2 times that in noncompacted soils. Lateral growth of roots prior to emergence can also be extensive, frequently reaching the extreme limits of the hill or bed that contains the seed piece. **Roots pruned by cultivation are rarely re-formed.** Because reduced root mass will detrimentally affect water and nutrient uptake, cultivation should be performed prior to emergence or shortly thereafter, to avoid root pruning and disturbing the development of the potato plant. A single cultivation after planting is sufficient to control most cool-season weed species. Cultivation at this point of the growing season will be performed "blind": since plants have not yet emerged, traditional navigators used on tractors are of little value, which explains why global positioning system (GPS) technology is now being utilized in planting and cultivating potato fields prior to emergence. Early preemergence cultivations are a good crop management tactic to guarantee that the developing crop and its root system will not be disturbed or damaged. The advantage of a GPS-based navigation system is that it can be used from planting to harvest, indirectly improving potato health by limiting mechanical injury to the crop at all growth stages.

Hill shape is an important component in the overall health management scheme of the crop. Broad, flat hills are generally preferred to narrow, peaked hills, especially when potatoes are produced under irrigation. Broad, flat hills tend to intercept irrigation water more readily than peaked hills. Peaked hills will shed irrigation water and allow soil in the tuber zone to remain dry, and dry soil can induce physiological tuber disorders (Chapter 23). This result is most likely to occur in light, sand-based soil that is low in organic matter, but it can happen in soil of any type. Cultivation using the Dammer Diker tillage system, which forms depressions between the rows, is useful on sandy soils; the depressions act as collection pools for water and nitrogen applied through the irrigation system, thereby preventing loss of water and nutrients due to runoff.

During cultivation, a **sidedress application of nitrogen** is often made. Split applications of N during the season have been shown to be an effective nutrient management strategy. This is particularly important in soil types where nitrate leaching is likely to occur. If sidedress applications are to be made, it is important that they be made at this time, to avoid the risk of root pruning, rather than delaying the application until later in crop development. It is not uncommon to apply 50–60% of the total N requirement of the crop between the preplant and sidedress applications, with 25–50% of that being applied as a sidedress or postplant application. The actual amounts of N applied during this growth stage (and during Growth Stage II) should be based on the yield goal for the crop and local potato production recommendations. It is most common to apply N at this stage in a form such as 28-0-0 (liquid) or urea (dry), but liquid blends of 28-0-0 and 10-34-0 are used to increase the availability of phosphorus to the developing crop in some potato production areas in the United States. Phosphorus availability is highly influenced by soil pH (Chapter 8), and in areas with either very high pH (above 7.5) or low pH (below 5.2) a blend of these two liquid fertilizers can be beneficial to crop health.

Most **herbicide applications** should be made after postplant cultivation and prior to emergence of the potato plants, to reduce the risk of injury to the developing potato crop. Herbicide injury can frequently exacerbate Rhizoctonia stem and stolon canker by slowing the development of the potato plant, thereby keeping it in a more susceptible, juvenile state for a longer time. Herbicides should be selected on the basis of their range of activity and the specific weed complex in the production area (Chapter 22). Many growers apply postplant N to the crop with herbicide, rather than applying N as a sidedress during cultivation, for two reasons. First, if the N applied is in the liquid form of ammonium sulfate or ammonium nitrate, the burn-down activity of the herbicide is accentuated and kills any emerged weeds not killed by cultivation. Second, applying N with the preemergence herbicide further reduces the risk of root pruning, which occurs in any field cultivation and sidedress application.

Growth Stage II: Vegetative Growth

Leaves and secondary stems begin to develop above ground, and photosynthesis begins, while roots continue to develop and stolons are initiated below ground during Growth Stage II (Fig. 2.1). Roots begin to absorb and translocate nutrients from the soil, since the leaves are beginning to transpire, providing the water flow needed for nutrients to move through the plant system. If **postemergence cultivation** is performed, it should be completed very early in this growth stage to reduce the risk of pruning roots. Sidedress or lay-by applications of fertilizer can also be made at this time, although the risk of damaging roots during this growth stage is much greater than during Growth Stage I. During Growth Stage II, the developing potato plant will become independent of the seed piece, since the energy required for further plant growth is supplied through photosynthesis.

Nonemergence of seed pieces can be due to soft rot seed piece decay, preemergence blackleg, Fusarium dry rot, Rhizoctonia stem canker, poor planter performance resulting in blind or misplaced seed pieces, or exposure of seed to the sprout inhibitor CIPC. Bud and sprout development is also variable in seed that is very young physiologically and chronologically.

Any problems with stand establishment as the potato plants are emerging should be investigated to determine the causes. This can only be done by walking the entire field and digging blank areas of the field. Nonemergence of seed pieces can be due to soft rot seed piece decay, preemergence blackleg, Fusarium dry rot, Rhizoctonia stem canker, poor planter performance resulting in blind or misplaced seed pieces, or exposure of seed to the sprout inhibitor chlorpropham (CIPC). Bud and sprout development is also variable in seed that is very young physiologically and chronologically. It is important to take stand counts to determine the actual number of plants that have emerged versus the amount of seed planted to determine whether to replant. Although at this point little can be done to ameliorate the situation, it is helpful for future potato production to know if critical errors were made or if management options were inadequate to deal with potential problems. If the number of emerged plants is less than 70% of the plant population expected, replanting the crop should be considered.

At this point in the development of the crop, the management of water and nutrients, particularly nitrogen, is critical. Fertilizer applied as a preplant broadcast, as a starter at planting, or as a sidedress during cultivation will ensure rapid and continuous growth of the developing crop. **Adequate nutrition** at this point will also aid in suppressing early blight, which is favored by nutrient deficiency (Chapter 21). Toward the end of this growth stage, petiole sampling can be initiated to determine the nutritional status of the crop. This is an important step, because it provides the opportunity to make early adjustments in the fertility program (Chapter 8). In some areas, it is common to take the first petiole sample approximately 30–35 days after emergence and analyze it to determine the concentration of all major and minor nutrients. However, in other areas, petiole sampling does not begin until tuber initiation (Growth Stage III). If a deficiency of a minor nutrient is detected, it must be dealt with early in the crop cycle if supplemental application of the nutrient is to have any benefit. Most deficiencies of minor nutrients can be corrected with chelated sprays. Deficiencies of major nutrients can also be corrected at this stage; however, they rarely appear at this stage, and applications of P and K at this time of the crop cycle are generally less effective than preplant broadcast applications or applications at planting. However, the application of P through the irrigation system can be an effective means of maintaining in-season sufficiency. If postemergence P applications are compatible with local recommendations, it is advisable to make these applications as early as possible during the growing season, while feeder roots are still present near the soil surface (Chapter 8), since P will rarely move more than 2 inches into the soil. Postemergence applications of N are usually delayed in indeterminate cultivars until after tuber initiation (Growth Stage III). Local recommendations should be followed in conducting and interpreting petiole tests.

Water management at this point in the season is critical for crop health but is frequently overlooked. Since the foliage is just beginning to develop, it is easy to become complacent and assume that the crop is not using much water. However, moisture loss from potato fields at this point is very real, and close attention must be paid to the demand of the crop and available soil moisture. In the absence of 100% ground cover, moisture loss is primarily due to evaporation from the soil rather than transpiration from the plants. Of course, potato plants in this stage of growth do use water, and it is easy for fields to get too dry. The key point is that adequate soil moisture is critical for continuous root development. If the roots fail to develop fully, the ability of the potato plant to continue growth is severely compromised by insufficient uptake of water and nutrients.

At this point in the development of the crop, moisture deficiency is more likely than excess moisture. Moisture in the top 12 inches of soil should not fall below 75% available soil moisture (Box 9.2). If the soil drops below 60–65% available soil moisture during this growth stage, root development and leaf expansion will be slowed and, in severe cases, cease. Lack of leaf expansion will limit photosynthesis, and impairment of root development will inhibit the plant's ability to take up water and nutrients. Photosynthesis is highly dependent on water and carbon dioxide. Potato plants growing under a soil moisture deficit will not only lack the water required for photosynthesis but also, in response to moisture stress, close their stomates to prevent further water loss, thus reducing their uptake of carbon dioxide. The combined effect of reduced water uptake and reduced carbon dioxide uptake is a decrease in the rate of photosynthesis, limiting further canopy development and ultimately reducing the amount of sugar that can be translocated and stored in the tubers as dry matter.

Field scouting to monitor the status of insect pests and beneficial insects should begin in earnest during this growth stage. Insect scouting involves walking the field, examining whole plants, particularly the undersides of leaves, and using a sweep net to determine pest status. Insect pests that can be troublesome during this growth stage vary with the potato production area and environmental conditions since planting. Unseasonably warm weather generally results in earlier emergence of insects. Key insect pests to scout for include the Colorado potato beetle, the potato leafhopper, psyllids, and flea beetles. Decisions regarding management of insects should be based on accurate estimates of their populations, derived from thorough and systematic field scouting throughout the season, and on action thresholds (Chapters 15 and 16).

Diseases can also be a production problem during this growth stage. However, unlike scouting for insect pests, disease monitoring is directed at determining the effectiveness of disease management tactics that have been already enacted in addition to determining the need for additional pesticide applications. There are disease forecasting systems for late blight and early blight that effectively aid in timing fungicide applications (Chapter 21). These programs are available for use as an on-farm version or through crop consulting services and universities. The initiation of foliar fungicide applications and their subsequent frequency can be scheduled by the use of these disease forecasting models. Depending on heat unit accumulations and region of production, effective stubby-root nematode management may require the application of a nematicide during Growth Stage II or Growth Stage III (Chapter 20).

When a field is scouted for late blight, efforts should be focused on areas most prone to disease development. These areas are usually located where the wind is obstructed or where moisture is likely to remain on foliage for a prolonged period. Areas near windbreaks, woodlots, low depressions, or the pivot of a center-pivot irrigation system are likely to remain wet longer and therefore are ideal locations in which to concentrate late blight scouting efforts.

The first symptoms of early blight can also be found during this growth stage. The early blight fungus is very aggressive on senescing foliage, and its presence during this time of the cropping season may be an indication of a nutrient deficiency, most likely nitrogen, resulting in higher susceptibility. The severity of early blight will increase unless adjustments are made to the nitrogen program.

Box 9.2

General Guidelines for Water Management in Irrigated Production of Potatoes

Growth Stage I: Sprout Development

Avoid planting in excessively wet or dry soil; the soil moisture profile should have a moderate amount of moisture, in the range of 75–85% available soil moisture. This range of moisture provides good planting conditions and adequate moisture for sprout development while avoiding compaction. If soils are excessively dry, fields should be irrigated prior to planting. Postplant irrigation is not advisable unless emergence is delayed by low soil temperatures and the soil becomes dry, or unless irrigation is needed to control soil erosion during high winds.

Growth Stage II: Vegetative Growth

As potato plants emerge and continue to grow, soil should be maintained at 75–85% available soil moisture. In areas of high rainfall and in fields with sandy soils, the percentage should be slightly lower to reduce the potential for nitrate leaching as a result of heavy rainfall.

Growth Stage III: Tuber Initiation

Soil should be maintained at 80–90% available soil moisture during tuber initiation. This amount of moisture is necessary for optimal tuber set and to reduce the potential for infection of newly formed tubers by the common scab pathogen. If the weather turns cold and soil temperatures are below 60°F, available soil moisture should be in the 70–80% range.

Growth Stage IV: Tuber Bulking

The highest demand for water occurs during this growth stage, and available soil moisture should be in the 80–90% range for most of this critical time in the crop's development. Water stress during bulking can have detrimental effects on yield and quality and, in the case of excess moisture, can also affect disease development, especially late blight, bacterial soft rot (aerial blackleg), and white mold. During the latter portion of this growth stage, water amounts can be reduced slightly, to the range of 75–85% available soil moisture, because wet soil can slow tuber bulking and predispose tubers to storage rot diseases, such as soft rot, pink rot, and leak.

Growth Stage V: Maturation

As natural senescence begins, water demand by the potato plant is reduced. Available soil moisture can decline to 70–75% before water is replaced through irrigation. This level of moisture will promote skin set. High soil moisture, in excess of 90% available soil moisture, can lead to storage rot due to soft rot, pink rot, and leak as a result of the enlargement of lenticels. Excessively dry conditions, with less than 65% available soil moisture, should also be avoided, because they can lead to dehydration, which increases susceptibility to blackspot bruise.

Growth Stage III: Tuber Initiation

Enlargement of stolon tips signifies tuber initiation during Growth Stage III. Most determinate cultivars flower. This is a relatively short growth stage, normally lasting 10–14 days. Moisture management is critical to future crop quality. A number of tuber disorders, such as translucent end, sugar end, and dark ends, can be initiated, depending on soil moisture (Box 9.1).

Common scab is caused by a filamentous bacterium that preferentially affects potato tubers growing in dry soil, usually with less than 70% available soil moisture. At this stage of growth, the potato plant begins to use a significant amount of water. A frequent management tactic is to maintain available soil moisture in the 80–85% range from tuber initiation until tubers are 1–1½ inches in diameter. Under these high soil moisture conditions, a microthin film of water forms on the surface of each developing tuber and limits infection by the common scab pathogen. However, high soil moisture content will favor infection by the pathogen causing powdery scab, a disease that can very easily be confused with common scab. It is important to know which disease is present in fields to manage the disease most effectively. Accurate identification can be done only by someone familiar with both diseases (Chapter 19).

Tuber disorders such as **translucent end** and **secondary growth** can be induced during this growth stage (Box 9.1) when soil moisture becomes too low, usually less than 60–65% available soil moisture. These disorders are not caused by low soil moisture alone but by an interaction between temperature and soil moisture status. If the rows are not closed during this growth stage (i.e., not at 100% ground cover), solar radiation will be intercepted by the soil rather than by the foliage, and dry soil will become warmer than moist soil. For this reason, it is important to maintain more than 80% available soil moisture in the tuber zone to minimize the development of translucent end and secondary growth.

Extremely high soil moisture, in soils approaching saturation, can also have deleterious effects on crop development at this stage. Tuber disorders such as **brown center** and **hollow heart** are favored by extremely wet soils, especially if the weather turns cold during this time (Box 9.1). Low soil temperatures are a primary factor in the development of brown center in the cultivar Russet Burbank. Soil temperatures below 55°F for 5–7 days during tuber initiation and early bulking can initiate brown center and hollow heart. Soil moisture can also affect the development of brown center; the higher the available soil moisture during tuber initiation, the higher the incidence of brown center (Chapter 23).

A few general guidelines can be used to reduce the risk of brown center and hollow heart in irrigated potatoes. If soil temperatures are below 60°F during tuber initiation, soil moisture levels above 70% available soil moisture should be avoided (Box 9.2). If weather conditions allow it, the soil moisture should be allowed to drop to 65% available soil moisture before the field is irrigated, to allow the soil to warm during this critical stage of tuber formation. It is critical for water in potato fields to be intensively managed. If weather conditions cause soil temperatures to be lower than normal, the soil should be kept on the dry side, but not below 65% available soil moisture, to mitigate brown center. Attention to the weather forecast is important to ensure that soil moisture is managed in case the air and soil temperatures rise, thus setting up a scenario for the induction of translucent end.

Postemergence **applications of N** usually begin once an adequate number of tubers have formed. These initial postemergence applications are usually quite substantial and are made depending on the yield goal, cultivar, petiole N status, and likelihood that nitrate leaching may have occurred as a result of excessive rainfall or irrigation after planting. Most potato cultivars take up N efficiently within 70–75 days after emergence; therefore, split applications of postemergence N should generally be made before or shortly after that time. It is important that the amount of N applied not be excessive and stimulate unnecessary foliage growth. Excessive foliage is extremely susceptible to lodging. It also provides a dense canopy impenetrable by wind, thus increasing the duration of leaf wetness, which favors the development of certain diseases, including late blight, early blight, Botrytis stem rot, bacterial stem rot, and white mold.

Foliar disease pressure is generally higher in potato crops grown under irrigation, because of the effects of frequent overhead water applications and a dense plant canopy. Overhead irrigation not only effectively disseminates some pathogens but also lengthens the daily wet period necessary for infection. Once the potato crop achieves 100% ground cover, the microenvironment within the potato canopy becomes very conducive to disease development. Systematic field scouting is necessary to continuously monitor the status of foliar diseases. On the basis of the results of scouting and disease forecasts, foliar fungicide applications to control early blight and late blight may be intensified or relaxed. There is no substitute for being in the field and having a detailed knowledge of disease incidence and severity. To achieve maximum control of some diseases, such as white mold, a foliar fungicide application at flowering may be require; this treatment should be based on field history and crop growth stage rather than field scouting (Chapter 21).

During **field scouting**, it is important to continue to dig plants throughout the field and examine their belowground and aboveground parts to assess their overall health. Any Rhizoctonia stem and stolon canker that will impact yield and quality will be evident by this stage of crop development. The *Rhizoctonia* fungus is preferentially aggressive on young, juvenile plant tissues, so any infection evident at this stage is likely to be the most economically damaging. Severe infection of the main stem can cause the whole plant to die. Less evident are stolon infections, which can cause a decrease in the number of tubers formed or (in less severe cases, when the stolon is not completely girdled) cause tuber malformations. At this stage in the development of the crop, little can be done to ameliorate this problem. When excessive stolons are girdled, resulting in a significant loss of tuber numbers, total yield can be affected. In such instances, adjustments to the fertility program, such as reducing the amount of postemergence nitrogen to reflect the reduction in yield potential, is all that can be done.

Potato crops should also be intensively scouted during this growth stage for the presence of **weeds**, especially prior to achieving 100% ground cover, in the event that a postemergence herbicide needs to be used to help rescue a field from a serious weed problem. Since many of the soil-applied herbicides are highly water soluble, they are susceptible to leaching after a heavy rainfall that dilutes the zone of weed control. The lack of weed control in potato fields has a number of negative impacts in the overall health of the developing potato crop. Only a few postemergence herbicides are registered for use on potato, and most have little or no burn-down capability (Chapter 22). A few herbicides, such as pendimethalin and metolachlor, must be applied early in the crop cycle (Growth Stage II or III) to minimize injury. These herbicides can control various common troublesome weeds in potato if they are applied early enough in the crop cycle. Careful consideration must be given to postemergence application of metribuzin, since it has been associated with an increase in the severity of black dot. *Selected trade names of herbicides used in potato production are given in Table 14.1.*

Growth Stage IV: Tuber Bulking

Tuber cells continue to expand and accumulate water, carbohydrates, and other nutrients during Growth Stage IV. Tubers are the dominant sink for carbohydrate deposition. Tuber yield and quality will be enhanced by uniform growing conditions and the elimination of biotic and abiotic stresses. It is important to maintain healthy foliage to maximize yield. **Foliar diseases** may need particular attention at this time to ensure that they do not inhibit the crop's photosynthetic capability. A number of **insect pests** also can be troubling at this point, including those that directly impede photosynthesis through the destruction of foliage (such as the Colorado potato beetle and leaf-feeding caterpillars) or the movement and deposition of carbohydrates (such as aphids, psyllids, and the potato leafhopper) and those that vector potato pathogens (various aphids and leafhoppers). Current-season spread of *Potato leafroll virus* by the green peach aphid can have devastating results in a cultivar such as Russet Burbank. Various noncolonizing leafhoppers that transmit phytoplasmas and related pathogens are important in some processing cultivars. The nutrient status of the crop going into this portion of the growing season can affect the development of some diseases, and water management can have profound effects on crop health. This growth stage is a critical time, one that demands a great deal of attention to ensure the proper management of crop health.

Adequate nutrition is vital for tubers to bulk to the cultivar's genetic potential. Petiole analyses begun in Growth Stage III should have identified any deficiencies in minor nutrients, and these should have been remedied. During Growth Stage IV, N is the only nutrient that can be effectively applied, although applications of P are made very early during this growth stage in some production areas as long as there are feeder roots near the soil surface (Chapter 8). Since tubers are the growth sink at this stage, it is important for nitrogen applications to be made in such a manner as to avoid stimulating excessive foliage growth. Excess foliage at this stage can favor foliar diseases, such as bacterial stem rot and late blight. Shifting potato plants from tuber bulking to vegetative growth or regrowth will invariably have detrimental effects on tuber yield and quality. The goal is to promote uniform growth of the crop, which will result in a nearly linear increase in tuber weight. Potato plants are much more likely to become deficient in N during this growth stage, making them more susceptible to early blight and to infection by weak pathogens, such as *Alternaria alternata* and *Botrytis cinerea.* Nutrient sufficiency ranges vary for different cultivars, but Table 8.1 provides a general target for producers to aim at. Local recommendations should be followed to account for regional differences in climate, soil type, cultivar, and end use of the potato crop.

Efficient water management is also important during this growth stage. Water use by the crop is generally quite high, and evapotranspiration rates of 0.25–0.40 inches per day are not uncommon. Obtaining and maintaining the desired level of available soil moisture require daily management decisions. **Avoiding plant water deficits** is the most common issue. Deficient soil moisture status, 60–65% available soil moisture, can dramatically affect crop health and tuber quality by increasing the severity of early dying and providing a soil environment that favors the development of common scab. Conversely, excessive soil moisture promotes nitrate leaching, which can lead to increased early blight susceptibility, and lush foliage, which favors some foliar diseases (Box 9.1). Under soil moisture conditions approaching saturation, the tuber lenticels will proliferate, or open, allowing entry of soft rot bacteria, thereby increasing the risk of rot in storage. Additionally, late in Growth Stage IV, excessive soil moisture can predispose tubers to infection by pathogens that cause pink rot, leak, and bacterial soft rot. Erratic soil moisture (too wet or too dry) favors a number of tuber disorders, such as internal brown spot, growth cracks, and secondary growths, all of which reduce marketability of the potato crop.

It is critical that the soil moisture status of each field be carefully monitored daily to decide whether water is deficient, adequate, or excessive. Although calculated and predicted evapotranspiration rates can be high during much of this growth stage, it is not unusual for a potato plant to use considerably less water during the latter half of this stage. Potato plants exposed to daytime temperatures above 95°F and nighttime temperatures above 70°F eventually shut down with very little growth or tuber bulking. These environmental conditions are very stressful to the potato plant, and it is not uncommon for water usage to dramatically decline if these conditions persist for more than 1–2 weeks. Under these conditions, continued irrigation increases the risk of over-irrigating fields, until the soil remains at or near saturation for an extended period. For this reason, the use of calculated or predicted evapotranspiration rates as the sole method of managing irrigation water is discouraged.

Foliar diseases can be influenced by the timing of overhead irrigation. Late blight and early blight are caused by pathogens that require free moisture on the leaf surface for infection. The early blight fungus requires alternating wet and dry conditions to sporulate (Chapter 21). For this reason, the time of day when a potato crop is irrigated can affect the development of these two diseases. However, the influence of irrigation timing varies among production areas and with local environmental conditions. For example, in the arid potato production areas of Idaho, Oregon, and Washington, dew formation occurs less frequently than it does in the more humid Midwest. In the **arid West**, the free moisture necessary for the late blight organism to infect is often supplied by overhead irrigation. If potato fields are irrigated at night when temperatures and wind speeds are lower, the wet period provided by irrigation is likely to be significantly longer than it is if the crop is irrigated during the day. Consequently, potato fields infected with late blight should be irrigated during the daylight hours to minimize pathogen development and infection. The converse is true in potato production areas in the **Midwest**, where dew forms frequently and is a major driver of late blight epidemics, which is why early blight is substantially more important in the Midwest than in the arid West. Since dew formation is a frequent occurrence, potato fields infected with late blight in this region should be watered at night, when the foliage is likely to be wet anyway, rather than during the day. Irrigating potato fields in the early morning may extend the wet period in portions of the field that are being irrigated at that time. These few hours of extended leaf wetness can frequently be the difference between infection occurring and infection not occurring. Altering the timing of irrigation in a field infected with late blight will have an effect on disease progress, but the effect is minor compared to the timing of a fungicide application. However, these minor effects may be the difference between a manageable disease and an unmanageable epidemic. Combined with an effective foliar fungicide program, timely irrigation can be an effective disease management tool.

Bacterial stem rot can be problematic in a number of potato cultivars, particularly those with an indeterminate growth habit or those prone to producing excessive foliage under high nitrogen fertilization. There is little that can be done chemically to control bacterial diseases at this time in the growing season. However, further spread and development of bacterial stem rots can be mitigated through irrigation management. Once these diseases are initiated, it is generally advisable to irrigate with higher water amounts less frequently, thereby allowing the foliage in the lower canopy to dry out between irrigations. Since bacteria require water to survive, water management is frequently the only strategy available to limit disease spread.

Most potato cultivars bulk much more rapidly during this growth stage in soil that is somewhat drier than would be desirable earlier in the growing season, such as during tuber initiation. For that reason, potato fields are managed on the "drier" side during the later portion of this growth stage, in the range of 75–85% available soil moisture. Maintaining this **soil moisture profile** accomplishes several disease management objectives, but most importantly it is an effective way to minimize the potential for tuber rots, Pythium leak, and pink rot. Lower soil moisture enables potato plants to draw moisture from a lower soil profile, thereby leaving additional field capacity for rain that may occur late in the season and prior to harvest. This is more important in the eastern and midwestern United States, where significant rainfall often occurs late in the growing season, than in the western United States. If soil is left at a higher soil moisture profile (more than 90% available soil moisture, for example), it does not take much rain to result in saturation of the soil, which promotes tuber rots.

As tubers continue to bulk and enlarge during this growth stage, there is an increased need for the diffusion of carbon dioxide (CO_2) across cell membranes and through the tuber lenticels into the air spaces of the soil. Tubers are living plants, and CO_2 naturally builds up in tuber tissue as a result of respiration. As tubers become larger, CO_2 takes longer to diffuse and escape through the lenticels. That is why **lenticels enlarge** and open readily during this growth stage in response to soil moisture. In soils approaching saturation (95–100% available soil moisture), a microthin film of water forms around the tuber, inhibiting the diffusion of CO_2 out of the tuber and the diffusion of O_2 into it. The result is that lenticels continue to enlarge and proliferate until the film of water is broken, thus allowing CO_2 to pass into the soil. Enlarged lenticels are important infection courts for soft rot bacteria, the pink rot organism, and (on rare occasions) the leak pathogen. The storability of a potato crop can be severely compromised if wet soil conditions are allowed to occur and persist during the latter part of Growth Stage IV and Growth Stage V.

Weekly **field scouting** must continue to determine the status of potential insect pests during this stage. A number of chewing and sucking insects can increase dramatically under favorable environmental conditions. Action thresholds for these pests have been determined in most production areas. The potato leafhopper, potato psyllid, green peach aphid, potato aphid, Colorado potato beetle, cabbage loopers, armyworms, and cutworms can significantly impact the ability of a potato plant to efficiently grow and the ability of tubers to bulk. In addition, the green peach aphid can transmit *Potato leafroll virus,* resulting in serious losses of tuber quality due to tuber net necrosis in some cultivars (Chapters 16 and 17). These pests can generally be controlled with insecticides once the pest reaches an action threshold. Weather conditions can affect nematodes, such as the stubby-root nematode, and an additional application of nematicide may be required for effective suppression during this growth stage.

It is during this stage that fields frequently receive **fungicide applications** to control foliar diseases, including late blight and early blight. However, decisions regarding fungicide application frequency should be made in concert with disease forecasting models and scouting reports to improve the efficiency of fungicides.

Growth Stage V: Tuber Maturation

Foliage and vines begin to yellow and eventually lose leaves and die or are killed with a desiccant during Growth Stage V. The rate of photosynthesis decreases, and tuber growth begins to slow as dry matter content reaches its maximum. Tuber skins begin to set as they mature. **Soil moisture** should continue to be reduced to ensure tuber health going into storage, declining to 70–75% available soil moisture before being replenished (Box 9.2). Excessive soil moisture must be avoided to minimize pink rot and leak (Box 9.1). Pink eye, a tuber disorder of unknown cause, can also develop during this stage of growth if the soil remains warm (above 70°F) and available soil moisture is high (above 95%). However, it is important that soil moisture not be depleted too much, because tubers that are desiccated are more susceptible to blackspot bruising during harvest (Chapter 23).

Since the potato canopy begins to retreat and ground cover diminishes as harvest approaches, sunlight hitting the ground can stimulate the germination of weed seeds, particularly those of grasses. **Grass management** is important to minimize bruising at harvest and to prevent tubers from being carried with the root mass and discharged from the harvester.

Vine desiccation is important for a number of potato market classes, not only to maintain appearance but also to minimize bruising and wound formation during harvest. A number of tuber rot pathogens, including those causing early blight tuber rot, Fusarium dry rot, and Pythium leak, all require a wound for successful tuber infection. Vine desiccation will improve skin set and help minimize the development of wounds during harvest and storage operations, thereby reducing the risk of infection by various tuber rot pathogens. Once tuber periderm is established, harvest can proceed.

CHAPTER 10

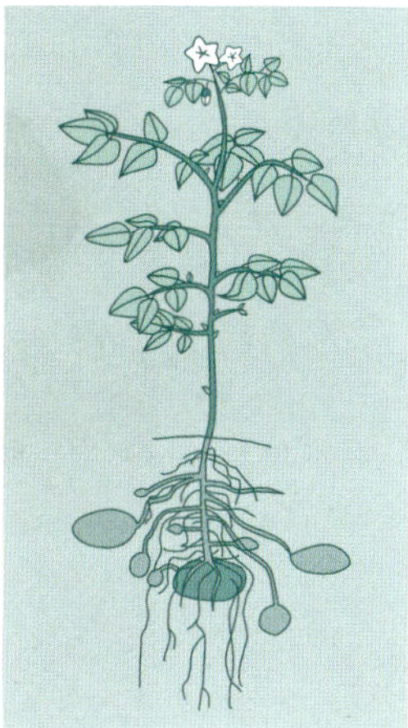

N. Richard Knowles
Department of Horticulture and Landscape Architecture
Washington State University, Pullman

Edwin S. Plissey
Department of Plant, Soil and Environmental Sciences
University of Maine, Orono

Maintaining Tuber Health During Harvest, Storage, and Post-Storage Handling

After a healthy potato crop has been grown, the next challenge is to limit the loss of tuber quality during harvest, storage, and post-storage handling. Potato tubers are living organisms, and tuber quality only deteriorates after harvest and through storage and delivery to the processor or consumer. The objective, therefore, is to implement handling and storage practices that preserve tuber quality for as long as possible following harvest. Achieving this objective is even more difficult if quality has been compromised by various stresses (e.g., temperature, diseases, and insects) during production.

This chapter presents optimal handling and storage practices for crops that are healthy at harvest. Some guidelines for handling and storage of "problem crops" will also be outlined. Maintaining tuber health through harvest, storage, and post-storage handling requires a thorough understanding of the crop history, to best predict its subsequent physiological behavior. Only then can effective management practices be developed for maintaining postharvest quality in relation to the condition of the crop prior to harvest and its end use as seed, processing, or fresh-market potatoes.

Tuber Quality

The holistic approach to maintaining tuber health during harvest, handling, and storage recognizes that preharvest factors (e.g., characteristics of the cultivar, growing conditions, agronomic practices, stress) and conditions during harvest and handling prior to storage affect the postharvest behavior of both seed and commercial potatoes (Fig. 10.1). Vine condition, soil and tuber temperature, and mechanical damage during harvesting can greatly affect crop quality and subsequent storability and thus must be considered in timing the harvest. In areas with a relatively short growing season, target yield and tuber size development are also important factors.

Decreases in tuber quality are evident at each stage of the supply chain from producer to consumer, and the causes of deterioration can be stage-specific. For example, mechanical injury during harvest and transport is a major cause of subsequent loss of quality. Bruising, improper curing, starch breakdown, sugar buildup, weight loss, decay, and sprouting affect quality during storage. Another cause of deterioration in wholesale and retail markets is greening of tubers exposed to light.

Economic returns for fresh-market potatoes are largely dictated by the tuber size profile, with premiums paid for specific size classes. In the market for premium clear- or foil-wrapped baking potatoes (Fig. 10.2), an absence of external defects is paramount. Regardless of the cultivar, blemish-free tubers with good skin set and good color command the highest market prices.

Processors require a constant supply of healthy, disease- and damage-free tubers of the proper size throughout the year. The ability to retain processing quality over a long storage period is thus desirable. While the preferred characteristics vary according to the type of processing (e.g., chips, fries, extruded by-products), a high ratio of processed product to raw product is desired, and this generally requires a high dry matter content. Low levels of reducing sugars (glucose and fructose) and a high starch content are essential in tubers used for chip and french fry processing. Processors often provide premiums to growers to encourage production and delivery of high-quality tubers.

For seed potatoes, a supply of disease-tested, unsprouted tubers of the correct size and physiological age that will produce vigorous, high-yielding plants in commercial production is of utmost importance.

From a postharvest perspective, the challenges are to preserve the characteristics that constitute quality in seed, fresh-market, and processing potatoes through proper management of handling and storage environments. To accomplish this, the process must begin before harvest and involve an integrated assessment of the impact of all facets of production on the health of a crop at harvest.

Preparing for Harvest

To attain the maximum postharvest quality and storability of the crop, decisions about the timing of the harvest should be based on consideration of several factors, including vine condition, crop maturity, target yield and grade, and soil and tuber temperatures. Killing vines at the end of the season

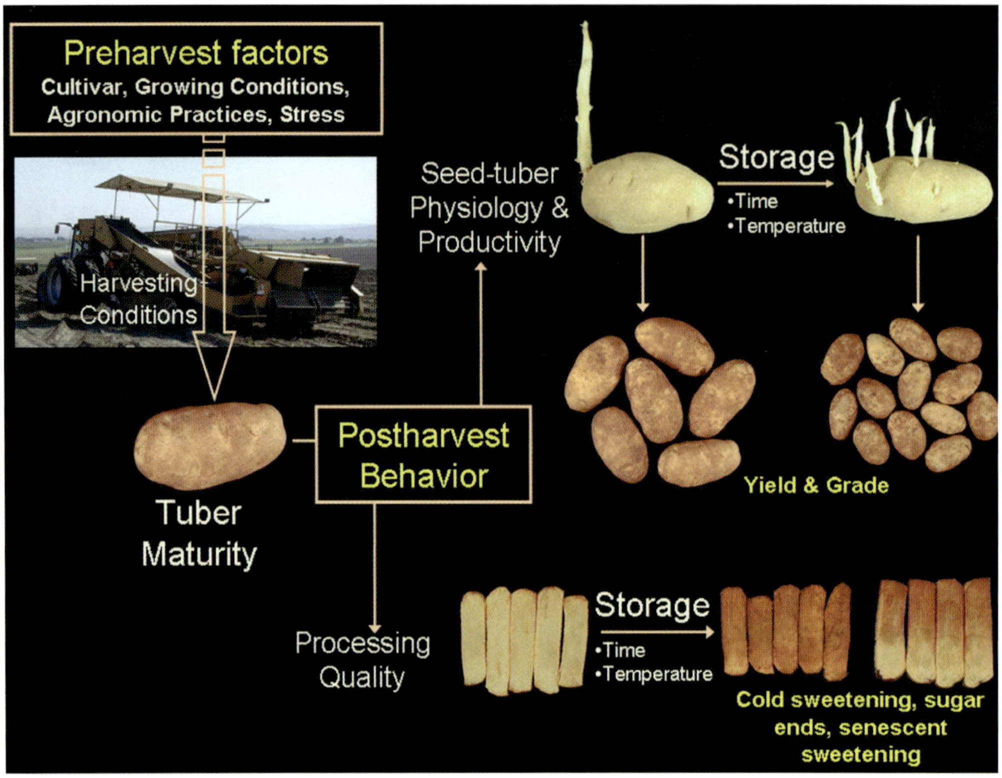

Fig. 10.1. Variables inherent in the production environment, together with conditions during harvest, affect tuber maturity, which subsequently affects the postharvest behavior and quality of seed and processing potatoes. A thorough understanding of crop history is needed in order to define the best management practices for maintaining postharvest quality.

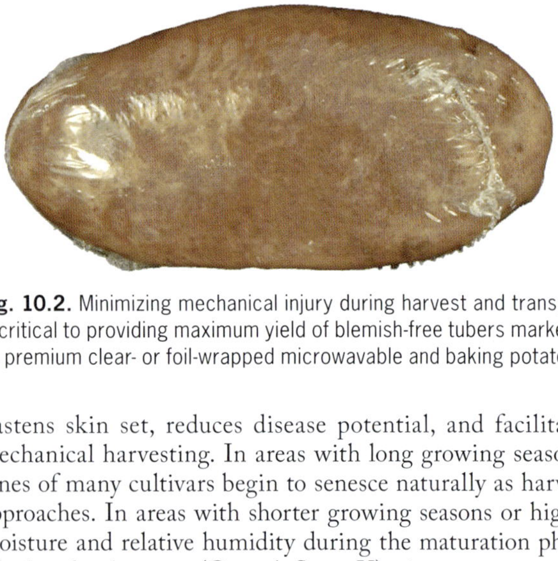

Fig. 10.2. Minimizing mechanical injury during harvest and transport is critical to providing maximum yield of blemish-free tubers marketed as premium clear- or foil-wrapped microwavable and baking potatoes.

hastens skin set, reduces disease potential, and facilitates mechanical harvesting. In areas with long growing seasons, vines of many cultivars begin to senesce naturally as harvest approaches. In areas with shorter growing seasons or higher moisture and relative humidity during the maturation phase of tuber development (Growth Stage V), vine senescence will be delayed, and managing vine condition and tuber maturity may be more challenging.

Managing Crop Maturity

Proper management of a crop prior to harvest minimizes mechanical damage to tubers during harvesting and handling. In some areas, the crop is harvested while the vines are still green and the tubers are immature. These potatoes are usually shipped directly to a fresh or processing market where they are used immediately, before the full effects of harvest damage develop. The lack of skin set associated with harvesting tubers from green vines results in increased skinning during harvest, which leads to desiccation and loss of fresh weight in storage. Immature tubers also have higher rates of respiration, which leads to dry matter loss and heat production during transit and storage.

In the major production areas of the United States and Canada, most of the harvested crop is stored before marketing. To preserve quality and maximize storability, growers must produce mature tubers that resist mechanical damage during harvest and handling operations. A mature tuber is one that has been managed to achieve full physiological maturity, either naturally or artificially. Physiologically mature tubers contain high levels of total solids (i.e., they have a high specific gravity and a high starch content) and low levels of reducing sugars and sucrose (Fig. 10.3) and have good skin set. Sucrose and glucose can be monitored at the end of the growing season to estimate relative maturity, which will dictate temperature management early in the storage season. Potatoes with a sucrose content greater than 0.15% of tuber fresh weight (approximately 0.8% of tuber dry weight) at harvest are prone to premature increases in reducing sugars (glucose and fructose) during storage, which eventually causes off-color problems during processing.

The timing of the harvest operation is critical in management for healthy, damage-free tubers. Early-maturing cultivars, such as Russet Norkotah, Norland, and Shepody, are frequently ready for low-damage harvest with minimal preharvest treatment. However, in many northern production

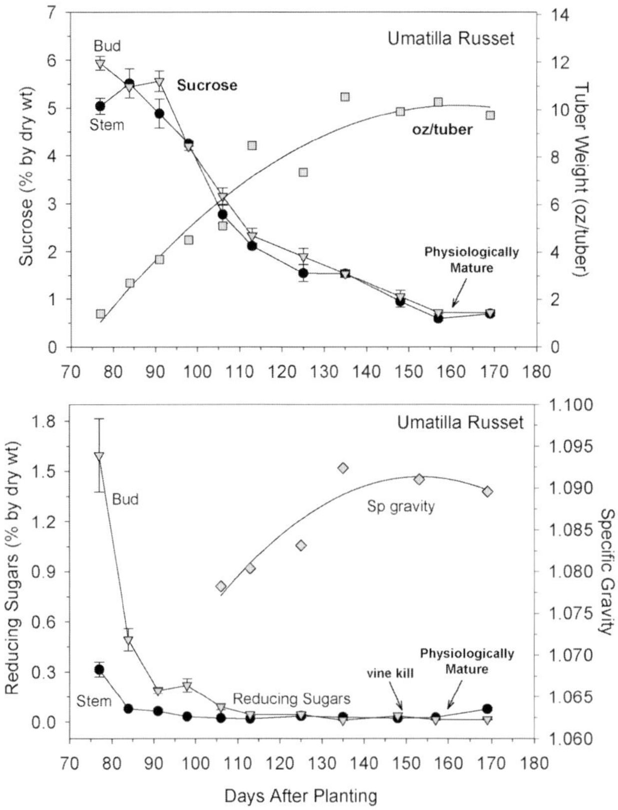

Fig. 10.3. A crop should be managed to produce physiologically mature tubers that contain high levels of total solids (i.e., have a high specific gravity) and low levels of sucrose and reducing sugars prior to harvest. Physiologically mature tubers with good skin set will maintain their processing quality the longest, under ideal storage conditions. Sucrose and glucose can be monitored to estimate tuber maturity. (Courtesy E. P. Driskill, Jr., and N. R. Knowles)

areas, damaging frost occurring before all fields are harvested often complicates planning for tuber maturity and a damage-free harvest, especially in crops of mid- or late-season cultivars. In these areas, early planting and early emergence are essential to allow time for crops to achieve full physiological maturity in a short growing season.

In most situations, careful management of fertility, irrigation, tillage, and vine killing is necessary to ensure the attainment of maturity. Nitrogen applied late in the season can delay maturity, reduce specific gravity, and raise the reducing sugar content. Nitrogen should be applied in amounts that do not exceed the crop's needs, and it should not be applied within 4–6 weeks of vine killing.

In irrigated production areas, the frequency and amount of irrigation should be reduced to achieve 70–75% available soil moisture during tuber maturation to encourage maturation and skin set. Excessively high soil moisture during maturation promotes pink rot, pink eye, and Pythium leak; it also causes enlargement of tuber lenticels, which often leads to increased bacterial soft rot in storage. Moreover, overly wet soil that remains caked on the surface of tubers will decrease the efficacy of chlorpropham (CIPC) and other sprout inhibitors applied later in storage. Soil and mud in the storage facility may clog vital air ducts and disrupt ventilation efficiency, leading to nonuniform movement of air through the bulk pile. Therefore, every effort should be made to ensure that potatoes going into storage are as clean as possible, and maintenance of ideal soil moisture conditions at harvest is the first step toward this goal. Where irrigation is available, excessively dry soil should be watered to prevent tuber dehydration, which increases the sensitivity of tubers to blackspot bruise during harvest (Chapter 23) and can lead to increased pressure bruising during storage. Soils with high clay content tend to form clods as they dry. Clods can be a significant source of bruising during harvest, and thus tillage and other field operations (e.g., irrigation) should be managed to minimize soil compaction and clod formation.

Vine Killing

Potato vines actively photosynthesize and translocate sucrose to developing tubers, where numerous carbohydrate interconversions take place. Vine killing halts this process, allowing soluble sugars to decline and solids (specific gravity) to stabilize in tubers during the maturation process prior to harvest (Fig. 10.3). It hastens maturity and facilitates periderm maturation and skin set (Box 10.1), thus minimizing certain kinds of mechanical injury during harvest and handling. A mature tuber with a well-set skin is considerably more resistant to injury during harvest, and infection by fungal and bacterial pathogens that enter tubers through wounds is minimized. For most cultivars, fully mature tubers can retain processing quality during prolonged storage.

Much loss due to disease in storage is directly attributable to failure to achieve proper vine kill and tuber maturation before harvest. Vine killing also kills weeds that might interfere with harvest, helps stop the spread of spores of the early and late blight pathogens from foliage to tubers at harvest (Chapter 21), and limits the late-season spread of viruses by aphids or mechanical means, which is of considerable importance in the production of seed potatoes.

Timely vine killing also improves the harvest operation by reducing the total vine mass moving through the harvester and by weakening the point where each tuber connects to its stolon. This allows easier separation of tubers from vines, thus reducing the loss of tubers from the harvester with the devining chains. Vine killing is also useful in controlling tuber size, where that is desirable.

The timing of vine killing is best determined by digging a few hills in several places across a field and examining tuber development. Glucose and sucrose monitoring should also be considered in assessing maturity status. In most production areas, vine killing is done by mechanical or chemical means (application of vine-desiccating compounds) or by a combination of these. Mechanical vine beaters can efficiently shred vines with minimal physical injury to tubers. Partial vine shredding or rolling of vines frequently enhances the efficacy of chemical desiccants. A drawback of very rapid vine killing is that it occasionally induces stem-end browning, a discoloration of the tuber vascular ring. This physiological disorder is usually more severe when chemical desiccants are applied at high temperatures when the soil is dry. Stem-end browning may reduce tuber grade sufficiently to affect the quality of french fries or chips.

Most potato cultivars require at least 10–14 days after application of a vine-killing treatment for tubers to mature sufficiently to resist bruising and skinning damage at harvest. During this period, the meristematic layer of cells, which

Box 10.1

Skin Set

The periderm (skin) of a tuber is composed of phellem (cork) cells, a meristematic layer called the phellogen (cork cambium), and the phelloderm. During growth, the phellogen produces multiple layers of cells toward the outer side of the tuber, which differentiate into phellem cells. The phellem cells become coated with suberin, a waxy, fatty substance, which seals the periderm and prevents water loss from the tuber. The phellogen is thus responsible for producing the protective skin of the tuber.

Skin set occurs during tuber maturation and is important in minimizing the susceptibility of tubers to skinning injury during harvest and handling. Green vines indicate that the phellogen is still meristematically active, producing periderm as the tubers continue to develop. Skin set begins with the cessation of meristematic activity by the phellogen, which occurs naturally as vines senesce, but the process is hastened by vine killing. During skin set, structural changes strengthen the cell walls of the phellogen as it becomes inactive. This process constitutes the maturation of the periderm, rendering the tuber resistant to skinning. Damage incurred during harvest and handling exposes underlying cells of the cortex, which will form new phellogen and suberized wound periderm, thus sealing off the underlying tuber tissues in the process of wound healing.

The progress of periderm maturation and skin set can easily be checked by digging a sample of tubers and determining the ease with which the skins can be "slipped" by hand. If the skin (the outer layer of suberized phellem cells) slides easily on the phellogen layer when pressure is applied with a sideways motion of the thumb, the phellogen layer is still active, and harvest should be delayed. Inability to slip the skin by hand indicates that the phellogen is no longer active as a meristem, periderm maturation is complete, and the skin has set.

The process of skin set is usually completed within 10–14 days of vine death. Skin set greatly enhances the resistance of tubers to skinning during harvesting. This in turn reduces the susceptibility of tubers to desiccation and infection by various pathogens.

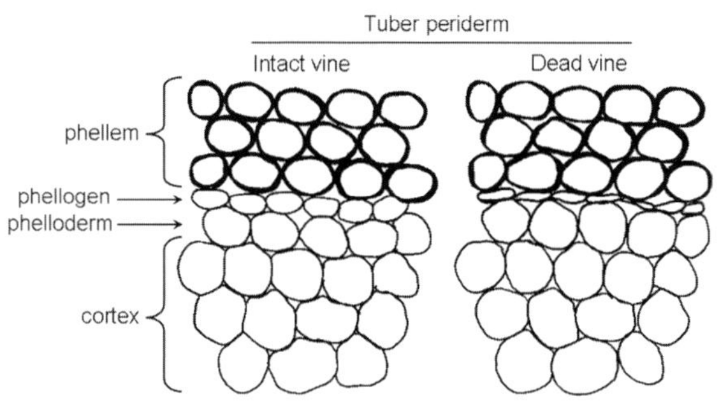

Organization of cells in the periderm (skin) of a potato tuber. During skin set, structural changes occur in the phellogen layer as the meristematic activity of this layer ceases under dead vines (right). (Drawing by G. N. M. Kumar)

produces the periderm (skin), becomes inactive, resulting in skin set. In fields where late blight or early blight is present, fungicide applications should be continued until all vines are dead. Cultivars that are susceptible to net necrosis, such as Russet Burbank, should also continue to receive insecticide applications to prevent late-season spread of *Potato leafroll virus* by aphids. Ideally, vines should be dead and dry before harvest begins. Delaying harvest more than 2 weeks after vines are completely dead is unwise, because tubers left in the ground are more likely to have increased levels of Rhizoctonia black scurf and silver scurf. Furthermore, some processing cultivars (e.g., Ranger Russet) will age and become overly mature if left too long under dead vines, especially in areas with relatively high temperatures at harvest, and this aging can lead to premature sweetening and loss of processing quality during storage. For seed crops, the accumulation of heat units by tubers left in the soil for a long period after vine killing contributes to aging, which can ultimately affect the productivity of the subsequent crop by altering the number of stems produced, tuber set, and tuber size.

Preharvest Storage Sanitation and Inspection

A clean storage facility is essential for preserving the health and quality of stored potatoes after harvest. Before harvest begins, the walls, plenum areas, air ducts, and floors of the storage should be thoroughly cleaned and sanitized (Chapter 7). After the storage is cleaned, all operating systems should be inspected carefully and repaired as needed (Box 10.2). The insulation, vapor barrier, fan and humidifier components, ductwork, sensory and regulatory components, and doors must all be functional and coordinated to ensure maintenance of the proper environment for long-term storage.

Preharvest Equipment Inspection

Mechanical damage to tubers is cumulative throughout the harvest and handling operations (Fig. 10.4). All harvest and handling equipment should be inspected before harvest to ensure that it is mechanically ready to perform the harvest with a minimum of damage to the crop. Significant bruise damage to more than 3–4% of the crop is excessive. Bruise damage can be considerably reduced by identifying problem areas and making "low-bruise" modifications to harvest equipment (Box 10.3). All harvesters, windrowers, bulk trucks, bin pilers, and other equipment used during harvest should be tested and serviced to ensure proper operation during periods of peak use. Worn and damaged parts should be replaced. Equipment should be regularly examined

Box 10.2

A Checklist for Preharvest Storage Inspection

Insulation

The insulation in potato storages should be inspected to be sure that it is intact in ceilings, walls, and doors. Adequate insulation must be installed in each component to avoid condensation and heat loss. The greater the expected difference between inside and outside temperatures during the storage season, the more insulation is needed. For most potato storages in northern areas, a minimum of R-30 in walls and R-40 in ceilings is recommended. Condensation on walls and ceilings and excessive heat loss along perimeter walls are often major factors leading to the breakdown of tubers in storage. These conditions are favorable for the development of bacterial soft rot and other damage that lowers processing quality and can result in rejection of entire lots.

Vapor Barrier

The vapor barrier serves two purposes: it maintains the high relative humidity needed to prevent water loss and shriveling of stored tubers, and it prevents water vapor from penetrating the insulation and thereby decreasing its ability to insulate. To be effective, it must be continuous over all interior storage walls. The preharvest inspection should include a thorough examination of the vapor barrier. The vapor barrier can be an integral component of the insulation. In modern storages, the insulation is specifically formulated and applied to create a textured surface that prevents condensation (Fig. 10.9E).

Fans and Humidifiers

Fans and humidifiers exchange air and add water vapor to the storage environment. Both functions are critical to the maintenance of tuber health in storage. Stored tubers require frequent air exchange through the pile to remove the undesirable products of respiration—carbon dioxide, heat, and excess water vapor. Fans should provide uniform temperature and even distribution of water vapor throughout the storage. The fans and humidifiers should be sized to deliver maximum output in the most difficult storage situations. Fan capacity is rated in cubic feet of air per minute (cfm). At least 1 cfm should be provided for each 100 pounds of stored potatoes. The relative humidity cell decks should be inspected and cleaned of mineral deposits. A recent development in fan motor control is the variable-frequency drive (VFD), by which the speed of ventilation fans can be adjusted according to the return air temperature, whenever outside air is available. VFDs prevent overventilation and can provide substantial energy savings, and they can cause less tuber shrinkage and prevent condensation more effectively than conventional fan control systems.

Before filling the storage, be certain that all bearings are properly lubricated and drive belts are in good repair. After the storage is filled, a smoke bomb may be used to check for proper air distribution through the storage pile. In storages equipped with only perimeter air distribution, a careful inspection must be made to ensure that the airflow is uniform throughout the facility and that it is removing excess condensation from the walls and ceilings. Consult with local advisors to determine procedures for optimal fan operation in specific storage facilities.

Duct System

The duct system includes a plenum chamber, which usually houses the fan and humidifier, and ducts, which distribute air uniformly to all storage areas. Ducts usually consist of galvanized culverts, wooden sections, or concrete troughs in the floor covered with boards. In newer designs (Fig. 10.9C) the air outlets are built into the concrete floor (Fig. 10.9D). Plugged or broken ducts interfere with airflow, and the quality of tubers may be affected in places where airflow through the pile is obstructed. During the preharvest inspection, make sure that all ducts are in good repair and are properly sized in relation to fan capacity, to provide uniform air distribution.

Doors

Storage doors must function as walls when the storage is in operation and thus must be equipped with adequate insulation and vapor barrier to maintain the desired environmental conditions in storage. Doors must seal tightly when closed, to avoid excessive heat loss during winter and to avoid heat gain from the outside air during warm periods. Because of their heavy use, doors are the storage component most often in need of maintenance and repair. The preharvest inspection is not complete until all doors have been repaired and function properly.

Sensors and Control Panels

In automated storages, the control panel and environmental sensors are critical to successful storage management (Fig. 10.9B). Improper installation, malfunction, or sensor failure can allow unfavorable environmental conditions to develop quickly, leading to subsequent damage to tuber health. All thermocouple wires should be inspected for proper placement and function. Be certain that automated controls function properly and that all storage systems are well integrated. Determine that dampers and damper motors function properly on demand from the control panel to maintain the storage environment within allowable tolerances.

Box 10.3

Equipment Modifications for a Low-Bruise Harvest

Considerable research has been done to identify the parts of harvest and handling equipment where the most bruise damage to tubers occurs during the harvest operation. The modifications described here can be made to existing equipment to reduce bruising. When new equipment is purchased, it is wise to look for these features as part of the original design.

Harvester

The harvester is the major source of mechanical damage to tubers (Fig. 10.4). Several modifications can be made that together reduce damage to tubers as they flow through the machine.

The digging blade on the harvester should be matched to the soil texture and be positioned so that the flow of soil and tubers is delivered up onto the digger chain, rather than bumping into the front of it (see the figure below). Hinged metal plates can be installed at the back of the digger blade to bridge the gap and rise from the blade to the primary conveyor.

With hooked chain on the primary conveyor, all links in the chain should be down links if bruise damage there is mainly due to pinching of tubers. However, if bruise damage on the primary conveyor results mainly from an inability to separate soil from tubers without excessive shaking or rollback (mostly a problem with round tubers), a configuration with one straight link and three or four down links provides positive tuber movement and increased soil separation. Ideally, all hooked chain on the harvester should be replaced with belted chain, which virtually eliminates pinching of tubers.

Other improvements that should be considered are hydraulically activated shakers on the primary bed, to improve operator control of soil separation, and full-width belted primary chain, particularly if a windrower is used with the harvester. When potatoes that have been windrowed into the furrow between rows are being picked up by the harvester, considerable damage to tubers is inflicted by the lag hooks of the twin-bed harvester chain. A full-span belted primary chain with an accompanying full-span lifting blade delivers windrowed potatoes onto the harvester with a minimum of bruising. Belting or rubberized deflectors should be installed along the sides of the conveyors to deflect tubers away from the gap between these areas and the edge of the conveyor.

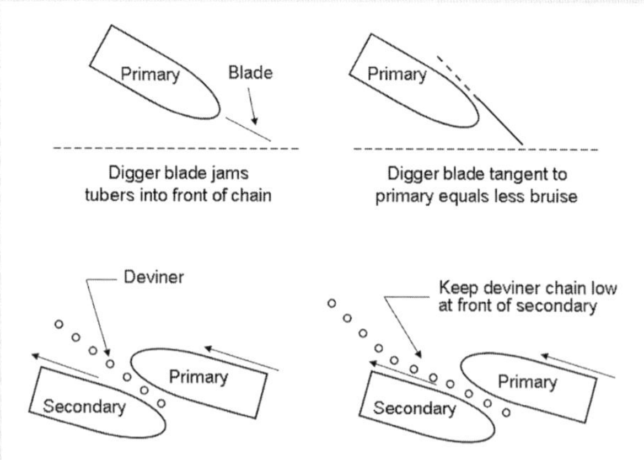

Position of the digger blade and deviner chain in relation to the primary and secondary conveyors for reduced tuber damage. (Adapted from G. M. Hyde, R. E. Thornton, and R. E. Hermanson, 1983, Reducing potato harvesting bruise, Extension Bulletin 1080, Cooperative Extension, College of Agriculture and Home Economics, Washington State University)

On harvesters with an overriding deviner chain, the deviner should operate in conjunction with the secondary conveyor. The deviner is frequently a major source of tuber damage on both harvesters and windrowers. If the first one or two sets of deviner chain rollers are removed, the chain runs flat on the secondary conveyor for the first few inches following the drop from the primary conveyor (see the figure below). Before this is done, the secondary conveyor speed and the deviner chain speed must be measured and adjusted so that they are the same. This rather inexpensive modification can result in a significant reduction in bruise damage on many harvesters.

The drop from one conveyor system to another can be a major source of bruise damage, and it is important to minimize the height of the drop. The drop from the secondary chain to the rear cross-conveyor is the largest on most harvesters. The 90° change in direction of the potato flow, in addition to the drop, increases the potential for damage at this point. The best padded chains available should be installed on the rear cross-conveyor to help reduce bruise damage. In addition, the upper end of the secondary conveyor should be modified by installation of a dogleg drive sprocket, which can decrease this drop to 6 inches or less. This modification is most easily achieved if belted chain is used on the secondary bed.

Modifications at both ends of the rear cross-elevator may significantly reduce bruise damage. A split secondary conveyor can be used, with the inside of the secondary extended over the rear cross-conveyor; the extension should be about one-third of the width of the rear cross. Similarly, the rear cross-conveyor can be extended over the side elevator, just enough to avoid damage at the inside corner, where the end of the rear cross meets the inside of the side elevator conveyor. Additional padding and chain edge deflectors can also be installed in the side elevator conveyor to reduce bruising.

Bruising due to rollback is a serious problem on the side elevator, because the angle of flow is rather steep. Rollback can be reduced by operating the conveyor at a speed that allows it to remain full of material. If this cannot be done, rollback can be reduced by installing a plastisol flight link every eight lengths of the side elevator chain. The plastisol material is stiffer at the wing ends of the flight and thus reduces rollback along the elevator edges. Another way to reduce rollback is to install a hugger belt on the side elevator conveyor.

The drop from the side elevator to the front cross-conveyor is also a key site for bruising, especially if many stones are present. This drop should be minimized as much as possible (see the figure below). Deflector belting is essential on the sorting table, to cover hooked or belted chain edges, where bruise damage often occurs. Exposed chain edges at the sorting table are also a safety hazard, as harvest personnel might catch their fingers there.

All areas of potato flow on the harvester should be examined carefully, and padding devices should be installed wherever possible to reduce tuber buffeting, which causes mechanical damage.

Harvester Chain Speed Calibration

An important low-bruise principle is that all conveyors on the harvester should be filled to capacity at all times during operation, to prevent bouncing, buffeting, and rollback of tubers. To accomplish this, the conveyor speed and forward ground speed must be coordinated, so that the volume of material handled on each conveyor is equal to its capacity.

The conveyor speed can be calculated from the chain's length and the number of revolutions per unit of time. First, measure the circumference of the conveyor with a flexible tape. Next, mark the chain with spray paint, and then time it as it makes one full revolution. The conveyor speed (in feet per second) is equal to the chain length (in feet) divided by the time taken for one revolution (in seconds). To express the speed in miles per hour, multiply the number of feet per second by 0.68. The desired speed of various harvester conveyors in relation to ground speed has been established for most areas and can be obtained from local advisors. Adjustments to conveyor speed should be based on these recommendations.

Details of harvester modifications and conveyor adjustment to minimize bruising have been outlined in several bulletins and in four videos prepared by the National Potato Anti-Bruise Committee of the Potato Association of America in cooperation with the University of Idaho and Washington State University (see Sources of Additional Information at the back of this book). Local advisors should be consulted to obtain these and other sources of information.

Bulk Trucks

There is considerable potential for damage to tubers during handling between the harvest operation and the warehouse (Fig. 10.4). Bulk truck bodies should be inspected, and any rough areas that could cause injury during handling should be repaired or padded to protect the crop. Unloading conveyors should be repaired or modified to minimize damage during unloading. Bulk trucks should be equipped with tarps or mechanical covers to protect potatoes from wind, rain, or sunburn during transit. A self-tarping mechanism should be installed to avoid the need for walking on the potatoes when the truck is being covered (see the figure below).

The drop from the conveyor should be limited to 6 inches.

Bin Piler

The bin piler is another significant source of bruise damage as the crop is moved into storage (Fig. 10.4). All rough or abrasive surfaces should be padded with protective foam or rubber cushioning. The piler should be modified to limit all drops to 6 inches or less. If necessary, the receiving end of the piler should be modified so that it can be raised up under the bulk truck's unloading belt sufficiently to minimize the drop of tubers into the hopper. All switches and control levers on a bin piler should be located conveniently for the operator, to ensure operation with minimal damage to the crop. The chain speed of the bin piler and the rate of unloading bulk trucks should be adjusted so that the bin piler conveyors remain filled to capacity at all times. To accommodate different unloading rates, a variable-speed control can be installed in the conveyor system.

When purchasing a new bin piler, look for features that will reduce bruising. Select a unit with a wide, shallow, well-padded hopper with sufficient capacity to prevent spill-out and with fingertip-adjustable height to minimize the drop from the truck. Rubberized draper chain or belted chain on the unloading conveyor is desirable, as it allows any remaining soil to sift out before the potatoes flow into the bin. The length of the boom conveyor should be about double the planned depth of the pile. To avoid rollback, the boom conveyor should never be operated at an angle greater than 45° from the floor. Hugger belts installed on the piler also minimize rollback. Additional features, such as a sorting table, a sizing chain area, a telescoping boom, variable conveyor speed, and multiple receivers, may make bin piling more efficient and may significantly reduce tuber damage during operation. Even-flow bins positioned in the receiving area between the unloading conveyor and the piler allow the conveyor to be kept fully loaded when the truck is not being unloaded.

A self-tarping mechanism eliminates walking on tubers in the truck bed during tarping.

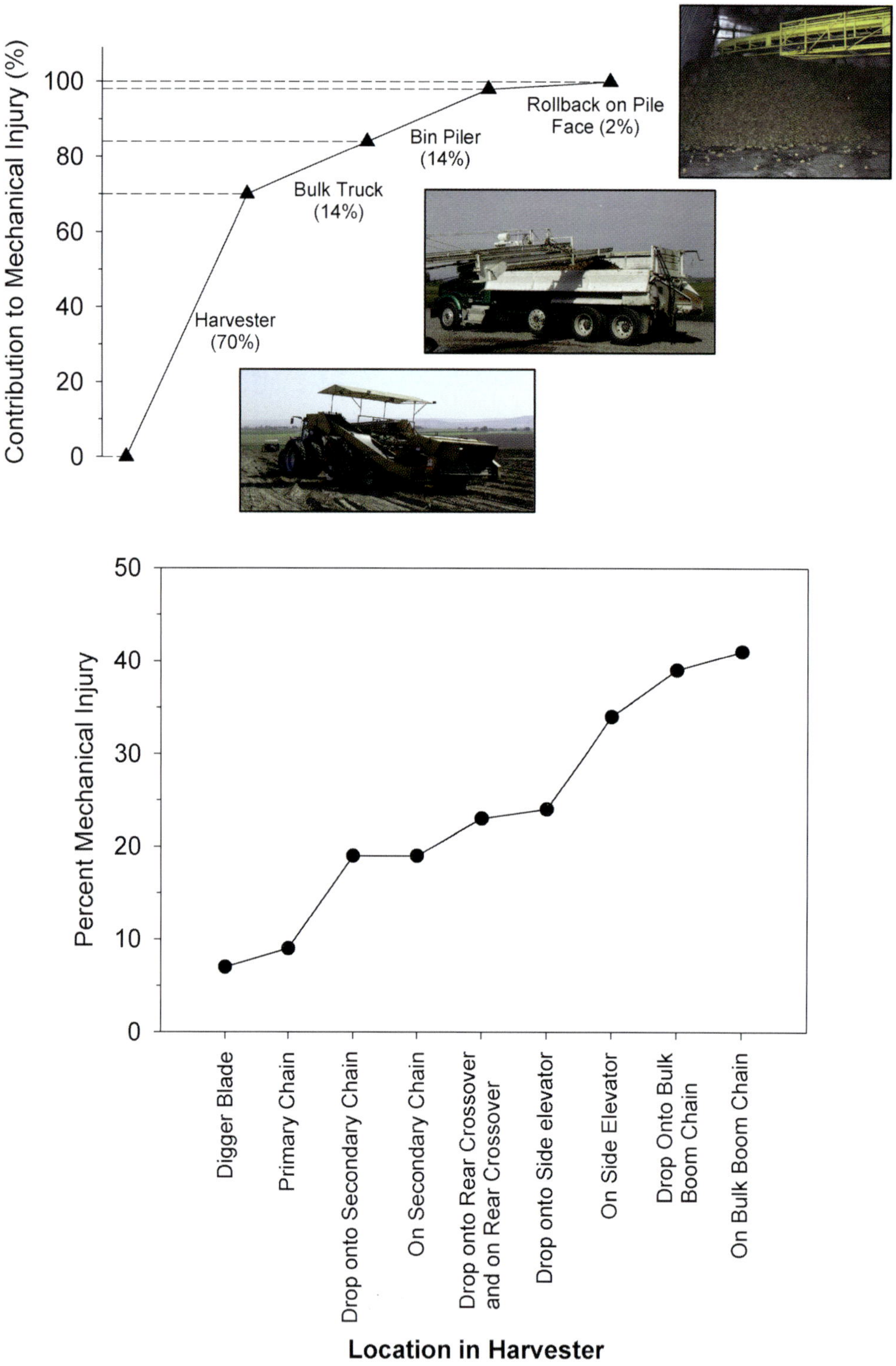

Fig. 10.4. Relative contributions of harvesting and handling equipment to mechanical injury of potato tubers. Mechanical injury is cumulative throughout the harvest and handling operations. Bruise damage to more than 3–4% of the crop is excessive. (Adapted from G. D. Kleinschmidt and M. K. Thornton, 1991, Bruise-free potatoes: Our goal, Bulletin 725, University of Idaho Cooperative Extension System)

for worn bearings. Gearboxes, transmissions, hydraulic lines, and reservoirs should be checked for leaks and repaired. Electrical systems should be inspected and repaired as needed. Rubberized chains that have become hard and brittle from being stored outside and exposed to bright sunlight should be replaced, because they will no longer protect tubers from bruising during harvest. Before use, all equipment should be thoroughly cleaned and sanitized to reduce the potential for contamination of freshly harvested tubers with disease organisms (Chapter 7).

Managing the Harvest Operation

Training Personnel in Low-Bruise Philosophy

Personnel involved in the potato harvest are just as important as the equipment in minimizing harvest damage. A properly trained crew working together with bruise control as a common objective is necessary to ensure the delivery of healthy tubers into the storage. The entire harvest staff must understand that even though the best management practices have successfully readied the crop for harvest, tubers remain very susceptible to skinning and bruising when handled. A team attitude should be developed, with each member of the crew constantly striving to protect quality at each step as potatoes move from the field to the storage bin.

A team attitude should be developed, with each member of the crew constantly striving to protect quality at each step as potatoes move from the field to the storage bin.

Operators of harvesters, windrowers, bulk trucks, and bin pilers are key staff in minimizing tuber damage during harvest. They should ensure that all drops are held to 6 inches or less. They should be made aware of the damage that can result from improper operation of their equipment, inattention, and neglect of maintenance. All staff should be trained to examine equipment during routine maintenance and lubrication shutdowns and to identify and correct any problems that may cause tuber damage.

Harvester and windrower operators should be instructed to maintain enough soil on the primary beds to cushion the tubers as they move across and ensure an even flow onto the secondary beds. Operators should be reminded that agitation of the primary bed chains should be held to the absolute minimum necessary to achieve adequate separation of tubers from soil. Bulk truck operators should be instructed to coordinate field operations closely with the harvester operator, to avoid dropping tubers into the truck from excessive heights or allowing the loading boom to plow into potatoes piled on the truck (Fig. 10.5). Once loading is complete, the load should be carefully covered with a tarp without walking on the tubers. Truck and bin piler operators must work together to match the flow of potatoes out of the truck with the capacity of the bin piler, keeping the piler conveyor filled to capacity at all times.

Monitoring Bruise Damage

Some bruise damage to potato tubers is obvious upon inspection, but fresh bruises often involve hairline cracks or internal shatter bruises that are difficult to detect visually. Tissue bruises that can develop into blackspot may also occur beneath unbroken skin and cannot be detected without peeling.

Several techniques can be used to monitor bruise damage at harvest and pinpoint operational and mechanical adjustments to minimize bruising. In large operations, managers frequently collect 20- to 30-pound test samples at regular intervals and determine the amount of damage as a percentage by weight. An autosampler on the front of the bin piler is ideal for random sampling of tubers entering the storage. The samples are then incubated, hand- or steam-peeled, and scored for bruising. Internal blackspot does not begin to develop in bruised tissues for 6–8 hours after the damage has occurred, and symptoms do not develop fully for a day or so. To detect blackspot bruising, tuber samples can be held at room temperature for at least 24 hours and then peeled. The holding period can be shortened to 8 hours if the tuber sample is held at 100°F. Blackspot is readily apparent as a bluish gray to black discoloration just under the skin (Chapter 23).

When tubers are being checked for blackspot, the first sample should come from the storage pile. If blackspot is found, samples should be collected at other harvesting and handling points to determine where the damage is occurring (Fig. 10.4). Large tubers are most susceptible to blackspot and should be selected for the first sample. If significant bruising is found, progressively smaller tubers should be tested.

Many processors include a bruise-damage clause in their contracts and penalize commercial producers for delivery of potatoes with excessive mechanical damage. Processors tra-

Fig. 10.5. Good coordination and communication between the harvester operator and the bulk truck operator are necessary to avoid dropping tubers into the truck from an excessive height or allowing the harvester's loading boom to plow into the pile of tubers on the truck. (Courtesy Washington State Potato Commission [top] and M. J. Pavek [bottom])

ditionally collect samples from delivered loads, hold them for 1 or 2 days, and then steam- or abrasion-peel them to detect blackspot bruises. A federal or state inspector is often present, with the approval of the producers, to make impartial judgments about the extent of bruise damage in the delivered potatoes.

Timing the Harvest

The timing of harvest can significantly affect tuber health throughout storage and marketing. In northern areas, the harvest should begin early enough to be completed before a hard frost is likely. For many cultivars, potatoes intended for long-term storage should not be harvested until the vines have been dead for 10–14 days, to allow for full skin set to occur. Immature tubers easily skin or "feather" during harvest and handling and are more difficult to handle and store. However, recent experience with **green vine harvest** of some cultivars shows that it can be done successfully with careful control of skinning, bruising, and initial weight loss during storage.

Soil moisture and temperature must be monitored closely once harvest has begun. Optimal harvest conditions are 60–65% available soil moisture and tuber pulp temperatures of 50–65°F. Tubers harvested from cold and wet soil are more susceptible to bruising, more difficult to cure, and more prone to breakdown in storage.

The amount of tuber hydration closely parallels soil moisture. Overhydrated tubers harvested from wet soil are highly sensitive to shatter bruise, especially when the pulp temperature is below 45°F (Fig. 10.6). Slightly dehydrated tubers taken from dry soil are less sensitive to shatter bruise but are highly sensitive to blackspot (Chapter 23). An intermediate level of tuber hydration is the best compromise to minimize overall bruise damage (Fig. 10.6). Soil moisture at harvest also has other effects. Dry, cloddy soil may result in more bruising, and overly wet soil is difficult to separate from tubers. Tubers from wet soil may also become oxygen-starved, and their lenticels may become enlarged; these conditions increase their susceptibility to internal blackheart (a physiological disorder caused by reduced oxygen) and decay. Ideally, the soil should contain enough moisture that it cushions the tubers until they reach the secondary conveyor on the harvester, at which point the soil should separate completely from the tubers.

Proper pulp temperature at harvest is critical. Harvest operations should not begin until tuber pulp temperatures are at least 45°F, and preferably 50–65°F. Tubers become much more sensitive to bruising at pulp temperatures below 45°F (Fig. 10.6). If the soil is frosty or the air temperature is near or below freezing, it is best to begin the harvest later in the day, when temperatures have risen, and then work into the evening. Soil temperatures are usually highest from 11 A.M. to 11 P.M. (Fig. 10.6).

Tuber pulp temperatures can also be too high, and harvest should be delayed if they are above 65°F. Tubers harvested with pulp temperatures of 70°F or higher are very susceptible to breakdown due to bacterial soft rot (Chapter 18) or Pythium leak (Chapter 19) in storage. Moreover, it is difficult to remove the field heat from these tubers once they are in storage. Thus, in warm weather, early morning may be the best time for harvest. Harvest scheduling is often unpredictable, with weather always a significant factor. Fall rains and early frosts often hamper operations. Harvest decisions should be based on soil and crop conditions, weather forecasts, long-term weather records, and grower experience. In spite of the best intentions, harvest conditions are not always optimal, and compromises sometimes must be made. However, if harvest operations can be conducted under favorable environmental conditions, a healthier crop can be delivered to storage.

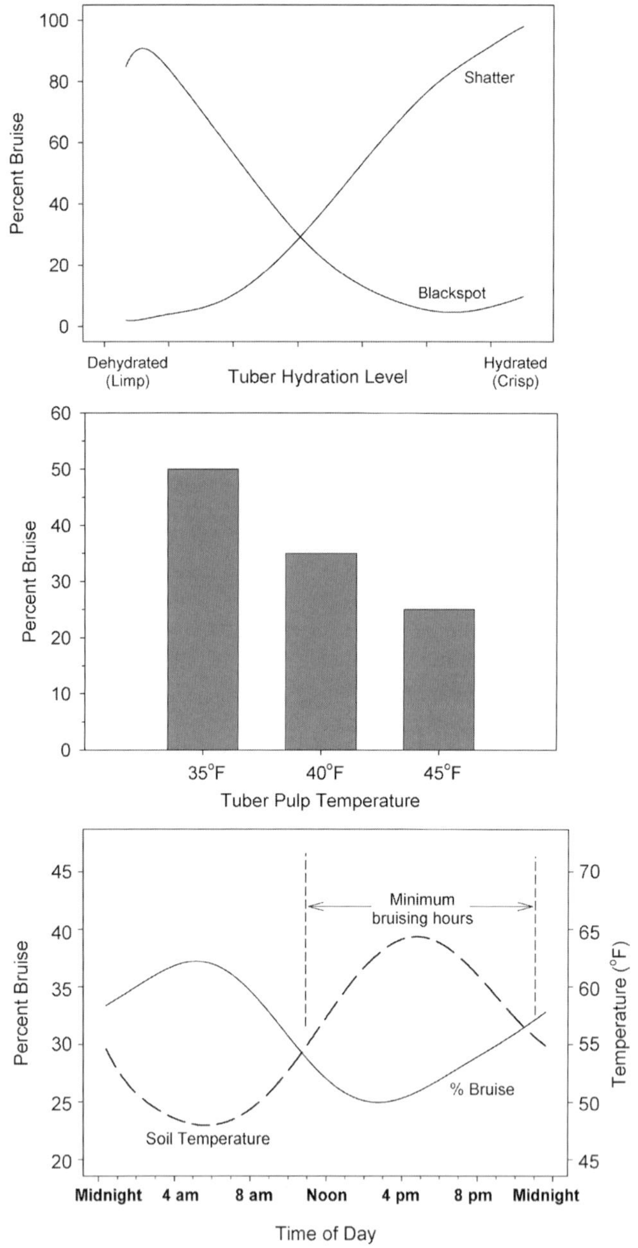

Fig. 10.6. Effects of tuber hydration, pulp temperature, and time of day on bruise injury to potatoes. Bruising is minimal when tuber pulp temperatures are 50–65°F. In northern production areas, tuber pulp temperatures will be highest when the soil has warmed, usually between 11:00 A.M. and 11:00 P.M. The timing of harvest should be planned accordingly, to minimize bruising. (Adapted from R. E. Thornton and H. Timm, 1990, Influence of fertilizer and irrigation management on tuber bruising, American Potato Journal 67:45–54 [top], and Kleinschmidt and Thornton, 1991 [middle and bottom])

Operating the Harvester

The most important principle in minimizing bruise damage during digging is that the volume of soil, tubers, and other material moving through the harvester should be held at capacity at all points in the machine. Operating the harvester with the conveyors filled below capacity allows tubers to bounce and roll excessively and thus increases bruise damage. Operating with the conveyors overfilled also increases bruising, due to excessive rollback.

The digger blade must be set deep enough to lift the maximum amount of tubers with minimal damage. If it is set too deep, the machine becomes overloaded with soil, and the result is poor separation of tubers. For each inch the blade is set too deep, the harvester handles about one additional ton of soil per minute for each row. If the blade runs too shallow, however, it cuts many tubers, and insufficient soil is carried up the primary conveyor. It is important to carry a layer of soil one-half to two-thirds of the way up the primary conveyor to cushion the tubers. Supplemental agitation of the primary conveyor should be closely monitored and reduced if tubers are bouncing. Conveyor agitation should be stopped if an appropriate cushion of soil cannot be maintained up the primary, as often occurs with dry, sandy soil.

To keep the machine filled to capacity with soil and tubers, forward speed and conveyor agitation must be adjusted as conditions change within the field. Some harvesters are equipped with a variable conveyor speed, allowing an additional adjustment. The harvester operator should visually judge changing conditions and adjust the forward speed by shifting gears, not by adjusting the engine speed or the blade depth, to maintain the desired amount of soil on the primary conveyor.

Windrowers are used routinely in some production areas. Two or more rows are dug with a windrower, and the tubers are placed on the ground between adjacent undug rows (Fig. 10.7). A harvester then follows and harvests these adjacent rows, picking up the windrowed tubers at the same time. This procedure not only increases the efficiency of the harvester but also helps lower overall bruising, by increasing the volume of tubers going through the machine, thus making it easier to keep the conveyors filled to capacity at all times.

Weather governs the management of windrowing operations. If tubers dug by the windrower are damp, they can be allowed to dry on the ground for 30–45 minutes before being harvested. Dry tubers, however, should be picked up immediately, to avoid dehydration. If pulp temperatures are near 50°F and the air temperature is below that, windrowed tubers should be picked up immediately to avoid further cooling. If pulp temperatures are closer to 65°F and the air temperature is above that, they should also be harvested immediately to avoid further warming. Thumbnail cracking (Chapter 23) can be accentuated while tubers are in windrows. Windrowed potatoes should not be left lying on warm, moist soil in bright sunlight; these conditions favor greening and sunscald in the field and the rapid development of Pythium leak in storage. Windrowed potatoes exposed to rain should be kept separate and marketed immediately, as they may not store well.

One of the most serious errors is to harvest potatoes from low, poorly drained areas of a field where water may have accumulated. Moisture levels are also usually excessive in the center of a pivot irrigation system, and planting within 80–100 feet of the pivot center should be avoided. Tubers harvested from this area are likely to be under increased disease pressure because of overwatering and should not be placed in storage, as they are much more susceptible to breakdown from blackheart, bacterial soft rot, Pythium leak, pink rot, and late blight. If harvested and mixed with healthy tubers in storage bins, tubers from overly wet areas of the field can be a major source of breakdown during storage. It may be best to leave tubers in overly wet areas, but if they are harvested, they should be kept separate and marketed immediately.

One of the most serious errors is to harvest potatoes from low, poorly drained areas of a field where water may have accumulated.

In areas where potatoes are grown in stony soils, some harvesters are equipped with air-vacuum devices to assist in separating tubers from stones as they pass by on the conveyor. The air-vac system takes advantage of the lower density of potatoes (a potato may weigh half as much as a stone of the

Fig. 10.7. A windrower digs two or more rows and places the tubers on the ground between adjacent undug rows. A harvester follows, harvesting these adjacent rows and picking up the windrowed tubers at the same time. (Courtesy N. Olsen)

same size) to lift them from among the stones and place them on the sorting or boom conveyor. With ideal soil moisture, harvesters equipped with air-vacs are 95–98% effective in separating potatoes from stones. Machines equipped with air-vacs have been shown to cause less bruise damage under these conditions than some conventionally equipped harvesters. Harvester operation must be monitored closely. The volume of stones plus tubers must be considered to determine the appropriate ratio of forward speed to conveyor speed needed to minimize tuber damage. After stones have been removed by an air-vac, the appropriate conveyor speed is related only to the volume of tubers. Because stony soils compound the potential for bruise damage during harvest, a management plan should be developed for removing stones from the field.

Placing the Crop in Storage

Bulk trucks should be covered with a tarp after loading, for several reasons. Chilling injury predisposes potatoes to breakdown in storage. Exposure to sun during transport may warm tubers excessively. Mild sunburn interferes with wound healing in storage and may promote the development of Pythium leak or Fusarium dry rot. It is especially important to protect harvested tubers from rain, which can greatly increase the potential for decay in storage.

To ensure tuber health in storage, unloading must be managed as carefully as harvesting. One person who is properly trained in low-bruise operation should be assigned to operate the bin piler. Personnel should not walk on the potatoes during unloading. The tubers should be fed out of the truck at an even rate, onto an unloading conveyor placed up under the truck unloading conveyor. A rubber chute is useful for buffering tuber drop onto the conveyor. The drop from the unloading conveyor to the receiving conveyor should not exceed 6 inches. Bin-loading conveyors must be operated fully loaded to prevent bouncing and rollback. To accomplish this, the conveyor speed must be matched to the rate of unloading the truck, and the speed of each successive conveyor must be matched to the delivery capacity of the preceding one. Hugger belts installed on bin-loading conveyors also help reduce bouncing and rollback (Box 10.3). When a truck has been unloaded, all conveyors should be shut off but left full. Evenflow bins positioned following an unloading conveyor allow the subsequent conveyors to be kept fully loaded when the truck is not being unloaded.

If significant disease or frost damage is evident or if excessive amounts of stones or debris are present, the tubers should be graded to remove damaged and diseased tubers, stones, and debris before being placed in storage. Potatoes should not be washed before they are placed in storage; washed potatoes do not store well. Crops containing over 1–2% blighted or soft-rotted tubers should be isolated from lots of healthy tubers and marketed quickly, as long-term storage of such crops is usually unsuccessful. If fungicides are applied as a fine mist, a low-volume application rate should be used, to avoid wetting the tubers before putting them into storage.

In bin piling, a progressive or **stepped piling procedure** minimizes rollback on the pile face (Fig. 10.8). An even flow of tubers should be maintained on the bin piler, which should be kept as full as possible to avoid rollback. High conveyor speed can cause potatoes to be flung against the storage structure and should be avoided. The boom should be kept close to the top of the pile, to minimize the drop. Short, frequent movements of the piler limit the height of the drop and evenly distribute any soil or debris carried onto the face of the pile. Pockets of soil within the pile prevent uniform airflow and may promote bacterial soft rot. In-floor air duct slots (Fig. 10.8) located under conveyors and bin pilers should be protected during the piling operation to prevent them from being filled with soil, which would interfere with air distribution after the pile is formed. Tubers should not be piled more than 16–18 feet deep, to avoid excessive pressure bruise, which is cultivar-dependent.

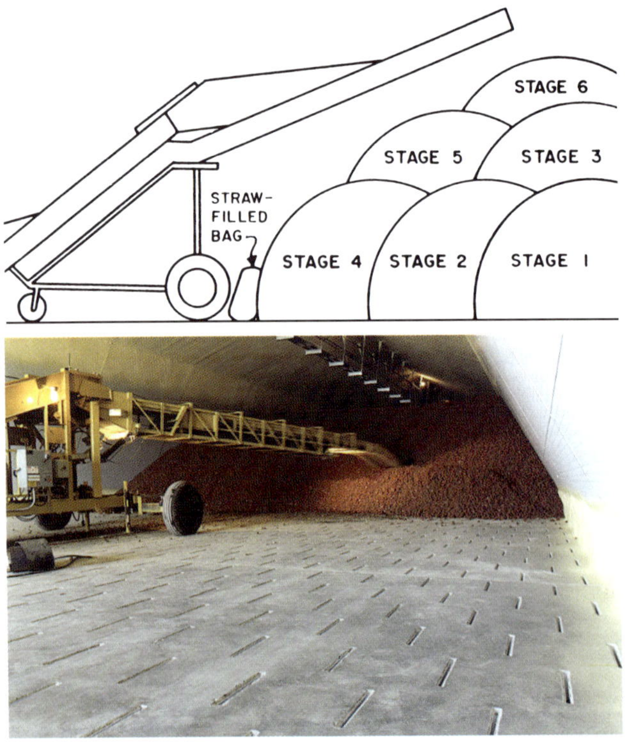

Fig. 10.8. A progressive or stepped bin-piling procedure minimizes the rollback of tubers on the pile face. (Drawing reprinted, by permission, from Kleinschmidt and Thornton, 1991. Photo courtesy R. Hesse, Suberizer, Inc.)

Maintaining Tuber Health Through the Storage Cycle

Even the best storage can only maintain the quality of the tubers put into it. Tuber health never improves in storage. Despite being dormant during much of the storage period, potato tubers utilize oxygen and produce heat, moisture, and carbon dioxide. A proper storage environment is essential for managing respiration, water loss, and sprouting by promoting wound healing and preventing disease development and progression, pressure bruise, loss of dry matter, and excessive accumulation of reducing sugars—all factors that reduce tuber quality.

Tubers undergo several distinct physiological stages during storage: (1) a curing period, or "sweat" period, during which harvest wounds heal, (2) a cooling period, when the pulp temperature is lowered to a level that is appropriate for the intended use of the tubers, and (3) a holding period, during which the tubers are dormant at first and respiration is low, and then dormancy ends and tubers are able to sprout. When sprouting occurs there is an increase in respiration and generation of heat and carbon dioxide. Modern storages (Fig. 10.9) are designed to provide the storage manager with the equipment required to control temperature, humidity, and air movement during each physiological stage (Table 10.1). Minimizing losses in tuber weight and quality during storage is critical to profitability. Precise temperature management during these stages is also essential for controlling the physiological aging of seed potatoes, which can have a tremendous impact on subsequent productivity (Chapter 7).

The Curing Period

Despite all precautions taken during harvest and handling, some injury to tubers will occur. The susceptibility of tubers to weight loss and infection by various pathogens is highest during the first weeks of storage, and up to half the shrinkage over an entire storage season can occur during this critical initial phase of storage. Environmental conditions must be managed to minimize weight loss and maximize the development of wound periderm during the curing period. Curing results in the production of a suberized wound periderm, which effectively protects underlying tissues from desiccation and invasion by pathogens during long-term storage. Proper wound healing at the beginning of storage is vital to

Fig. 10.9. Modern potato storage facilities (**A**) are equipped with control panels and environmental sensors (**B**) to control temperature, humidity, and airflow. Recent designs have a duct system (**C**) with air outlets built into the concrete floor (**D**). Dampers (**C**) can be selectively closed to direct air to specific areas of the pile. Minimizing condensation is paramount and depends on adequate insulation. Ceiling condensation can be reduced with small circulating fans between the top of the pile and the ceiling, but fans are not necessary in state-of-the-art storages where a vapor barrier is an integral component of the insulation (**E**), which is designed to prevent condensation. (**B–E** courtesy R. Hesse, Suberizer, Inc.)

maintaining tuber health and minimizing losses later in storage (Chapter 7).

Wound healing can be completed in a few weeks under ideal conditions, but many storage bins are filled over a period of several weeks. The period during which a bin is being filled should be as short as possible but should not exceed 7–10 days, to minimize the period of exposure of tubers to the high temperatures needed for curing. Freshly dug potatoes have a high rate of respiration, which contributes substantially to the initial heat load and potential weight loss in the first days of storage. Therefore, as a storage bin is being filled, the immediate concerns are to remove field heat and stabilize the temperature of the bulk pile as quickly as possible. Ideally, the temperature should be lowered to between 50 and 60°F within days of entering the storage by maximizing the movement of cooled and humidified air through the pile. The curing period begins when the first potatoes are placed in the bin and ends when wound healing is complete for the last potatoes placed there, usually a total of 3–5 weeks. Crop management during this time sets the stage for the entire storage season.

Wound healing and suberization are greatly affected by temperature; oxygen and carbon dioxide levels; and relative humidity. Wound healing is most rapid at about 70°F, but a compromise temperature of 50–60°F during the curing period is recommended, to minimize the potential for desiccation and limit tuber decay. Most pathogen activity declines at lower temperatures. If the crop is generally healthy, the **curing temperature** can safely be maintained at 55–60°F for 2–3 weeks. However, if any significant amount of tuber disease is present, a curing temperature nearer to 50°F is preferable. Tuber pulp temperatures should be monitored continually with a pulp thermometer or with thermocouples placed in tubers within the pile. Wireless temperature sensors can be placed at strategic locations within the pile during bin filling to monitor temperatures during curing and throughout the season. The temperature of the return air supply should also be monitored, since it indicates conditions at the top of the pile.

Air movement within the pile is essential during the curing period. Wounded potatoes have high rates of respiration, and wound healing is an aerobic (oxygen-requiring) process. Evenly distributed airflow is necessary for controlling temperature and providing oxygen while removing carbon dioxide and excess moisture. The ventilation systems in modern potato storages are equipped to provide up to 25 cubic feet per minute (cfm) per ton of potatoes (Fig. 10.9). The airflow and humidification should be sufficient for cooling without causing tuber desiccation. To prevent weight loss and promote rapid formation of suberized wound periderm, the **relative humidity** of the supply air must be kept above 95%. It is, however, very important that condensation not develop on the tubers. Free water on tuber surfaces inhibits the exchange of oxygen and carbon dioxide and promotes bacterial soft rot.

Since relative humidity and air temperature are interrelated, differences between the supply air temperature and tuber pulp temperature must be monitored closely. If saturated air much colder than the tubers is forced through the pile, it will be warmed by the tubers, and the relative humidity will drop considerably, resulting in tuber dehydration and increased pressure bruising. Conversely, if saturated air warmer than the tubers is introduced, condensation will occur on cold tuber surfaces, favoring the development of bacterial soft rot. The best practice is to monitor the pulp temperature of stored tubers carefully and **keep the supply air just a few degrees lower than the tuber temperature** until the pile reaches the desired curing temperature. After that, the supply air should be adjusted to maintain a temperature differential of 0.5–2°F from the bottom to the top of the pile. This is accomplished most efficiently and economically by adjusting fan speed to reduce ventilation when the temperature differential is small and increase ventilation when the temperature differential increases toward the high end of the optimum range. Fans equipped with variable-frequency drive are especially efficient for this purpose.

Tubers with pulp temperatures below 50°F should not be placed in storage, if possible. Cold tubers must be warmed to wound-healing temperatures (55–60°F), and condensation on cold tuber surfaces during the warming process is the major concern. Condensation (free water) will enhance tuber decay, especially during this critical period before wound healing has occurred. Cold tubers should be warmed with dry air (no humidity added). If possible, allow the cold tubers to warm naturally, using limited airflow if no field frost

Table 10.1. Environmental management for maintaining potato health in storage

	Curing period	Cooling period	Holding period	Marketing period
Temperature	Maintain tuber pulp temperature at • 55–60°F if tubers are healthy • 50°F if tuber decay is present	Rapidly cool seed and fresh-market tubers to the appropriate holding temperature Slowly cool processing tubers, lowering pulp temperature by 2–3°F per week	Maintain tuber pulp temperature at • 38–40°F for seed • 38–50°F for fresh market • 44–50°F for french fry processing • 50–55°F for chip processing	Warm slowly to 50–55°F over several weeks
Relative humidity	Maintain at 95–99%	Maintain at 95–99%	Maintain at 90–95%	Maintain at 90–95%
Ventilation	Supply at high rates to remove field heat, stabilize pile temperature, reduce CO_2 buildup, and provide O_2 for wound healing	Supply at high rates to control cooling and maintain a temperature differential of 0.5–2°F from the bottom to the top of the pile during cooling	Supply at reduced rates, adjusted as necessary, to supply O_2, remove CO_2, and maintain a temperature differential of 0.5–2°F from the bottom to the top of the pile	Supply at reduced rates to allow heat of respiration to raise pulp temperature to 50–55°F and thus minimize bruising during removal from storage

is present. Once healing temperatures have been reached, the pile should be managed as described above.

It is critical to be aware of the amount of disease in each lot of tubers coming into storage to properly manage the curing period for each lot. **Storing tubers with a high level of disease** is difficult. If more than 1–2% of the tubers have symptoms of bacterial soft rot, late blight, pink rot, or leak, or if a similar proportion is frost-damaged, they should be cooled to below 50°F immediately after being placed in storage (Box 10.4). The problems of weight loss, subsequent pressure bruise, and reducing sugar buildup must be considered secondary to potential losses from decay. In this situation, the use of humidifiers should be avoided, and fans should be operated continuously to dry the diseased tubers and prevent the buildup of free moisture in the pile. In warm weather, no attempt should be made to store a heavily diseased crop unless pulp temperatures throughout the pile can be brought down and maintained below 50°F. If tubers at the top of the pile cannot be cooled properly, decay will start there, and the entire pile may "melt down" and be lost. It is preferable either to grade tubers in this condition carefully and market them straight from the field or to delay harvest until the weather

Box 10.4

Storing Problem Potatoes

J. P. McMorran, A. R. Mosley, and M. I. Vales
Department of Crop and Soil Science
Oregon State University, Corvallis

Blighted, frozen, wet, or otherwise compromised potatoes call for extraordinary storage measures. Common storage rules must sometimes be bypassed to save the crop. The following guidelines for harvesting and postharvest handling should be considered in managing problem potatoes.

- Market straight from the field, if feasible; if not, sell at the earliest reasonable opportunity.
- Avoid problems by harvesting dry, sound potatoes with a pulp temperature between 45 and 65°F.
- Where potatoes are frozen or blighted, wait a few days, if possible, to allow symptoms to fully develop before harvest.
- Leave rotten, frozen potatoes and debris in the field if possible. This may call for additional people on the harvester.
- Sort tubers as they are placed in storage. Provide sufficient light, people, and time to do the job properly.
- Prepare the storage and ensure that it is at the proper temperature, with the air delivery and control systems in good order. Provide an adequate flow rate (25 cfm per ton) in all areas of the storage. Add portable systems to otherwise airless storages. Good air movement is absolutely essential for storing problem potatoes.
- Cure the tubers at the lowest possible temperature (50°F), or eliminate the curing period altogether, depending on the condition of the crop. Problem potatoes are usually wet and infected with decay organisms and therefore should be cooled and dried as quickly as possible.
- Quickly cool the pile to the final storage temperature (about 38°F for seed, 42°F for table stock, 45°F for french fry processing, and 50°F for potato chips). It may be necessary to cool and hold processing potatoes well below the levels typically recommended for frozen processing (45°F) and chips (50°F).
- Do not humidify. The potatoes are probably much too wet already.
- Run fans continuously until the crop is dry and decay is under control. Running fans does not necessarily call for ventilation. Recirculate air through the potatoes at all times during the problem period, even when outside air is not being introduced.
- Keep the pile as shallow as possible, to promote air movement and easy removal of potatoes in hot spots. Rotting tubers and dirt sometimes form barriers to air movement.
- Monitor the storage daily. Thermometers suspended at various depths in the pile provide a good indication of the average temperature. Infrared guns are helpful in locating hot spots before they begin to sink and spread.
- Do not expose cold potatoes to warm outside air. A layer of free water will condense on the tubers. Try to use air no warmer than about 5°F above the desired tuber temperature. Water on tubers tends to suffocate them, and it favors soft rot bacteria.
- Do not expose tubers to air at or below the freezing point.
- When unloading the storage, do not wash dry seed potatoes unless they are covered with dried slime. Washing will probably help wet slimy seed. If seed is washed, use sprays rather than a dip tank. Try to use multiple nozzles, so that all surfaces are washed clean. Misting washed potatoes with a 10% sodium hypochlorite solution is recommended (check the labeling for your area). Add 1 gallon of bleach to 9 gallons of water and mix well before applying in a well-ventilated area. Problem seed that requires washing should be planted as soon as possible, if soil conditions are suitable.

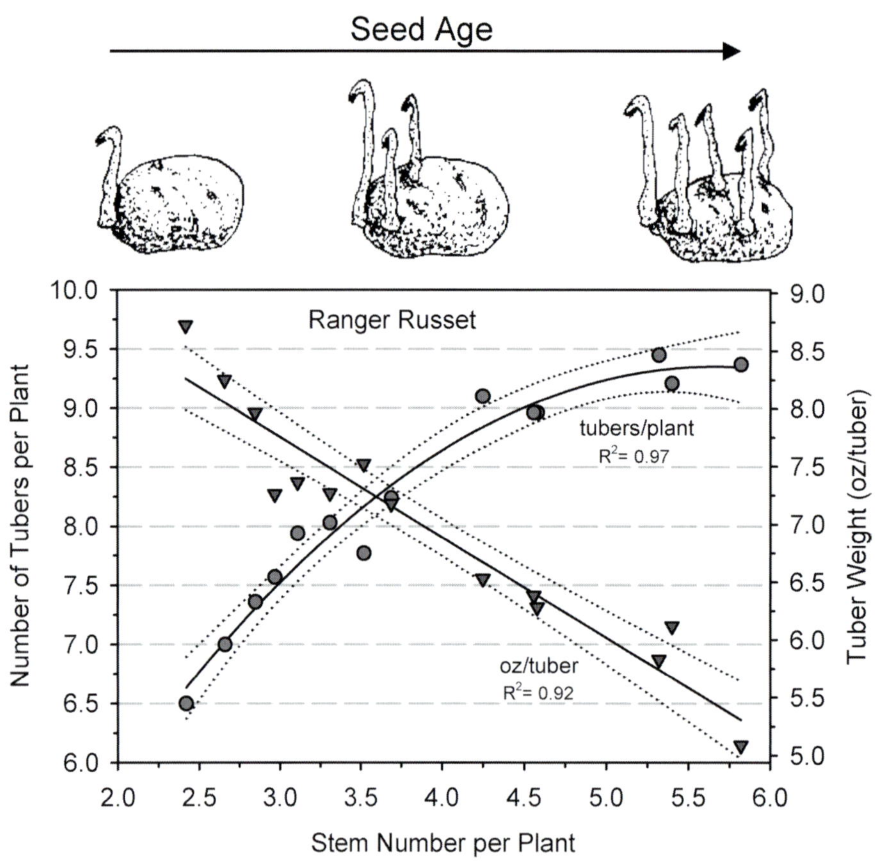

Fig. 10.10. Stem number and tuber set of plants grown from Ranger Russet seed tubers and weight of tubers produced over a 150-day growing season in the Columbia Basin of Washington. The stem number is indicative of the physiological age of the seed tuber, which increases with the accumulation of heat units during handling and storage. The physiological age in turn affects the tuber size distribution (Table 10.2). (Drawing by G. N. M. Kumar. Adapted from N. R. Knowles and L. O. Knowles, 2006, Manipulating stem number, tuber set, and yield relationships for northern- and southern-grown potato seed lots, Crop Science 46:284–296)

permits rapid cooling in storage. Delayed harvest will also allow time for infected tubers to decay totally in the field, so that they won't be harvested.

If more than 2% of the tubers have symptoms of bacterial soft rot, late blight, pink rot, or leak, or if a similar proportion is frost-damaged, they should be cooled to below 50°F immediately after being placed in storage.

The Cooling Period

Storage temperatures should be lowered to holding levels after the curing period (Table 10.1). Potatoes to be sold as **seed tubers** should be cooled to 38–40°F if a minimum number of stems per hill is desired in the subsequent crop. Higher temperatures can stimulate physiological aging of seed, which may affect stem number, tuber set, tuber size, and yield in the succeeding crop. Seed tuber age can have a tremendous impact on productivity. While the biochemical basis of seed age is not fully understood, research has shown that the accumulation of heat units above 39°F results in early dormancy break and reduced apical dominance. This leads to the production of multiple sprouts and thus multiple stems in the progeny crop (Box 7.2). The resulting effects on productivity are cultivar-dependent and can be either positive or negative, depending on length of the growing season and market requirements for tuber size of seed, fresh, and processing tubers.

After vine kill, the accumulation of heat units by a particular seed lot is additive and need not be continuous. High pulp temperatures after vine kill, as tubers set their skin and mature, and during harvest, transport, wound healing, storage, warm-up prior to transport, and post-storage transport and handling at the cutting facility all contribute to physiological age and variation in the productive potential of seed lots.

Aged seed tubers produce more stems per seed piece, which in turn affects tuber set and size development in a predictable manner (Fig. 10.10). The number of tubers set per hill increases with increasing stems per seed piece. This increases competition among tubers within a hill for carbohydrate and other resources during growth, resulting in smaller tubers. In areas with long growing seasons, such as the Columbia Basin of Washington, yield of marketable U.S. No. 1 tubers and total yield remain relatively constant over a wide range of stem numbers, and thus only the tuber size distribution will be affected (Table 10.2). In other growing regions, both tuber yield and size distribution are affected by seed tuber age. Therefore, the optimum seed tuber age for a specific yield and tuber size distribution will differ according to the growing conditions unique to various geographic areas of production.

Seed growers should keep records of the degree-days accumulated by their crops, starting at vine kill and continuing through storage to final sale. This will help to answer questions that may arise later concerning performance of the commercial crop. Daily heat units are calculated by subtracting a base temperature of 39°F from the average daily tuber temperature; the sum of the daily heat units for each day in a specified period (e.g., from vine kill through storage), expressed in degree-days, represents the total accumulation of heat units by the crop. Maintaining a constant number of degree-days in seed lots from year to year will decrease year-to-year variability in seed performance and better enable commercial growers to optimize management (e.g., in-row spacing) to get the best yield and tuber size distribution possible, within the constraints of their growing region.

Tubers for **fresh consumption** can be stored at 38–50°F. Lower temperatures delay dormancy break, inhibit sprouting, and lengthen the overall storage period. However, a gradual conversion of starch to reducing sugars at low temperatures (38–42°F) will cause tissues to turn off-color and lead to an undesirable waxy texture in many fresh-market cultivars. Optimal cooking and culinary quality for fresh-market potatoes is maintained at 44–48°F. In contrast to processing potatoes, seed and fresh-market potatoes may be cooled relatively rapidly by careful introduction of outside air or by refrigeration. Care must be taken to maintain high relative humidity in the cooling air, to prevent desiccation. Precautions regarding temperature differences and condensation during the curing period apply here as well. The supply air temperature should never be more than 2°F lower than the tuber pulp temperature during cooling.

Table 10.2. Yield of potato cultivars Russet Burbank and Ranger Russet in relation to number stems per seed piece in the Columbia Basin of Washington State[a,b]

	Average stem number of:			
	Russet Burbank seed pieces		Ranger Russet seed pieces	
	3.2	5.4	3.2	5.4
Total yield (tons/acre)	30.9	28.7	31.5	31.0
Marketable yield (tons/acre)[c]	28.8	27.2	31.5	31.9
U.S. No. 1 (tons/acre)	25.8	22.1	28.4	26.6
Tuber grade[d]				
<4 ounces (%)	10.8	18.5	10.5	16.3
4–6 ounces (%)	17.4	23.4	14.4	22.4
6–10 ounces (%)	32.8	32.9	29.5	31.7
10–12 ounces (%)	12.1	9.1	11.1	9.9
12–14 ounces (%)	8.4	5.8	9.4	7.2
>14 ounces (%)	18.3	10.4	25.8	11.1
Tubers per plant	7.7	9.0	7.6	9.2
Average tuber weight (ounces)	7.0	5.7	7.7	6.1
Tubers per acre	130,200	156,500	139,000	168,500

[a] Adapted from Knowles and Knowles, 2006.

[b] Russet Burbank and Ranger Russet seed tubers were stored at temperatures above 39°F for various periods directly following harvest, to produce different numbers of stems per seed piece. Tuber set and tuber size distribution are affected by the number of stems per seed piece, which increases with the physiological age of the seed tuber. In areas with a long growing season, such as the Columbia Basin, tuber size distribution is affected more than yield.

[c] Total yield minus culls.

[d] Percentage of total marketable yield.

Tubers for processing must be cooled more slowly than seed and fresh-market potatoes, preferably by no more than 2–3°F per week. Rapid cooling to holding temperatures can induce a subsequent buildup of reducing sugars, which results in unacceptable darkening of chips and french fries during processing. To produce acceptable chip fry color, reducing sugars must not be allowed to accumulate during the curing and storage periods (Fig. 10.11).

The final holding temperature for processing potatoes is determined by the cultivar and the intended use. The region of production can have a significant effect on storability, because of different environmental and agronomic conditions in different production areas. Therefore, for a particular cultivar, the ideal storage temperature for preservation of processing quality may vary somewhat among geographic areas. Growers are encouraged to consult local crop advisors to determine cultivar-specific holding temperatures for locally grown crops. General guidelines for storage temperatures follow.

Tubers stored for processing into french fries generally should be cooled to no lower than 44°F, although individual processors may have specific requirements (usually between 44 and 48°F), depending on the cultivar and the estimated storage duration. The tolerance for sugar-induced color formation is lower in tubers processed into chips than in tubers processed into french fries. Hence, potatoes for chip processing must be stored at a slightly higher temperature, to avoid the accumulation of reducing sugars. Generally, healthy chip potatoes harvested at soil temperatures of 55–70°F may be stored at 48–50°F with minimal concern about reducing sugars. Chipping potatoes harvested during cooler weather may require long-term holding at 50–55°F to ensure acceptable fry color. Chipping cultivars have been developed that recondition well after storage at lower temperatures. The prevailing temperature and rainfall during production and soil temperature at harvest may affect the chipping performance of potatoes removed from storage.

Carbon dioxide (CO_2) buildup in storage can adversely affect tuber quality. The source of CO_2 is tuber respiration, which increases with temperature and varies with the physiological state of the tubers. Immature tubers, wounded tubers, and tubers that have been stressed (e.g., by disease, cold, or heat) respire at a higher rate than mature, suberized, unstressed tubers. Carbon dioxide production by tubers is highest during the curing period and lowest during the holding period when tubers are dormant; it increases as tubers emerge from dormancy and begin to sprout. More frequent ventilation with outside air during periods of high respiration rate is needed to prevent CO_2 buildup in storage. Clogged air ducts and excess soil can restrict ventilation through the bulk pile, so that CO_2 can accumulate in localized pockets, leading to a loss of tuber quality.

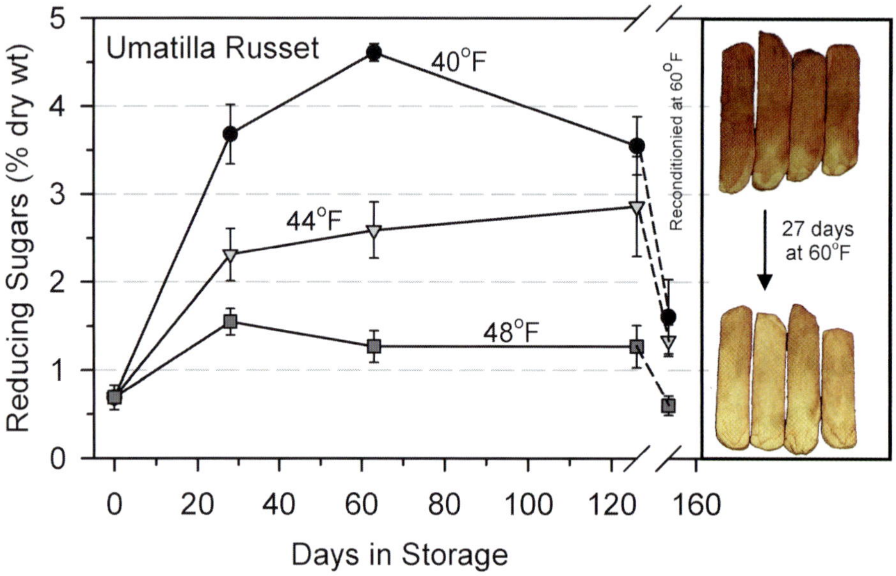

Fig. 10.11. Effect of storage temperature on the concentration of reducing sugars (glucose and fructose) in Umatilla Russet tubers. Low-temperature sweetening results in unacceptable darkening of fries and chips during processing. The reducing sugar content falls when the storage temperature is raised to 60°F to recondition the tubers (from day 126 to day 153, dashed line), restoring processing quality. (Adapted from N. R. Knowles and L. O. Knowles, 2002, Storage management and processing quality of Umatilla Russet potatoes, pages 103–110 in: Proceedings of 41st Annual Washington State Potato Conference, Washington State Potato Commission, Moses Lake)

Tuber respiration results in CO_2 levels of 0.1 to 4–6% in commercial storages. Carbon dioxide levels as high as 4% have been shown to promote soft rot and stimulate the growth of *Fusarium* and other fungal pathogens. Tuber glucose and fructose content may also increase if CO_2 is maintained at 3–4%; however, CO_2-induced sweetening is cultivar-dependent and reversible. Sugar levels usually fall when CO_2 levels return to normal. In addition to causing sweetening and favoring decay, high CO_2 can inhibit suberization, promote physiological disorders (such as blackheart), and stimulate sprouting.

To prevent CO_2-induced loss of quality, ventilation should be managed to maintain CO_2 concentration below about 0.5% (5,000 ppm). This can be particularly challenging in circumstances where respiratory heat produced by the tubers is used to maintain storage temperature (e.g., in regions with cold winters) or where refrigeration is required to maintain temperature. Both situations necessitate reduced ventilation with outside air, which leads to CO_2 buildup. State-of-the-art storages incorporate CO_2 monitors to help determine the frequency of ventilation with outside air, thus enhancing the ability to provide an optimum storage atmosphere for maintaining tuber quality.

The Holding Period: Dormancy

After the appropriate long-term storage temperature has been reached, the frequency of **ventilation** can be reduced, if desired. During the holding period, it is necessary to provide enough air movement through the pile to supply oxygen for tuber respiration, remove carbon dioxide and excess heat, and maintain a constant temperature. The temperature should be maintained within 2°F of the desired holding temperature, and the variation from the top to the bottom of the pile should be no more than 2°F, and preferably less. The ventilation systems in modern storages are capable of maintaining a variation of 0.5°F throughout the holding period. The relative humidity should be kept at 90–95% throughout the holding period, to minimize weight loss and pressure bruising.

Condensation is a problem in some storages. Excess moisture dripping from the ceiling onto the top of the pile creates wet areas, favoring tuber rots. Adequate insulation is the key to minimizing condensation (Fig. 10.9). Condensation will develop in an insufficiently insulated storage even when the outside air is only moderately cool. If necessary, ceiling condensation can be reduced by a constant flow of air between the top of the pile and ceiling, maintained by small circulating fans. Outside air warmer than the potatoes in the pile should never be introduced into the storage, as this can result in condensation directly on the surfaces of tubers.

During the holding period, most cultivars are in a dormant state, characterized by very low respiration and the production of only minimal heat and water vapor. The length of dormancy is a cultivar characteristic. Early-maturing cultivars (such as Russet Norkotah, Norland, and Superior) often have a short dormancy, whereas late-maturing cultivars (such as Russet Burbank) have a longer dormancy. The length of dormancy also depends on tuber maturity at harvest and storage temperature. Ranger Russet, Umatilla Russet, and Russet Burbank will remain dormant for approximately 60, 120, and 135 days, respectively, at 45°F, but only 50, 80, and 120 days, respectively, at 48°F. Knowledge of cultivar-dependent differences in dormancy length will facilitate effective timing of application of sprout inhibitors.

Inhibition of sprouting is required to maintain quality in tubers stored beyond their natural dormant period. The most effective chemical registered for postharvest sprout control is chlorpropham (CIPC). It is applied by licensed applicators as a thermal aerosol fog, which is evenly distributed throughout a bulk pile by the storage ventilation system, usually operating at reduced airflow. After application, the air is recirculated within the storage to facilitate even distribution and deposition of the product on the tubers. Recirculation continues until the fog has settled on the tubers, which normally occurs within 2–12 hours. The storage should then be ventilated with outside air. Longer durations of exposure with no outside ventilation (e.g., more than 24 hours) will lead to a buildup of carbon dioxide, which can induce sweetening and loss of quality of processing potatoes.

Proper timing of treatment is critical to tuber health and effective sprout control. Chlorpropham inhibits cell division. Therefore, it should not be applied during the curing period, as it will inhibit wound healing. For maximum efficacy, CIPC should be applied after curing and well before tubers emerge from dormancy and the sprouts begin to "peek." In the Pacific Northwest, the first treatment normally occurs in November or December. A well-balanced, evenly distributed airflow is necessary during application. Any tuber buds not adequately covered by the CIPC aerosol may enlarge and produce oppressed, rosette-type sprout clusters, which may grow inward to produce ingrown or internal sprouts, especially if tubers have been subjected to desiccation resulting in increased pressure bruising. Since thorough distribution of the material is essential, an inspection of the ventilation and duct system (Box 10.2) by the custom applicator before potatoes are placed in the storage is advisable. Measures taken to ensure uniform airflow through the pile before the storage is filled will prevent serious problems later. This sprout inhibitor should never be applied to seed potatoes, and bins treated with CIPC should never be used the following spring to store incoming seed potatoes (Chapter 7).

Other registered sprout inhibitors include the substituted naphthalenes 1,4-dimethylnaphthalene and diisopropylnaphthalene. The mode of action of these chemicals is not entirely understood, but it is likely that they delay sprouting through hormonal action. These compounds are more aptly referred to as sprout suppressants rather than inhibitors, as their effects are temporary. Multiple applications are required to suppress sprouting through a normal storage season. Diisopropylnaphthalene acts synergistically with chlorpropham, allowing chlorpropham to be used at lower rates while maintaining the degree of sprout inhibition achieved by applying it once at a higher rate.

The Holding Period: Dormancy Broken

Stored tubers begin to break dormancy during the holding period. The time when this begins is determined by the dormancy characteristics of the cultivar, tuber maturity at harvest, the storage temperature, and whether sprout inhibitors have been applied. As tubers break dormancy, the respiration rate increases, and internal changes lead to rapid cell division in growing points, resulting in sprouting. The rise in metabolic activity results in increased oxygen demand, generation of heat, and emission of carbon dioxide and water vapor.

Proper storage management becomes critical after dormancy is broken, because tuber health may easily be jeopardized. Maintaining a uniform low temperature throughout the storage facility can significantly extend dormancy in most cultivars. Refrigerated storages are most valuable at this time. For potatoes not treated with a sprout inhibitor, especially seed potatoes, careful management is critical to **avoid sprouting** when bins are frequently opened to market tubers. Premature sprouting enhances desiccation and weight loss and accelerates the aging of seed tubers, potentially resulting in significant loss of vigor in the crop grown from them.

Alternative sprout inhibitors that physically "burn" developing sprouts can be used to limit further sprout growth once dormancy is broken. These suppressants include essential oils such as peppermint, spearmint, and clove. Sprouts are most susceptible to these compounds at the peeping to ⅛-inch-long sprouting stage. Forced evaporation or wick applications are more effective than thermal or cold aerosol fogs of the mint oils. The efficacy of these volatile compounds is somewhat dependent on maintaining a sufficiently high concentration of them in the atmosphere surrounding the tubers, which generally requires frequent reapplication (every 2–3 weeks). Clove oil (eugenol), marketed under the trade name Biox-C, is best applied as a thermal aerosol fog. As plant derivatives, these compounds are approved by the National Organic Program of the U.S. Department of Agriculture for the control of sprouting of organically grown potato crops.

Throughout the holding period, the stored tubers should be monitored closely for changes in temperature or humidity causing condensation of moisture on rafters, floor wetness, wet spots or sunken areas in the pile (Fig. 10.12), objectionable odors, and the presence of fruit flies. All of these are signs of **tuber decay** within the pile. Most modern storages feature adjustable airflow dampers at duct entrances to allow extra ventilation air to be delivered to trouble spots in the storage bin (Box 10.2 and Fig. 10.9). Generally, once a pocket of breakdown occurs in a bin, the prospects for continuing long-term storage are greatly reduced. If decay is detected, the storage humidity should be lowered and the potatoes graded and marketed promptly.

Depending on the cultivar, the duration of storage, and the storage temperature, quality may be compromised by a buildup in **reducing sugars**, which cause an unacceptable darkening of potato chips and french fries during processing. The accumulation of reducing sugars due to advanced tuber age, referred to as senescent sweetening, is irreversible. Cultivars differ in the time of onset of senescent sweetening. For example, under optimal storage conditions (48°F and 95% relative humidity) for processing, Umatilla Russet can begin to sweeten irreversibly 200 days after harvest, while Russet Burbank generally will not undergo senescent sweetening for 300 days or more. The expected duration of storage should thus be matched to the cultivar to avoid loss of processing quality.

Prior to senescent sweetening, tubers that have accumulated reducing sugars as a result of exposure to low temperatures during the storage cycle can be reconditioned to lower the sugar concentration before processing (Fig. 10.11). **Reconditioning** involves raising tuber pulp temperatures at the end of the storage season, in order to lower the reducing sugar content. Storage temperatures of 57–65°F for 3–4 weeks at the end of the storage period stimulate the loss of reducing sugars by increasing respiration and partial conversion of sugars back to starch. The decrease in the reducing sugar content improves the processing quality of the tubers (Fig. 10.11).

The temperature and condition of the pile must be monitored closely during reconditioning. Raising the pile temperature at the end of a prolonged storage season will stimulate sprouting, respiration, and thus weight loss, and it

Fig. 10.12. During the holding period, stored tubers should be monitored closely for wet spots, sunken areas, and objectionable odors, which indicate the presence of disease or decay within the pile.

favors the development of disease and decay. Reconditioning is risky and should be used only as a last-ditch effort to improve processing quality after careful consideration of the risks and potential benefits. Moreover, the extent to which processing quality can be improved through reconditioning is cultivar-dependent. Storage managers are encouraged to seek advice from local extension specialists or consultants in determining the potential for improvement through reconditioning.

The preferred method for maintaining processing quality is to identify the market for which tubers are being stored at the beginning of the storage season and impose temperature and relative humidity regimes that are optimal for the cultivar and the end use. If tuber health can be maintained during storage, then reconditioning, with all its inherent risks, will not be necessary.

Maintaining Tuber Health During Post-Storage Handling

To minimize mechanical damage to tubers while they are being removed from storage for marketing, it is essential to raise the **pulp temperature** gradually to 50–55°F over a period of several weeks. This can be accomplished by restricting ventilation and allowing the heat of respiration to increase the temperature of the pile. Handling cold tubers usually results in a high incidence of shatter bruise or internal blackspot. Rough handling will stress tubers and may cause the accumulation of a significant quantity of reducing sugars in chipping potatoes, resulting in unacceptable fry color. Fresh-market potatoes may develop internal blackspot, after-cooking darkening, or other surface defects that lead to excessive waste during peeling. French fry processors often reject roughly handled loads or impose heavy penalties for accepting them.

Once desired tuber pulp temperatures have been reached and removal from storage begins, **bruise control** is a major concern. The same principles of low-bruise handling applied when loading the storage must be applied when unloading. Careful operation of bulk unloading equipment is essential. Dumping heights in hoppers and trucks should be 6 inches or less. If a bin loader is used to move potatoes into a truck, load through the back to minimize the drop as loading is begun, and then continually adjust the loader as loading progresses. As in filling the storage, bruising is minimized by maintaining the outflow of potatoes at capacity on all conveyors and by padding sharp or rough surfaces on all equipment.

Prior to use, all equipment should be cleaned and sanitized to prevent the spread of bacterial pathogens. If a **fluming system** is used to transport tubers in water, it must be cleaned and sanitized frequently, and the flume water must be changed often. During fluming, tubers should not be submerged more than an inch or two, because the water pressure may force bacteria into lenticels. Clean water should be used for washing fresh-market potatoes on the packing line, and the wash water should not be recirculated. It is very important that washed tubers be well dried before packaging, as a film of moisture on the surface of tubers favors the development of bacterial soft rot in the bag, further down in the marketing chain. Ventilated bags with holes or perforations help to ensure complete drying.

Removing **seed tubers** from storage requires intense precaution to maintain seed quality. If off-farm trucks are used to transport bulk or bagged seed, they must be thoroughly cleaned and properly sanitized before loading.

After packaging, bagged seed or fresh-market potatoes are damaged mainly at the time of loading and unloading. **Bagged potatoes** are likely to suffer severe shatter bruising when dropped onto a hard surface from a height greater than 30 inches. When potatoes are being shipped, the outside temperature should be watched carefully. If it is below freezing, precautions must be taken to prevent frost damage during transit.

Proper **disposal of waste potatoes** is an important management practice to ensure the health of future potato crops. A plan for proper disposal must be in place before cull potatoes are generated in the packing-out operation. Cull potatoes should never be dumped outside in piles where some tubers will survive freezing temperatures. Cull piles are notorious sources of spores of the late blight pathogen, which form on sprouts from surviving tubers. They also serve as reservoirs of some viruses. Acceptable waste disposal methods include using culls as livestock feed, composting them, and spreading them in the winter to freeze on land not intended for potato production the following season. An expensive and less desirable method is to bury cull potatoes at least 6 feet deep. Contamination of groundwater may result from this method.

Mechanical damage continues to be a major factor in potato health management, not only because of the direct results of injury, but also because of the introduction of disease organisms. Healthy seed tubers are critical to the potato industry—they are the vital link from one season to the next. Repeat orders at the processing plant and in the fresh-market supply chain depend on tuber health. The potato grower, along with employees, field advisors, and creditors, must maintain vigilance to ensure that a tuber health management program is followed through harvest, storage, and marketing. Continued profitability in potato farming depends on the ability of the potato health manager to accomplish this objective.

CHAPTER 11

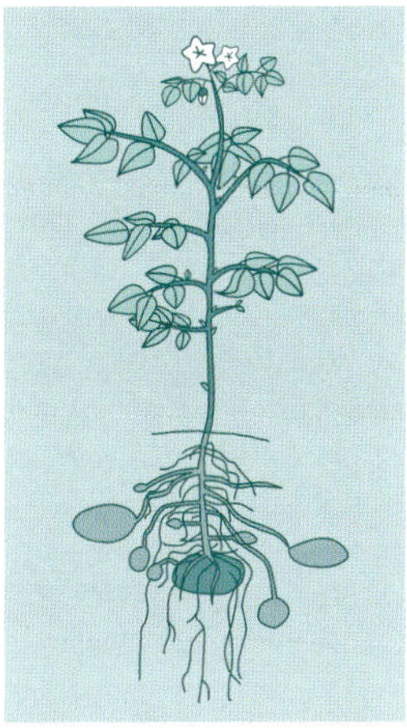

Bryan G. Hopkins
Department of Plant and Animal Sciences
Brigham Young University, Provo, Utah

Ronda E. Hirnyck
Department of Plant, Soil and Entomological Sciences
University of Idaho, Boise

Organic Potato Production

Organic agriculture is described by the U.S. National Organic Standards Board as "an ecological production management system that promotes and enhances biodiversity, biological cycles and soil biological activity. It is based on minimal use of off-farm inputs and on management practices that restore, maintain and enhance ecological harmony. 'Organic' is a labeling term that denotes products produced under the authority of the Organic Foods Production Act. The principal guidelines for organic production are to use materials and practices that enhance the ecological balance of natural systems and that integrate the parts of the farming system into an ecological whole." In practice, a crop is certified as organic if a certifying agency approves the field in which it is grown and the inputs used to produce it.

The word *organic* has more than one definition, which can cause confusion. Chemicals are classified as organic (containing carbon) or inorganic (not containing carbon). However, many organic chemicals, by this classification, such as urea fertilizer and most pesticides, are not permitted in organic production of crops. The distinction is important, to avoid application of an organic chemical that is not approved for organic farming by the state or regional certifying agency.

Organic production should generally implement many of the same management practices recommended for conventional production to enable plants to be healthy and maximize yield. However, some practices are more important for organic growers than conventional growers, because there are fewer options for preventing or correcting problems in organic production. General guidelines for organic production of potatoes are presented in this chapter. However, organic producers are required to consult the appropriate organizations and certifying agencies for standards, guidelines, and approved inputs in both the production area and the market destination. The U.S. Department of Agriculture, through its National Organic Program, provides certifying standards for organic crops produced and sold in the United States, with interpretation and product approval conducted by certification agencies (generally, state departments of agriculture). However, other government, consumer, and grower organizations may have more or less stringent standards, which should be observed when applicable.

Field Selection and Crop Rotation

Selecting a field with a history and likelihood of minimal pest and disease pressure is extremely important for organic production, since most pesticides available for conventional production cannot be used in organic production. A vital consideration in selecting a field for organic production is the **number of years since the last potato crop**. Many of the insect pests, weeds, and soilborne pathogens affecting potato increase in proportion to the frequency of potato crops in a field (Chapters 15 and 19–22). A minimum of 4 years between potato crops in a field is recommended for organic production, but 7 years or more is better. If a nonapproved pesticide has been applied to a field, a specified time interval must elapse (3 years is required in the United States), following the application, before the field can meet organic certification standards.

Certain **rotational crops** serve as hosts for pests and pathogens that attack potato. It may be necessary to exclude such crops from rotations for organic potato production. For example, dry bean, sunflower, and canola are susceptible to white mold, and growing these crops in rotation with potato can increase the severity of white mold in the potato crop. Pasture grasses serve as hosts of wireworms and should be avoided altogether as a rotational crop if the field has a history of wireworms. Clover, a common source of nitrogen in organic systems, is also a host of wireworms.

Green manure crops incorporated into the rotation reduce soil erosion, add carbon to the soil, and capture and recycle residual nutrients. Biofumigants released from certain green manure crops can be useful in suppressing some weeds, insect pests, and pathogens, including nematodes (Chapters 3 and 20). These natural chemical biofumigants are released from the decaying crop if it is incorporated into the soil while still green. Green manure crops can be planted during the summer, fall, or spring immediately prior to potato planting. The biofumigation effect will not be realized if the crop does not produce several tons of biomass per acre. The crop needs about 8 weeks of sustained growth with adequate water and nutrients, particularly nitrogen. Weeds in green manure can also serve as hosts of certain pests and pathogens, potentially diminishing the biofumigation effect (Chapter 5).

Field isolation is important when selecting a field for organic production. Certain insects and disease-causing agents (such as the Colorado potato beetle, *Potato leafroll virus,* and the late blight pathogen) can overwinter on potato and then infect neighboring fields. Organic potato fields should be separated by at least 400 yards from other potato fields and any site where volunteer potatoes are growing. A separation of this distance will restrict the introduction of some pests and pathogens. However, winds can disperse airborne pathogens many miles, and extreme isolation from other potato fields is often a benefit. Field isolation not only helps protect the organic crop but also reduces the risk that pests and pathogens may spread from "untreated" organic fields to conventional fields in the vicinity. Field isolation also helps prevent application of pesticides to the wrong field by aerial applicators, resulting in the loss of organic certification. All refuse potato tubers and cull piles within several miles of organic and conventional fields should be destroyed.

Fields with superior soil and water resources (Chapter 5) should be selected. Potatoes have a relatively high nutrient requirement, and unfortunately they also have a shallow rooting system. Many fertilizer materials used in organic production are slow to release their nutrients, and nutrient deficiencies can develop if the soil fertility levels are low. Soil should be tested for nutrients (Chapter 8). **Avoid soils with nutrient deficiencies** or toxicities and problems with salts, sodium, and pH. Similarly, irrigation water should be tested for salt, sodium, calcium, magnesium, potassium, nitrate, chloride, sulfate, bicarbonate, and boron. Wastewater, such as livestock lagoon water, is often used for irrigation and fertilization and should be tested for possible toxins, such as copper, which is commonly found in animal footbaths.

The **climate** is an important factor in organic potato production. There is an obvious advantage for organic producers if certain pests and pathogens, such as tuber moth, mites, and the late blight pathogen, are not present or are rarely a problem in an area used for organic production. Many pathogens require moisture to complete the infection process, and growing potatoes in a semiarid climate can be an advantage for organic production. The length of the growing season is another factor. In a long growing season, pests have more time to get a foothold and damage plants and tubers. Warm, long-season climates with mild winters may also favor the survival of some pests and pathogens between potato crops. A short season also results in a lower yield, but the premium paid for organic potato and rotational crops may be sufficient to offset the financial effect of a reduced yield.

Soil and Nutrient Management

The condition of the soil is extremely important in growing a healthy crop (Chapters 3 and 5). Potatoes grown in healthy soil are better able to compete with weeds, insect pests, and pathogens. One of the most difficult aspects of organic production is nutrient management; however, most nutrients can be supplied by commercially available organic fertilizers with guaranteed analysis, composted plant and animal wastes, and legumes.

Fertilizers, especially organic materials, applied at high rates cannot immediately compensate for poor soil fertility. Many **organic fertilizers release nutrients slowly**, at rates that are difficult to predict. If the fertility of a soil has been built up over a period of years, the soil will provide nutrients in relatively high concentrations along the entire length of the root system over the course of a season. For most nutrients, soil test levels in organic fields should be in the upper end of the range of values considered sufficient for potato production. However, some nutrients (notably nitrogen, phosphorus, boron, and copper) can accumulate to excessive levels, with potentially negative impacts on soil and water quality.

Fertilizers applied at high rates cannot immediately compensate for poor soil fertility.

Nitrogen is the mineral nutrient of greatest concentration in plants, followed closely by potassium. Phosphorus, chloride, sulfur, calcium, and magnesium are present in much smaller quantities, and the remaining micronutrients (zinc, iron, manganese, copper, boron, and molybdenum) are present in very minute amounts in plants (Chapter 8). Organic producers generally use manure, compost, some animal byproducts (blood meal and fish meal), some mined minerals (sodium nitrate, rock phosphate, potassium chloride, sulfur, calcium carbonate, magnesium carbonate, soluble boron, and magnesium, zinc, iron, manganese, and copper sulfates), and various other materials approved for organic production.

Organic standards prohibit the use of raw waste products, such as manure, within either 90 or 120 days (depending on regulations in the growing area) of potato harvest. However, thoroughly composted waste materials can be exempt if approval is received from the certifying agency. **Composting** is achieved by adding ample, but not excessive, water and periodically mixing the material over several months of warm weather. The compost is complete when the internal temperature of the pile subsides and the odor is less pungent and more like that of soil. However, compost is not certified organic if certification procedures are not followed.

Organic fertilizers, especially composted animal waste and plant residues, are low in nitrogen (N), compared to the other nutrients. This imbalance can result in environmental contamination after long-term use. Initially, waste material used as fertilizer should be analyzed to determine its nutrient concentrations and applied on the basis of the N requirement of the crop. The application of enough compost to meet the N requirement is likely to supply certain other nutrients in excess of the amounts required by the crop (as verified by laboratory analysis) and thus is a good method of building soil fertility. However, this practice will need to be altered once the phosphorus (P) content of the soil reaches an environmentally sensitive threshold established by national or local regulatory agencies. In switching from application of an organic fertilizer based on the N requirement of the crop to application based on the P requirement, the overall rate of application is greatly reduced. Some of the N needed by the crop will be supplied at this reduced rate, but the remainder of the N requirement will have to be met by a fertilizer with a high concentration of N and a low concentration of P. For example, if an organic fertilizer such as manure is applied at a rate based on N need, the amount applied will be fairly large. It would require 20 tons of manure per acre, containing 10 pounds of N per ton, to supply the N for a crop requiring 200 pounds of N per acre. However, continued use of this material would result in a buildup of other nutrients, especially P. Eventually the soil P level will reach an excessively high level, and runoff of water and soil from the field will be potentially hazardous to surface waters. Therefore, the application rate eventually has to be reduced to meet the P removal rate. Typically this reduces the application of organic fertilizer to 20–30% of the amount applied on the basis of the N requirement only. In the example given above, only 4–6 tons per acre could be applied, instead of 20 tons. The remainder of the N requirement would have to be supplied by some other fertilizer low in phosphorus (few organically approved materials fit this description, and cost is often high) or with legumes.

Legumes, which include crops such as bean, pea, soybean, clover, and alfalfa, convert atmospheric nitrogen into a form available to plants. Most legumes will supply 40–60 pounds of N per acre to the following crop (about 25% of the requirement of a potato crop). Nearly 100% of a potato crop's need for N will be met following a dense, healthy alfalfa crop. Nitrogen is also supplied to a crop in the second year after alfalfa, usually in amounts of 30–50 pounds per acre. The nitrogen credit is proportionally reduced as the alfalfa stand thins and weeds become more dominant. Alfalfa should be plowed down early in the fall prior to potato production, to maximize potato yields and to speed the rate of decomposition of the crowns, which can be problematic for potato planting and harvesting.

Many potato cultivars benefit from being "spoon-fed" nitrogen throughout the growing season, with 50–75% being provided after tuber initiation. The other nutrients may also be supplied in this way, although tuber yield and quality are not reduced if the entire required amount of these nutrients is supplied to the plants early in the season. Organic fertilizer materials generally decompose slowly and release nutrients throughout the growing season, often coinciding with plant need. Petiole and soil samples should be taken every 7–10 days to determine the nutrient status of the soil-plant system. Nutrient concentrations in the plant should not drop below deficiency levels (Table 8.1).

Excessive amounts of some nutrients, especially nitrogen, are detrimental. Fertilizers of guaranteed analysis can be added to meet the needs of the plant without providing excess amounts of nutrients. However, it is more difficult to meet the nutritional needs of the plant by applying composted organic fertilizers, because of the variability of nutrient concentrations in the compost pile, difficulty in achieving uniform application of the material, and difficulty in predicting the release rates. Composted materials should be well mixed and tested by a reliable lab. Typically, 30–50% of the organic nitrogen and sulfur from manure will become available to plants during the first year after application, but only about 10% from completely composted manure will become available. Other dry organic materials would most likely have similar rates. About 60–80% of the phosphorus and micronutrients and 90–95% of the potassium in these materials will become available to plants during the first year after application, regardless of stage of decomposition. Incorporation of these materials into the soil several months prior to planting will result in earlier release of nutrients and reduce the likelihood of excessive nitrogen release during the late season, when it can be detrimental to tuber bulking. The application of organic materials with high carbon contents, such as straw, wood shavings, and other fibrous materials, in late fall or in spring before planting should be avoided. A carbon-induced nitrogen deficiency is more likely to occur as the interval between application and plant growth shortens. It is better to work straw into the soil in late summer or early fall than in late fall or early spring prior to planting.

Building soil fertility to high levels will meet plant needs for most nutrients throughout the season. Deficiencies identified from in-season soil and petiole analysis can be corrected if an organically approved material is applied in a form that is immediately available to the plant. **In-season nitrogen monitoring** is common and effective in potato production, but this tool is difficult to use in an organic system, because of difficulty in predicting the release rate of many organic fertilizers. Nevertheless, in-season sampling should be conducted, and nutrients should be applied accordingly, but the rate of application of nitrogen should be reduced (generally to 25–75% of the rate used in a nonorganic crop) to compensate for the anticipated release of nitrogen from slowly decomposing organic materials applied earlier. Generally, slowly decomposing organic fertilizers will not supply nutrients soon enough to correct in-season deficiencies. However, commercial nutrient sources that are both organically approved and immediately available to plants can be used for this purpose.

The **physical properties of the soil** are just as important as its chemical properties. Management of physical properties of the soil in an organic system is largely the same as in a conventional growing system. Adding large quantities of organic matter to the soil generally results in improved physical properties, but heavy equipment used in application can cause soil compaction (Chapter 5).

Biological properties of the soil are also important (Chapter 3). A healthy soil has active populations of microbes, earthworms, and other organisms. These soil organisms have requirements similar to those of plants, including adequate temperature, nutrients, water, oxygen, and pH. In general, the best situation is to stimulate a diverse and substantial community of soil organisms by planting an appropriate crop sequence and adding organic matter to the soil, so that organisms that are pests and pathogens of potatoes have more competition from beneficial soil microbes. Organic production has an advantage in that broad-spectrum pesticides that kill both pests and beneficial organisms are not applied to soil.

Managing Weeds

One of the best defenses against weeds in potato fields is vigorously growing potato plants. It is important to establish **canopy cover** over the ground as quickly as possible, to provide shading, which will delay weed germination and slow weed growth. Rapid development of canopy cover is facilitated by ample nutrients and water, friable and oxygenated soil, healthy and adequately sized seed, uniform seed spacing, proper planting depth, and minimal damage from cultivation,

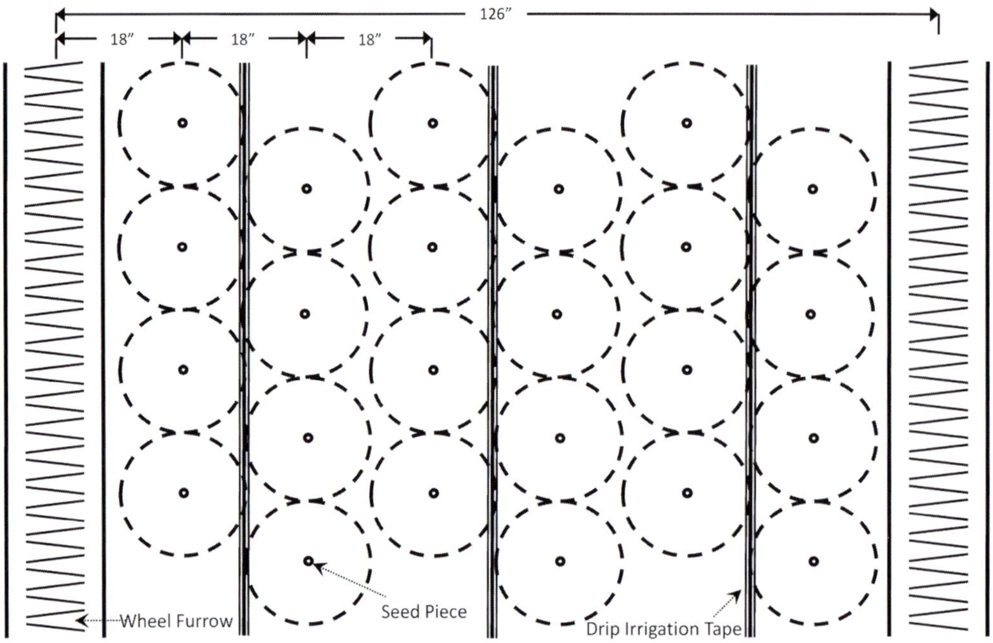

Fig. 11.1. Potato planting bed with 18-inch alternating grid spacing. The circles represent canopies of individual potato plants. The bed receives drip irrigation, with each line of drip tape irrigating two rows of potatoes. This planting system results in rapid canopy closure and efficient utilization of sunlight, soil, nutrients, and water.

pathogens, and pests. Warming seed to the point of sprout initiation and planting in relatively warm soil also combat weed pressure by favoring rapid canopy establishment.

Potatoes can be planted in **flat beds**, rather than in the traditional hills, with the in-row and across-row spacing adjusted so that plants are equidistant in an alternating grid pattern, to facilitate more uniform spacing and rapid canopy cover (Fig. 11.1). Bed planting will result in canopy closure 10–20 days earlier than traditional hill planting. It also more effectively utilizes sunlight and soil resources. Initially, bed planting is challenging, as it requires major changes in planting, cultivating, and in some cases harvest equipment. Plant spacing and depth of planting are as important in bed planting as in conventional hill planting. The optimum spacing of the cultivar Russet Burbank is 18–20 inches between plants in an alternating grid pattern. The distance between plants can be decreased to reduce tuber size, for the production of some red potatoes and other specialty cultivars that sometimes require smaller tubers.

It is important to establish canopy cover over the ground as quickly as possible, to provide shading, which will delay weed germination and slow weed growth.

Traditional, cost-effective herbicides are not available for organic production, and either cultivation or weed matting is needed to manage weeds. **Cultivation** is most effective when done immediately prior to potato emergence. Small plants (4–6 inches) can be cultivated without much damage, but late cultivation will damage potato roots, slow canopy development, and reduce yields (Chapter 22). Yield loss from minor to moderate weed competition is less than that from late cultivation.

Flaming, mulching, and roguing are other organic methods of weed management. Early-season weeds can be destroyed by **flaming** immediately prior to or shortly after potato emergence. However, high energy prices have nearly eliminated flaming as a weed control option. Minor damage to emerging potato plants from flaming will not negatively impact tuber production. The use of plastic or plant fiber sheeting can minimize weed pressure, but this option is very labor-intensive and expensive. The most common method of organic weed control is **roguing** (hand removal).

A few organically approved herbicides, such as acetic acid, are available for use on potatoes. However, all of these herbicides are burn-down products which will also damage potatoes and should be used only on very young potato plants. Some annual weeds (e.g., foxtail, pigweed, lambsquarters, and velvetleaf) can be suppressed, and the perennial Canada thistle may be temporarily burned down with acetic acid. Suppression of thistle is inadequate with 5–7% acetic acid solutions (household vinegar is 5%), but suppression is usually improved with 10–20% solutions, which can create a severe hazard to the applicator. Application to potato plants immediately after emergence may result in burn-down, but plants will generally regrow without a negative impact on tubers.

Weed competition is one of the more serious problems in organic potato production, and rigorous weed management efforts are needed year-round in potato fields, field borders, and neighboring fields. **Weed management in the years prior to potato production** is important in reducing the weed seed bank. The techniques used for weed management in potatoes can generally be used for rotational crops, although cultivation is not possible in many perennials and annuals that are seeded close to one another. Frequent cutting of perennial crops, such as alfalfa, can effectively suppress some weeds. Green manure crops such as mustards or winter peas planted in late summer through early spring and incorporated before potato planting can be an effective weed management strategy in some potato-growing areas. It is also important to minimize the number and type of weed seeds entering the field. Planting certified weed-free seed of rotational crops and minimizing the amount of soil on seed potatoes will help reduce the weed seed bank and minimize the risk of introduction of weed species not already present. In addition, filters should be installed on surface irrigation water to remove weed seeds.

Managing Insects and Rodents

The length of time between potato crops is very important in suppressing populations of soil-dwelling pests. Certain other crops or weeds serving as hosts in previous seasons also favor soil insect populations. Potatoes should not be planted in fields previously cropped with pasture grass having a history of **wireworm** within the past 6 years. Clover, cereal grains, corn, and grass weeds are also wireworm hosts and should be avoided in fields where wireworm is a known problem. Alfalfa, which is not a wireworm host, can be planted instead. During cooler fall and spring weather, tillage that results in drying of the soil may help to reduce wireworm populations by desiccation; it also exposes wireworms to attack by birds, rodents, and other predators. Fallowing a field with tillage when soils are cool and pupae are active close to the soil surface is another tactic for reducing wireworm populations (Chapter 15).

The **Colorado potato beetle** is a serious insect pest in organic production. Favored hosts of the beetle (e.g., nightshade, tomato, or volunteer potato) should not be grown in an organic potato field for 1–2 years before production. In areas with high rainfall, the beetle can overwinter in lush native vegetation bordering potato fields, and overwintering populations of beetles can walk to nearby potato plants. Field isolation of at least 400 yards aids in managing beetles from these sources. This proximity recommendation is generally not practical for fields with borders of lush native plants, but beetles in these areas can be managed by trenching, as outlined in Chapter 15. Beetles walking toward a potato field can be trapped in a trench from which they cannot escape if the sides are steep.

Spinosad is a selective insecticide that provides modest control of the Colorado potato beetle while preserving some beneficial insects. Several strains of *Bacillus thuringiensis* (Bt) suppress beetle populations when applied to small larvae. Azadirachtin is also labeled for beetles. Alternatively, a trap crop can be planted near the edge of a potato field, and the Colorado potato beetle and other insects can be vacuumed, flamed, or sprayed to minimize the number moving into the field. Conventional pesticides can be used on trap crops if the distance from the organic field is sufficient to meet organic certification standards, although this is arguably out of harmony with the spirit of organic production. Another option for Colorado potato beetle control is "beetle banks," which are strips of native vegetation near or in potato fields that provide a breeding ground for beetle species that prey on the Colorado potato beetle. Predatory insects, such as the carabid beetle, can be obtained from commercial insectaries and released in potato fields to augment native populations of predators. Other natural predators include lady beetles, green lacewings, pirate bugs, predatory mites, spiders, and ground beetles.

The **green peach aphid**, a vector of *Potato leafroll virus*, is transported with bedding plants and other nursery stock. Isolation from these and other hosts, such as peach trees, can help reduce aphid infestation. Nightshade and volunteer potatoes are favored hosts of the green peach aphid, and management of these weeds will help reduce viral infections (Chapter 16). Azadirachtin is labeled for aphid management. Stylet-Oil, a mineral oil product, is labeled for management of *Potato virus Y* (PVY) in organically grown potatoes; it interferes with the transmission of the virus by aphids (Chapter 17).

Damage due to the **potato tuber moth** can be minimized by planting at a proper depth and by hilling and providing proper irrigation to avoid soil cracking (Chapter 15). Rolling the soil between senescence and vine kill can seal cracks in the soil. A biological insecticide containing a baculovirus is approved for tuber moth control in some regions; it can be sprayed on seed tubers in storage and on vegetation during the growing season.

The **potato leafhopper** is a serious pest in certain areas and must be suppressed or the crop will be destroyed. Many of the control measures listed for the Colorado potato beetle will also suppress leafhoppers. If leafhopper populations become excessive, pyrethrin products and PyGanic 50 EC are labeled for control. Some forms of Diatect (a combination of pyrethrum and diatomaceous earth) are labeled for use in organic production and have some suppressive effects on leafhoppers. However, these products are broad-spectrum pesticides and will reduce populations of beneficial insects as well as pests.

There are few choices of organically approved **insecticides**, which are typically not as toxic and not as broad in spectrum as those used in nonorganic production. Spinosad and *Bacillus thuringiensis* (Bt) are certified organic and labeled for use on potatoes to suppress insect populations, although certain strains of Bt are not organically approved. Other organically approved insecticides include azadirachtin, pyrethrum, rotenone, cryolite, elemental sulfur, lime sulfur, natural oils, and insecticidal soaps. These may also help suppress insects, but less information is available on their effectiveness in potato production. As pesticides, they must be registered for use on potatoes in the production site. It is important that insect management measures minimize the suppression of beneficial biological populations. Therefore, avoiding excessive application of insecticides is critical to preserve natural enemies of potato pests important for biological control. The objective is to treat only when economic loss is likely (Chapters 1 and 4). Potato plants can tolerate a certain amount of injury before net returns are impacted. Typically, defoliation from insects doesn't negatively impact tuber yield and quality unless the canopy is reduced to less than full cover. However, there is a low tolerance for insects that damage tubers and those that transmit viruses (Chapters 15 and 16). *Selected trade names of insecticides are given in Table 14.4.*

Rodent management can be achieved through the use of sulfur dioxide or vitamin D_3 pellets. Some organic growers have been known to house, water, and partially feed cats in an effort to minimize vole damage. Applying garlic through drip lines has been shown to repel rodents.

Managing Diseases

Managing disease in organically grown potatoes relies heavily on sanitation practices, to reduce initial inoculum levels before planting, and cultural practices, to reduce the rate of disease development during the growing season and in storage. **Sanitation practices** include biofumigation, soil solarization, effective crop rotation, field selection, eliminating cross-contamination, and the use of certified seed tubers. Practices that **reduce disease rates** include planting disease-resistant cultivars (partial resistance), establishing plant spacing and row orientation to reduce the duration of leaf wetness, irrigation management, and chemical control. Tuber rots are reduced by harvesting mature tubers with good skin set (Chapter 10) and by proper regulation of temperature, humidity, and airflow in storage. A biological fungicide containing the bacterium *Pseudomonas syringae* (Bio-Save) has been reported to reduce Fusarium dry rot and silver scurf when applied to tubers before they are placed in storage.

Biofumigation, in which substantial amounts of fresh green manure are incorporated into the soil, can effectively suppress nematode populations (Chapter 20). Soil solarization will reduce levels of nematodes and other soilborne pathogens, including *Verticillium dahliae* (Chapter 20). Allowing a long period of time between potato crops can also reduce some soilborne pathogens; however, the crops grown in rotation should be nonhosts, and weed hosts must be controlled. Fields with moderate to high levels of parasitic nematodes or with a history of diseases caused by soilborne pathogens, such as Verticillium wilt (Chapter 19), should be avoided for organic production.

Avoid cross-contaminating fields with infested equipment. Steam-wash and sanitize all equipment entering the field to minimize the transference of pests and pathogens from other fields (Chapter 7). Certain soaps, hypochlorite, peroxide, and other materials approved for organic production are available for sanitizing seed-handling and field equipment.

Select **disease-resistant cultivars**, when available. For example, the cultivar Defender is resistant to late blight and moderately resistant to early blight; tubers of Umatilla are moderately resistant to late blight, even though the foliage is susceptible; Ranger Russet is moderately resistant to Verticillium wilt; and tubers of some cultivars are resistant to scab (Chapter 6). If the crop is to be stored for several months, select a cultivar with a long dormancy period. For example, Ranger Russet will store for only about half the time of most other russet cultivars. White- and red-skinned potatoes tend to store for shorter periods than most russets.

Plant only **certified seed tubers** for organic production. Organic-grown seed is required, if available. Inspect the seed crop both in the field and in storage and examine certification records to select seed lots with low levels of disease (Chapter 6). Early generations of certified seed lots generally have lower disease levels than later generations and should be considered for organic production. Whole, uncut seed tubers should be planted, to minimize potential infection sites and the spread of pathogens. If seed tubers are cut, the cutting blades should be sanitized frequently. Seed tubers should be examined and tested to estimate the potential severity of key diseases (Box 7.1). There are three main tactics for reducing seed piece decay and establishing a uniform stand: select high-quality seed, handle it carefully, and encourage quick emergence by properly warming the seed tubers and planting them in warm soil (Chapter 7).

Control the amount, frequency, and timing of **overhead irrigation** to reduce the duration of wetness of foliage, since leaf wetness favors foliar diseases such as late blight and white mold (Chapter 9). Several foliar diseases can be reduced in crops grown with sub-, furrow, or drip irrigation rather than overhead irrigation. However, dry foliage may favor powdery mildew and mites under some conditions. Organic fields need to be thoroughly and regularly monitored for diseases and pests (Chapter 9).

Elemental sulfur and copper **fungicides** have been used for centuries as a means of disease management. Copper sulfate and copper hydroxide are used for early blight and late blight management in organic production. However, copper can accumulate in soil over time and can build up to levels that are toxic to potatoes and rotational crops and to any organisms that consume them. Because of the potential toxicity, copper fungicides have been removed from the list of products approved for organic production in the European Union. However, these fungicides remain approved for organic production in most regions, and management options for avoiding toxicity are available.

Other materials that are listed by the Organic Materials Review Institute and have potential use in disease management include hydrated lime, lime sulfur, potassium bicarbonate, hydrogen peroxide, and various natural oils. Although these materials have been approved for use, there is little information on their efficacy in potatoes. Trials for their effectiveness should be conducted prior to use. Check with certifying agencies to ensure that products are registered for control of the pathogens being targeted.

Managing the Crop During Harvest and Storage

None of the foliar desiccants or senescence-promoting chemicals approved for conventional production of potatoes are certified for organic production. Organic growers can roll the vines if the canopy is not senescing suitably and the cultivar is known to bulk more rapidly with modest senescence (most cultivars yield best with 50% senescence at the time of vine kill). Rolling the soil helps to reduce cracks, through which pathogen spores, moths, and sunlight can reach tubers.

Mechanical chopping is a management option if vine kill is necessary for timely harvest and adequate tuber skin development. Early vine desiccation may be necessary in areas where late blight is present (Chapter 21). Avoid harvesting in wet weather, and avoid harvesting areas of fields with known tuber infections and areas that are excessively wet.

Storage conditions need to be ideal for organically grown potatoes. Fewer options for disease management and sprout suppression in storage are available in organic production than in conventional production. Therefore, special care in sanitizing the storage with steam and disinfectants is particularly important in organic production. The crop should be inspected and infected tubers removed both on the harvester and on the storage piling conveyers. This practice is important for all potato growers, but organic growers should exercise increased vigilance, having fewer options for storage pathogen management.

The sprout control chemicals commonly used in conventional production of potatoes are not certified for organic production. However, the USDA National Organic Program has approved essential oils from certain herbs and spices, such as clove and various mint oils, for the suppression of sprout development. These compounds are volatile plant derivates that damage emerging sprouts. Hydrogen peroxide may also inhibit sprouting, but some formulations contain adjuvants that are not certified for organic production. Clove oil and *Bacillus subtilis* may both suppress silver scurf, which is a common storage problem. *Pseudomonas syringae* (Bio-Save) may suppress silver scurf and dry rot in storage.

Checklist for Organic Potato Production

This checklist is parallel to the checklist in Chapter 2, with emphasis on management practices that are unique to organic potato production and practices that are more important in organic production than in conventional production. Not every practice listed below is necessary, and some practices can be substituted for others.

Management in Years Prior to Growing Potatoes

Select a field with good soil and water resources, free of toxicities and deficiencies. Control volunteer potatoes and weeds closely related to potato (e.g., nightshade). Select rotational crops that are not hosts of the most serious potato pests. Avoid compacting the soil. Incorporate large quantities of crop residues. Ensure that only clean equipment enters the field.

☐ Isolate the field from peach and other *Prunus* spp., bedding plants, cull piles, and other potatoes grown in the same or a previous season.

☐ Select a field where there has not been a potato crop for at least 7 years.

☐ Select a field where there has not been a rotational crop that is a host of a pest or pathogen that is of particular concern in the growing area.

☐ Use certified weed-free seed in rotational crops. Filter surface irrigation water to remove weed seeds.

☐ Use organic methods to manage weeds year-round in the organic field, neighboring fields, and field borders to reduce the weed seed bank.

☐ Plant perennial legumes (such as alfalfa) for at least 2 years immediately prior to potato production, to reduce many annual weeds and fix nitrogen into the soil.

☐ Plant and incorporate an appropriate green manure crop.

Management Prior to Planting

Apply ample (not excessive) amounts of manure or certified organic fertilizers. Select certified potato seed. Handle seed with care to minimize damage and the spread of pathogens.

☐ Select a cultivar that is resistant to foliar, root, and tuber pathogens.

☐ Select a cultivar with a long dormancy period, if the crop is to be stored.

☐ Select limited-generation and certified seed tubers with superior results in off-season testing.

☐ Use organically grown seed when available.

☐ Sanitize seed-handling equipment at least daily and before handling each new seed lot.

☐ Plant whole seed tubers (single-drop planting), rather than cut tubers.

☐ Treat seed with a baculovirus insecticide if tuber moth is a serious concern.

☐ Apply preplant organic fertilizer on the basis of soil test results and yield potential.

Management During the Growing Season

PLANTING

Plant uniformly in moist soil (75–85% available soil moisture), at an appropriate soil temperature (optimum of 50–55°F), and at an appropriate depth.

☐ Remove diseased, small, or damaged seed.

☐ Plant a trap crop at or near the edge of the field, and periodically flame, vacuum, or spray insects (e.g., Colorado potato beetle and aphids).

GROWTH STAGES I–V

Promote rapid canopy cover. Cultivate weeds before potato plants are more than 4–6 inches tall. Irrigate to prevent the soil moisture content from dropping below 70% available soil moisture (Chapter 9).

☐ Apply sulfur dioxide or vitamin D_3 to control rodents, if necessary.

☐ Form plastic-lined trenches in which to drown Colorado potato beetles walking toward the field from lush border vegetation.

☐ Apply spinosad, *Bacillus thuringiensis* (Bt), or other certified organic insecticides.

☐ Consider the release of predatory insects obtained from commercial insectaries to augment natural populations of beneficial organisms.

☐ Consider installing insectaries, or "beetle banks," in strips of native vegetation close to or within the field.

☐ Cultivate or flame weeds prior to or just after potato emergence.

☐ Consider plastic or plant fiber sheeting as a weed barrier.

☐ Rogue diseased plants and weeds.

☐ Apply copper hydroxide, elemental sulfur, or copper sulfate to suppress some foliar diseases and insects, as determined by forecasting models or action thresholds.

☐ Spray plants with a baculovirus insecticide if tuber moth is a serious concern.

Management During Harvest

Harvest with adequate soil moisture (60–65% available soil moisture) and when tuber pulp temperatures are 50–65°F (Chapter 10). Avoid bruising and frost damage. Remove as much soil as possible from tubers while they are on the harvester.

☐ For most cultivars in which the tuber bulking rate increases with senescence, roll the vines and soil several days before vine kill.

☐ Chop vines mechanically at least 18 days prior to harvest if the cultivar and market destination benefit from this practice.

☐ Roll soil in conjunction with vine chopping if soil cracking is a problem.

☐ Remove diseased and damaged tubers on the harvester.

Management During Storage

Avoid damaging tubers during handling. Remove diseased and damaged tubers entering storage. Keep the pile height less than 16–18 feet. Cure the tubers for 14–21 days at 50–55°F with high air volume and high humidity. Then gradually reduce the temperature to an appropriate level, keeping it as low as is recommended for the cultivar and the destination market (Chapter 10).

☐ Clean and disinfect storages with steam and organically approved soaps, hypochlorite, or peroxide.

☐ Consider application of hydrogen peroxide or peroxyacetic acid for pathogen suppression.

☐ Apply spearmint, peppermint, or clove oil once the dormancy period is past, and repeat the application every 2–3 weeks to suppress sprout development. Alternatively, use hydrogen peroxide to suppress sprouts.

☐ Consider application of clove oil or *Bacillus subtilis* for suppression of silver scurf.

☐ Apply *Pseudomonas syringae* for suppression of silver scurf and dry rot.

CHAPTER 12

Thomas F. Cummings
Department of Plant Pathology
Washington State University, Pullman

Robert E. Thornton
Department of Horticulture and Landscape Architecture
Washington State University, Pullman

Managing Potatoes in the Home Garden

Home gardens provide affordable and nutritious food for millions of North Americans. Potatoes in the home garden offer advantages to those willing to put in the time to grow this exceptional vegetable. One advantage is that various cultivars and tuber sizes, from baby reds to large bakers, can be grown in a garden plot and be readily available for different uses. Additionally, growing potatoes, like all gardening activities, provides physical and emotional benefits.

Holistic health management is needed in the home garden just as in commercial agriculture. Many of the same principles explained in other chapters of this book apply to potato production in small garden plots as well as large corporate farms. In addition, potatoes in the home garden can be hosts of insect pests and pathogens that affect potato production in commercial agriculture. Some insects and pathogens survive and build up on hosts in gardens or associated with gardens and can be dispersed to commercial potato plantings. Potato, tomato, pepper, petunia, and other plants in the nightshade family (Solanaceae), including weeds, are of greatest concern for potato production. For example, tomato plants propagated in greenhouses and then transplanted to home gardens have been the source of initial inoculum of the late blight pathogen in outbreaks of the disease in commercial potato production as well as in home gardens. Other pests and diseases that can move from gardens to commercial potato production areas are aphids, leafhoppers, and viruses.

Site Preparation and Nutrients

Potatoes are adapted to a wide range of soil conditions (Chapter 5). Select a site that has reasonable drainage, plenty of sun, and soil that can be cultivated. Do not use a site where lime or large amounts of wood ashes have been incorporated, as both may decrease tuber quality. Potatoes are generally hardy and don't require large amounts of soil amendments, but they may benefit from the incorporation of well-decomposed compost, especially in a clay, sandy, or high-mineral soil or at a site where water conservation is important. A thorough 6- to 8-inch preplant tillage is needed for the garden plot. If any compost or fertilizer is applied, it is incorporated during tillage. Fresh organic material that is not decomposed, such as cow manure, is generally low in nutrients and may not be biologically active to release nutrients for the current crop. In addition, such material may adversely alter the soil pH, and it may contain weed seeds and plant pathogens. The best organic fertilizers for potatoes, such as in chicken manure, have high nutrient value and should be well composted or processed (Chapter 11). Such fertilizers tend to release adequate amounts of nutrients to the potato plant throughout the season.

Soil testing is generally not needed for growing potatoes in a home garden, but a soil test is advisable if there is concern that a site may have a nutrient deficiency or nutrient imbalance. Fertilizer requirements vary according to many factors, such as the potato cultivar, the soil type, and the length of the growing season. However, fine tuning of fertilizer amendments is usually not needed for potatoes in the home garden. Nitrogen is most often the limiting factor in obtaining a successful potato crop. Incorporation of organic or inorganic fertilizer at a rate of 4–5 pounds of actual nitrogen per 1,000 square feet is sufficient for most garden soils. The potato plant also requires phosphorus, potassium, and several micronutrients (Chapter 8). These nutrients can be adequately supplied by common formulations of NPK (nitrogen, phosphorus, and potassium) fertilizer, such as 10-10-10, 16-16-16, or 16-8-12, or by organic amendments, if the approximate nutrient content of the amendments is known. With organic amendments or incorporation of a green manure crop, the rate of application of inorganic fertilizer can be reduced by one-third to one-half. If plants begin to show yellowing due to nitrogen deficiency, a water-soluble fertilizer can be applied through an in-place irrigation system or a hose-end sprayer, at a rate of ½ pound of nitrogen per 1,000 square feet. Overapplying nitrogen can result in excessive plant growth, which is detrimental to tuber growth and favors some foliar diseases (Chapters 9 and 21). It is often better to let the plant die off naturally if yellowing occurs after tubers have reached a desirable size.

Planting and Culture

The seedbed should be smoothed after the soil has been tilled and fertilizer and soil amendments have been incorporated. The soil should be moist before planting, but not wet to the point where water can be squeezed out of a soil ball by hand. The optimum soil temperature for planting is 50–55°F. Planting when soil temperature is in this range and is increasing will ensure quick emergence. Planting in soil at temperatures below the optimum range does not result in quicker-emerging plants or earlier or larger yields, because plants from seed pieces planted in soil at the optimum temperature quickly catch up and surpass those planted in cooler soil.

Only **certified seed tubers** should be used for planting. Carefully handle seed tubers to **avoid bruising**, which is a major cause of poor stand establishment. The three main management tactics of planting are to use quality seed, handle it carefully, and encourage **quick emergence** (Chapter 7). Seed tubers can be cut into **seed pieces** of approximately 2 ounces. Each seed piece should have one or two eyes. There is no need to dry or treat seed pieces once they are cut. They can be planted immediately, 3–4 inches below the surface. Seed pieces can be spaced 6–12 inches apart within a row. Generally, plants that are more closely spaced will provide more but smaller tubers, which are desirable for small table reds; plants grown from more widely spaced seed pieces will yield fewer but larger tubers, which are usually desirable for large bakers. The rows should be 24–36 inches apart.

Hilling is a method of culturing potatoes in which soil from between rows is mounded around the bases of newly emerged plants, before they are touching one another within the row. The soil can be pulled with a rake or hoe under the leaves of the plant. The hill should be more flat on top than pointed, with enough soil added so that the original seed pieces are 6–8 inches below the top of the hill. There are many other innovative ways to culture potatoes, involving straw, tires, and planters, which are described elsewhere, in books, Internet sources, and extension publications.

Managing Weeds

Weeds in potatoes in the home garden can be managed by hand pulling or light cultivation with a hoe. However, the planting bed needs to be properly prepared by preplant tillage to remove established weeds. After planting, weeds are most efficiently removed when small. Care should be taken not to damage the potato plants or disturb their roots within hills (Chapter 22). Herbicides are not recommended for the home garden, because of the potential damage to potatoes and the risk of drift to other sensitive plants in the garden. Potato plants generally outcompete most weeds after the potato leaves begin to touch between rows (row close). Row close usually occurs about 60 days after planting, depending on temperatures, row spacing, the cultivar, and general growing conditions.

> **Weeds are most efficiently removed when small.**

Managing Insect Pests and Diseases

The first step in managing insect pests and diseases in the home garden is to plant only **certified potato seed tubers** (Box 12.1). Do not plant tubers produced in a garden the previous year or purchased from a grocery store. Certified seed tubers have been disease tested, and planting only certified seed will restrict the introduction of many viruses, bacteria, fungi, nematodes, and insect pests (Chapter 6). **Rotating** areas of a garden plot to plants from other plant families and disposing of potato, tomato, and eggplant stems at the end of the growing season will reduce the incidence of Verticillium wilt and other diseases in potatoes subsequently grown there. Biofumigation, through the incorporation of a substantial amount of fresh green manure into the soil, and soil solarization can effectively suppress nematodes and other soilborne pathogens (Chapter 20). Potato cultivars resistant to common scab, Verticillium wilt, and other diseases can be selected in areas where these diseases are important (Chapters 6 and 19).

Problems in the home garden must be **correctly diagnosed** to be effectively managed. Many problems in home gardens are associated with abiotic factors, such as temperature extremes (frost damage and high-temperature leaf scorch),

Box 12.1

Key Management Tactics for Growing Potatoes in the Home Garden

- Potatoes are adapted to many soil conditions. Do not apply lime or wood ashes to garden soil.
- Plant only seed tubers from sources identified as certified by a state or province. Do not plant tubers produced in a garden the previous season or obtained from a grocery store.
- Plant when the soil warms. The optimum soil temperature is 50–55°F.
- Control weeds by hand pulling or cultivation with a hoe while they are small. Herbicide use in the home garden is not recommended.
- Keep potato plants vigorous through proper cultural practices.
- Harvest potatoes when they have reached the desired size.
- Store healthy tubers in a cool (ideally, 40–50°F), dark, dry environment.

moisture extremes (including improper watering), and misuse of fertilizer. Such stresses may increase susceptibility to damage due to insect pests, favor some plant diseases, and increase tuber defects due to physiological disorders (Chapter 23). Several insect pests and diseases can be minimized by maintaining plant vigor through proper cultural practices.

Management of irrigation water, including the frequency of irrigation, time of irrigation, and amount of water applied, especially from sprinklers (Chapter 9), is extremely important for disease management in the home garden. Verticillium wilt and powdery scab are favored by moist soils during Growth Stages II and III, whereas common scab can be managed partially by maintaining high soil moisture during this period (Chapter 19). Detailed information on managing insect pests and diseases can be obtained in Chapters 15–21 of this book. Common insect pests and diseases of potato encountered in the home garden and management practices to address them are briefly described below.

> **Many problems in home gardens are associated with abiotic factors, such as temperature extremes (frost damage and high-temperature leaf scorch), moisture extremes (including improper watering), and misuse of fertilizer.**

Potato flea beetles are tiny, round, brown to black insects that jump like a flea when disturbed. They create many small holes on potato leaflets. It is best to manage them early, since a few beetles can rapidly increase to a population that will completely defoliate plants. Effective insecticides are registered for home use. Instructions on product labels and bee protection warnings should be followed closely (Chapter 15).

Potato leafhoppers are tiny, lime green, elongated, fast-moving insects generally found on the underside of leaves. Apply a registered insecticide if more than one or two leafhoppers per plant are observed (Chapter 16).

Colorado potato beetle adults are oval beetles with alternating dark and yellow stripes running lengthwise on their backs. The larvae are sluglike and orange to red, with two rows of black spots on the sides. The larva is the most destructive feeding stage and can defoliate plants. For the home garden, this pest can be effectively managed when first observed by simply crushing adult beetles by hand while wearing gloves. The biocontrol agent *Bacillus thuringiensis* (Bt) is also effective, as are other insecticides registered for home use (Chapter 15).

Spider mites are small mites, visible with a hand lens. They produce webbing on the underside of leaves and cause yellowing, bronzing, or stippling of leaves. Insecticidal soaps can reduce mite populations (Chapter 15).

Late blight lesions begin as water-soaked spots, quickly increasing in size, on leaves and stems. A white, downy growth develops at the margins of lesions on the underside of leaves. The disease rapidly spreads among potato and tomato plants during moist weather. Fungicides are available for garden production and are best applied either before wet weather, if inoculum of the late blight pathogen is in the area, or when symptoms first appear. Reducing or eliminating overhead irrigation, to shorten periods of wetness of plant foliage, can help to control late blight. Irrigation at midday is less conducive to late blight development than irrigation early in the morning or in the evening. Rogue all diseased plant tissue. Check tomatoes for late blight, since infected transplants can be a source of inoculum of the pathogen. Tuber refuse from previous potato crops should be completely destroyed or buried at such a depth that it will not produce shoots that reach the soil surface before plants emerge in the spring (Chapter 21).

Early blight lesions are oval to angular spots, often containing dark, concentric, targetlike rings. They usually appear on older leaflets. Maintaining nitrogen fertility is critical in preventing damage from early blight. Fungicides are available and are effective when the disease begins early in the season. The early blight pathogen can be spread to potatoes from other plants in the nightshade family, such as tomato and pepper (Chapter 21).

Virus diseases cause various symptoms, including stunting of plants, leaf rolling, leaf wrinkling, and mosaic (indistinct yellow and green patterns on leaves). To minimize the risk of virus diseases, plant only certified seed. Remove and destroy infected plants, and control insect vectors, such as aphids, if they are present (Chapters 16 and 17).

Common scab lesions occur on the tuber surface and are tan to dark brown and usually rough-textured. Fungicides will not control common scab. Grow resistant cultivars (Table 6.2), and enhance the microbial community of the garden soil by incorporating green plant manures (Chapter 3). Do not apply wood ashes, fresh manure, or lime to potato beds. Manage irrigation water (Chapters 9 and 19). Common scab is a cosmetic problem, because the lesions can be peeled off and the tubers eaten. Smooth-skinned cultivars are more susceptible to common scab than russet-skinned cultivars (Chapter 19).

Harvest and Storage

The best time to harvest potato tubers varies and depends on the size of tubers desired. Tubers can be harvested from less than 3 months to 6 months after planting, depending on the cultivar and the environment. If small tubers are desired, they can be harvested after plants flower and when the rows of potato plants completely touch one another. For full-size or large tubers, delay harvest until the plant dies naturally or before the first freeze of the year. Periodically dig single-hill plants to determine tuber size and to obtain fresh potatoes for eating throughout the season.

Tubers from the home garden can be stored for an extended time. After the vines die, tubers can often be left in the ground until needed as long as the temperature does not fall to 30°F or below for several days. Tubers should not be left in wet soil for any period of time. For longer storage after harvest, select only the cleanest undamaged tubers that have been in the ground for at least 14 days after the vines have died. This will ensure that the skin has set and thickened.

Burlap or net bags or partially opened boxes will let the tubers "breathe" and reduce tuber rots; do not store tubers in tightly enclosed containers or closed plastic bags (Chapter 10). Tubers should be stored in a cool, dry environment, ideally 40–50°F, with no light. The cool environment ensures longer storage and reduces sprouting. A basement or garage that is cool but not exposed to freezing can be a satisfactory storage area. Do not store wet tubers. It is important to check tubers for rot every couple of weeks and to remove diseased tubers immediately.

CHAPTER 13

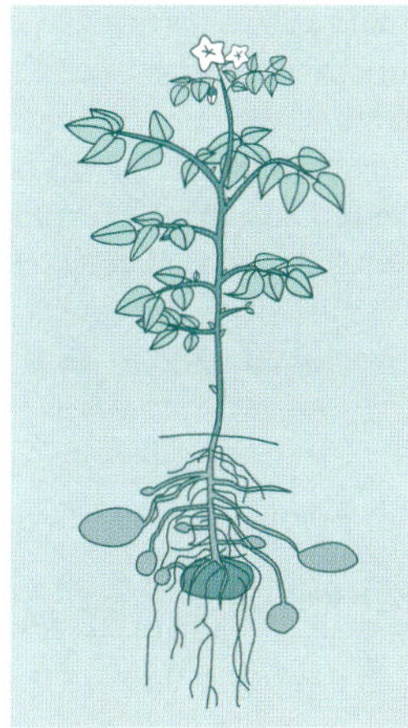

Philip B. Hamm
Department of Botany and Plant Pathology
Oregon State University, Hermiston

Rick A. Boydston
U.S. Department of Agriculture, Agricultural Research Service
Prosser, Washington

Casey W. Hoy
Department of Entomology
Ohio State University, Wooster

Walter R. Stevenson
Department of Plant Pathology
University of Wisconsin, Madison

Pamela J. S. Hutchinson
Aberdeen Research and Extension Center
University of Idaho, Aberdeen

Applying Pesticides

Pesticides are an important component of many pest and disease management strategies in potatoes. The selection of an appropriate pesticide and the correct timing, effective rate, and appropriate method of application must be considered for effective pest management. The appropriate method of application is the focus of this chapter.

Pesticides can generally be applied three ways: (1) by ground applicator, (2) by airplane or helicopter, or (3) by chemigation, in which pesticides are applied through an irrigation system. Each of these methods has inherent characteristics that need to be recognized and understood so that the most efficacious and cost-effective benefit can be obtained. These methods may be influenced differently by environmental conditions, such as wind speed, humidity, rain, and air temperature. Furthermore, the application target may include crop foliage (contact insecticide or fungicide), weed foliage (above or below the potato canopy), pests on or in the foliage (contact insecticide), the soil (herbicide, insecticide, or soil fumigant), soil around the seed piece in the furrow (Chapter 7), or a combination of these.

Whenever pesticides are used as part of a pest management program, forethought must be given to which application method is most appropriate. In addition, the method used may reduce the risk of selecting for and increasing the number of pest biotypes that are resistant to the applied pesticide. This is especially important with the newer site-specific pesticides, which may be applied multiple times during the growing season, or those that have a long period of residual activity (Chapter 14).

Application Methods

No application method works in all circumstances, and in some cases products are not labeled for all application methods or production regions. Each pesticide use situation is unique, and knowing the characteristics of each application method will help in selecting the best method and timing for the specific task. For example, although the amount of actual product applied per unit area may not differ among application methods, the amount of carrier volume varies immensely. The amount of water used in aerial application is commonly 3–7 gallons per acre (GPA), whereas chemigation may use 2,700–6,500 GPA. This major difference in the carrier volume can affect the amount of pesticide that is deposited and remains on plant surfaces, compared with the amount that moves through the canopy to the soil, and in the case of soil fumigation it dictates the depth to which the material moves downward in the soil. Moreover, carrier volume has a direct bearing on the amount of active ingredient present, plant coverage, and possibly the level of target pest control achieved.

There are three ways to consider pesticide coverage: (1) the percentage of the target pest population (primarily insects and weeds) contacted at the time of the application, (2) the area of initial coverage on the plant (crop or weed) or soil, and (3) the area of the plant or soil subsequently covered by redistribution of the product by irrigation, rain, or cultivation. These views of pesticide coverage are fundamentally different, but all can provide successful control of target organisms, particularly with an understanding of how the application method relates to coverage and the performance of the pesticide.

Regardless of the application method, proper operation and calibration of the application equipment is essential to achieving correct pesticide placement, coverage, and application rate. Before using air or ground equipment, check for proper nozzle selection and spacing to ensure uniform and complete coverage. Each nozzle should be inspected to ensure that all are the same size, and each should be individually calibrated. A nozzle should deliver within 10% of the rated amount specified in the manufacturer's instruction booklet. For ground applications, boom height and placement and the speedometer or tachometer should be checked for accuracy under actual field conditions. Inspect center-pivot systems to ensure that accessory pumps, backflow valves, and sprinklers are functioning properly. Sprinklers need to be appropriately positioned. The system should be calibrated to ensure that the correct amount of water is being applied. Misapplication of pesticides due to poor sprayer calibration and lack of a proper overlap in aerial or ground application or poorly working sprinklers during chemigation can result in crop damage and poor pest control. The primary goal of the application of pesticides is safe, efficient, economical, and effective control of the target pest. This can be achieved by ensuring the best coverage possible while maintaining the highest concentration of the active ingredient in the targeted area—sometimes competing goals.

> **Regardless of the application method, proper operation and calibration of the application equipment is essential to achieving correct pesticide placement, coverage, and application rate.**

Ground Application

Ground application includes three general placement areas: (1) aboveground, (2) in-furrow, and (3) at a specific depth or distance from the row, when shanks are used for placement.

Aboveground Application. Various aboveground applicators are available, including broadcast, band, and directed-spray applicators, but for the most part they behave similarly. Regardless of which is used, a precise, constant speed in the field must be maintained in order to deliver the correct pesticide dose. Unless a monitor system is used to adjust the spray volume or pesticide rate, constant speed is especially important in sloping or contoured fields, as changing gears or engine speed can result in significant variations in the application rate.

A ground applicator (Fig. 13.1) usually applies 20–200 GPA of water carrier (aboveground, broadcast), while moving typically at speeds of 5 MPH, with the boom pressure set at 100–400 pounds per square inch (psi) for fungicides and insecticides and 20–50 psi for herbicides. The major types of ground sprayers are **air-blast sprayers**, which use forced air at relatively high velocity to drive the spray into the canopy, and **boom sprayers**, which force the spray through nozzles of various kinds onto the canopy. The **air-assisted boom sprayer** is a relatively new development, adding a cur-

Fig. 13.1. Ground applicator applying pesticide. (Courtesy G. Secor)

Fig. 13.2. Skip in fungicide application at the edge of a field, allowing late blight to become established. (Courtesy D. Preston)

Fig. 13.3. Nozzle positioned directly behind where the seed piece drops during planting. Material applied by this method will coat the seed piece before it is covered with soil. (Courtesy G. Secor)

tain of forced air around the output from a boom, with the goal of improving penetration into the canopy. Both types of applicators behave similarly. Regardless of the equipment being used, the amount of water, pressure, ground speed, and nozzle and sprayer types should be carefully considered and balanced to create the kind of product deposition needed for managing a specific pest with the appropriate pesticide. Carrier volume can greatly impact the effectiveness of post-emergence herbicides. For example, application volumes and pressures at the high end of the recommended range are usually necessary for complete coverage of a dense weed population.

Ground application has many **advantages** (Box 13.1). Careful ground application results in product distribution throughout most of the potato canopy at a high concentration of active ingredient, even when the canopy has closed between rows. Furthermore, if the appropriate nozzles are selected and pressure is adjusted to minimize fine droplets, pesticides can be dispersed when there is a wind of moderate velocity, with minimal product loss due to drift. Ground application is preferred in situations where off-site drift must be as low as possible. In addition, there is less likelihood of skips with ground application conducted after row mark-out or planting, as long as the number of rows to be sprayed matches the effective spray swath from the boom (Fig. 13.2) and nozzles are frequently examined for proper output and pattern.

Disadvantages of ground application include the high initial cost of equipment, longer time required for application over a large area, crop damage from equipment wheels after row closure, inability to apply product when soil moisture levels are high, inability to apply product when wind speed is too high or the wind is from an undesirable direction, potential spread of pathogens or weeds on equipment from infested areas, soil compaction in wheel tracks, and inability to spray during a prolonged period of wet weather. Most preemergence herbicides require mechanical or sprinkler incorporation, and some require incorporation immediately after application. If ground application equipment is owned by the grower, application timing may be governed more by need, whereas custom application timings can be influenced by equipment availability.

In-Furrow Application. Pesticides applied in the seed furrow are typically applied either at the side of, on top of, or beneath the seed piece in relatively low volumes of water at relatively low pressure, and with large droplet size (Fig. 13.3).

Box 13.1

Comparison of Pesticide Application Methods Used in Potato Crops

	Ground application	Aerial application	Chemigation
Initial amount of product in canopy[a]	High	Moderate	Low
Initial coverage throughout the canopy[b]	Very good	Moderate	Excellent
Droplet or deposition concentration[c]	Moderate to high	Very high	Very low
Redistribution required[d]	Some	Yes	No
Amount of product moved to the soil surface at application[e]	Low to moderate	Low	High
Amount of product in canopy after 7 days[f]	Moderate	Moderate	Low
Carrier (water) volume	Moderate	Low	Very high
Amount of time required for application[g]	Moderate	Short	Long
Soil compaction or mechanical damage to crop	High	None	Low
Application under wet field conditions	Poor	Excellent	Good
Off-target drift potential	Low	High	Moderate
Ease of treating small areas of a field	Easy	Difficult	Difficult
Amount of labor required for application	High	Low	Low
Cost[h]	High	Moderate	Low

Comparisons are based on typical results of air and ground application by an experienced applicator with modern, well-maintained and calibrated equipment and chemigation with 0.15 inch of water per acre (4,100 gallons per acre), using the same application rates (amount of active ingredient per acre), where irrigation water is applied frequently following closure between rows. The method used may depend on the target pest: air or ground application of protectant fungicides may provide better long-term protection, while other methods may be more efficacious for systemic insecticides or soil-active herbicides. Methods of application of soil fumigants are not included in this table.

[a] Total amount of active ingredient in the canopy following application.
[b] Initial pesticide coverage throughout the canopy following application.
[c] Concentration of active ingredient found in spots of residue or on leaves following application.
[d] Redistribution of the pesticide after application, in rain or irrigation water, from a higher to a lower location in the canopy.
[e] Amount of product that moves entirely through the potato canopy at application and reaches the soil surface.
[f] Overall amount of a contact pesticide in the potato canopy 7 days after application.
[g] Relative amount of time required to apply a pesticide, per acre.
[h] General expense, including the cost of time and equipment, when all three application methods are available.

Granular products also can be placed in the furrow just before or after the seed piece is dropped, so the product is below or above the seed piece, whichever is most appropriate for pesticide uptake. In-furrow products tend to target soil-dwelling insects or pathogens that attack the seed piece and roots. Some products are systemic materials that can be taken up by the plant through the roots and moved to the shoots, where they are useful for protection against insects feeding on the potato foliage. The key to successful in-furrow application is placement for maximum exposure to belowground pests or efficient uptake by potato plants. Excessive irrigation or a high amount of rainfall prior to root development can move water-soluble products downward in the soil profile, reducing their availability to the developing roots or shoots and contaminating ground water. In-furrow products are generally not impacted by environmental factors such as wind, but poor application can occur because of improper calibration, plugged or worn nozzles, or wet soil.

Broadcast Application and Incorporation into Soil. Pesticides can also be broadcast on the soil surface and then incorporated to a prescribed depth. Since product placement and uniform distribution are important to ensure adequate pest management, preparation for mechanical incorporation should include an inspection of all incorporation implements to ensure that the teeth or tines are working properly and will incorporate the material consistently at the targeted depth. If a product is applied after hills are formed, confirmation that the teeth or tines are at the proper row spacing is necessary. The entire tractor-drawn assembly should be field-tested for proper depth and spacing before application. Some products, such as metam-sodium and metam-potassium, require soil sealing after incorporation, so careful adherence to the product label is important. If an overhead sprinkler irrigation system is used for incorporation, it should be inspected and calibrated to ensure uniform distribution of the desired water volume for proper product placement. Product label recommendations should be followed for approved irrigation system type and water amount, especially in relation to soil type and moisture during incorporation.

Aerial Application

Several different types of aircraft are used to apply pesticides. As with ground application, the effectiveness of aerial application is usually due not to the type of aircraft used (Fig. 13.4) but to properly functioning and calibrated equipment and environmental conditions during application. The water carrier volume used in aerial application typically ranges from 3–7 GPA, although it can be as high as 10 GPA. Depending on the nozzle type, high carrier volumes can result in more visible spots of product on the leaf. These spots are highly concentrated deposits of the pesticide and represent the drops that reach the plant canopy. As the water volume increases, the concentration of pesticide in these spots decreases, but the number of spots increases. When the potato canopy is closed, aerial application usually results in more product applied on the upper canopy than on the middle and lower canopy.

Fig. 13.4. Aerial application of pesticide. (Courtesy D. A. Johnson)

Advantages of aerial application over chemigation include the ability to apply higher levels of more concentrated product to the potato canopy (Box 13.1). Large acreages can be treated from the air in a relatively short time. In addition, because there is no ground contact, aerial application does not cause problems with soil compaction, soil moisture, and crop damage from equipment moving in the field.

Disadvantages of aerial application include limited availability of equipment (if not owned and operated by the grower), which could prevent timely applications; higher cost of application, compared with chemigation; and the potential for large areas to remain untreated as a result of skips or malfunctioning nozzles (Fig. 13.5). The recent addition of global positioning systems has been beneficial to help avoid gaps between spray swaths and application to the wrong field.

Other disadvantages of aerial application include difficulty of application near field obstructions, such as power lines, trees, public or private buildings, and public roads. The efficiency of aerial application may be reduced by stagnant air masses or drift due to high winds. Also, since aerial application deposits a disproportionate amount of product in the upper canopy, redistribution (particularly of fungicides) throughout the canopy is dependent on subsequent rain or ir-

Fig. 13.5. Infrared aerial photograph showing late blight infection (dark areas) where skips occurred during aerial application of fungicide. Healthy plants appear red.

rigation. Water dislodges the product and moves it across leaf and stem surfaces and downward in the canopy. Protection of the lower canopy from diseases or insects may be actually improved after application followed by irrigation or light rain. Formulation, adjuvants, and product solubility may play a crucial role in effective coverage in aerial application of herbicides and contact insecticides and fungicides.

The amount of product that reaches the canopy is affected by the temperature, humidity, speed of aircraft, droplet size, nozzle pressure, irregularities in field topography, and distance from the aircraft to the canopy. In general, more product reaches the canopy when the air is cool or the humidity is high, droplet size is large, and application is made close to the canopy. Even under ideal conditions, the initial canopy coverage by aerial application (before pesticide redistribution) is most often less than coverage by ground application or chemigation (Box 13.1). Aerial herbicide application coverage on emerged weeds may be insufficient because of low water volumes, particularly in high-density weed populations.

Chemigation

Chemigation is the injection of a pesticide into the water stream of an irrigation system. Chemigation in a potato crop is most commonly done through a center-pivot irrigation system (Fig. 13.6), although wheel-line and solid-set/hand-line systems are also used. A properly designed center-pivot system can provide much more even application and distribution of a product than wheel or hand lines. Chemigation by center-pivot systems usually involves high water volumes, sometimes more than 6,000 GPA (0.23 inches per acre), but typical volumes are 2,000–4,100 GPA (0.08–0.15 inches per acre). These large water volumes are the result of balancing nozzle size, operation pressure, and pivot speed required to apply the desired irrigation volume over a designated period. The system must first satisfy the water needs of the crop; application of pesticides is a secondary function. The faster a center-pivot system can travel in a complete revolution, the lower the amount of water applied per revolution. The amount of water applied during chemigation influences the amount of product remaining on the potato canopy or weed foliage; more water leaves less product, since the product can be washed off the canopy or foliage. Chemigation using solid-set or hand-line systems generally requires the injection of the pesticide near the end of the watering period, to reduce loss from runoff. Volatilization can also be influenced by how a pesticide is applied. Volatilization loss increases as air temperature increases but can be reduced when drop nozzles on center-pivot systems are placed as close to the ground as possible.

Regardless of the amount of irrigation water, chemigation usually leaves less product on the potato canopy or weed foliage than ground or aerial application. Substantially less water (50–100 GPA) can be applied by chemigation when a second, modified boom is attached to the center-pivot system, and only then does chemigation become comparable to ground or aerial application in carrier volume and the amount of product remaining on the potato canopy or weed foliage (Fig. 13.7).

Advantages of chemigation include the low cost of application per acre treated, uniform application of pesticides throughout the canopy or soil, and (depending on the nozzles) the ability to apply pesticides in winds of low to moderate speeds. Unlike products applied by air, products applied by chemigation do not need subsequent rainfall or irrigation for redistribution throughout the potato canopy or for complete coverage of a dense weed population, since uniform application occurs during chemigation. In addition, products requiring soil incorporation do not need a subsequent mechanical operation, rainfall, or irrigation. Soil moisture adequate for herbicide activation (availability in the soil water solution for uptake by the targeted weed species) can be supplied during chemigation. Some products may be "safened" for a potato crop when applied by chemigation rather than aerial or ground application, since the relatively high volume of water used in chemigation can wash the product off potato plants and onto the soil.

Disadvantages of chemigation include the greater amount of time required, compared with aerial or ground application; skips due to malfunctioning equipment (Fig. 13.8); and washoff during application, due to the large amount of water used, so that less product remains on the potato canopy or weed foliage than in other methods of application. Contact herbicides, contact fungicides, and contact insecticides may not be as effective when applied by chemigation as in ground

Fig. 13.6. Chemigation, in which pesticide is applied by injection into the water stream of an irrigation system, in this case a center-pivot system. (Courtesy D. A. Johnson)

Fig. 13.7. Secondary boom added to a center-pivot irrigation system to apply pesticide in a reduced volume of water. (Courtesy D. A. Johnson)

Fig. 13.8. Infrared aerial photograph showing infected plants (dark areas) where skips occurred as a result of a plugged pump during the application of a fungicide by chemigation. Healthy plants appear red.

or aerial application, since foliar retention is often critical. **Wheel lines** or **hand lines** can be used to apply pesticides, but two limitations should be recognized: (1) these methods usually apply water and product unevenly, and (2) regardless of when products are injected into the system, proper application may not be achieved. For instance, if the product is injected at the beginning or throughout a complete irrigation set, which typically occurs for extended periods of 6–12 hours, the product will be substantially diluted and washed from the potato canopy. If the product is injected at the end of an irrigation set, it may emerge almost immediately from the nozzles closest to the inject system, but, depending on the length of the line, a long time may be required for the product to reach the last nozzles. The difficulty then becomes a question of timing, determining when and how long to inject the product to ensure a proper application rate along the entire length of the irrigation line.

Wheel- or hand-line irrigation systems can be used effectively if the irrigation set time period is reduced so that the irrigation amount and subsequent product washoff is reduced. Special attention must be given to chemigation application of volatile products, such as the herbicide EPTC (Eptam and others). Loss of EPTC is greater from wet soil surfaces than from dry soil. If EPTC is injected into the chemigation system during the last portion of an irrigation set, loss will be great from the already wet soil surface. By comparison, if EPTC is injected into the system at the beginning of the set, it will encounter initially dry soil, some adsorption will occur, and volatilization losses will be less than if the herbicide was applied to the wet soil surface at the end of the set. When a water-soluble herbicide, such as rimsulfuron (Matrix) or metribuzin (Sencor and others), is applied early in the irrigation period and the appropriate amount of irrigation water is used, the herbicide will be moved into the top 1–2 inches of soil—the zone where weeds are germinating—and effectiveness is maximized. However, if the herbicide is applied at the beginning of the chemigation set and too much water is subsequently applied, the herbicide may move below the zone of maximum effectiveness and into the potato root uptake zone, possibly increasing the risk of crop injury. Always carefully read pesticide labels for specific application information.

Soil Fumigant Applications

Soil fumigants are pesticides that are applied to the soil weeks prior to planting potatoes, targeting soilborne fungi, weeds, and soil-dwelling insects and nematodes (Chapters 5 and 20). The most commonly used fumigants in potato production are metam-sodium and 1,3-dichloropropene (1,3-D). Metam-sodium and metam-potassium are usually applied by chemigation but can be applied by shanks that place the product at a specific depth in the soil. When applied by chemigation, the amount of irrigation water generally dictates the depth of fumigant penetration into the soil. Shank-applied metam-sodium and metam-potassium are injected into the soil profile with limited amounts of water at one or more depths. Shank application requires sealing the soil surface, to reduce the amount of product that escapes from the "chimney" produced by the shanks as they move through the soil. Soil fumigants should not be applied by chemigation during windy periods or air inversions or at high ambient air temperatures, to reduce the risk of volatilization and drift. Regardless of the application method, minimum soil temperature and moisture requirements must be met for proper and effective application.

The fumigant 1,3-D is primarily used to control soilborne nematodes. This material is always applied by shank, usually at a depth of 18–24 inches (Fig. 13.9). It should not be applied unless the residue of the previous crop has been incorporated and soil temperature and moisture requirements are met. Soil sealing is also required. A common treatment in some potato-producing areas is the application of both metam-sodium and 1,3-D, at reduced rates of both products. When both are used, 1,3-D is always applied first, followed not less than 1 week later by metam-sodium. The rates of both products in this sequential combination can be reduced without lowering efficacy. Metam-sodium, metam-potassium, and 1,3-D are restricted-use pesticides, requiring special precautions for handling and use.

Alternating Aerial Application and Chemigation

Alternating aerial application and chemigation utilizes the advantages of both methods to reduce grower expense where repeated fungicide applications are required. Aerial application provides a high concentration of pesticide on

Fig. 13.9. Long shanks used for application of 1,3-dichloropropene (1,3-D).

the upper canopy, whereas chemigation helps to maintain an adequate level throughout the canopy until the pesticide is applied again by air. This program has been effective where new foliar growth must be protected and frequent pesticide applications are necessary for control, but at a reduced cost, compared with the exclusive use of aerial application.

Alternating aerial application and chemigation utilizes the advantages of both methods to reduce grower expense where repeated fungicide applications are required.

What Application Method Should Be Used?

The application method or methods selected should be based on the desired distribution and concentration of the pesticide relative to the target organism.

Systemic Pesticides

Coverage throughout the potato canopy or over an entire weed plant is less critical for systemic pesticides, which can move through the leaf or roots and into the plant. In contrast, contact pesticides do not move and therefore require complete coverage to protect the potato plant, ensure consumption by a targeted insect, or effectively control a weed.

Fungicides

The distribution of foliar fungicides throughout the canopy is often an important factor. Late blight, for example, often begins within the canopy, and fungicide coverage throughout the canopy is important in managing the disease. **Ground application** distributes fungicides throughout the canopy at effective concentrations, and this is the commonly used application method in many production areas where late blight is a regular problem.

Aerial application of fungicides is common in the large production areas of the semiarid western United States and in the Midwest. It is particularly useful during periods of wet weather, which may preclude the use of ground equipment. This method provides effective coverage of the upper canopy and, with the correct nozzle selection, equipment calibration, and proper adjuvant, provides adequate but reduced levels, compared with ground application, in the middle and lower canopy. During the week after an aerial fungicide application, a very high concentration of the material is typically present in the upper potato canopy, but with rain or sprinkler irrigation the fungicide is moved downward, improving the protection of the lower areas. Regardless of the application method used, reapplication of

Fig. 13.10. Fiddleneck symptom of injury due to clopyralid (Stinger) carryover or drift.

Fig. 13.11. Chlorotic new growth produced as a result of injury due to glyphosate (Roundup) drift.

Fig. 13.12. Boat- or lance-shaped, elongated, light-colored leaves produced as a result of injury due to imazethapyr (Pursuit) carryover or imazamethabenz (Assert) drift or carryover.

fungicide is required to ensure and maintain effective fungicide concentration on new foliar growth and at all positions in the potato canopy.

Fungicides applied by **chemigation** provide immediate protection throughout the potato canopy, but when repeated irrigation is required every few days, little fungicide remains in the canopy 1 week after application. These conditions also provide a favorable environment for some foliar pathogens, and complete canopy coverage with a high concentration of product is obtained with weekly ground applications.

Insecticides

Thorough coverage throughout the canopy, or at least sections of the canopy where the target insect pest occurs, is critical when contact insecticides are applied. Contact between the insecticide the insect is often needed, and application can be improved by high application pressure, resulting in very small droplets, which flow across leaf and stem surfaces and contact insects that rise above the surface of the plant. However, as droplet size decreases, the risk of drift increases. On the other hand, some insecticides must be consumed by insects as they eat foliage, making thorough coverage of both the upper and the lower leaf surfaces an important factor. Knowing the pesticide mode of action and the habits and life cycle of the target pest is important in selecting the appropriate application method for a given situation.

Fig. 13.13. Veinal chlorosis of older leaves caused by injury due to metribuzin (Sencor) in a plant of the cultivar Shepody, which is sensitive to this herbicide. This is unlike injury due to atrazine (AAtrex) carryover, which causes interveinal chlorosis.

Fig. 13.14. Potato leaf distortion from postemergence application of dimethenamid-P. Similar symptoms occur in injury due to postemergence application of metolachlor.

Fig. 13.15. Necrotic areas on leaves caused by injury due to root uptake of sulfentrazone (Spartan).

Fig. 13.16. Dumbbell-shaped, curved tubers produced as a result of injury due to chlorsulfuron (Glean) carryover.

Fig. 13.17. Popcorn effect, in which small tubers with many knobs are formed, the typical symptom of injury due to flumetsulam (Python or Broadstrike), sulfometuron (Oust), or imazapyr (Arsenal) drift at tuber initiation.

Fig. 13.18. Multiple sprouts from each eye produced as a result of injury due to glyphosate (Roundup) drift. (Courtesy W. T. Cobb)

Fig. 13.19. Growth cracks and folds typical of injury due to imazethapyr (Pursuit) carryover or thifensulfuron-methyl (Harmony) or nicosulfuron (Accent) drift.

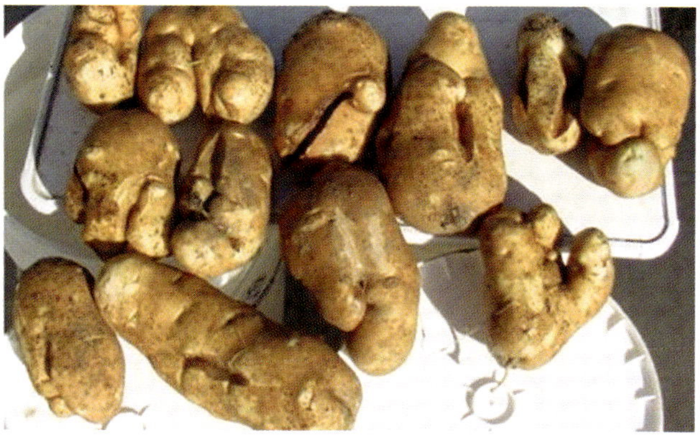

Fig. 13.20. Growth cracks and folds in Ranger Russet tubers, caused by injury due to sulfometuron (Oust).

Herbicides

Ground application is usually the best method for contact herbicides applied to emerged weeds requiring good coverage, especially when the weeds are growing prostrate rather than erect. **Chemigation** is often the most convenient and best choice for the application of preemergence, soil-active herbicides needing placement in the weed seed germination

> **Potatoes are extremely sensitive to small amounts of most imidazolinone, sulfonylurea, benzoic acid, and picolinic acid herbicides, which might be present in the spray tank from a previous application to a tolerant crop.**

zone (1–2 inches deep) and requiring "activation" by water for availability to and uptake by targeted weeds. **Aerial application** of systemic herbicides that may not require complete coverage can be effective in managing emerged weeds, if the weed density is not so high as to prevent herbicide distribution to small weeds located beneath tall plants. Most preemergence herbicides are not active against emerged weeds so a timely hilling operation can reduce weeds that have emerged since planting.

Since herbicides have biological activity on plants, the risk of injury to the potato crop due to overapplication or misapplication is greater with herbicides than with other pesticides. Herbicide selectivity between crop and weeds is usually metabolism-based and can be rate-dependent, so overapplication or application to stressed potato plants can result in crop damage (Figs. 13.10–13.20). Potatoes are extremely sensitive to small amounts of most imidazolinone, sulfonylurea, benzoic acid, and picolinic acid herbicides, which might be present in the spray tank from a previous application to a tolerant crop. When ground or aerial application equipment is used, spray tanks, booms, injection systems, and nozzles should be thoroughly cleaned after each use, to ensure that no herbicide residues are present that could damage the potato crop. In addition, some preemergence herbicides, such as flumioxazin, should not be applied to emerged potato plants, and an adequate amount of soil should cover any vegetative portion of the sprouting potato plant, to reduce the risk of crop injury.

CHAPTER 14

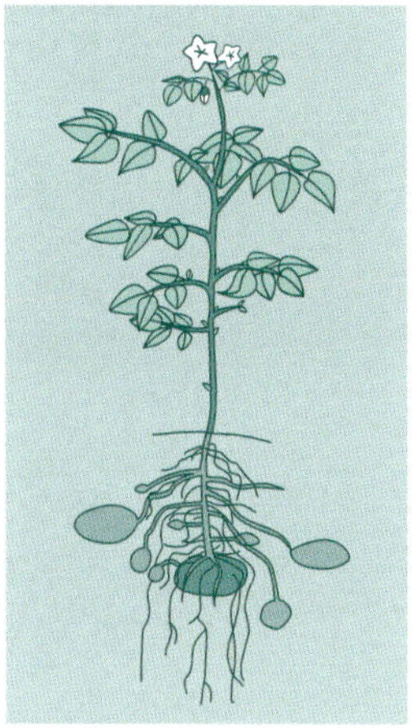

Philip B. Hamm
Department of Botany and Plant Pathology
Oregon State University, Hermiston

Casey W. Hoy
Department of Entomology
Ohio State University, Wooster

Pamela J. S. Hutchinson
Aberdeen Research and Extension Center
University of Idaho, Aberdeen

Walter R. Stevenson
Department of Plant Pathology
University of Wisconsin, Madison

Rick A. Boydston
U.S. Department of Agriculture, Agricultural Research Service
Prosser, Washington

Juan Manuel Alvarez
Department of Plant, Soil and Entomological Sciences
University of Idaho, Aberdeen

Andrei Alyokhin
Department of Biological Sciences
University of Maine, Orono

Galen Dively
Department of Entomology
University of Maryland, College Park

Neil C. Gudmestad
Department of Plant Pathology
North Dakota State University, Fargo

William W. Kirk
Department of Plant Pathology
Michigan State University, East Lansing

Managing Pesticide Resistance

Pesticide resistance is the inherited ability of a pest or pathogen population to survive a pesticide to which the original population was susceptible. It occurs in populations of insects, plants, fungi, and bacteria. Principles governing the development and management of pesticide resistance are similar regardless of the type of pest or pathogen. Pesticides have been grouped by mode of action—the sequence of events by which a pesticide controls a pest. Populations of insects, weeds, and pathogens vary genetically. The term **biotype** is used to denote an individual or groups of organisms within a population that differ in one or a few genetic traits, such as resistance or susceptibility to a particular pesticide. Most pest populations contain a small number of biotypes that are naturally resistant to a given pesticide or pesticides with the same mode of action. There is no evidence that pesticides cause alterations in individual organisms leading to resistance. The development of resistance in a field is a process of selection. Resistant pest biotypes that are not managed produce progeny that may also be genetically resistant to the pesticide, so resistant biotype numbers increase in a population when the same pesticide or pesticides with the same mode of action are repeatedly used (Fig. 14.1). Pest or pathogen populations resistant to different pesticides with the same mode of action are termed **cross-resistant. Multiple resistance** occurs when a pest or pathogen population is resistant to two or more pesticides with different modes of action (Tables 14.1–14.5). The

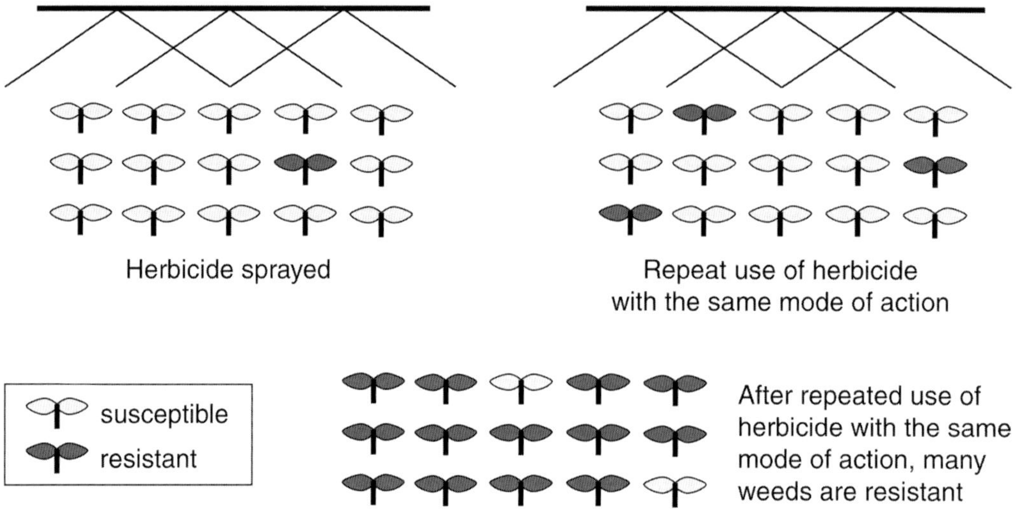

Fig. 14.1. Selection and increase of a resistant biotype in a pest population when the same pesticide or pesticides with the same mode of action are used repeatedly.

Table 14.1. Herbicides used in potato production, classified by mode of action[a,b]

Group	Mode of action	Chemical family	Common names	Trade names	Resistance risk
1	Inhibition of acetyl coenzyme A carboxylase (ACCase)	Cyclohexanedione	Clethodim Sethoxydim	Select Poast	Medium to high
2	Inhibition of acetolactate synthase	Sulfonylurea	Rimsulfuron	Matrix	High
3	Inhibition of microtubule assembly	Dinitroaniline	Pendimethalin Trifluralin	Prowl Treflan	Low to medium
5	Inhibition of photosynthesis at Photosystem II Site A	Triazinone	Metribuzin	Sencor	High
7	Inhibition of photosynthesis at Photosystem II Site A (binding behavior different from Group 5)	Urea	Linuron	Lorox	Medium to high
8	Inhibition of lipid synthesis (not ACCase inhibition)	Thiocarbamate	EPTC	Eptam	Low
9	Inhibition of EPSP synthase	Glycine	Glyphosate	Roundup	Low
10	Inhibition of glutamine synthetase	Phosphinic acid	Glufosinate-ammonium	Rely	Medium
14	Inhibition of protoporphyrinogen oxidase (protox)	*N*-Phenylphthalimide Phenylpyrazole Triazolinone	Flumioxazin Pyraflufen-ethyl Carfentrazone-ethyl	Chateau ET Aim	Low
15	Inhibition of cell division (very long chain fatty acids)	Chloroacetamide	Dimethenamid-P Metolachlor S-Metolachlor	Outlook Stalwart Dual Magnum, Dual II Magnum	Low
22	Photosystem I electron diversion	Bipyridilium	Diquat Paraquat	Reglone Gramoxone	Medium

[a]Adapted from Herbicide Resistance Action Committee, 2005, Classification of Herbicides According to Mode of Action (http://www.plantprotection.org/hrac/). Group numbers are according to the approved classification system of the Weed Science Society of America (C. A. Mallory-Smith and E. J. Retzinger, Jr., 2003, Revised classification of herbicides by site of action for weed resistance management strategies, Weed Technology 17:605–619).

[b]The list of herbicides in this table is not comprehensive. The trade names listed are examples only. Mention of products in this table does not constitute an endorsement.

terms *insensitivity* and *tolerance* are frequently used as synonyms for *resistance* in discussions of pesticide resistance.

Resistance to insecticides involves some unique characteristics of insects not shared by weeds and pathogens. Insects disperse actively, have multiple life stages in which they may or may not be exposed to insecticides, and vary considerably in gene passage to their progeny. Insect behavior, particularly active dispersal among populations and between fields and natural refugia, can determine the level of exposure to insecticides and the chances of mating with susceptible in-

Table 14.2. Herbicides used in small grain production, classified by mode of action[a,b]

Group	Mode of action	Chemical family	Common names	Trade names	Resistance risk
1	Inhibition of acetyl coenzyme A carboxylase (ACCase)	Cyclohexanedione	Tralkoxydim	Achieve	Medium to high
		Aryloxyphenoxy propionate	Clodinafop Diclofop Fenoxaprop	Discover Hoelon Puma	
2	Inhibition of acetolactate synthase (ALS)	Imidazolinone	Imazamethabenz Imazamox	Assert Beyond	High
		Sulfonylurea	Chlorsulfuron Mesosulfuron Metsulfuron-methyl Prosulfuron Sulfosulfuron Thifensulfuron Thifensulfuron + tribenuron Triasulfuron Tribenuron	Finesse, Glean Osprey Ally Escort Peak Maverick Harmony GT Harmony-Extra Amber Express	
		Sulfonylamino-carbonyltrazolinone	Flucarbazone-sodium Propoxycarbazone	Everest Olympus	
4	Action like indoleacetic acid (synthetic auxins)	Phenoxy	2,4-D, MCPA	Many products	Low
		Benzoic acid	Dicamba	Banvel	
		Carboxylic acid	Clopyralid Fluroxypyr	Stinger, Curtail Starane	
		Quinaline carboxylic acid	Quinclorac	Paramount	
6	Inhibition of photosynthesis at Photo-system II Site B	Phenylpyridazine	Pyridate	Tough	Medium to high
		Nitrile	Bromoxynil	Buctril	
6/4	Inhibition of photosynthesis at Photo-system II Site B + synthetic auxin	Nitrile + phenoxy	Bromoxynil + MCPA	Bronate	Low
7	Inhibition of photosynthesis at Photo-system II Site A (binding behavior different from Group 5)	Urea	Diuron	Karmex	Medium to high
8	Inhibition of lipid synthesis (not ACCase inhibition)	Thiocarbamate	Triallate	Far-Go	Low
14	Inhibition of protoporphyrinogen oxidase	Triazolinone	Carfentrazone-ethyl	Aim	Low
26	Unknown	None	Difenzoquat	Avenge	Medium

[a]Adapted from Herbicide Resistance Action Committee, 2005, Classification of Herbicides According to Mode of Action (http://www.plantprotection.org/hrac/). Group numbers are according to the approved classification system of the Weed Science Society of America (C. A. Mallory Smith and E. J. Retzinger, Jr., 2003, Revised classification of herbicides by site of action for weed resistance management strategies, Weed Technology 17:605–619).

[b]The list of herbicides in this table is not comprehensive. The trade names listed are examples only. Mention of products in this table does not constitute an endorsement.

dividuals, both critical factors influencing the development of resistance. Colorado potato beetles do not move very far away from fields to overwinter, and thus resistant beetles tend to mate with each other and can be exposed again to an insecticide with the same mode of action in the same field the following year. In contrast, potato leafhoppers migrate long distances each year and infest many noncrop plants, so that there is a considerable mixing of leafhopper populations, and only a small portion of a population will be exposed again to an insecticide with the same mode of action used on potato. Insect eggs and pupae are rarely affected by insecticides, whereas adult, larvae, or nymph feeding stages of most potato insects can be affected. Exceptions include potato flea beetle larvae, which feed on roots and tubers and may not be exposed to foliar insecticides, and potato tuberworm adults, which do not feed on potatoes. Aphids reproduce asexually during the growing season, each aphid being a genetic clone of its mother. Most other potato insect pests reproduce sexually, so the offspring can have traits from either parent, and those traits can be dominant and expressed or recessive and hidden.

The development of resistance in a field is a process of selection.

Colorado potato beetle populations are notorious for developing resistance to a wide variety of insecticides and are

Table 14.3. Herbicides used in sugar beet production, classified by mode of action[a,b]

Group	Mode of action	Chemical family	Common names	Trade names	Resistance risk
1	Inhibition of acetyl coenzyme A carboxylase (ACCase)	Cyclohexanedione	Clethodim	Select	Medium to high
		Aryloxyphenoxy propionate	Quizalofop	Assure II	
2	Inhibition of acetolactate synthase	Sulfonylurea	Triflusulfuron	UpBeet	High
4	Action like indoleacetic acid (synthetic auxins)	Carboxylic acid	Clopyralid	Stinger	Low
5	Inhibition of photosynthesis at Photosystem II Site A	Phenylcarbamate	Desmedipham Phenmedipham Desmedipham + phenmedipham	Betanex Betanel Betamix	High
		Pyridazinone	Pyrazon	Pyramin	
5/16	Inhibition of photosynthesis at Photosystem II Site A + unknown	Phenylcarbamate + benzofuran	Desmedipham + phenmedipham + ethofumesate	Progress	Medium
8	Inhibition of lipid synthesis (not ACCase inhibition)	Thiocarbamate	Cycloate	Ro-Neet	Low
9	Inhibition of EPSP synthase	None	Glyphosate	Roundup	Low
15	Inhibition of cell division (very long chain fatty acids)	Chloroacetamide	Dimethenamid-P	Outlook	Low
16	Unknown	Benzofuran	Ethofumesate	Nortron	Low

[a]Adapted from Herbicide Resistance Action Committee, 2005, Classification of Herbicides According to Mode of Action (http://www.plantprotection.org/hrac/). Group numbers are according to the approved classification system of the Weed Science Society of America (C. A. Mallory-Smith and E. J. Retzinger, Jr., 2003, Revised classification of herbicides by site of action for weed resistance management strategies, Weed Technology 17:605–619).
[b]The list of herbicides in this table is not comprehensive. The trade names listed are examples only. Mention of products in this table does not constitute an endorsement.

by far of most concern for insecticide resistance. The high incidence of resistance is due to high reproductive potential of the Colorado potato beetle, low gene flow among populations, and a probable predisposition to develop resistance, as a result of the coevolution of the beetle and its host plants in the family Solanaceae, which have high concentrations of toxic glycoalkaloids in their foliage. The first case of Colorado potato beetle resistance to a synthetic insecticide was observed in 1952. Since then, over 40 different compounds belonging to eight different chemical classes have failed against new strains of this pest. In some cases, insecticide effectiveness was lost within the first year of use. Resistance mechanisms in the Colorado potato beetle are highly diverse, even within a relatively narrow geographical area. Currently, beetles in various locations in the eastern United States show cross-resistance to organophosphates and carbamates and multiple resistance to these two insecticides and pyrethroids; therefore, mixtures of these insecticides are not effective. Colorado potato beetle populations can develop resistance to all components of an insecticide mixture, and the resultant resistance will be more permanent and difficult to manage than resistance to an individual compound. The majority of Colorado potato beetle populations are currently susceptible to neonicotinoid and spinosad insecticides. However, instances of neonicotinoid resistance in commercial potato farms have been reported since 2001 in New York, Delaware, and southern Maine. The number of production areas reporting resistance continues to expand.

Nearly 300 **herbicide-resistant** grass and broadleaf weed biotypes have been recorded in over 50 countries worldwide. Herbicide-resistant populations of weed species, such as common lambsquarters, kochia, redroot pigweed, Russian thistle, and wild oat, to name a few, have developed in potato production areas. The spread of resistant weeds from one production area to another is affected by weed seed dispersal mechanisms, pollination methods, and equipment sanitation practices.

The distribution of weeds resistant to herbicides with a given mode of action has changed over time. For instance, weed resistance to the group of herbicides inhibiting the enzyme acetolactate synthase (ALS), such as imidazolinones and sulfonylureas, increased from 22 to 30% over the period 1984–2005. More populations of weed species are resistant to ALS-inhibiting herbicides than to any other herbicide group. This phenomenon is due to many factors. Since 1984, commercial availability of herbicides with this mode of action has increased, and they are currently in widespread use on many crops. ALS-inhibiting herbicides have excellent activity on sensitive biotypes at the rates commonly used, which in turn exert high selection pressure for resistant biotypes. ALS-inhibiting herbicides have a single-site mode of action, affecting the ALS enzyme. In most cases, ALS-inhibiting herbicide resistance is caused by an altered ALS enzyme in a weed biotype, and this alteration can easily occur through several point mutations within the gene encoding ALS, resulting in an herbicide-resistant ALS enzyme. The ALS gene is nuclear inherited, so the herbicide resistance is transmitted by both seed and pollen.

In the past, season-long weed control was a highly desirable herbicide trait. In relation to resistance development, however, this characteristic is currently not as desirable, since experience has shown that the longer an herbicide remains

Table 14.4. Insecticides classified by mode of action[a,b]

Group	Sub-group	Mode of action classification	Chemical family	Common names	Trade names	Resistance risk
1	A	Acetylcholine esterase inhibitors	Carbamate	Aldicarb, oxamyl, methomyl, carbaryl, carbofuran	Temik, Vydate, Lannate, Sevin, Furadan	Low to high
	B		Organophosphate	Dimethoate, diazinon, disulfoton, ethoprop, malathion, parathion, phosmet, phorate, methamidophos, azinphos-methyl	Dimethoate, Diazinon, Di-Syston, Mocap, Malathion, Methyl Parathion, Imidan, Penncap-M, Thimet Monitor, Guthion	Medium to high
2	A	GABA-gated chloride channel antagonists	Cyclodiene organochlorine	Endosulfan	Thiodan	Medium
3		Sodium channel modulators	Pyrethroid, pyrethrin	Esfenvalerate, cyfluthrin, permethrin, bifenthrin, cypermethrin, others	Asana, Baythroid, Pounce, Ambush, Leverage	High
4	A	Nicotinic acetylcholine receptor agonists/antagonists	Neonicotinoid	Thiamethoxam, imidacloprid, clothianidin, acetamiprid, dinotefuran	Platinum, Admire, Cruiser, Gaucho, Genesis, Leverage, Actara, Provado, Venom	Low
5		Nicotinic acetylcholine receptor agonists (not group 4)	Spinosyn	Spinosad	Success, SpinTor, Entrust	Medium
6		Chloride channel activators	Avermectin	Abamectin	Agri-mek	Low
9	A	Compounds of unknown or	Cryolite	Sodium fluoaluminate	Kryocide	Low
	B	nonspecific mode of action	Pymetrozine	Pymetrozine	Fulfill	?[c]
	C	(selective feeding blockers)	Flonicamid	Flonicamid	Turbine	?[c]
11	B2	Microbial disruptors of insect midgut membranes (includes	*Bacillus thuringiensis* var. *kurstaki*	Btk	Javelin, Dipel	Low
	C	transgenic crops expressing *Bacillus thuringiensis* toxins)	*Bacillus thuringiensis* var. *tenebrionensis*	Btt	M-One, Novador	Low
12	C	Inhibitors of magnesium-stimulated ATPase	Propargite	Propargite	Comite	Low to medium
15		Inhibitors of chitin biosynthesis, type 0	Benzoylurea	Novaluron	Rimon	Low
18	B	Ecdysone agonist/molting disruptor	Azadirachtin	Azadirachtin	Azadirect, Ecozin	Low
22		Voltage-dependent sodium channel blocker	Indoxacarb	Indoxacarb	Avaunt	Low
25		Neuronal inhibitors (unknown mode of action)	Bifenazate	Bifenazate	Acramite	?[c]

[a] Adapted from Insecticide Resistance Action Committee, 2007, IRAC Mode of Action Classification, version 5.3 (http://www.irac-online.org/crop_protection/moa.asp). Not all compounds with the same mode of action have been shown to be cross-resistant. Different resistance mechanisms that are not linked to the target site of action, such as enhanced metabolism, may be common for chemicals belonging to the same group. Alternating compounds from different subgroups within a group with the same mode of action may be an acceptable practice in an insecticide resistance management strategy.

[b] The list of insecticides in this table is not comprehensive. The trade names listed are examples only. Mention of products in this table does not constitute an endorsement.

[c] Knowledge of the mode of action or toxicology of this material is not yet sufficient to estimate resistance risk.

active in the soil, the longer the selection pressure for resistant biotypes lasts. In other words, using a persistent herbicide would be similar to repeatedly using herbicides with the same mode of action during the time when that herbicide persists.

Fungicide resistance is also an important production issue. Several instances have occurred since the introduction of fungicides with single-site modes of action in the early 1970s, including loss of efficacy of thiabendazole (Group 1) used to manage Fusarium dry rot and silver scurf and metalaxyl (Group 4) used to manage late blight, pink rot, and Pythium leak (Table 14.5). More recently, the early blight fungus, *Alternaria solani*, has become less sensitive to chemicals in the quinone outside inhibitor (Q_oI) fungicide group (Group 11). All of these fungicides have a **single-site mode of action**; that is, they affect single, well-defined biochemical processes of specific pathogens.

The **Q_oI fungicide group** includes strobilurin fungicides, such as azoxystrobin, pyraclostrobin, and trifloxystrobin, and materials with similar modes of action, such as famoxadone and fenamidone (Table 14.5). All members of this group affect the same metabolic site in fungal pathogens (the Q_o site

Table 14.5. Fungicides classified by mode of action[a,b]

Group	Site of action	Chemical family	Common names	Trade names	Resistance risk
1	Mitosis: β-tubulin assembly	Benzimidazole	Thiabendazole	Mertect	High
		Thiophanate	Thiophanate, thiophanate-methyl	Tops, Topsin M	High
2	NADH cytochrome c reductase in lipid peroxidation (proposed)	Dicarboximide	Iprodione	Rovral	Medium to high
4	RNA polymerase I	Phenylamide	Metalaxyl Mefenoxam	Ridomil Ridomil Gold, Ultra Flourish	High
7	Complex II of fungal respiration	Carboxamide	Boscalid Flutolanil	Endura Moncut, Moncoat	Medium
9	Methionine biosynthesis (proposed)	Anilino-pyrimidine	Pyrimethanil	Scala	Medium
11	Complex III of fungal respiration: Q_oI site	Strobilurin:			
		Methoxy acrylate	Azoxystrobin	Quadris, Amistar, Quadris Opti	High
		Methoxy carbamate	Pyraclostrobin	Headline	
		Oximino acetate	Trifloxystrobin	Gem	
		Ozazolidine dione	Famoxadone	Tanos (also contains cymoxanil)	
		Imidazolinone	Fenamidone	Reason	
12	MAP protein kinase in osmotic signal transduction	Phenylpyrrole	Fludioxonil	Maxim	Low to medium
14	Lipid peroxidation (proposed)	Aromatic hydrocarbon	Quintozene (PCNB)	Blocker	Low to medium
15	Cell wall synthesis	Cinnamic acid	Dimethomorph	Acrobat, Forum	Low to medium
21	Complex III of fungal respiration: Q_iI site	Cyanoimidazole	Cyazofamid	Ranman	Medium to high
22	Mitosis: β-tubulin assembly	Benzamide	Zoxamide	Gavel (also contains mancozeb)	Low to medium
27	Unknown	Cyanoacetamide oxime	Cymoxanil	Curzate	Low to medium
28	Cell membrane permeability, fatty acids (proposed)	Carbamate	Propamocarb hydrochloride	Previcur Flex	Low to medium
29	Uncouplers of oxidative phosphorylation	2,6-Dinitroaniline	Fluazinam	Omega	Low
30	Inhibitors of oxidative phosphorylation	Organotin compounds	Triphenyltin hydroxide	SuperTin	Low to medium
M	Multiple-site contact	M1: copper	Copper hydroxide	Kocide, Champ	Low
M	Multiple-site contact	M3: ethylenebis-(dithiocarbamate) (EBDC)	Mancozeb Maneb Metiram	Dithane, Manzate, Penncozeb Maneb Polyram	Low
M	Multiple-site contact	M4: phthalimide	Captan	Captan	Low
M	Multiple-site contact	M5: phthalonitrile	Chlorothalonil	Bravo, Echo, Equus	Low

[a] Adapted from Fungicide Resistance Action Committee (FRAC), 2006, FRAC Code List 1: Fungicides Sorted by FRAC Code (http://www.frac.info/frac/). In 2001, the U.S. Environmental Protection Agency proposed a pesticide labeling scheme that added the FRAC mode of action classification to fungicide labels. In this scheme, all registered pesticides are classified by mode of action, and each mode of action is assigned a group symbol (number or letter). There are currently over 40 groups, and more are added as new modes of action are discovered or identified. Letters are assigned when the mode of action is unknown (U) or when the mode of action involves multiple-site contact (M). Products in 15 groups plus several groups with multiple-site contact are registered for use on potatoes in the United States and are currently available for disease management.

[b] The list of fungicides in this table is not comprehensive. The trade names listed are examples only. Mention of products in this table does not constitute an endorsement.

of the cytochrome bc_1 complex in mitochondria) and inhibit respiration. Pathogens exhibiting resistance to one member of this group have varying levels of resistance to other members of the group. Residual contact fungicides, such as chlorothalonil and the ethylenebis dithiocarbamates (e.g., maneb and mancozeb), have multiple sites of action (affect many general biochemical processes) and are therefore less prone to selection of resistance in the target pathogen.

Recognizing Resistance

Pesticide resistance is first recognized when expectations of pest or disease management in the crop are not met, even though a pesticide that normally reduces the pest or pathogen population has been properly applied at the labeled rate. Efficacy may gradually decrease over time, or resistance may appear suddenly, with significant loss of control. Resistance to a pesticide may be recognized by the following:

- No apparent problems with pesticide application but poor reduction of the pest population
- Reduced populations of other species that are normally sensitive to the pesticide
- History of managing the same pest species in the same field with the same pesticide or pesticides with the same mode of action
- Repeated use of one pesticide or pesticides with the same mode of action

Failure to manage an insect, weed, or plant pathogen does not necessarily mean that resistance has developed, as management failures can result from causes other than resistance, including

- Use of a pesticide that is not normally effective against the pests that are present
- Poor calibration of spray equipment
- Ineffective application methods, such as inadequate pressure or volume, resulting in poor coverage (Chapter 13)
- Application skips
- Failure to adjust the pH of spray water when required
- Excessive rainfall following application
- Adverse or stressful environmental conditions
- Heavy pest pressure
- Poor timing of application, such as application when a pest is in a less susceptible life stage or incorrect timing in relation to preemergence or postemergence of weeds
- Rapid reinfestation of fields by mobile insect pests

Insecticide resistance usually has been developing for some time before field control problems become apparent. Monitoring programs for insect resistance often detect shifts in susceptibility or increased tolerance in populations that have been treated in the past but are still adequately controlled by the insecticide. Such programs are well established, particularly for the Colorado potato beetle, and are critical in detecting early stages of resistance in time to allow the implementation of resistance management practices. If pesticide resistance is suspected, contact a university or industry expert for advice on how to collect and submit samples to be tested to verify resistance.

Factors Affecting the Development of Pesticide Resistance

Factors that affect the development of pesticide resistance include the types of pesticides used; frequency of use; whether pesticides are applied alone, tank-mixed, or alternated with other products with different modes of action; quality of application; genetic diversity of the target pest or pathogen; initial frequency, reproductive potential, and fitness of resistant biotypes; weather conditions that favor increases in pathogen and pest populations; and crop rotation. Factors that affect the spread of pesticide-resistant biotypes from one area to another include local and long-distance weather patterns that influence the spread of insects, weed seeds, or fungal spores by air currents.

Resistance to some pesticide groups has occurred more frequently than resistance to others, and some pest and pathogen species are more prone to developing resistance than others. As previously shown with the example of resistance to ALS-inhibiting herbicides, selection pressure imposed by pesticides with a single-site mode of action is more likely to lead to the development of resistant pest populations than that imposed by pesticides with a multiple-site mode of action. A member of a pest population might require only one specific mutation to be able to circumvent the activity of a pesticide with a single-site mode of action. In contrast, resistance to multiple-site pesticides is much less likely to occur, because it requires several specific mutations in exactly the right combination.

Strategies for Proactive Management of Pesticide Resistance

Loss of pesticide efficacy due to the development of resistance has led to management failures and increased production risks and costs. Therefore, proactive resistance management is critical for the retention of satisfactory pesticide efficacy in potato cropping systems. The general goal of a pesticide resistance management plan is to minimize selection pressure on the pest or pathogen population. Integrated crop and pest management strategies will help reduce the risk of selecting for pesticide-resistant biotypes in pest and pathogen populations. Knowing the mode of action and pesticide class is the key to planning a management strategy that will reduce the potential for development of resistant pest and pathogen populations (Tables 14.1–14.5). To slow the development of resistant populations, pesticides having different modes of action should be rotated from year to year, and combinations of pesticides with different modes of action for controlling the same pest species should be used during a single growing season. For insects, it is additionally important to conserve susceptible individuals in the population.

Pesticide resistance is much easier to prevent than to cure. Most of the practices listed below are designed to reduce selection pressure, thereby limiting the development of resistance. The following are proactive pesticide resistance management tactics:

- Use integrated pest management practices, including the selection of cultivars with resistance to pests and diseases.
- Practice good crop hygiene by eliminating sources of primary pest populations or initial inoculum, e.g., by management of volunteer plants, proper tuber refuse disposal, management of alternate hosts, and selecting an appropriate rotation interval between potato crops.
- Minimize the number of pesticide applications, and use action thresholds or disease forecasting models to apply pesticides only when needed.
- Rotate pesticides that have different modes of action, or use formulated mixtures or tank mixes of pesticides having different modes of action. For effective resistance management, mixing partners must be active against the target pathogen.
- Use pesticides that are designed to optimize pest or disease management at labeled rates.
- When pest or disease pressure is high, avoid using pesticides with a single-site mode of action, such as Group 11 fungicides (Table 14.5).

- Follow label guidelines for application of all pesticides. The labels of most recently registered herbicides include detailed information on resistance management as well as the pesticide mode of action or group code. For example, carboxamide fungicides (e.g., boscalid) belong to Group 7; Q_oI fungicides belong to Group 11; mancozeb belongs to Group M3; and chlorothalonil belongs to Group M5 (Table 14.5).
- Monitor fields after pesticide applications to detect surviving pests that could be resistant biotypes.
- Keep accurate records of past pesticide use and unexpected changes in pest populations or disease occurrence in individual fields.
- Be alert for control failures, and report control difficulties to university or industry representatives, so that the possibility of resistance can be monitored and evaluated.
- Work cooperatively, both locally and regionally, to ensure there is an effective area-wide resistance management program. Local and regional cooperation in resistance management is essential.

Strategies Specifically for Weeds

- Integrate weed management practices, including cultivation, use of competitive crops and crop cultivars, biocontrol, and properly selected herbicides.
- Where possible, extend crop rotation intervals to avoid a buildup of dominant weed species associated with particular crops.
- In crop rotation, use diverse crops that allow various weed management options, including the use of herbicides with different modes of action (Tables 14.1–14.3).
- Rotate summer annual crops with winter annual or perennial crops to allow different timing of tillage and herbicide applications.
- Utilize proper fertilization and irrigation practices to ensure competitive crops.
- Practice good hygiene by eliminating new sources of weed seed; for example, plant certified seed in rotation years when potatoes are not grown, manage weeds in field borders and rights of way, clean equipment between fields, and screen weed seed from irrigation water.
- Correctly identify weeds in a given field to select the appropriate herbicides and other management methods.
- Prevent seed production from weed escapes; hand-weed, cultivate, mow, burn, remove. To control the escapes before seed production, do not use an herbicide with the same mode of action as the herbicide that allowed the escapes.
- Map weed infestations in fields, and monitor changes, weed shifts, and appearance of new weeds.

Strategies Specifically for Insects

- Use compounds that can retain populations of predators and parasitoids that will reduce pest populations, even those with resistance to insecticides.
- Establish untreated refuge areas, where susceptible insects can survive without selection pressure. Commonly recommended refuge areas range from 5 to 20% of potato acreage. Refuges are designed to conserve susceptible insects that will dilute resistance after it occurs. Refuges for the susceptible Colorado potato beetle should be located near fields with resistant beetles, so that resistant beetles will mate with susceptible beetles. The perimeter barrier treatment discussed in Chapter 15 is another way to create an untreated area within fields as a refuge for insecticide-susceptible insects. Within such refuges, management would have to be minimal and preferably would not include the use of insecticides. Because resistant genotypes are usually incompletely dominant, the progeny of crosses between resistant and susceptible genotypes are usually susceptible to doses of insecticides applied at the labeled rate. Using the full label rate of an insecticide outside the refuge can therefore help kill the offspring of crosses between resistant and susceptible genotypes and reduce the resistant gene frequency in the population.
- A fundamental strategy for resistance management is to rotate chemical classes (Table 14.4) from one insect generation to the next. Pests with multiple generations in the same year, particularly the Colorado potato beetle, are important in considering insecticide rotation schemes. Consider the relative advantages of a given class of insecticides for both early- and late-season pests, and choose a time for applying an insecticide belonging to a given chemical class when it will do the most good. The use of neonicotinoids in the North Central states is a good example. Early use of these materials is very effective for first-generation Colorado potato beetles, but late-season use is more effective for both second-generation potato beetles and aphids. A number of alternative insecticides could be used early in the growing season to manage first-generation potato beetles, preserving the opportunity to use neonicotinoids late in the season for second-generation beetles and aphids. To obtain information on specific insecticides and rates of application, obtain current recommendations for your area from university or industry representatives.

Strategies Specifically for Diseases

- Group 11 fungicides (Table 14.5) should be used no more than two times in a single season. Fungicides belonging to other mode-of-action groups also have product-specific use statements on their labels. This information needs to be incorporated into effective disease and resistance management plans.
- Apply fungicides prior to plant infection, and avoid curative treatments (application after infection has occurred) with fungicides belonging to the Q_oI group and other site-specific fungicide groups.
- Use effective multiple-site fungicides less prone to fungicide resistance as mixing partners, especially late in the growing season, when disease pressure is often greatest.
- Avoid using fungicides with a high risk of resistance development alone in potato seed production. Pathogens that develop resistance in seed production fields will spread to every area where the seed is shipped.

What to Do in Cases of Confirmed Resistance

Resistance in Weeds

- Limit the movement of resistant populations from field to field by cleaning equipment before moving between fields.

- Avoid using herbicides with the same mode of action as the herbicide to which resistance has been confirmed, when possible. If an herbicide with the same mode of action must be used to manage weed species other than the resistant weed population, tank-mix the product with an herbicide having a different mode of action that is active against the resistant weed population.
- If the crop in the field with a resistant weed population is to be grazed or cut for feed, be careful not to transfer resistant seed in manure.
- In cases where resistance has been confirmed before the end of a growing season, eradicate the remaining weeds to limit the production of resistant weed seed.
- Seek advice from university and industry experts to assist in the long-term planning of weed management in these fields. One strategy is to change the crop rotation sequence to allow use of herbicides effective against the target weed in crops other than potatoes (Tables 14.2 and 14.3).
- Once resistant weed biotype numbers are at a level that requires management, integrated weed management practices that implement herbicide resistance management strategies can help ensure satisfactory weed reduction economical weed management and potato production.

Resistance in Insects

- Rotate fields as far away as possible from field sites where resistant insect populations have been detected.
- Avoid using a neonicotinoid at planting time on farms where resistance to insecticides in this class has been detected. Monitor fields, use action thresholds, and apply foliar insecticides with different modes of action (Table 14.4).
- Keep abreast of the situation in the neighbor's fields, and cooperate in a community effort to reduce selection pressure in the area where resistance has been detected. One strategy is to plan crop and insecticide rotation schemes at the community level, to reduce the buildup of a population of resistant insects and their exposure to the same insecticide.
- Seek help from university or industry experts on collecting and submitting samples of insects to track changes in resistance levels. It is important to know whether the population has reverted back from resistant to susceptible levels following mitigation practices.

Resistance in Pathogens

- Avoid using fungicides having the same mode of action or belonging to the same fungicide group (Table 14.5). Since many plant pathogens may be easily dispersed over a broad planting area, all growers in a production region should be advised and adhere to a resistance management plan.
- Avoid using highly efficacious fungicides with single-site modes of action late in the growing season, especially in curative treatments after disease symptoms have been observed.
- All growers in a production region should use tank mixes or alternate fungicides having different modes of action.
- Seek help from university or industry experts to monitor resistant pathogen populations. Report observations on reduced levels of management, so that management tactics can be adjusted.

CHAPTER 15

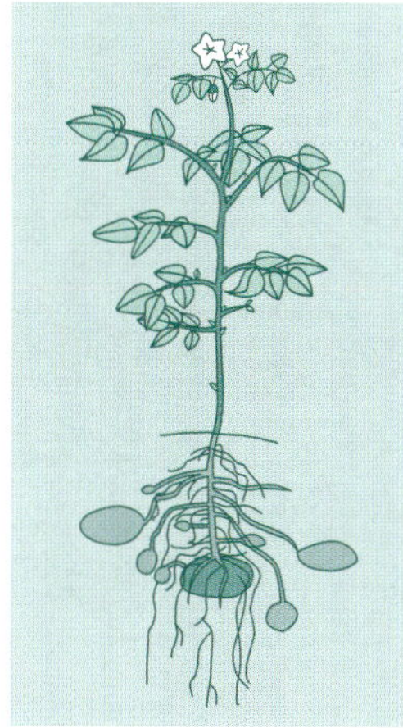

Casey W. Hoy
Department of Entomology
Ohio State University, Wooster

Gilles Boiteau
Agriculture and Agri-Food Canada
Fredericton, New Brunswick

Andrei Alyokhin
Department of Biological Sciences
University of Maine, Orono

Galen Dively
Department of Entomology
University of Maryland, College Park

Juan Manuel Alvarez
Department of Plant, Soil and Entomological Sciences
University of Idaho, Aberdeen

Managing Insect and Mite Pests

Most insects, mites, and spiders found in potato fields are either benign or beneficial and provide important ecological services in the food web, including nutrient cycling, feeding on weeds, or feeding on insects and other organisms in the soil and on plants. Only a few species feed on the potato plant itself and impact crop health either as a result of their feeding damage or by transmitting plant pathogens.

Managing insect populations to protect potato health requires a sequence of decisions every year. Many of these decisions can be anticipated following the three steps outlined in Chapter 1: (1) identification of historical problems, (2) identification of manageable factors, and (3) selection of management alternatives. Key elements of a successful integrated pest management plan include forecasting pest presence, monitoring pest populations, understanding and estimating threshold population densities, and combining multiple tactics to economically manage pests. An understanding of variation in the associated techniques through the growing season and geographic variation across North America can be used to tailor insect management to a holistic potato crop health management program for any particular location.

Overview of the Most Significant Pests

The **Colorado potato beetle** is the most important defoliating insect pest of potato (Fig. 15.1). None of the control techniques developed during the past 140 years has provided long-term protection of potato crops from this pest. The high reproductive potential of the beetle, sustained feeding by larvae and adults (Figs. 15.2 and 15.3), a diverse and flexible life history, and a remarkable ability to develop insecticide resistance make management of this pest a challenging task.

Four species of **flea beetles** are found on potato (Fig. 15.4). The two most common, the potato flea beetle (Fig. 15.5) and the tuber flea beetle, overwinter as adults and often are the first pests to appear on potato plants in the spring in northern potato-growing regions. Adult feeding results in small, round "shotholes" in the leaves (Fig. 15.6), which rarely affect yield. However, extensive leaf feeding by adults may defoliate plants and reduce tuber growth. Severely damaged leaves turn brown and die. Larvae can reduce tuber quality by feeding on the tuber surface. The damage can be severe, because the larvae can create tunnels that extend up to ½ inch directly into the tuber, making the tuber unsuitable for processing (Fig. 15.7). Larvae of the western potato flea beetle, however, burrow just under the peel and seldom penetrate over ¼ inch. Flea beetles can transmit potato pathogens, such as the spindle tuber viroid and bacteria, and leaf wounds caused by their feeding may allow additional pathogens to enter the plant (Chapter 18).

Larval stages of the **European corn borer** (Fig. 15.8) damage potato foliage by boring into the stems (Fig. 15.9), interfering with the movement of plant nutrients and water and opening entry routes for soft rot bacteria. European corn borer damage to potato is usually sporadic, depending on the timing of injury relative to tuber bulking. Damage can be severe if heavy infestations occur during early tuber development. Yield reduction results from breakage of the weakened stems and from the introduction of soft-rotting bacteria and fungi through wounds.

Cutworms are soil-dwelling caterpillars that damage potato foliage or tubers, especially those that are exposed at the soil surface or are accessible through cracks in the soil. The variegated cutworm (caterpillars with a line of small, pale spots down the back, which are more distinct toward the

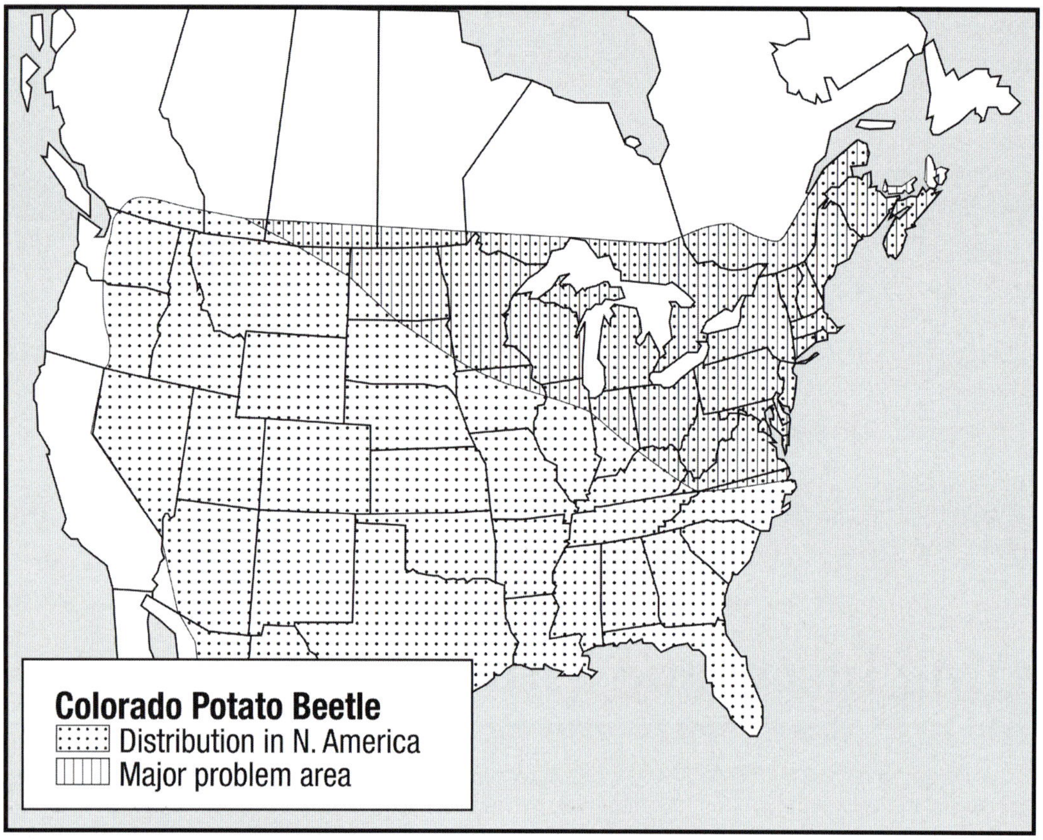

Fig. 15.1. Distribution of the Colorado potato beetle in North America. (© 2007 Minister of Public Works and Government Services Canada, used by permission)

Fig. 15.2. Colorado potato beetle adult, larva, and egg mass.

Fig. 15.3. Colorado potato beetle larvae and damage to a potato plant.

Colorado Potato Beetle

Colorado potato beetles overwinter in the adult stage in the soil in and around potato fields. Eggs are deposited on potato foliage in the spring, after adults have found potato plants and fed for about a week. Larvae pass through four life stages and then burrow into the soil to pupate. This species has one to three generations per year, depending on climate. Alternate hosts include solanaceous weeds, such as nightshade, horsenettle, and buffalobur.

head), one of numerous species known as climbing cutworms, feeds on potato stems and foliage at night, returning to the soil during the day. The black cutworm (gray to black upper half and distinctly greasy appearance) tends to remain in the soil, feeding on potato tubers. Other cutworms, such as the spotted cutworm (pairs of black oblique marks on top of last four segments), the army cutworm (gray body with darker top-lateral and spiracle stripes), and the red-backed cutworm (top often distinctly reddish, bordered with dark bands), also feed at night and can defoliate potato plants. The western yellow-striped armyworm feeds during the day and, like the army cutworm, may migrate in large numbers into potato fields from adjacent crops, particularly grains. Significant

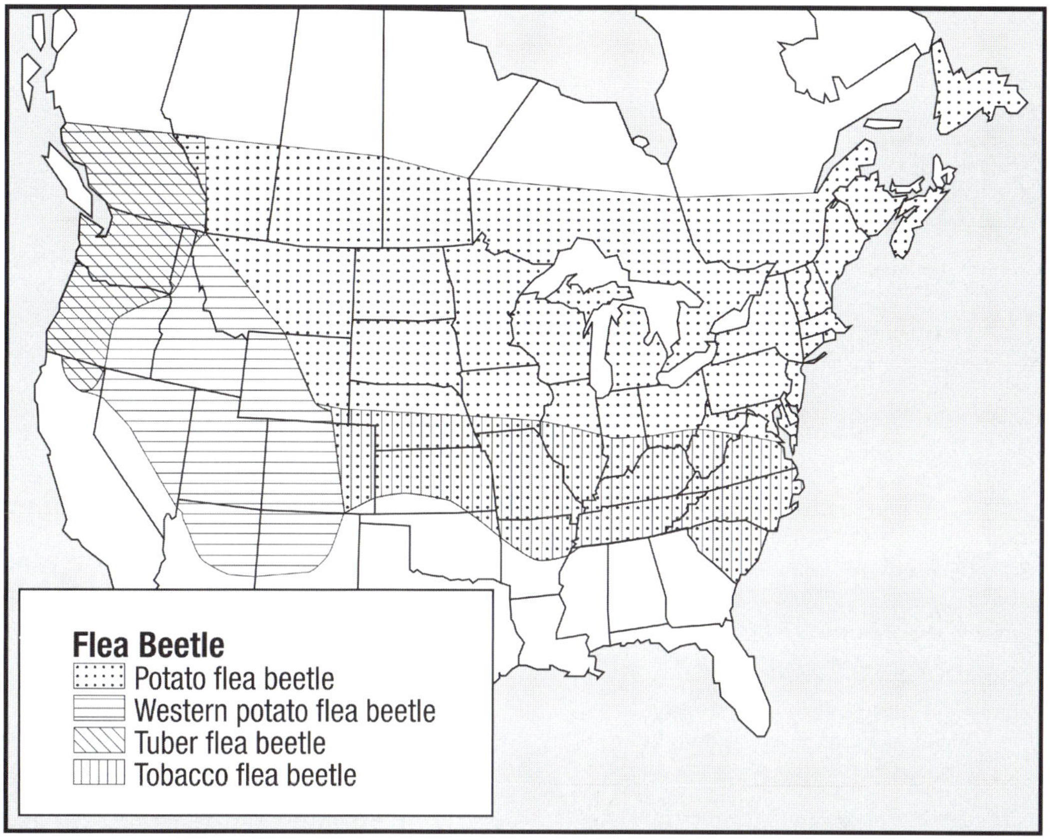

Fig. 15.4. Distribution of common species of flea beetles in North America. (© 2007 Minister of Public Works and Government Services Canada, used by permission)

Fig. 15.5. Adult potato flea beetle. (Ken Gray image courtesy of Oregon State University, used by permission)

Fig. 15.6. Shotholes and necrotic spots from adult flea beetles feeding on potato leaflets. (Courtesy D. A. Johnson)

Flea Beetles

Flea beetles are small black beetles with enlarged hind legs specialized for jumping. The characteristic feeding damage on foliage looks like the result of a shotgun blast. Flea beetles overwinter as adults in the soil and lay eggs in cracks in the soil around the base of plants in the spring. The larvae develop in the roots and can scar the tubers. Adults emerge again in summer and typically resume feeding.

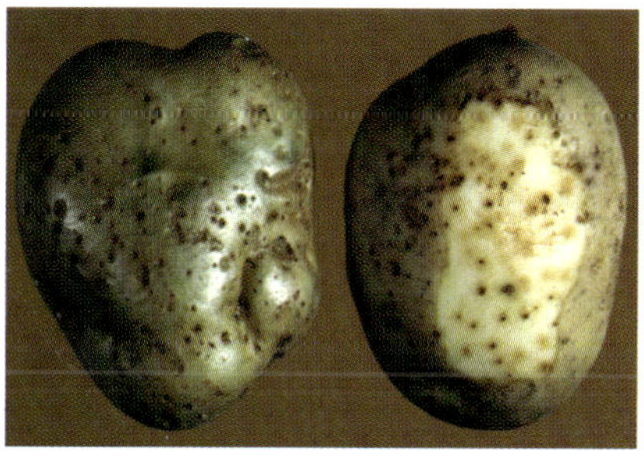

Fig. 15.7. Tunneling by flea beetle larvae in potato tubers. (Photo by Ray Kriner courtesy of Rutgers University, used by permission)

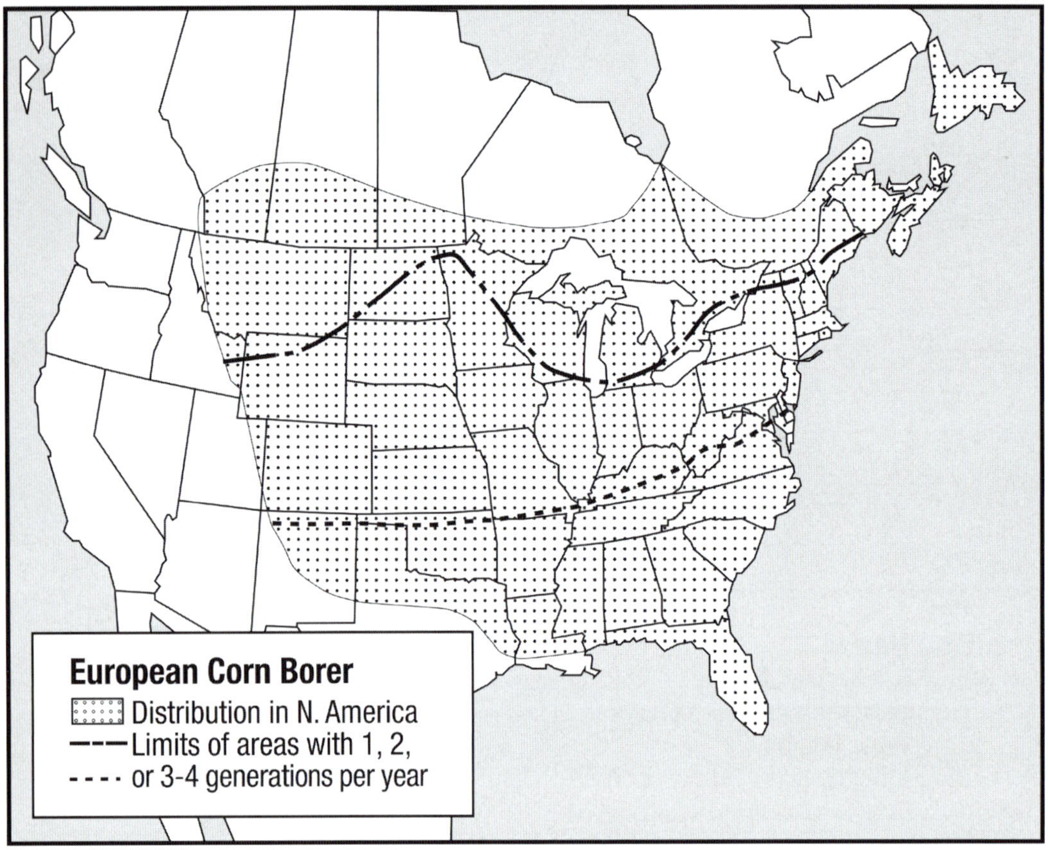

Fig. 15.8. Distribution of the European corn borer on potatoes in North America. (© 2007 Minister of Public Works and Government Services Canada, used by permission)

Fig. 15.9. European corn borer larva in a potato stem.

European Corn Borer and Cutworms

European corn borers and cutworms are caterpillar pests. Corn borers infest corn, snap beans, and other crops besides potato. Egg masses are white and flattened against the leaf. Larvae bore into stems, which tend to break at the point of entry and wilt beyond the area of feeding damage. Cutworm moths also lay eggs on many plants besides potato. The larvae spend daylight hours in the soil and feed at night.

damage by cutworms is sporadic but can be extensive when their population density is high.

Plant bugs, in the genus *Lygus*, occasionally damage potato plants by sucking sap from the young growth at the tips of stems and injecting a toxin as they feed. Common species include the tarnished plant bug (*L. lineolaris*), the western tarnished plant bug (*L. hesperus*), and the pale legume bug (*L. elisus*). They overwinter in debris in fields or at field margins. The insects are strong flyers and move from field to field. They usually move into a potato field just after an adjoining field has been harvested or has matured. The damage is typically most severe at field margins.

The **potato tuberworm** has been recorded in California since 1856 and several times over the last two decades in the Columbia Basin of Oregon and Washington, but it is not a pest elsewhere in the United States or Canada (Fig. 15.10). Long-distance movement of the tuberworm is probably due to the movement of infested tubers rather than the flight of the moths (Fig. 15.11). The adult female moth lays 150–200 eggs on the underside of potato leaves, on stems, and in the eyes of exposed tubers in the field or in storage. Economic damage results primarily from feeding by larvae on tubers (Figs. 15.12 and 15.13). In the field and in storage, the larvae excavate tunnels throughout the potato tuber, often leaving mounds of frass near tunnel entrances. In addition to the physical damage to tubers, the tunnels allow bacteria and fungi to enter tubers. Tuberworm larvae also damage foliage, burrowing through leaf petioles and creating transparent leaf

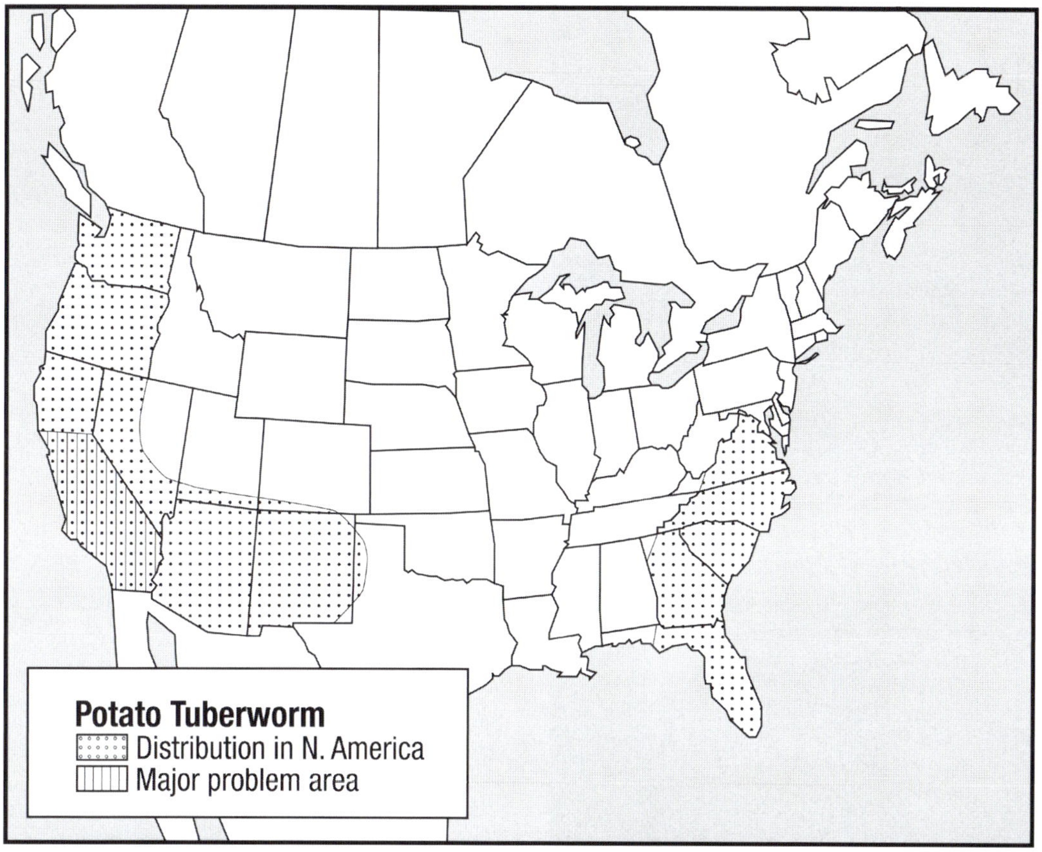

Fig. 15.10. Distribution of the potato tuberworm in North America. (© 2007 Minister of Public Works and Government Services Canada, used by permission)

Fig. 15.11. Adult potato tuberworm moth. (Courtesy A. Jensen)

Fig. 15.12. Tunnel damage caused by the potato tuberworm on the exterior of a potato tuber. (Courtesy A. Jensen)

blisters. Infested plants can be recognized by mines made by larvae in the leaves and stems and by webbing on adjacent leaves.

Wireworm is the common name for the soil-dwelling larvae of several species of click beetles (Fig. 15.14). The adults cause no damage to potato, but the larvae feed on potato roots and burrow into tubers (Fig. 15.15). Nearly 40 species from 12 genera attack potato, but only a few are economically important (Table 15.1 and Fig. 15.16). Many of these are native species, but several species introduced from Europe have become serious pests, such as *Agriotes obscurus*, *A. lineatus*, and *A. sputator*. The combination of wireworm species present in a given location varies widely across North America and can also change over time, because of changes in cultural practices and use of insecticides. Potato crop losses due to wireworm damage vary from as high as 25% in some Florida fields to less than 5% in northern Maine.

Fig. 15.13. Tunnel damage caused by the potato tuberworm in the interior of a potato tuber. (Courtesy A. Jensen)

Fig. 15.14. Wireworm larvae. (Courtesy A. Jensen)

Fig. 15.15. Wireworm tunnels in a potato tuber. (Courtesy A. Jensen)

White grubs are the larvae of several species in the beetle family that can burrow and feed in potato tubers (Fig. 15.17), especially in potato following pasture or sod in a rotation. Two species of white grubs that frequently damage potatoes in the western United States are the carrot beetle (*Ligyrus gibbosus*) and the tenlined June beetle (*Polyphylla decemlineata*). Both species are more abundant in sandy soils where grass sod or large quantities of organic matter, such as manure, have been plowed into the soil before potatoes are planted. The carrot beetle has an annual life cycle, whereas the tenlined June beetle spends 2–3 years in the soil as a grub. The Japanese beetle and other scarab species attack potato tubers in the eastern United States.

Table 15.1. Wireworm species that are common pests of potatoes in the United States[a,b]

Region	Species
Pacific Northwest	Great Basin wireworm (*Ctenicera pruinina*)
	Ctenicera lobata
	Pacific Coast wireworm (*Limonius canus*)
	*Sugarbeet wireworm (*Limonius californicus*)
	Limonius infuscatus
	*Dusky wireworm (*Agriotes obscurus*)
	Agriotes sparsus
	*Lined wireworm (*Agriotes lineatus*)
Central	*Wheat wireworm (*Agriotes mancus*)
	Corn wireworm (*Melanotus communis*)
	Melanotus depressus
	Pacific Coast wireworm (*Limonius canus*)
	*Sugarbeet wireworm (*Limonius californicus*)
	Limonius dubitans
	Limonius subauratus
	Hemicrepidius memnonius
	**Ctenicera destructor*
Northeast	*Wheat wireworm (*Agriotes mancus*)
	*Eastern wireworm (*Hypnoides abbreviatus*)
	Hemicrepidius decoloratus
	*Dusky wireworm (*Agriotes obscurus*)
	*Lined wireworm (*Agriotes lineatus*)
Mid-Atlantic	Corn wireworm (*Melanotus communis*)
	Conoderus lividus
	Conoderus vespertinus
Southeast	Corn wireworm (*Melanotus communis*)
	Conoderus rudis
	Conoderus amplicollis
	Conoderus falli
	Conoderus vespertinus
	Conoderus scissus

[a] Adapted from T. P. Kuhar, J. Speese, J. Whalen, J. M. Alvarez, A. Alyokhin, G. Ghidiu, and M. R. Spellman, 2003, Current status of insecticidal control of wireworms in potatoes, Pesticide Outlook 14:265–267.

[b] Species marked with an asterisk are also found in Canada.

Wireworms

Wireworms can have up to a 7-year life cycle from egg to adult. Species with multiyear life cycles, such as the corn wireworm and the wheat wireworm, hatch from eggs laid in grasslands or in a cereal crop, feed on grass roots, overwinter in the soil, and then attack potato the following year. Their burrowing in tubers severely reduces quality. Wireworms with a 1-year life cycle, such as the abbreviated wireworm, pupate in the spring or early summer, emerge, and then search for grasslands. In potato fields with grassy weeds, they can oviposit and infest tubers later in the year.

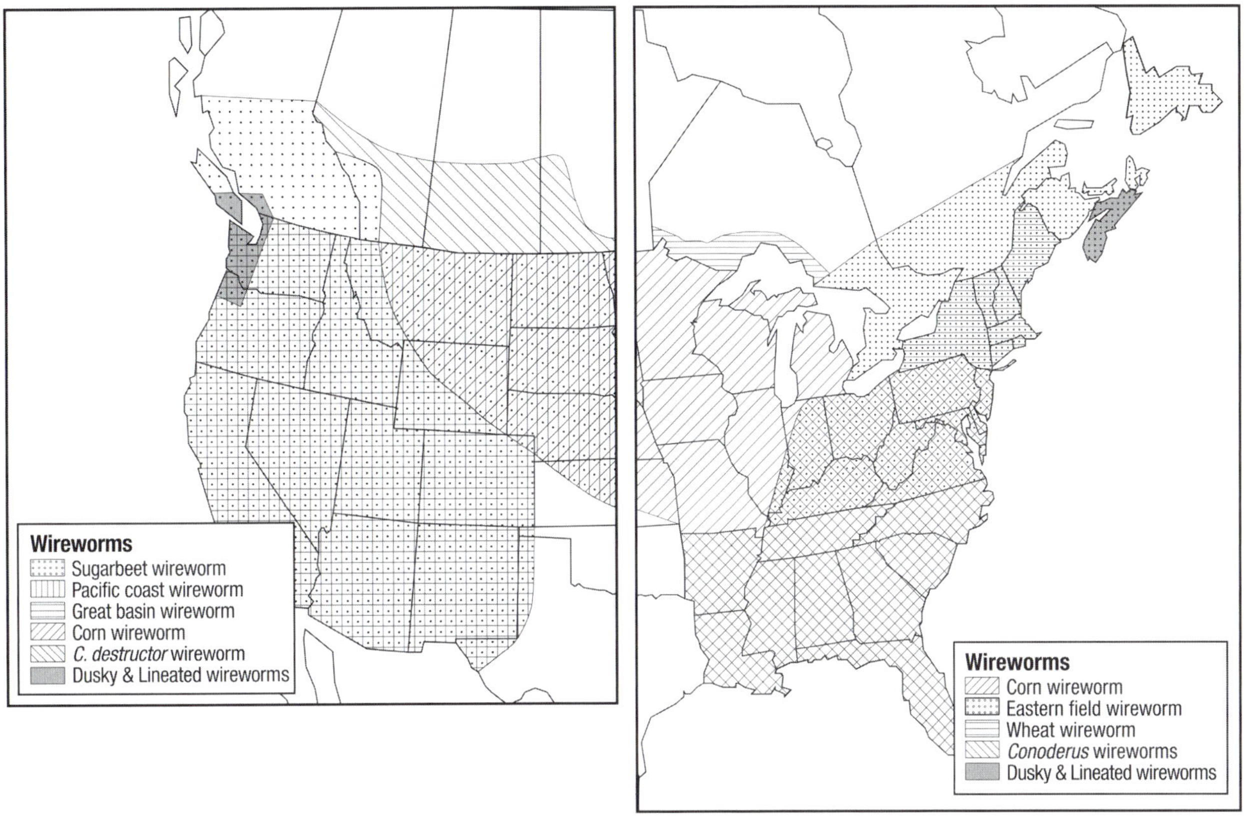

Fig. 15.16. Distribution of common species of wireworms in North America. (© 2007 Minister of Public Works and Government Services Canada, used by permission)

Fig. 15.17. White grub larva in a potato tuber.

White Grubs

White grubs are the larvae of beetles in the family Scarabaeidae, including June beetles, chafers, and Japanese beetles. Like wireworms, they normally feed on the roots of grasses but will also feed on potato tubers. The adults lay their eggs in grasses, typically the year before potatoes are planted. The larvae overwinter in the soil and resume feeding the following summer.

Spider mites that colonize potatoes are typically the twospotted spider mite (*Tetranychus urticae*). Problems occasionally occur in the Pacific Northwest and usually occur downwind from infested bean, corn, alfalfa, or clover seed fields or along dusty roads. Spider mites prefer stressed plants. They damage plants by puncturing the leaf tissue with their mouthparts to extract the plant juices. The injury appears as stippling or blotching on leaflets, followed by a change in leaf color progressing from yellowish to brown. In severe infestations, browning progresses rapidly across the field. Spider mites overwinter as adults in the soil or in debris in fields, fence rows, or field margins. The adult female emerges and lays eggs on the underside of plant foliage in late spring. A female can lay 20 eggs per day and a total of 300 eggs during her lifetime. The eggs hatch in 3–5 days. During hot weather, the young develop to adults in 7–9 days. The female spins a fine web over the leaf, which apparently protects the eggs and mites from rain and predators. When populations become severely crowded, mites climb to the top of a plant or fence post, secrete a strand of silk, and "parachute" to a new location. This is the reason sudden infestations commonly develop downwind from previously infested fields.

Management Prior to Planting

Pest Dispersal and Crop Rotation for Pest Avoidance

Selecting field locations on the basis of an understanding of the dispersal of pests from previous to present crops can both minimize pest damage and improve the predictability of insect pest problems.

Colorado potato beetle and **flea beetle** can overwinter in or near potato fields and move to new potato plantings during spring. Locating current-season potato fields away from fields cropped to potatoes the previous year has proven to be the best preventive strategy for Colorado potato beetle management. Rotated fields are colonized later by overwintered beetles and at lower densities than nonrotated fields. Adult offspring of late-arriving beetles may emerge after the day length is shortened enough that beetles stop reproducing, reducing the risks of a second generation of larvae; second-generation larvae and adult offspring of late-arriving beetles have less time to prepare to survive the winter. Fewer remedial treatments are required in rotated fields than in nonrotated fields, and the reduction of potato beetle infestation increases with the distance between a potato field and the nearest potato planting the prior year. Crop rotation also works well against tuber flea beetles.

For **wireworms**, the spread of infestations from field to field is slow, and larvae tend to have a multiyear life cycle, so that the more important consideration is whether other crops in the potato rotation serve as hosts of these pests, not how far away the nearest previous potato field was located. Wireworms can be reduced by including alfalfa but avoiding clovers and grasses in rotation with potatoes. Because dry soil increases wireworm mortality, fallowing a field will reduce wireworm numbers in dry climates, but the control achieved must be weighed against the income lost from missing a crop year.

Insect pest populations can be reduced by managing habitat around potato fields. Overwintered **Colorado potato beetle** adults need to feed to regain their ability to fly, and thus the control of volunteer potato plants and solanaceous weeds, such as horsenettle and nightshade, is important, as they are an early food source for the emerging adults. This tactic does not provide complete control, but it may reduce or delay the spread of the beetle population from overwintering sites outside the field.

Monitoring Soil Insect Pests

Problems due to **wireworms** and **white grubs** can be anticipated from the crop history of a field. These species prefer grasslands, and problems occur more often when potatoes follow cereal crops, sod, pasture, or a grass cover crop. Numerous soil factors (moisture and texture) and agronomic factors (previous dryland farming, use of a field as pasture, or soil recently brought under cultivation) affect wireworm populations. Most wireworms that attack potatoes require 3–4 years in the soil under favorable conditions to complete their life cycles, although a few species have an annual life cycle. Dissemination of wireworms occurs mainly by flights of the adults, which then deposit eggs in soil under grasses.

Because wireworm populations are erratic, the reliability of sampling and treatment thresholds are lower than for foliar pests. The most accurate way to assess the need for control is to examine field and rotation histories. If past crops in a field have sustained economic damage, there may be a need to treat. The need for control can be estimated from the previous occurrence of tuber defects in the field or in similar rotations in the area. If a field history is not available, soil sampling can give a rough indication of the need for control and will help detect high-density populations. However, soil sampling may fail to detect low-density populations that could still cause damage.

Soil can be monitored for wireworms by direct observation, baiting, or digging and sifting. Monitoring is best done before planting potatoes in fields recently rotated from sod or pasture, but it can also be done in the fall, before the first frost, when larvae move down in the soil profile. The simplest way to determine the presence or absence of wireworms is to follow the plow and look for them in early spring or fall when soil temperatures are above 50°F and wireworms are near the surface. Digging and sifting the soil to inspect for wireworms and white grubs is more direct but also much more labor intensive. Soil samples should be taken at random from at least 20 locations per field. Samples should be at least 10–13 inches deep, and in a hot summer it may be necessary to sample down to 18 inches. If the soil is friable and easily sifted, ¼-inch hardware cloth can be used to sift the samples and extract the wireworms; otherwise, it may be necessary to sort through each sample by hand.

Baiting methods use buried food sources, such as carrots, shelled corn, or potato tubers, to attract wireworms from the surrounding soil and concentrate them for counting. Bait systems are based on the principle that wireworms are attracted by CO_2 given off by germinating seeds. Therefore, this technique is affected by factors such as soil humidity and temperature and the presence of alternative food sources that can also generate CO_2. Baits are not effective in soils that are too dry, wet, or cold for wireworms to be active near the surface. Bait bags can be constructed by placing about 1 ounce of flour, germinating corn or grain seeds, rolled oats, or carrot pieces in a mesh bag or netted stocking at a depth of 4 inches into the soil and marking each bait station with a flag. Oatmeal baits are as attractive as germinating seed baits to both the Pacific Coast wireworm and the sugarbeet wireworm and are easier to process. For reliable results, 12–20 bait stations should be used per acre. After 3 or 4 days, the baits can be uncovered and the wireworms counted. The use of baits is usually limited to determining the absence or presence of larvae and can give a rough indication of the need for control. A 3- by 3-foot piece of plastic mulch is sometimes used to warm the soil and increase wireworm activity around the bait, allowing sampling earlier in the season, when the surrounding soil is are still cool. Even with the plastic mulch, the soil temperature should be at least 40°F at a depth of 10 inches below the soil surface when bait sampling for wireworms.

The precision of sampling by any of these methods tends to be low, and the associated **action thresholds** should be considered to be rough guidelines for management. Furthermore, action thresholds vary from one region to another, so local advisors should be consulted to determine an appropriate threshold for the area and specific situation. In eastern potato production areas of the United States and Canada, if more than eight wireworms are found per 20 soil samples, the field should either be treated with a soil insecticide before potatoes are planted or not be used for potato production. In Atlantic Canada, if an average of one or more wireworms is encountered per baiting station, damage to potatoes can be severe, and control is recommended. However, control of wireworms is difficult, as the soil insecticides used are often not completely effective in preventing damage. If insecticides are applied in the seed furrow, weeds or cover crops at the soil surface should be treated with herbicide well before the soil is worked, or the decaying plant material will attract wireworms and keep them away from the insecticide-treated area. By the time wireworms have spread from decaying plant material in the soil to the seed potatoes, there may not be sufficient insecticidal residual activity to protect the tubers. Therefore, it is preferable to anticipate potential problems due to wireworms and avoid them through crop rotation.

Time of Planting for Pest Avoidance

Some of the most damaging insect pest problems result from an unfortunate match between an increasing pest population and a susceptible crop stage. The pattern of occurrence typically is known for common pests and can be used to manipulate the synchrony of the pests with the crop by adjusting the planting date. For example, in regions of North America with a long growing season, early and late plantings can reduce the risk of second- or third-generation larval populations of **Colorado potato beetle**. An early planting, especially of an early-maturing cultivar, can reduce or eliminate the risk of second-generation Colorado potato beetle larvae, because the crop is usually harvested by the time the second generation peaks. With early planting, the beetle infestation may also occur after the bloom stage and close to harvest, when the crop is much more tolerant of defoliation without loss of yield. Overwintered adult potato beetles may never infest a late-planted crop, having already dispersed to find host plants elsewhere. Those that do infest may be sufficiently delayed that the next generation of adults emerges when a short-day photoperiod keeps them from reproducing.

Fertility and Insect Pests

Organic matter and microbial activity can help maintain a nutrient balance in plants between growth, reproduction, and defense against insect herbivores. Synthetic fertilizers, which provide nutrients to plants more quickly than the relatively slow release from soil microbial activity, may shift this balance toward growth and reproduction, rather than defense against insects. Most of the research in this area has focused on Colorado potato beetle and has demonstrated effects such as lower reproduction, higher mortality, slower development, and lower foliage consumption on manure-amended plots.

Management During Planting

Cultivar Selection for Insect Tolerance or Resistance

Although no commercially available potato cultivars are completely resistant to any insect pest, some variation in cultivar susceptibility to insect damage is common. Host plant resistance to insect pests is an important component of many successful integrated pest management programs. Certain potato cultivars are more susceptible to **wireworm** damage than others. For example, Shepody sustains heavy damage, Irish Cobbler and Superior sustain relatively low damage, and Atlantic, Russet Norkotah, and Norgold sustain intermediate levels.

Rapidly maturing cultivars will be more useful for either an early or a late planting strategy. Later-maturing cultivars also tend to produce greater biomass above the soil and may be better able to tolerate a given amount of insect defoliation as a result. For example, early-maturing cultivars, such as Superior, are less tolerant of late-season feeding injury by **flea beetles** than later-maturing cultivars, such as Russet Burbank.

Systemic Insecticides and Resistance Management

Seed and soil treatments with systemic insecticides have become a standard preventive approach for controlling Colorado potato beetle, aphids, and leafhoppers. These insecticides are taken up by plant roots or absorbed through the leaves if applied as a foliar spray. The systemic delivery of an insecticide maximizes plant coverage and significantly increases insecticide persistence in potato foliage. Unfortunately, whole-field applications of systemic insecticides also create strong selection pressure favoring insecticide resistance in insect populations. Colorado potato beetle, in particular, is notorious for developing resistance to a wide variety of chemicals (Chapter 14).

Trap Crops and Barriers

Growing a trap crop can draw insects away from the main portion of a production system and aggregate them in a relatively small area, where they can be effectively suppressed by chemical or physical means. Planting a border of approximately 5% of the total field area with potatoes about 10 days before planting the rest of the field, followed by controlling the beetles on the plants in the border, has been shown to reduce field-wide **Colorado potato beetle** pressure throughout the season. This strategy exploits the typical behavior of overwintered beetles, which initially invade fields at the field edges, primarily by walking, and orient themselves toward larger plants. Overwintered potato beetles move through the field border to the rest of the field within a few days, so they must be controlled in the perimeter with a timely application of an insecticide.

An approach that does not require earlier planting at the border of the field is to treat the perimeter with a systemic neonicotinoid insecticide applied in-furrow at planting in rotated fields that are relatively close to the previous year's potato fields. This **perimeter barrier** is established by treating the end sections of the middle rows (usually the first 100 feet from the field edge) and a number of rows running the length of the field along each side. The treated rows serve as barriers to Colorado potato beetles that walk in from the field edges, because the neonicotinoid insecticide kills them after very little feeding and before they move on. This strategy creates an untreated area in the middle of the field and thus can reduce the amount of systemic insecticide used. The untreated area can serve as a refuge for insecticide-susceptible insects (Chapter 14). Implementation of the perimeter treatment technology is very simple and does not require special skills or major modifications in other crop management practices. All that is needed is placing flags to mark the inner boundary of the perimeter barrier and then shutting off the in-furrow insecticide applicator on the planter each time the tractor enters the flagged area.

Another trapping approach is to place a **plastic-lined trench** around the perimeter of a potato field to trap spring adults as they disperse from overwintering sites. This type of barrier can reduce the population of adult Colorado potato beetles by about 50% and oviposition by 70%. The slippery dust-covered sides of the plastic, laid at a 60° angle, are on average 95% effective at retaining the captured beetles, and 40–90% of summer adults walking to overwintering sites may be trapped. Long-term use of trenches may therefore reduce populations in successive years. Plastic-lined trenching devices are available commercially, as are extruded plastic traps that can be removed and reinstalled to allow machinery to enter the field. The best control is achieved when all field boundaries are trenched. This tactic is best suited for locations where a large percentage of the Colorado potato beetle population walks to a new potato field from a nearby overwintering site. The cost of establishing and maintaining trenches can be high, but the cost per unit of area decreases as the field size increases, and it replaces the cost of chemical treatments to control dispersing fall adults.

Another partial barrier to insect pests is **straw mulch**, which has several advantages: it can increase the time required by Colorado potato beetles to find potatoes; it decreases the likelihood that flying beetles will locate potato plants; it increases the proportion of beetles leaving the area by flight; and it increases predation on eggs and larvae of Colorado potato beetles and other insect pests. As a result, it reduces larval populations of the Colorado potato beetle and may delay the time at which they reach their peak density. A 2- to 4-inch layer of straw has produced as much as a 2.5- to 5-fold decrease in potato defoliation, but results have been variable. Mulching is also known to reduce aphid densities and virus transmission in potato fields (Chapters 16 and 17), but it may also encourage the development of plant diseases in some years.

Management During Growth Stages I to III: Sprouting to Tuber Initiation

Monitoring Foliar Feeders

Potato fields should be sampled regularly to estimate pest populations as the crop becomes established and progresses toward bloom. As shoots emerge, each field should be checked weekly for colonizing **Colorado potato beetles** and damage. Although systemic insecticides applied as seed treatments or at planting can control Colorado potato beetles, it is still important to scout fields to detect the time when control by systemic insecticides breaks down and to inspect for other pests. Where damaging infestations of Colorado potato beetle are not an annual occurrence, the best option for control is to scout the crop after emergence and apply foliar insecticides as needed. Knowledge of the field history and whether any potatoes or other host crops were grown nearby will help determine where to sample initially. Early-season populations are often concentrated in areas of the field nearest to where potatoes were previously grown. Check these areas first to assess whether adults are dispersing into the field, and survey for damage at all times while walking through the field. Although individual potato plants can be distinguished early in the season, within a few weeks it becomes difficult to determine which stems belong to which plant. For this reason, sampling tends to focus on leaves or single stems from the soil level to the tip or on all of the foliage above a chosen length of row when viewed from directly above. Thresholds vary in the relevant sample unit (e.g., number per stem, number per compound leaf, number per foot of row, and percentage of infested stems), and the sampling scheme should match accordingly. The sampled leaves or stems should be chosen at random rather than according to where insects are suspected, because insects tend to move after feeding, and their presence does not always match the location of damage. Several leaves or stems should be examined at each of at least 10 sites spread throughout each field. Additional sites should be sampled if the field exceeds 40 acres. At each site, record the number of adult beetles, egg masses, young larvae (less than 3⁄16 inch), and older larvae (3⁄16 inch or more) per plant stem, and estimate the percentage of leaf area lost to feeding injury. Sampling for additional insect pests can be accomplished while sampling for the Colorado potato beetle.

To monitor the **European corn borer**, keep track of the number of moths captured by the nearest blacklight or pheromone trap, to aid in timing field visits to assess initial infestation levels. When blacklight catches exceed 10 European corn borer moths per night, record the number of plant stems with wilted, flagged leaves at the end, which indicate the presence of corn borer galleries. The same plants used for potato beetle sampling can be checked for corn borer injury. Another method is to begin scouting when adult moths appear, after approximately 200–300 degree-days, with a base temperature of 10°C. Larvae are predicted after 400–800 degree-days. An efficient way to monitor male moth activity where European corn borers tend to be a problem is to place pheromone traps 100 feet apart in weedy areas around the edges of or within a potato field. Walking through weedy areas, headlands, and ditches to observe European corn borer moths flying out of the vegetation is another monitoring practice. When adult moths appear, scouting for egg masses should begin. Regular scouting is required to accurately identify the timing of egg hatch. Monitoring egg masses should continue for a few weeks as moths continue to lay eggs (Table 15.2).

Cutworms are difficult to monitor, because they usually spend the daylight hours in the soil. If recent damage is found on foliage and stems with no visible signs of insects on the plants, dig 6–8 inches into the soil around the base of the damaged plants to inspect for cutworm larvae.

Spider mite infestations can be detected by placing a white paper sheet underneath a suspected infested plant and shaking the leaves over the paper. Small dots moving on the paper may indicate the presence of spider mites. They can also be observed with the help of a hand lens. Adult spider mites are 0.04 inches long and yellow, with a dark spot on either side of the body and eight legs. Nymphs look similar but are smaller and have only six legs. Eggs are clear spheres found in feeding areas. Several effective miticides are available when chemical control becomes necessary.

Flea beetle feeding damage is easier to monitor than the beetles themselves, because the beetles are extremely active and are easily disturbed when approached. The fourth terminal leaflet from the growing tip of at least 30 plants per field should be inspected, and the size of the area with flea beetle

Table 15.2. Examples of action thresholds for key insect pests of potatoes in various growing regions[a]

Pest	Atlantic Canada	East	North Central	West
Colorado potato beetle	1.5 Colorado potato beetle equivalents per plant	10% defoliation and five adults or 40 small larvae or 15 large larvae on 10 plants	1–10 beetle units per plant	2–14% of plants infested with larvae
European corn borer	Egg masses on 10–25% of plants	No consistent relationship between infested stems and yield	One egg mass per 25 leaves	. . .
Flea beetles	50 feeding holes per leaf or 10% defoliation	15 feeding holes per leaf	50 beetles per 25 sweeps	10% defoliation
Wireworms	One wireworm per bait station	Eight wireworms per 20 soil samples[b]	One wireworm per bait station	Average of one wireworm per bait station

[a] Action thresholds vary throughout North America and within regions. Refer to local guidelines, and adapt thresholds to growing conditions.
[b] Each soil sample is 1 square foot × 10–13 inches deep or up to 18 inches deep in hot weather.

shothole damage should be estimated by holding it up to the light. In the past, broad-spectrum insecticides applied to control Colorado potato beetles and aphids usually kept flea beetle populations in check. With increased dependence on systemic insecticides for the control of the Colorado potato beetle, flea beetle populations requiring control occur more frequently. In the eastern United States and Atlantic Canada, monitoring is recommended in June and then again from late July through August (Table 15.2).

Damage and Action Thresholds

Yield reductions resulting from insect feeding on foliage and stems depend on the cultivar, plant growth stage, growing conditions, and ability of the crop to tolerate damage or compensate with additional growth. The potato plant can compensate for fairly extensive defoliation if it occurs in the earliest growth stages and can either tolerate or compensate for extensive damage late in the period of tuber bulking as harvest approaches, if growing conditions are adequate. From early row closure (between rows) to the bloom period, when tubers are being initiated, insect defoliation can reduce the number of tubers formed, with a significant effect on yield. Because the plants are relatively small during the early growth stages, it takes fewer insects to cause a given percentage of defoliation at this time. Therefore, action thresholds for insect pests tend to remain relatively low until the end of the bloom period and then increase (Table 15.2). The relationship between insect feeding and yield is variable, so that actual growing conditions and cultivar characteristics must be taken into account in addition to the locally recommended threshold. Once the threshold has been exceeded, an insecticide treatment typically is the tactic required to suppress the population (Table 15.3). Treatment with therapeutic insecticides, however, can lead to insecticide resistance, can adversely affect biological control, and can be expensive, and the other management tactics and biorational insecticides described below should be considered first.

A range of threshold values has been calculated for the **Colorado potato beetle** on potato. In many regions, thresholds are now expressed in beetle units or beetle equivalents, which take into account the differences in feeding damage expected from this insect different life stages. Formulae vary in the weights given to different life stages. On Prince Edward Island, for example, the formula is spring adults + (small larvae × 0.125) + (large larvae × 0.33) + (summer adults × 0.625); for Ohio, adults + large larvae + (0.5 × small larvae). The recommended economic threshold for Prince Edward Island is 1.5 Colorado potato beetle equivalents (CPBE) per plant, but the threshold is reduced to 0.75–1.0 CPBE if the crop is stressed. Ohio thresholds range from 1 beetle unit per plant at early bloom to 10 beetles per plant in the late growth stages. In the mid-Atlantic area, treatment is recommended when defoliation reaches 10% and the population density exceeds five adults per 10 plants, 40 small larvae per 10 plants, or 15 large larvae per 10 plants. A combination of any two stages at more than half the threshold level also warrants control action, if at least 10% defoliation has occurred. In each case, the increase in feeding damage from later growth stages is taken into account. The given thresholds serve as a guide and should be adjusted upward when growing conditions are good and downward when conditions are poor.

The relationship between the number of **flea beetle** shotholes per leaflet and potato yield is quite variable, and thus recommended action thresholds are variable (Table 15.2). In Atlantic Canada, an action threshold of 50 shotholes per leaflet is considered conservative, but an action threshold of 15 is recommended in Maine. The recommendation on Prince Edward Island is to estimate the area of defoliation caused by shotholes: less than 5% of leaf surface with shotholes is considered low, 5–10% is moderate, and more than 10% is high. Management action is recommended if the average defoliation is greater than 10%. If the plants are stressed by drought or Colorado potato beetle damage, less flea beetle damage can be tolerated before management is needed.

A relationship between direct damage by the **European corn borer** and yield loss has not been established. Reduction in yield following a heavy infestation of corn borer larvae is suspected of resulting from wind damage, water stress, and secondary infection by soft rot bacteria. Generally, no management of European corn borer is necessary for early-season cultivars or seed fields, because the damage has not been shown to impact yield. The action threshold for European corn borer in Atlantic Canada is egg masses on 20–25% of plants checked; a stressed crop has a lower threshold of 10%. Management measures should be initiated when egg masses reach peak egg hatch.

Physical Controls

Physical controls use heat, suction, or mechanical action to kill insect pests or prevent infestations. A propane flamer,

Table 15.3. Insecticides for potato health management[a]

Timing and insecticide[b]	Wire-worms	White grubs	Flea beetles	Colorado potato beetle	Cut-worms	European corn borer	Plant bugs	Potato tuber-worm	Spider mites
Seed piece treatment									
Neonicotinoids[c]	...	...	×	×[d]	...	...	...	...	...
Preplant soil or in-furrow treatments									
Aldicarb	...	×	×	×	...	...	...	...	...
Diazinon	×	×	×	...	...	...	...	...	...
Phorate	×	...	×	...	...	...	...	...	...
Ethoprop	×	×	...	...	...	...	...	...	...
Disulfoton	...	×	×	...	...	...	...	...	...
Neonicotinoids[c]	...	...	×	×[d]	...	...	...	...	...
Foliar treatments									
Pyrethroids	...	...	×	×[d]	×	×	×	×	...
Phosmet	...	...	×	×[d]	...	...	...	...	...
Methamidophos	...	...	...	×[d]	...	...	...	×	...
Oxamyl	...	...	...	×[d]	...	...	...	...	...
Propargite	...	...	...	...	...	...	...	...	×
Neonicotinoids[c]	...	...	×	×[d]	...	...	...	...	...
Novaluron	...	...	...	×	×	×	...	×	...
Abamectin	...	...	...	×	...	...	...	...	×
Indoxacarb	...	...	...	×	...	×	...	...	...
Spinosad	...	...	...	×	×	×	...	...	...
Spiromesifen	...	...	...	...	...	...	...	...	×
Bacillus thuringiensis	...	...	...	×	×	...	...	×	...
Cryolite	...	...	...	×	...	...	...	...	...
Azadirachtin	...	...	...	×	...	...	...	...	...
Rotenone	...	...	...	×	×	...	...	...	...

[a] Check local recommendations for the most up-to-date information, particularly regarding resistance in Colorado potato beetle populations, and always follow label directions.
[b] Classes of insecticides and selected trade names and are given in Table 14.4.
[c] Use either at planting or as a foliar treatment, but not both.
[d] Populations may be resistant. Check local recommendations.

normally sold for nonchemical weed control, can reduce the number of overwintering **Colorado potato beetle** adults by about 80%. Proper timing is essential; a flamer is best used on warm, sunny, windless days when adults are present in great numbers on small plants. The use of a propane flamer is not recommended on plants more than 8–10 inches high, because the canopy absorbs enough of the heat to reduce the effectiveness of the treatment. A flamer can be used to kill beetles in trap crops, but the treatment must be repeated frequently, and multiple passes with the flamer can lower tuber yield and cause soil compaction. Some success has been achieved using different types of insect collectors. The vacuum collector uses forced air and suction to remove all stages of the Colorado potato beetle from foliage, suck them up, and pulverize them. Documented removal with vacuum collectors includes 39% of small larvae, 27% of large larvae, and 48% of adults. Cultivating the soil mechanically destroys eggs and larvae of **flea beetles** and reduces egg laying, because females prefer to lay eggs in crusted soil. Physical tactics must be integrated with other management measures to provide the level of pest reduction required in most commercial plantings.

Conserving and Augmenting Biological Controls

Natural enemies of insect pests need to be considered when control measures at the early stages of crop development are selected, as the consequences of reducing their populations could develop later. For example, many insect and mite predators attack spider mites, but because some insecticides kill these predators, mite outbreaks often follow treatments for aphids or other insect pests. Maintenance of effective natural enemies can reduce production costs by decreasing pesticide use. Natural enemies of potato insect pests include arthropods (mostly other insects and spiders), birds, fungi, and nematodes.

Several species of arthropods attack the **Colorado potato beetle**. The twospotted stink bug (*Perillus bioculatus*) and the spined soldier bug (*Podisus maculiventris*) prey on beetle larvae. The tenspotted lady beetle (*Coleomegilla maculata*) preys on eggs and small larvae. Adults of the ground beetle *Lebia grandis* feed on Colorado potato beetle eggs and larvae, and *L. grandis* larvae parasitize Colorado potato beetle pupae. The tachinid flies *Myiopharus doryphorae* and *M. aberrans* parasitize beetle larvae and adults. Several generalist predators, such as ground beetles other than *L. grandis*, spiders, and harvestmen (daddy longlegs), also occasionally feed on Colorado potato beetle eggs. *Pterostichus melanarius*, an introduced European species, is the dominant ground beetle in the Pacific Northwest. It has been observed feeding on larval stages of Colorado potato beetle and can also eat **aphids**, such as the green peach aphid, potato aphid, and rose-grain aphid.

The **European corn borer** has an even larger complex of natural enemies. Some of these are native to North America, while others were deliberately introduced for pest management. Generalist arthropod predators, such as lady beetles and green lacewings, feed on eggs and young larvae that have not yet burrowed inside the plants, and they also attack **potato tuberworms**. Birds (mostly downy woodpecker and crow) eat larger larvae living inside stems. Numerous parasitoid species attack eggs and larvae at different stages of development.

Populations of natural enemies of potato insect pests often build up considerably in potato fields in the absence of broad-spectrum pesticides. However, their populations are usually insufficient for bringing all insect pests under complete control. One possible way to achieve better suppression of pest populations is releasing natural enemies obtained from a commercial supplier. Mass releases of predatory stink bugs will reduce **Colorado potato beetle** numbers and protect potato plants with the same level of efficacy as insecticides, but supplies are limited for commercial field releases. Entomopathogenic nematodes have the potential to reduce Colorado potato beetle numbers and plant damage. Similarly, mass releases of the egg parasitoid *Trichogramma ostriniae* can result in significant control of the **European corn borer**. Beneficial organisms may be purchased from a number of commercial suppliers, but the cost-benefit ratio of the augmentative biological control should be evaluated (Chapter 4).

Biorational Insecticides

Because natural enemies of insect pests are difficult to augment but very helpful when preserved, insecticides that have minimal effects on natural enemies deserve special attention. Biorational insecticides are those that have little effect on nontarget organisms and are often of biological origin themselves. *Beauveria bassiana,* a naturally occurring fungus, is capable of causing fairly high mortality in **Colorado potato beetles**. Exposure to sublethal concentrations of imidacloprid has been shown to increase the susceptibility of the beetle to this fungus. *B. bassiana* also has been observed to infect the **European corn borer**, but its impact on borer populations has not been thoroughly quantified. Spores of this fungus are commercially available for application with standard spray equipment and have been used by organic farmers for many years. Although less effective than most chemicals (including organically certified ones), applications of commercial formulations of *B. bassiana* can help control the Colorado potato beetle. Commercial formulations of entomopathogenic nematodes are and available for application to the soil with standard spray equipment. Suppression of Colorado potato beetle pupae has been demonstrated, but the results have not been consistent, and the cost of the nematodes is high.

Commercial formulations of *Bacillus thuringiensis* var. *tenebrionis* (Btt) have been very successful in controlling the Colorado potato beetle while preserving its natural enemies. Btt is a bacterial insecticide typically consisting of both spores and a protein produced with the spores that destroys the beetle digestive system. The *tenebrionis* strain is effective only against some species of leaf-eating beetles, including the Colorado potato beetle. It is different from *B. thuringiensis* var. *kurstaki,* which is commonly used for control of caterpillar pests. Btt must be consumed by beetle larvae to be effective. Maximum efficacy depends on very careful timing of applications to coincide with the hatching of egg masses and feeding of the smallest larvae, but Btt can equal synthetic insecticides in reduction of populations and especially in preventing feeding damage. A more recently available insecticide of fungal origin is spinosad, which is very effective for control of the Colorado potato beetle and caterpillar pests of potato, but it can also be toxic to their natural enemies.

Management During Growth Stages IV and V: Tuber Bulking and Maturation

Wireworms and **white grubs** burrow into developing tubers, causing reduced tuber quality for processing and fresh-market use during the period of tuber bulking. In some years, up to 45% of the total potato harvest in some fields in the Northwest has been downgraded from U.S. No. 1 to No. 2 because of wireworm injury. Potato crop losses due to wireworm damage vary, from as high as 25% in some Idaho and Florida fields to less than 5% in northern Maine. Temperature affects wireworm populations in the field as well as their biology and seasonal pattern of vertical movement in the soil. At higher temperatures larvae are very active, feeding voraciously and growing rapidly. Movement from the soil surface in the summer is a response to high soil temperatures or a decrease in soil moisture. When soil surface temperatures reach 85°F or higher, wireworms move downward in the soil. In irrigated fields with complete foliage cover, the soil may remain below this temperature. In potatoes, much of the feeding damage to tubers seems to occur late in the growing season at lower soil temperatures.

Almost all tuber damage due to the **potato tuberworm** occurs after vines begin to die or after chemical vine desiccation. Any effort to reduce the exposure of tubers to egg-laying potato tuberworm moths or larvae will reduce crop damage. These insects will not burrow through more than 2 inches of soil to reach tubers. Therefore, extra hilling to increase soil coverage can reduce tuber infestations. Soil cracking also provides entrance for the potato tuberworm. Timely irrigation will reduce soil cracking. Sprinkler irrigation seals the soil and reduces soil cracking better than furrow irrigation. The number of potato tuberworm moths during the season gives an indication of field infestations. Commercially available pheromone traps can be used to monitor male moth activity. Sticky traps need to be located on the edges of the field where the pheromone lure is less likely to get wet. Traps should be suspended 12 inches above the ground or the crop canopy. Thresholds for potato tuberworm based on moth catches in pheromone traps have not been determined, but early detection is the key to eradicating new infestations of this pest and preventing its spread.

Although pests often disperse from potato crops late in the season, they can occasionally be concentrated and killed, so that the population in a subsequent crop will be lower. With the right timing, late-season trap crops for the **Colorado potato beetle** can be generated by skipping four to eight rows of potatoes every 100–200 feet when applying a vine killer. Adult beetles remaining in the field will concentrate in these few remaining green rows within days, and they can then be destroyed to reduce the beetle population in the following year. The beetles can be sprayed or destroyed mechanically with a flail chopper or propane flamer, either of which would accomplish the vine killing at about the same time as the rest of the field. Adult beetles can also be trapped in plastic-lined trenches at the perimeter of the field as they move into bordering refuge areas late in the season.

Management During Harvest and Storage

Mature tubers should be harvested as soon as possible to reduce exposure to insect pests such as the **potato tuberworm**. Leaving tubers on the soil surface or in windrows overnight can encourage potato tuberworm infestations and should be avoided. Potato vines and tubers on and in the ground should be destroyed after harvest, to destroy any pests inside them. For example, **European corn borer** larvae can overwinter in potato vines, so the vines should be chopped and disked. Plowing a dry field early in the fall can also disrupt the pupal cases of **wireworms**.

Where the **potato tuberworm** is a problem, tuber storage during the winter will influence future populations. The pupae of tuber moths overwinter in tubers in the ground, cull piles, storages, and other locations where the temperature is above freezing. The warmer conditions associated with stored potatoes allow the potato tuber moth to breed year-round. In addition, moth populations can be maintained in storage sheds, because of direct contact between infested and uninfested tubers. Larvae will pupate in dead potato leaves, in soil, and on stored potato tubers. If none of these are available, tuberworms will seek other protected places for pupation, such as crevices in walls, floors, and crates.

Once tubers are harvested and store[illegible]uld be frequently checked for insect infestations. This is especially true for potato storages in the Columbia Basin of Washington and Oregon. Recommendations for preventing tuberworm infestations in storage include detecting the first invading insects, so that insecticides or other controls can be applied before they become established. Adult moths can be detected early with pheromone traps, and potatoes can be inspected for larval infestation, although the insect tunnels are very small and very difficult to see early in an infestation. Suspicious tubers should be cut carefully to check for signs of tuberworm damage and the presence of larvae or pupae. Eliminating the damaged tubers and treating the remainder with an insecticide registered for use in storage facilities is critical. Pyrethroids are effective for seed potatoes, whereas biological products such as *Bacillus thuringiensis* var. *kurstaki* and granulosis virus are safe on potatoes destined for consumption and have provided good control in many parts of the world. Screens placed over potential entry points serve as barriers to adult moths. All used sacks and bins should be cleaned of plant residues.

Management in Years Prior to Planting

The end of the growing season is the time to update or develop a multiyear integrated insect pest management plan for future crops (Box 15.1). A significant proportion of the alternative insect control techniques available against potato pests must be administered before estimates of insect densities can be obtained through crop monitoring. Application of

Box 15.1

Key Tactics for a Holistic Approach to the Management of Selected Insect Pests of Potatoes in North America

Pest	Before planting	At planting	Emergence to bloom	Maturation and harvest
Colorado potato beetle	Crop history Rotation plan Planting date Soil fertility	Systemic insecticides[a] Trap crops Barriers	Monitoring Action thresholds[b] Foliar insecticides Physical controls Biological controls Biorational insecticides	Control of dispersing adults with trap crops or physical controls
European corn borer	Crop history	. . .	Monitoring Action thresholds[b] Foliar insecticides Biological controls	Destruction of crop vines
Flea beetles	Rotation plan	Cultivar selection Soil insecticides	Monitoring Action thresholds[b] Foliar insecticides	. . .
Wireworms	Crop history Rotation plan Soil monitoring Action thresholds[b]	Cultivar selection Soil insecticides	. . .	Plowing

[a] Growers should consult local guidelines for insecticide resistance management.
[b] Growers should refer to local extension guidelines and consider current growing conditions in selecting action thresholds.

these tactics is based on pest history and forecasting population densities. The forecast is based on pest population estimates in the region from the crop year that just ended. After harvest is the time to plan land allocation on the farm or in the region to maximize the use of sustainable insect pest management control methods. Planning early will ensure that preventive insect control techniques are ready to implement before the next crop is planted.

CHAPTER 16

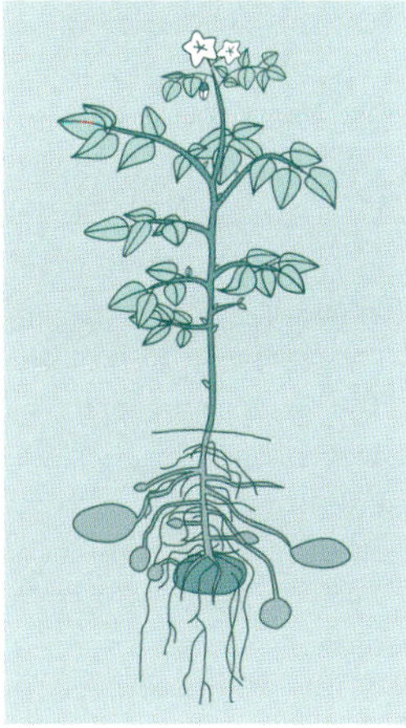

David W. Ragsdale
Department of Entomology
University of Minnesota, St. Paul

Edward B. Radcliffe
Department of Entomology
University of Minnesota, St. Paul

Kathy L. Flanders
Department of Entomology and Plant Pathology
Auburn University, Auburn, Alabama

Managing Aphids and Leafhoppers

Aphids and leafhoppers have sucking mouthparts, and in feeding they can affect the health of a potato crop directly by sap removal and indirectly by transmitting viruses and phytoplasmas and depositing salivary toxins. As vectors of pathogens, aphids and leafhoppers are key pests in seed potato production, because tuber-borne pathogens can limit the marketability of tubers intended for seed and can reduce yield in subsequent crops. Transmission of plant pathogens to potatoes during the growing season can reduce yield, especially if transmission occurs during vegetative growth, but viruses and phytoplasmas can also cause necrosis or discoloration of tuber vascular tissues, which can affect the marketability of the crop. This chapter focuses on aphids and leafhoppers found on spring-planted crops in the major potato-growing regions of North America.

Aphids

Aphids can injure potato plants directly by removing sap when feeding in vascular (phloem) tissue. However, aphid populations usually do not reach levels at which such feeding damage occurs. Aphids cause more damage in their role as vectors, transmitting viruses that cause diseases in potato.

Fig. 16.1. Green peach aphid (*Myzus persicae*). (Courtesy D. N. Ferro)

Identification and Biology

Aphid species that colonize potato differ in the amount of feeding injury they cause and the particular viruses they can spread. In North America, five aphid species commonly colonize potato: the green peach aphid, *Myzus persicae* (Fig. 16.1); the potato aphid, *Macrosiphum euphorbiae* (Fig. 16.2); the buckthorn aphid, *Aphis nasturtii;* the foxglove aphid, *Aulacorthum solani;* and the melon aphid (or cotton aphid), *Aphis gossypii.* In the northeastern United States and Canada, the green peach aphid, potato aphid, and buckthorn aphid are the most common, but the melon aphid can also occur in damaging populations. In the Midwest, the green peach and potato aphids typically predominate, but at times the buckthorn aphid can be as common. The buckthorn aphid is typically more sensitive to insecticides; thus, applications targeting other pests will eliminate this species early in the season. The foxglove aphid is generally the least common in both the Northeast and the Midwest. In the Pacific North-

Fig. 16.2. Potato aphid (*Macrosiphum euphorbiae*). (Courtesy A. Jensen)

west and the intermountain region, the green peach aphid is the predominant species, and the potato aphid is the next most common.

Most aphid species have both winged and wingless forms. Wingless aphids are by far the predominant form colonizing potato during most of the year. Wingless forms of the five most important potato-colonizing aphid species can be differentiated in the field by observing specimens with a 10× hand lens. The key in Box 16.1 can be used to distinguish these species by their overall body shape and color, the length of the legs and cornicles, and the shape of the head, cauda, and cornicles (Fig. 16.3).

In regions too cold for continuous outdoor reproduction, potato-colonizing aphids alternate between a sexual generation, laying overwintering eggs on one species (the primary host), and generations reproducing asexually the rest of the year, generally on a much broader range of host plants (secondary hosts). The primary host ranges of all potato-colonizing aphids are highly restricted, in some cases limited to a single species, generally a woody species. The secondary host ranges are much broader, and secondary hosts are used for much of the growing season. In spring, wingless aphids called stem mothers hatch from eggs on the primary host (Fig. 16.4). They and all their descendants in the spring and summer reproduce without mating, giving birth to live young. Asexual reproduction allows exceptionally short generation times (7–10 days). This type of reproduction, coupled with the ability to reproduce at relatively cool temperatures, facilitates rapid population increases. Under favorable environmental conditions, an aphid population may double in as

Box 16.1

Key to Wingless Forms of Five Aphid Species Commonly Found on Potato

1a Body outline egg- or teardrop-shaped; cauda short . . . 2

1b Body outline elongate; antennal tubercles large and pointing outward; cauda long and pointed; cornicles longer than the distance between their bases; legs prominent; body green, yellow, or pink, sometimes with a darker dorsal stripe (Fig. 16.2); aphids highly mobile . . . . **potato aphid**

2a Antennal tubercles prominent; body thick 3

2b Antennal tubercles absent; body flattened 4

3a Body pear-shaped, widest at base of cornicles; antennal tubercles prominent, with almost parallel sides; cornicles tapered, with dark tip and prominent flanges at the tip; body light yellow green to dark green, with dark areas around base of cornicles; legs and antennae with dark joints . . . . **foxglove aphid**

3b Body egg-shaped, almost the same width from base of middle legs to base of cornicles; antennal tubercles prominent and pointing inward; cornicles unevenly swollen; body light green to almost translucent, pink, or peach; legs and cornicles the same color as body (Fig. 16.1) . . . . **green peach aphid**

4a Cornicles pale or dusky; some hairs on hind femur long and fine, the longest exceeding the diameter of the femur at its base; body opaque lemon yellow to green, turning black in autumn . . . . **buckthorn aphid**

4b Cornicles dark; hairs on femur all rather short; color variable, most commonly light green mottled with darker green, but ranging from dark green to pale yellow . . . . **melon aphid**

Key based on information in M. E. MacGillivray, 1979, Aphids Infesting Potatoes in Canada: Life Cycle and Field Key, Publ. 1678, Agriculture Canada, Ottawa, Ontario, and R. L. Blackman and V. F. Eastop, 2000, Aphids on the World's Crops: An Identification and Information Guide, 2nd ed., John Wiley and Sons, New York. Drawings adapted from MacGillivray, 1979.

little as 2 days. The offspring of stem mothers are initially all wingless. By the second or third generation on the primary host, the offspring are mostly winged and referred to as *spring migrants*. Spring migrants, as their name implies, leave the primary host in search of suitable secondary hosts (typically annual plants). It is during this migration that plant viruses can be acquired from perennial weeds, volunteer potato plants, and other sources, and potato crops can be inoculated with them. As spring progresses, the primary host usually becomes unsuitable for further aphid survival, and aphids are longer found on that host. In mild climates where herbaceous plants are available during the winter, aphids may continue to reproduce asexually.

Spring migrants can be transported long distances on low-level jet streams. **Green peach aphid** migrants have been known to travel more than 900 miles, and long-distance transport occurs primarily at altitudes of 2,900 to 49,000 feet. As spring migrants begin searching for host plants, it appears that they cannot distinguish a host from a nonhost plant at a distance but instead land on plants and then feed for short periods, typically in epidermal tissues, sampling the sap before accepting the plant as a reproductive host. Aphids remain on a secondary host plant only if they receive the proper feeding stimuli when sap sampling. Movement across the landscape occurs at relatively low altitudes and in short flights of a few yards. During host finding, aphids move among plants apparently at random, stopping to feed to determine the suitability of the plants. Sap sampling facilitates the acquisition and transmission of nonpersistent viruses. In fact, many of the aphids associated with epidemics of *Potato virus Y* are species that never colonize potato. Transient aphids present a challenge to the seed potato producer, because conventional insecticides have proven ineffective in preventing virus transmission by aphids that do not colonize potato.

Once an acceptable host has been found, spring migrants settle and begin to reproduce. The offspring of winged aphids are initially all wingless, but a small proportion of each succeeding generation is winged. Several environmental factors and changes in host plant quality can trigger aphids to produce primarily winged offspring, which collectively are termed *summer migrants*. Summer migrants leave the host plant in search of other secondary hosts. As the season progresses, day length shortens, and daily low temperatures decline, aphids enter the sexual phase of their life cycle and produce *fall migrants*, which include both male and female offspring. Fall migrants seek out primary hosts, on which the females asexually give birth to wingless females. These wingless females then mate with winged males, and instead of giving birth to live offspring they produce fertilized, overwintering eggs, which are deposited near buds on the primary host.

The **buckthorn aphid** and the **potato aphid**, like most aphids, alternate between primary and secondary host plants and produce sexual forms in the fall. The primary hosts of these species are, respectively, common buckthorn (*Rhamnus cathartica*) and wild and cultivated roses (*Rosa* spp.). The foxglove aphid and the green peach aphid have varied reproductive strategies, which are highly dependent upon the climate and the availability of host plants. They are both pests in greenhouses. These aphids may go through the typical life cycle and alternate between primary and secondary hosts, or they may overwinter as wingless forms in greenhouses and sheltered areas where the foliage of secondary hosts remains green throughout the winter.

The life cycle of the **green peach aphid** varies between and within potato-growing regions. Spring migration in western states involves a mixture of migrants from primary hosts—peach (*Prunus persica*), apricot (*P. armeniaca*), and occasionally other *Prunus* spp.—and migrants from secondary hosts that persist through the winter. Peach trees and weeds on the orchard floor appear to be the most important sources of populations of true spring migrants (second- or third-

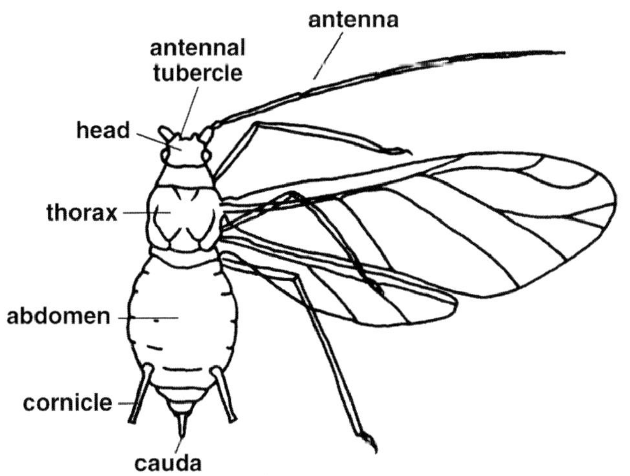

Fig. 16.3. Anatomical characteristics noted in the identification of aphid species. These characters are more pronounced in winged aphids, as are color and melanization. The cornicles are tubelike structures protruding from each side of the posterior abdomen. The cauda is the aphid's "tail," projecting from the tip of the abdomen. The antennae arise from protuberances called the antennal tubercles, on the front of the head.

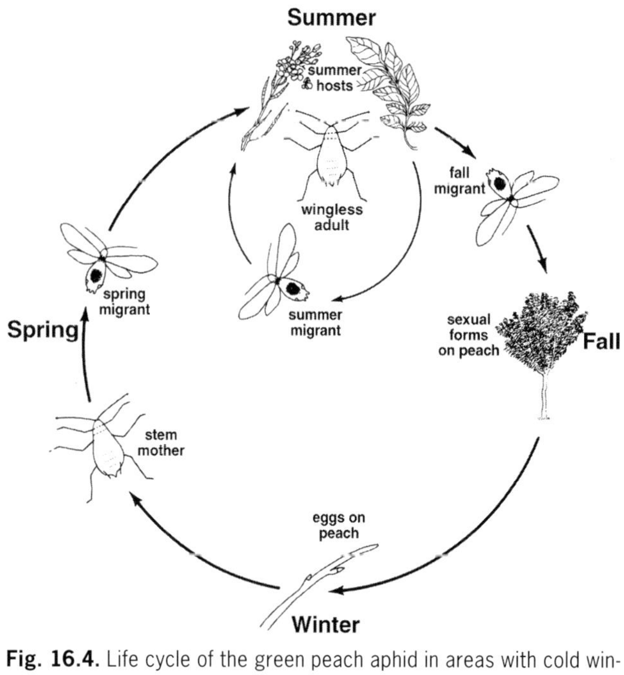

Fig. 16.4. Life cycle of the green peach aphid in areas with cold winters. (Reprinted, by permission, from Western Regional IPM Project, 1986, Integrated Pest Management for Potatoes in the Western United States, Publ. 3316, University of California, Division of Agriculture and Natural Resources, Oakland)

generation offspring of stem mothers). A single backyard peach tree may produce enough winged green peach aphids to cause an economically damaging infestation of 500 acres of potato! In regions with mild climates, such as the Pacific Northwest, winged migrants also develop from aphid populations that survive the winter as adults. These aphids may reproduce asexually on plants in sheltered areas, such as weeds in drainage ditches, or on winter cabbage and other *Brassica* crops.

In areas with severe winters, greenhouse-grown bedding plants may be an important source of initial populations of the green peach aphid in the spring. The green peach aphid is not known to overwinter in the northern Great Plains, and it rarely overwinters in New Brunswick. In these areas, spring migrants must immigrate from distant locations or originate on bedding plants. In parts of Maine, green peach aphid eggs overwinter on Canada plum (*Prunus nigra*), but long-distance migration may also be an important source of spring populations. In the Yakima Valley of Washington, overwintering on peach is more common, and green peach aphid populations are larger than those found on Canada plum in Maine. Winged forms of the green peach aphid colonize almost 1,000 species of secondary host plants, including potato, along with many other cultivated crops and common annual and perennial weeds.

The **potato aphid** overwinters in the egg stage in all potato-growing areas. Winged potato aphids are produced in June and colonize potato, tomato, groundcherry (*Physalis* spp.), and hairy nightshade (*Solanum sarrachoides*), among other species. When a large number of potato aphids feed on the upper portion of a potato plant, they can cause a condition known as toproll in some cultivars.

In Maine, the **foxglove aphid** typically remains year-round on its primary hosts, hawkweed (*Hieracium* spp.) and foxglove (*Digitalis purpurea*). In the summer, a small portion of the population migrates to other hosts, including potato. In New Brunswick, this aphid alternates between primary and secondary hosts, but it survives the winter primarily as females reproducing asexually in greenhouses. In New Jersey, the foxglove aphid exists year-round as asexually reproducing females, which overwinter on both primary and secondary hosts. In all regions, potato is a secondary host, and few winged forms are ever produced. The foxglove aphid is found mainly on older leaves and in moist, shady areas, apparently requiring high relative humidity. This aphid injects toxins while feeding, so it can cause economic injury at lower populations than other aphid species colonizing potato.

The **buckthorn aphid** is most often the earliest colonizer of potato. Spring migrants move from buckthorn to potato in June. Summer migrants are commonly caught in suction traps in Manitoba throughout much of the growing season. Fall migrants return to buckthorn, beginning in August. Populations of buckthorn aphids rarely reach high densities on potato, but because this aphid colonizes potato during early vegetative stages, virus transmission is of particular concern.

The production of winged aphids is influenced by several environmental factors. For the green peach aphid, the most important factor in the production of winged forms is crowding, but changes in host quality are also important; for example, senescing plants can trigger the production of winged aphids. High temperatures generally inhibit the production of winged forms. Shortening day length is a critical factor regulating the production of sexual forms and fall migrants. The photoperiod is usually not a factor in the production of summer migrants.

The length of time required for development varies for winged and wingless forms and is dependent upon temperature and food quality. Aphids can be considered cool-weather pests, and maximum population growth rates often occur under cool conditions, typically at temperatures below 70°F. For the green peach aphid, higher optimal temperatures have been recorded, which is perhaps why this species is such an important pest under a range of environmental conditions. At optimal temperatures, development from birth to the first reproduction of wingless aphids can take less than a week, depending on the species. All aphids molt four times before becoming adults. On average, the total lifetime production of a single wingless female ranges from 60 to over 100 offspring. Aphids attain maximum reproduction at relatively low temperatures; this characteristic favors aphid development, since predators and parasites typically require higher temperatures to attain optimal growth rates.

Several **other species of aphids** can be found on potato, including the black bean aphid, *Aphis fabae;* the mottled arum aphid, *Aulacorthum circumflexum;* the shallot aphid, *Myzus ascalonicus;* the bulb and potato aphid, *Rhopalosiphoninus latysiphon;* the rice root aphid, *Rhopalosiphum rufiabdominalis;* and *Smynthurodes betae.* All but the last two are potato virus vectors.

The **melon aphid** has developed resistance to many insecticides, and outbreaks of this aphid have occurred in the northeastern United States and the Canadian Maritime Provinces. Insecticides that are highly effective against the green peach aphid and the potato aphid may not be effective against the melon aphid, so care must be taken when identifying aphids on potato.

Transmission of Viruses by Aphids

Successful potato production is dependent upon planting healthy seed tubers. The development of seed stock in which virus infection levels are held below set tolerances is one of the principal goals of seed certification programs (Chapter 6).

Six potato viruses commonly found in North America are transmitted by aphids: *Potato leafroll virus* (PLRV), multiple strains of *Potato virus Y* (PVY), *Potato virus A* (PVA), *Potato virus S* (PVS), *Potato virus M* (PVM), and *Alfalfa mosaic virus* (AMV) (Chapter 17). PLRV and PVY are the most important of these in limiting potato yield and quality.

PLRV and PVY are transmitted by several aphid species, but the most efficient vector of both viruses throughout North America is the green peach aphid. This aphid is also the most abundant species colonizing potato. Both winged and wingless green peach aphids transmit viruses.

The potato aphid is considered a vector of both PLRV and PVY, but it is much less efficient than the green peach aphid. Some potato aphid populations have been shown to be unable to transmit PLRV in laboratory tests.

The buckthorn aphid is a relatively inefficient vector of PLRV, but it can be important in spreading PVY, PVA, and PVS.

The pea aphid, *Acyrthosiphon pisum,* does not colonize potato, but it is considered the key vector of AMV, and in some regions transmission of PVY has been attributed to this species.

The soybean aphid, *Aphis glycines,* became established in North America in 2000 and has since spread to all major soybean-growing regions in the northern United States and Canada. This aphid is capable of acquiring PVY and transmitting it to potato. In the Northeast and the Midwest, soybean aphid populations can reach extremely high densities, and summer migrants are common. The role of the soybean aphid in PVY epidemiology is yet to be determined.

The role of the foxglove aphid and the melon aphid in the epidemiology of potato viruses is unclear. The sporadic nature of outbreaks of these aphids makes it unlikely that they play a major role annually, but since they are competent vectors of PVY, their role cannot be completely ignored.

In addition to differences in transmission efficiency among aphid species, there can be large differences in efficiency of virus transmission among populations within the same species.

Aphids transmit viruses in two basic ways (Table 17.1). In **nonpersistent transmission**, the aphid remains infective for a brief period after the acquisition of a virus, transmitting the virus with its stylet. Viruses transmitted in this manner are called stylet-borne viruses. In **persistent transmission**, the virus circulates through the gut of the aphid, eventually reaching the salivary glands, so that the aphid can transmit the virus in its saliva for many days. Viruses transmitted in this manner are called circulative viruses. PLRV is the only potato virus transmitted by aphids in a persistent, circulative manner. All other aphid-transmitted viruses infecting potato are transmitted nonpersistently as stylet-borne viruses.

Aphids can acquire PLRV only while feeding from vascular (phloem) tissues of an infected plant. After an aphid ingests PLRV-infested plant sap, virus particles pass through the aphid's gut wall and into the body cavity. From there, virus particles are transported across the salivary gland membrane. As the aphid feeds, virus particles are injected into the plant along with saliva. Only aphids that accept potato as a host and feed on vascular tissue can acquire PLRV. Typically, acquisition by a virus-free aphid takes several minutes to hours of phloem feeding. It then takes hours to 1 or 2 days following acquisition before the aphid is capable of transmitting the virus. PLRV does not replicate within the aphid, but the insect retains the virus throughout much of its life (hence the term *persistent*). Unlike other potato viruses, PLRV cannot be transmitted mechanically, by machinery or by plant-to-plant contact; it is entirely dependent on aphids to move it from infected to healthy plants.

In **nonpersistent transmission**, virus particles are confined to the mouthparts and the front part of the aphid's gut. An aphid can infect only one plant with a nonpersistently transmitted virus, or at most a few plants, before losing the ability to transmit (hence the term *nonpersistent*). A virus-free aphid feeding on an infected plant can acquire virus particles from epidermal tissues after sap sampling for as little as a few seconds. Once the virus has been acquired, the aphid can immediately transmit it to other plants.

Nonpersistently transmitted stylet-borne viruses, such as PVY, PVA, PVM, PVS, and AMV, are concentrated in plant epidermal tissues. Since winged aphids search for host plants by sap sampling, they will readily transmit these nonpersistent viruses to potato, regardless of whether the aphids can colonize and reproduce on potato.

Population Monitoring and Thresholds

The concept of economic thresholds has not been widely applied to aphid management in potato production, because the complex relationships among aphid species, potato viruses, potato cultivars, and factors causing crop stress are difficult to quantify. Nevertheless, thresholds have been proposed for aphid management in processing, fresh-market, and seed potato production. These thresholds vary by region and with the intended use of the crop (Table 16.1).

Action thresholds for aphid management are more difficult to establish for seed potato production than for fresh-market or processing potatoes. European seed potato producers have long relied on various systems of monitoring aphid flights to time insecticide applications or to recom-

Table 16.1. Sampling method and action threshold for colonizing aphids on processing potatoes, table stock, and seed potatoes in different production regions

	Sampling method	Action threshold	Reference[a]
Processing potatoes and table stock			
Northern Great Plains	Leaf samples from the middle to lower parts of plants (10 leaves at each of five locations per field, for a total of 50 leaves per field)	Early in the season, an average of 15 wingless aphids per 50 leaves After bloom, an average of 100 wingless aphids per 50 leaves	Sexson et al., 2005
California	*Potato leafroll virus* (PLRV) symptomology	5% of plants with PLRV symptoms	Bacon et al., 1976
Idaho	Leaf samples from the middle to lower parts of plants (10 leaves at each of five locations per field, for a total of 50 leaves per field)	In south central Idaho, 10 wingless aphids per 50 leaves for two consecutive weeks In southwestern Idaho, 40 wingless aphids per 50 leaves	Byrne and Bishop, 1979
Seed potatoes			
Northern Great Plains	Leaf samples from the middle to lower parts of plants (100 leaves, randomly collected throughout the seed field)	In PLRV-susceptible cultivars, 3–10 wingless aphids per 100 leaves In PLRV-resistant cultivars, 30–100 wingless aphids per 100 leaves	Flanders et al., 1991 DiFonzo et al., 1995
Idaho	Beat sheet or leaf sample	First detection of green peach aphids	Mowry, 2001

[a] Complete references are given in Sources of Additional Information.

mend early vine killing and harvest. Similar methods have been tried in North America. In New Brunswick, a vine-kill date is established for seed potatoes after a cumulative capture of five green peach aphids in any one yellow pan trap. This approach is impractical, however, when pan trap captures exceed the threshold before an economically acceptable tuber yield has been obtained—a situation that occurs often in the northern Great Plains and the Pacific Northwest. In the event of early-season flight activity, insecticide applications are the only recourse against the green peach aphid. If the winged aphids have come from virus-free plants, pan trap captures are of no significance, except to indicate increasing numbers of aphids feeding on the crop. This information could be obtained more readily by counting aphids on sampled foliage.

For PVY, sources of inoculum in the seed crop appear to be more important than any weed or adjacent crop source. Clearly, when a seed lot is known to have some level of virus infection, early roguing of infected plants along with preventing the buildup of a colonizing aphid population on the crop is essential.

An action threshold of three to 10 wingless aphids per 100 leaves will reduce the spread of PLRV within a field. Lower thresholds have resulted in excessive applications of insecticide, with minimal overall gain in cumulative aphid control, while also increasing pressure for insecticide resistance and decimating natural enemies. However, this aphid density is essentially at the level of detection for most crop professionals. In practical terms, the first detection of the green peach aphid in a seed potato field is likely to indicate field population levels that are above this action threshold. This has been suggested in Idaho as the threshold for the green peach aphid to prevent excessive levels of tuber net necrosis.

Leaf counts are the simplest means of monitoring populations of wingless aphids within a field to determine the need for treatment or to evaluate the success of implemented management measures. For sampling wingless green peach aphids, selecting a random sample of leaves from the lower portion of the plant is the most efficient method. Sampling by sweep net is not a useful technique for monitoring aphid populations, because few are collected, and those that are collected may be crushed and unrecognizable. In the Pacific Northwest, beat sheets have been used to dislodge winged green peach aphids as a means of detecting when summer migrants (winged adults) are landing on the crop.

Leafhoppers

Leafhoppers are wedge-shaped insects, broadest in the head and thoracic regions and tapering along their wings. They are strong fliers and are much more mobile than aphids. Unlike aphids, the key leafhopper pest of potato, the **potato leafhopper**, *Empoasca fabae* (Figs. 16.5 and 16.6), is important mainly because of the direct feeding damage it causes. Other leafhopper species cause little direct feeding damage and are considered pests because of their role as vectors of plant pathogens. The potato leafhopper has long been recognized as a major pest of potato in eastern North America. Its full importance was only appreciated with the introduction of DDT and parathion, the first truly effective insecticides used on potatoes. Between 1946 and 1950, average potato yields in the United States increased by more than 60%, with most of the improvement being directly attributable to potato leafhopper management.

Prolonged feeding by the potato leafhopper causes a condition known as hopperburn. The first conspicuous symptom of injury is a brown, triangular lesion at the tip of an infested leaflet. As symptoms develop, lesions spread backward and inward from the margin, eventually destroying the entire leaflet. Injury to potato plants, however, occurs even before the first symptoms of hopperburn are visible. A subtle curling of the leaflet and paling of the veins may signal the onset of injury, but these symptoms can easily pass undetected. Some cultivars, such as Russet Burbank, may show little visible hopperburn, yet their yields are as severely limited as those of cultivars that readily show symptoms. Commercial cultivars at present do not appreciably differ in their ability to withstand potato leafhopper infestation. Both adults and nymphs are injurious, but late-instar nymphs can reduce yields more than an equal number of adults.

Potato leafhoppers must feed in the phloem of the plant. Damage from feeding is the result of the disruption of phloem tissues, which transport products of photosynthesis within the plant. A chemical in potato leafhopper saliva left behind after feeding induces swelling and replication of certain phloem cells, which eventually crush the phloem tissue. Potato leafhopper injury is thus irreversible. For this reason,

Fig. 16.5. Potato leafhopper adult (*Empoasca fabae*). (Courtesy D. N. Ferro)

Fig. 16.6. Potato leafhopper adult (left) and nymph (right). (Courtesy A. Al-Dawood)

the economic threshold for this pest is as low as one nymph per 10 leaves.

Potato plants in all growth stages are susceptible to injury from the potato leafhopper, but infestations are most damaging during early tuber bulking. Yield is inversely related to leafhopper population, and the relationship is effectively linear, because it takes so few leafhoppers to cause economic injury.

Identification and Biology

The species of leafhoppers that commonly colonize potato vary among production regions. Damaging populations of the **potato leafhopper** are limited to areas east of the Rocky Mountains. This species occurs throughout most of the western United States but is of little importance there. In Canada, it is the most important insect pest of potato in eastern Ontario, but it is rare in the Maritime Provinces and areas west of Manitoba. The **southern garden leafhopper**, *Empoasca solana,* is a minor pest of potato in the southern United States.

Other *Empoasca* spp. occasionally limit potato yields in the western United States, but insecticides used to manage other pests usually keep their populations low. These leafhopper species feed only on structural leaf tissues and typically cause small, white, stippled areas on leaves. They cause much less damage than the potato leafhopper, which has a different feeding mechanism. The **intermountain leafhopper**, *E. filamenta,* is a potato pest in the high arid regions between the Rocky Mountains and the Sierra Nevada. The **western potato leafhopper**, *E. abrupta,* and the **arid potato leafhopper**, *E. arida,* are pests along the Pacific slope.

Three other species—the **aster leafhopper** (six-spotted leafhopper), *Macrosteles quadrilineatus;* the **beet leafhopper**, *Circulifer tenellus* (Fig. 16.7); and the **clover leafhopper**, *Aceratagallia sanguinolenta*—are significant because they act as vectors of potato pathogens. None of these insects injure the crop directly. The aster leafhopper and the clover leafhopper are present throughout most of North America. The beet leafhopper is primarily a desert species. It occurs in the largest numbers in the western United States and is of little importance east of the Rocky Mountains, although it has been found from eastern South Dakota to Florida. The beet leafhopper has recently been associated with the transmission of a phytoplasma, known as the beet leafhopper–transmitted virescence agent, which has caused substantial tuber quality losses in the Pacific Northwest. Management programs for this insect are not yet fully developed but will likely include low-density thresholds requiring early and possibly frequent application of foliar insecticide to prevent tuber quality losses. Caution should be exercised if this pest becomes the target of routine spray applications, because experience from potato production regions in eastern North America and the Midwest has shown that repeated application of insecticides to control potato leafhoppers can induce green peach aphid outbreaks.

All of these leafhopper species have broad host ranges. The potato leafhopper, the aster leafhopper, and the beet leafhopper each feed on over 100 species of plants, and the clover leafhopper feeds on at least 38 species. The **potato leafhopper** and the **clover leafhopper** are commonly found on legumes, the **beet leafhopper** is commonly found on crucifers (members of the mustard family) and plants in the family Chenopodiaceae (the goosefoot family, including lambsquarters), and the **aster leafhopper** is commonly found on small grain cereals. Each of these species maintains populations on a wide range of weeds, some of which are alternate hosts of potato pathogens. The **intermountain leafhopper** is found mainly on potato, bean, and sugar beet. The **western potato leafhopper** and the **arid potato leafhopper** are found on many truck crops in California. The western potato leafhopper prefers cucurbits (members of the gourd family, including melons, cucumber, and squash), whereas the arid potato leafhopper favors potato and sugar beet.

Fig. 16.7. Beet leafhopper (*Circulifer tenellus*), vector of a phytoplasma known as beet leafhopper–transmitted virescence agent. (Courtesy A. Jensen)

Adults of the four leafhopper species commonly found on potato can readily be distinguished by close observation: the **potato leafhopper** by its color (a vivid lime green) and size (⅛ inch); the **clover leafhopper** by its robust body; the **aster leafhopper** by a distinctive pattern of six spots on the top of its head, differentiating it from the **beet leafhopper**, which is highly variable in its color patterns.

Leafhoppers pass through three developmental stages: egg, nymph (five instars), and adult. On potato, they usually complete several generations per year. The population density depends on the temperature and the date of arrival on the crop. In northern states, the potato leafhopper can complete two to four generations per year on potato. On all hosts, the beet leafhopper may produce three to five generations per year, and the clover leafhopper two or three. The aster leafhopper may produce as many as five generations per year, but unlike other potato-colonizing leafhoppers, this species does not reproduce on potato.

The leafhoppers that attack potato all overwinter as adults, often at great distances from potato production areas. The aster leafhopper can also overwinter locally as eggs, which are the source of aster leafhopper populations in the Northeast, although they appear to contribute little to the overall population in the North Central states.

Potato leafhopper adults overwinter in a permanent breeding area along the Gulf Coast from southern Louisiana to northern Florida, especially in the Mississippi Delta region. In spring, as potato leafhopper populations increase, flying adults may be caught in updrafts and transported on upper-level airstreams to distant sites throughout the eastern two-thirds of the United States and in southern Canada (Fig. 16.8). Wisconsin often receives the brunt of these long-distance invasions. Weather conditions that permit transport into the Midwest are a high-pressure system over the East Coast, a low-pressure system over the Great Plains, a north-south front moving east, a stationary east-west front over the Midwest, and precipitation in the fallout area. Influxes of potato leafhoppers into Illinois have been associated with southerly winds of 36 hours or more in duration and accompanying precipitation. Potato leafhoppers are carried to the eastern United States when weather conditions result in upper-level airflows extending from the Gulf Coast to the east. They often arrive in northern regions before potatoes have emerged. Often, it is the progeny of these migrants that make the first significant invasion of potato. There is a common belief that potato leafhoppers move into potato after alfalfa is cut, but in Minnesota the largest increases in population density on potato are related to leafhopper maturation on undisturbed hosts. In summer, potato leafhopper adults normally live 30–40 days, but some live over 90 days, so that generations overlap. Females lay an average of two or three eggs per day and can lay as many as 200 over a lifetime. Leafhopper eggs are inserted into leaf veins, where they are not exposed to contact insecticides. At summer temperatures, eggs hatch in about 10 days. Nymphs are nearly white when they first hatch and then turn pale yellowish green, but in later instars their color is similar to that of the adult leafhopper. They require about 12 days to develop into adults. New adults begin laying eggs when they are 6 days old.

The preferred hosts of the **aster leafhopper** are grasses. Large populations of this species can build up in small grains, especially oat and wheat. When weather conditions are right for long-distance transport, large influxes of aster leafhoppers can move into the upper Midwest from maturing cereals in the South Central states (Fig. 16.8). In the North Central states, immigrating aster leafhoppers accumulate on small grains, mainly oat. As cereals ripen, the aster leafhopper moves to other hosts, including potato. It will feed and lay eggs on potato, but those eggs fail to hatch. In the West, the aster leafhopper commonly overwinters in desert areas on Russian thistle (*Salsola* spp.). In California, it overwinters as adults on various desert plants.

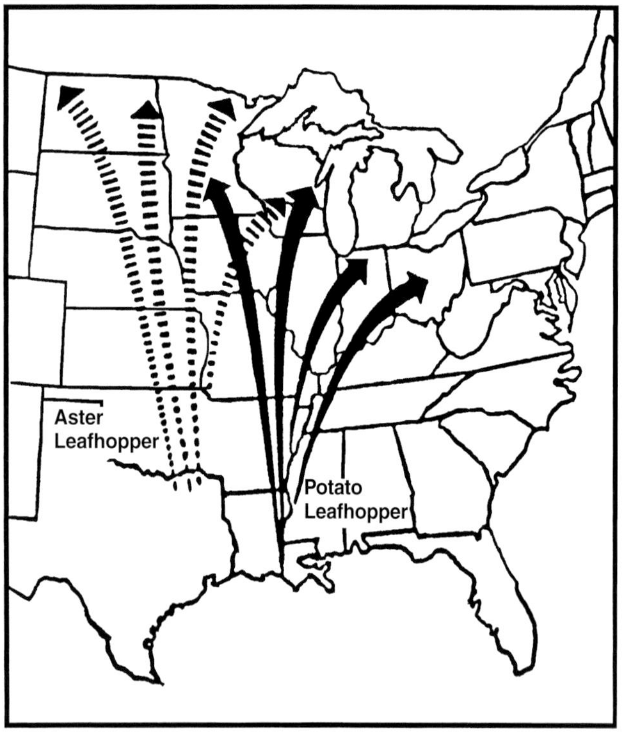

Fig. 16.8. Long-distance migration routes of the aster and potato leafhoppers. (Adapted from J. T. Medler, 1962, Long-range displacement of Homoptera in the central United States, pages 30–35 in: Proceedings of the 11th International Congress of Entomology)

In northern areas, the **clover leafhopper** overwinters as adults in and under residues of host plants, such as alfalfa. In central and southern areas, adults emerge on any warm day. They have been found in every month of the year in southern New York. Alfalfa, because of its predominance as a forage crop, is now the major source of clover leafhopper populations, but red clover (*Trifolium pratense*) and crimson clover (*T. incarnatum*) are its preferred hosts. On clovers, eggs of the clover leafhopper hatch after 5–12 days, and nymphs complete development in 25–30 days. The nymphs are creamy white with heavy dark spots and bands.

The **beet leafhopper**, in the western states, can overwinter in the immediate area of its summer hosts or can migrate seasonally. In localities where long-distance migration is characteristic, the beet leafhopper matures and leaves its summer hosts each fall, returning to permanent breeding sites. Seven such sites have been identified in the western deserts. The specific wild hosts used by breeding beet leafhoppers vary with the site, season, and amount of rainfall. Common wild hosts are Russian thistle in winter and wild mustards in spring and fall. On a suitable host plant, the beet leafhopper inserts its eggs in rows of two to five, mainly in leaf veins. Each female is capable of laying 300–400 eggs, which hatch in about 5 days when the average temperature is 100°F, but in about 40 days at 60°F. Nymphs are white when small, becoming spotted red and brown as they mature. The time required for development to the adult stage ranges from 13 days at 95°F to 75 days at 65°F. The length of the pre-egg-laying adult stage varies from 5 days at 90°F to 13 days at 68°F.

Transmission of Viruses and Phytoplasmas by Leafhoppers

Some leafhopper species are important to potato health not because of their feeding but because they transmit plant pathogens. Historically, the **aster leafhopper** has been considered the most common vector, because of its role in transmitting the phytoplasma associated with **aster yellows** (AY). The beet leafhopper–transmitted virescence agent is an emerging problem in the Pacific Northwest. Both pathogens cause a plant disease known as **potato purple top**, which affects the processing quality of daughter tubers. Reducing sugars accumulate in infected tubers, resulting in discoloration of chips and french fries produced from them.

The aster leafhopper acquires the AY phytoplasma by feeding on an infected host plant during dispersal from ripening small grains to potato. The phytoplasma cannot be acquired by feeding on infected potatoes, and thus roguing AY-infected potatoes is of no value in disease management. More than 350 plant species have been identified as hosts of the AY phytoplasma. Once the phytoplasma has been acquired by the leafhopper, replication must occur within the leafhopper's body before the insect can transmit the phytoplasma to a healthy plant. An infective leafhopper must feed on a susceptible plant for several hours for transmission to occur. The long incubation period and the long feeding period required for the transmission of the AY phytoplasma afford opportunities to reduce the spread of purple top by controlling the vector. The transmission cycle of the virescence agent transmitted by the beet leafhopper is not understood in detail, but it is probably similar to the transmission of the AY phytoplasma by the aster leafhopper.

The **clover leafhopper** is a vector of *Potato yellow dwarf virus* (PYDV). This virus was once more prevalent but now is of minor importance in potato production. Red clover and crimson clover are hosts of PYDV and are preferred hosts of the clover leafhopper. Alfalfa, now grown much more extensively than the clovers, is less preferred by the clover leafhopper and is immune to PYDV. The virus is present throughout the range of the clover leafhopper, but historically outbreaks of potato yellow dwarf have mostly been confined to the northeastern United States and Alberta. Circumstantial evidence suggests that the spread of PYDV is associated with drought that forces leafhoppers from clover to potato.

Population Monitoring and Action Thresholds

Potato leafhopper populations are relatively easy to monitor. Adults are best sampled by sweep net, and nymphs by examining randomly picked leaves from the middle of the plant. The number of samples required to satisfactorily estimate the size of a population depends on the population density and the required degree of sampling precision. In Minnesota and Wisconsin, 25 sweeps across two rows with a net 15 inches in diameter or a sample of 35 midplant leaves per field generally gives population estimates of sufficient accuracy for management decisions. In Minnesota, treatment is recommended when the population density exceeds one nymph per 10 leaves. Because potato leafhopper pressure is typically more severe in Wisconsin, and adult leafhoppers moving into the crop can rapidly cause substantial damage, sweep net sampling for potato leafhopper adults is a common technique. In practice, if producers observe a large and sudden influx of adult leafhoppers, the recommendation in Wisconsin is to spray at the first appearance of nymphs. If the number of adults caught is less than one per sweep, spraying can be delayed until the number of nymphs reaches 10 per 25 leaves. Advice on appropriate threshold values for specific production regions should be obtained from local advisors before any treatment is applied.

Management Strategies

There are three circumstances in which it is critical that management measures be specifically directed against **aphids** that colonize potatoes: (1) in seed tuber production, to eliminate virus vectors, (2) in the production of cultivars susceptible to tuber net necrosis (such as Russet Burbank), to prevent the spread of PLRV within a field, and (3) in an aphid population outbreak induced by previously applied insecticides. The appropriateness of different management tactics for dealing with aphids and aphid-borne viruses varies with the intended use of the crop, the cultivar, and the aphid species and virus diseases that must be controlled. For example, control measures directed specifically against aphids are seldom necessary for fresh-market tubers, at least for cultivars whose tuber quality is not adversely affected by current-season PLRV infection. In contrast, aphid control measures are essential in the production of seed potatoes.

Even when virus spread is the critical consideration, management must be tailored to the specific situation, because each vector-virus relationship presents a somewhat different problem. For example, insecticides are much more likely to be effective in preventing the spread of PLRV than in preventing the spread of nonpersistent viruses. The continuous presence of insecticide residues sufficient to kill aphids may not preclude the transmission of nonpersistent viruses. In general, insecticides are of little or no value for the control of nonpersistent viruses. A complicating factor is the new strains of PVY that are becoming more common, especially PVY^{NTN}, which causes tuber necrosis (different from PLRV-induced net necrosis). The first line of defense against nonpersistent viruses is seed certification programs that limit inoculum in seed crops (Box 16.2). In-season strategies that focus on vector management and interference with vector transmission are far less effective. Often these strategies do not provide sufficient control when inoculum levels in the seed exceed tolerance limits set by seed certification agencies.

Cultural Tactics

In the Pacific Northwest, various cultural practices have been proposed for green peach aphid management, based on the observation that the most vulnerable period in the aphid life cycle is passed on the primary host, peach trees. These strategies include applying chemical defoliants to peach trees in the fall to deny foliage to fall migrants, maintaining tree bands to provide shelters for beneficial insects, and pruning trees to remove most overwintering eggs. When peach trees are pruned, the twigs that are removed must be destroyed, or the eggs can still hatch, and the aphids then develop on orchard weeds that can serve as spring hosts. In orchards where ground cover is not maintained, weeds can be important sources of aphid populations in the spring. In the Yakima Valley of Washington, wingless forms can overwinter on weeds near irrigation ditches and warm-water springs.

Host Plant Resistance

Host resistance in potato cultivars is currently not a useful management measure. Currently available potato cultivars do not differ in susceptibility to aphids. Even if aphid-resistant cultivars were developed, without virus resistance they would probably only slow the spread of PLRV and have no effect on nonpersistent viruses. There are efforts to introduce genes from wild potato species that are highly resistant to aphids and to the viruses they transmit, but agronomically accepted cultivars that have both aphid resistance and virus resistance are currently not commercially available. In some potato species, this resistance is associated with glandular, sticky hairs on the stems and leaves, which are effective barriers against a wide range of pests, including aphids, leafhoppers, other small insects, and early instars of the Colorado potato beetle. Transgenic plants that have constructs limiting viral

Box 16.2

Key Management Tactics for Aphids and Leafhoppers

Aphids and Aphid-Transmitted Viruses

Proper aphid identification is needed before applying insecticide.

Green peach, potato, and melon aphids can be highly resistant to most insecticides. Use the appropriate insecticide for adequate management.

Insecticides targeting other insects (the Colorado potato beetle and leafhoppers) in combination with frequent fungicide applications can cause a rapid increase in the green peach aphid population.

For potato virus management, always plant the cleanest (virus-free) seed available.

***Potato leafroll virus* management**

In seed potatoes and cultivars susceptible to net necrosis, maintain aphid density below an economic threshold of three to 10 wingless aphids per 100 leaves.

Insecticide can control vectors and thus prevent the spread of the virus.

***Potato virus Y* management**

Insecticides provide no reduction in the spread of the virus.

Mineral oils applied weekly are effective in reducing the spread of the virus.

Crop borders can intercept aphids and thus reduce the spread of the virus into potato fields.

Leafhoppers and Leafhopper-Transmitted Pathogens

Proper leafhopper identification is needed before applying insecticide.

The action threshold for the potato leafhopper is one nymph per 10 leaves.

Monitor with a sweep net to detect a sudden increase in adult potato leafhoppers.

Apply insecticide if the potato leafhopper population density exceeds one adult per sweep and nymphs are present.

Proper identification is needed to determine the risk of transmission of pathogens.

The beet leafhopper and the aster leafhopper are the most important disease vectors.

For the potato leafhopper only, use reduced rates of insecticides, to avoid causing a rapid increase in the green peach aphid population.

replication in the plant have been produced but are currently not accepted in the marketplace.

Insecticides

Many insecticides registered for use on potato do not give satisfactory control of potato-colonizing **aphids**. Insecticide-induced outbreaks of the green peach aphid have commonly occurred, primarily because the insecticide used was one to which this aphid was already resistant. As a rule of thumb, the green peach aphid should be considered resistant to most organophosphate, carbamate, and pyrethroid insecticides (Chapter 14). The potato aphid is typically resistant to many of the same insecticides as the green peach aphid, although there are fewer reports of resistance in the potato aphid worldwide, probably because it colonizes fewer crops. The melon aphid has also developed resistance to many different insecticides. Care must be taken to properly identify which aphid is colonizing a potato crop, so that appropriate insecticides can be selected.

A new class of insecticides, the neonicotinoids, has been available in the United States since 1995. In general, neonicotinoids are active at low rates (ounces per acre), are highly persistent (persisting more than 70 days) when used as soil-applied or seed treatment systemics, provide a high degree of aphid control, and are less disruptive to natural enemies than foliar-applied insecticides. The first neonicotinoid registered for use was imidacloprid; it was followed by thiamethoxam. These compounds can be formulated as seed treatments, soil-applied systemics, or foliar-applied contact insecticides. Neonicotinoids break down rapidly with exposure to ultraviolet light and thus have a much shorter half-life when applied to foliage than when applied to seed or incorporated into the soil.

Another novel insecticide that gives acceptable control of the green peach aphid is pymetrozine (class pyridine azomethine). Insecticide-resistant green peach aphids are readily controlled with this material. Once exposed to pymetrozine, aphids immediately withdraw their stylets from the plant and never again attempt to feed, but it can take several days following application for aphid numbers to decline. To date, there has not been a report of resistance to insecticides in the neonicotinyl and pyridine azomethine classes in any potato-colonizing aphid. *Selected trade names of insecticides are given in Table 14.4.*

Insecticides applied for aphid management are of inconsistent benefit in protecting against the spread of potato viruses, even when they prevent the reproduction of aphids within the field. **Soil-applied systemic insecticides** tend to provide good early to midseason control of aphids and can limit the spread of PLRV within a field, but they are of virtually no value in limiting the spread of nonpersistent viruses, such as PVY.

Mineral oil sprays are valuable in reducing the transmission of nonpersistent viruses. Mineral oil apparently interferes with the transmission of these viruses during aphid probing. This technique is not entirely effective, however, since some vectors, such as the potato aphid, prefer to probe newly expanding leaves, which may not be protected. The efficacy of mineral oil rarely exceeds a 60% reduction in the spread of PVY, compared to the spread of the virus in an untreated control. Mineral oils and other crop protection chemicals for

PVY management can never be as effective as eliminating inoculum through seed certification programs and should be recognized as being of only supplemental benefit. Coverage is critical when using mineral oils. When the crop is actively growing, weekly application of oil is needed to ensure coverage of new growth. Application of 1–2% mineral oil in 100–125 gallons of water per acre is generally safe for the crop, but the risk of phytotoxicity increases if mineral oil is mixed with fungicides containing copper or zinc or applied at high temperatures (over 90°F). Recent research in North Dakota and Minnesota has shown that mineral oil can be tank-mixed with fungicides without compromising the control of late blight and early blight.

Mineral oils and other crop protection chemicals for *Potato virus Y* management can never be as effective as eliminating inoculum through seed certification programs and should be recognized as being of only supplemental benefit.

The insecticides most commonly used for potato **leafhopper** management are pyrethroids, although virtually any organophosphate or carbamate will provide acceptable control. Neonicotinoids applied as soil systemics at planting provide some level of potato leafhopper management but typically do not persist long enough to prevent colonization 4–6 weeks after crop emergence. It is unknown whether soil-applied neonicotinoids provide any disease reduction, but phorate used at planting has been shown to reduce purple top spread by the aster leafhopper. The standard industry practice for leafhopper management on fresh-market potatoes is to apply foliar sprays, but this can increase selective pressure for insecticide resistance in other pests, such as the green peach aphid and the Colorado potato beetle (Chapter 14).

Insecticide Resistance and Leafhopper Management

Because the potato leafhopper is a migratory insect with a very wide host range but is not economically important in its overwintering range, there is little or no selection for insecticide resistance. The real concern with potato leafhopper management is to avoid causing a flaring of the green peach aphid population in the course of treating for the leafhopper. One strategy advocated in the North Central region is the use of insecticide for potato leafhopper management at a rate below the labeled rate. Although it is common practice to caution against the use of below-label rates, the unique biology of potato leafhoppers makes the reduced rate appropriate. Even if resistance were to develop in one season on potato, the resistant individuals would not persist into the next growing season. The benefit of application rates lower than the labeled rate for potato, to avoid causing a flaring of the green peach aphid population, appears to outweigh any risks associated with exposing potato leafhoppers to sublethal doses of insecticide. It is unacceptable to apply any neonicotinoid or other insecticide that provides acceptable control of aphids or Colorado potato beetles at rates lower than the labeled rate, as this could result in the development of resistance in these nontarget insects.

Natural Enemies

Aphids on potato grown with minimum use of insecticides are typically low-density pests, since their populations are effectively regulated by natural enemies. In Maine, aphid predators were shown to be typically more beneficial than aphid parasites. Of the predatory insects, ladybird beetles, lacewings, minute pirate bugs, hover flies, damsel bugs, and seed bugs are the most important. Numerous species of parasitic wasps attack aphids on potato, and this group of natural enemies appears to play a major role in maintaining the green peach aphid at low density in the northern Great Plains. Several insect-pathogenic fungi, including *Pandora neoaphidis, Entomophthora planchoniana, Zoophthora radicans,* and *Conidiobolus obscurus,* attack aphids on potato. The fungicides commonly used on potato can be highly detrimental to these fungi. However, commercial potato production is seldom practical without routine use of broad-spectrum fungicides. Populations of green peach aphids often resurge only when fungicides are combined with multiple foliar applications of broad-spectrum insecticides.

Naturally occurring biological control agents can contribute substantially to the regulation of aphid population densities. When foliar sprays are used for aphid control, it is important to select only products that are effective aphicides and to exercise restraint in the frequency of application, to avoid causing aphid populations to resurge.

Most **leafhoppers** have few effective natural enemies. An insect-pathogenic fungus, *Erynia radicans* (*Zoophthora radicans*), commonly infects the potato leafhopper in Wisconsin and Minnesota, but it is rare in Illinois. Temperatures may be limiting there, since the fungus does not germinate above 86°F. The beet leafhopper is susceptible to the same pathogen, but the aster leafhopper is not. This fungus appears to be very diverse and may actually comprise several species. *E. radicans* seems to be a valuable biological control agent. Most fungal pathogens are limited by various factors: some have strict temperature requirements, some can germinate and cause infection only at high humidity, some are readily disseminated only when the host population density is high, and some are sensitive to commonly applied broad-spectrum fungicides. At the present time, biological control is not a viable management option for leafhoppers.

Summary

Selection of the most appropriate management techniques for aphids depends on the production locality, the cultivars grown, and the intended use of the crop. It is essential to carefully analyze the need for aphid management and the limitations of various management options. Consultation with local agricultural advisors should be useful in devising a strategy that addresses regional needs for aphid control and can deal effectively with management limitations.

CHAPTER 17

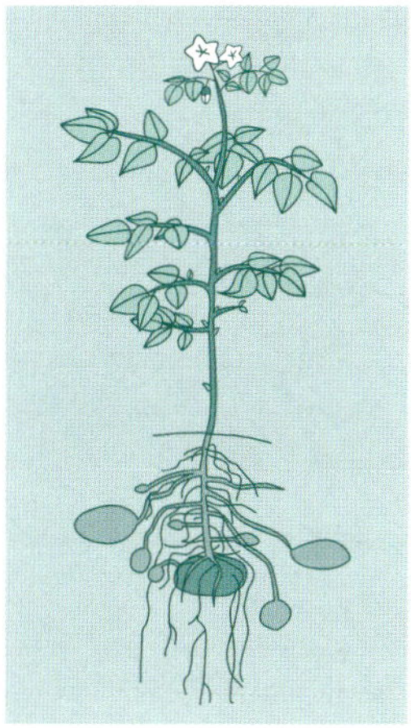

James M. Crosslin
U.S. Department of Agriculture, Agricultural Research Service
Prosser, Washington

Philip B. Hamm
Department of Botany and Plant Pathology
Oregon State University, Hermiston

Keith S. Pike
Department of Entomology
Washington State University, Prosser

Thomas M. Mowry
Parma Research and Extension Center
University of Idaho, Parma

Phillip Nolte
Idaho Falls Research and Extension Center
University of Idaho, Idaho Falls

Hassan Mojtahedi
U.S. Department of Agriculture, Agricultural Research Service
Prosser, Washington

Managing Diseases Caused by Viruses, Viroids, and Phytoplasmas

Viruses are submicroscopic, noncellular pathogens that require a living host for replication. In simplest form, plant-pathogenic viruses contain one or more segments of genetic material, either RNA or DNA, surrounded by a protein coat. Viroids are small pieces of RNA, similar to viruses but lacking a protein coat. Phytoplasmas are one-celled living organisms related to bacteria, but unlike bacteria they lack a rigid cell wall. Potato diseases caused by viruses, viroids, and phytoplasmas result in significant reductions in yield and tuber quality. The extent of these losses is often not noticed by the producer.

The primary means of dispersal of viruses are infected seed tubers and vectors, such as aphids and nematodes. Virus infection of the seed piece usually results in infection of the daughter plant and, ultimately, the daughter tubers. Many of the potato viruses are transmitted by aphids, which can rapidly spread viruses from initial infection sites to other plants in the same or nearby fields. The two most important aphid-transmitted viruses are *Potato virus Y* and *Potato leafroll virus. Potato virus X* also passes from seed pieces to daughter plants, and it can be spread mechanically from plant to plant, without an insect vector. *Potato virus A, Potato virus M, Potato virus S,* and *Alfalfa mosaic virus* are also tuber-perpetuated and aphid-transmitted, but they are less common than *Potato virus Y* and *Potato leafroll virus* in most potato-growing regions. *Potato virus T* passes from one generation to the next in true potato seed. *Tomato spotted wilt virus* is vectored by thrips and may reduce tuber quality. *Tobacco rattle virus* is transmitted by soil-inhabiting nematodes and can severely reduce tuber quality. *Potato mop-top virus* is transmitted by the soil-inhabiting powdery scab fungus, *Spongospora subterranea* f. sp. *subterranea.*

The only viroid important in potato is *Potato spindle tuber viroid.* This pathogen is easily spread mechanically and can be transmitted through true potato seed.

Most phytoplasmas that infect potato are transmitted by leafhoppers.

Modes of vector transmission and methods of detection of virus, viroid, and phytoplasma pathogens of potato are summarized in Table 17.1.

Diseases Caused by Viruses

Potato leafroll virus (PLRV) is a serious threat to potatoes wherever the crop is grown. As the name implies, the primary symptom is a distinct rolling of the leaflets (Fig. 17.1). Other symptoms may include yellowing or yellow purple discolorations of foliage, shortened internodes, an upright growth habit, and production of aerial tubers. The initial source of virus in a field is infected seed tubers or volunteers. The virus can also be introduced into a field by migrating winged aphids that have acquired it from outside sources, such as volunteer potatoes or infected weeds. As plants with primary infections emerge and grow, wingless aphids, especially the green peach aphid, *Myzus persicae,* can acquire the virus and transmit it to other plants. Other aphid species, including the potato aphid, also transmit PLRV, but the green

Table 17.1. Vector, mode of transmission, and diagnostic methods for detection of viruses, viroid, and phytoplasmas causing potato diseases

Pathogen	Vector	Mode of transmission	Diagnosis by mechanical inoculation	ELISA[a] available	Molecular diagnostic procedures[b]
Potato leafroll virus (PLRV)	Aphids	Persistent	No	Yes	RT-PCR
Potato virus Y (PVY)	Aphids	Nonpersistent	Yes	Yes	RT-PCR
Tobacco rattle virus (TRV)	Stubby-root nematodes	Persistent[c]	Yes	Yes	RT-PCR
Potato virus X (PVX)	None	. . .	Yes	Yes	RT-PCR
Potato mop-top virus (PMTV)	Powdery scab fungus	Infected zoospores	Yes	Yes	RT-PCR
Potato virus A (PVA), *Potato virus S* (PVS), *Potato virus M* (PVM), *Alfalfa mosaic virus* (AMV)	Aphids	Nonpersistent	Yes	Yes	RT-PCR
Potato spindle tuber viroid (PSTVd)	None[d]	. . .	Yes	No	HA, RT-PCR
Phytoplasmas	Leafhoppers	Persistent	No	No	HA, PCR

[a] Enzyme-linked immunosorbent assay.
[b] Hybridization assays (HA), polymerase chain reaction (PCR), or reverse transcription PCR (RT-PCR).
[c] The virus is lost when the nematode molts. Adult nematodes may retain infectivity for extended periods.
[d] Some transmission of PSTVd by aphids may occur in mixed infections with PLRV.

peach aphid is the most efficient vector (Chapter 16). PLRV is persistent in aphids; once they acquire the virus, they can transmit it for a long period, usually for life. From midseason to late in the season, winged aphids develop in the potato crop and can rapidly spread PLRV over relatively large areas unless they are managed.

In addition to significant yield reductions, current-season infections of some susceptible cultivars, such as Russet Burbank and Shepody, also cause internal tuber symptoms termed **net necrosis** (Fig. 17.2). Net necrosis is a dark, somewhat fibrous discoloration in and near the vascular tissue of tubers, usually concentrated at the stem end. The quality of a crop can be downgraded because of the internal symptoms, and severely affected crops may be rejected. Tuber symptoms are not always obvious at harvest but can become pronounced after 1–2 months in storage. Stored tubers of susceptible cultivars should be periodically monitored for symptom development. Net necrosis may be more pronounced in tubers stored at 50°F than those stored at 38–42°F.

Primary infections of PLRV in foliage from tuber-borne virus can be readily identified by enzyme-linked immunosorbent assay (ELISA). Infections occurring during the growing season may be more difficult to identify, as the symptoms may appear late or the virus concentration may be low. Sprouts on daughter tubers are a reliable source of tissue for detection of PLRV by ELISA. A molecular method, reverse transcription polymerase chain reaction (RT-PCR), can also detect the virus.

Potato virus Y (PVY), like PLRV, occurs worldwide wherever potatoes are grown. PVY is also transmitted by aphids, but in a nonpersistent (stylet-borne) manner. This means that an aphid can transmit the virus in a matter of minutes after acquiring it but does not remain infective for an extended period or after feeding on uninfected plants. Because aphids can acquire and transmit the virus quickly, before insecticides can kill them, insecticides are not effective in controlling the spread of PVY. The green peach aphid is the most efficient vector of PVY, but several other aphids, including the potato aphid and the soybean aphid, have been reported to transmit the virus (Chapter 16). In some situations, aphids that do not colonize potato may be responsible for considerable spread of the virus. PVY can be of economic importance in tomato, pepper, and a few other vegetable crops, in addition to potato.

PVY occurs as numerous strains or biological types, many of which overlap in certain characteristics. Traditionally these viruses have been designated as the ordinary strain (PVY^O), the tobacco necrotic strain (PVY^N), the tuber necrotic strain

Fig. 17.1. Chlorosis and rolling of leaflets caused by *Potato leafroll virus.* (Courtesy D. A. Johnson)

Fig. 17.2. Net necrosis caused by *Potato leafroll virus* in a tuber of the cultivar Green Mountain. (Reprinted from W. R. Stevenson, R. Loria, G. D. Franc, and D. P. Weingartner, eds., 2001, Compendium of Potato Diseases, 2nd ed., American Phytopathological Society, St. Paul, MN)

(PVY^{NTN}), and the stipple-streak strain (PVY^{C}). The PVY^{N} strains were unknown in North America until approximately 1990, when an outbreak in Canada resulted in implementation of control and quarantine procedures. In about 2002, reports of PVY^{N} in the western United States also appeared, and by 2005 PVY^{N} strains were commonly found across the United States and Canada. Sequencing the genetic material (RNA) of various PVY isolates has identified strains that have characteristics of both PVY^{O} and PVY^{N}; these recombinant viruses have been called $PVY^{N:O}$. Some, if not all, of these $PVY^{N:O}$ isolates also cause tuber symptoms and therefore overlap in symptomatology with the classical PVY^{NTN} isolates. The widely used ELISA tests for detection and differentiation of PVY strains identify the $PVY^{N:O}$ isolates as PVY^{O} and not as necrosis-inducing strains. This has allowed $PVY^{N:O}$ to be unrecognized during certification testing, and it has become widely established in seed and commercial fields in the United States and Canada. PVY isolates that have characteristics of both PVY^{O} and PVY^{N} have also been reported in many areas in Europe. PVY^{C} causes necrotic symptoms and dwarfing in potato cultivars carrying the *Nc* resistance gene. This virus strain is not commonly found in the United States.

Infections with a PVY strain can significantly reduce yield, and virus strains that cause tuber symptoms also reduce tuber quality. Foliar symptoms vary somewhat by strain but include mottling, yellowing, stunting, and leaf drop (Fig. 17.3). In many cultivars, infections with PVY^{N} strains produce very mild foliar symptoms that may be difficult to diagnose solely on the basis of visual symptoms. A diagnostic test such as ELISA is often needed to detect these infections. The symptoms of PVY^{O} are more pronounced in some cultivars than others. In Ranger Russet, for example, PVY^{O} causes a pronounced leaf drop, whereas in Russet Burbank this strain produces a mild mosaic. The tuber symptoms induced by PVY^{NTN} and some $PVY^{N:O}$ isolates include surface cracks, rings, and arcs that may penetrate a short distance into the tuber flesh. There may also be light brown internal rings, arcs, and blotches deep within the tuber flesh, in the absence of external symptoms. Cultivars such as Yukon Gold, Alturas, Ranger Russet, and some red-skinned cultivars appear particularly susceptible to tuber necrosis produced by these viruses. Russet Norkotah, Shepody, Gem, and Gem Russet frequently develop very mild foliar symptoms of PVY but still suffer significant yield losses. These last four cultivars, particularly Russet Norkotah and Shepody, are considered virus reservoirs, because of the difficulty in visually identifying infected plants in seed-growing areas.

Fig. 17.3. Foliar symptoms caused by a recombinant strain of *Potato virus Y* ($PVY^{N:O}$) in Ranger Russet.

PVY infections in foliage can be readily detected by ELISA. However, detection in dormant tubers by ELISA is more problematic. Tubers are usually tested by initiating sprouting and then subjecting the sprouts to ELISA. The three serotypes of the virus, O, N, and C, can be identified by strain-specific monoclonal antibodies. The PVY^{O} and $PVY^{N:O}$ strains cannot be differentiated by ELISA. Alternatively, RT-PCR can be used for detection and specific identification of the various PVY strains.

Tobacco rattle virus (TRV) is of economic importance in the Pacific Northwest, some of the Rocky Mountain states, and Florida, usually in areas with sandy soils. Elsewhere in the United States its occurrence is sporadic. The virus is transmitted by stubby-root nematodes (Chapter 20), in the genera *Paratrichodorus* and *Trichodorus*. In the western United States, the primary vector identified to date is *P. allius*, although *P. teres* was recently reported in Washington. In other regions of the United States, *P. minor* has been reported as a vector of TRV. In Europe, several additional species of *Paratrichodorus* and *Trichodorus* are vectors of the virus. Potatoes are the most economically important host of TRV in the United States. Many floral bulb crops, such as narcissus and tulips, can also suffer losses from this virus. Spinach and some other vegetables can be affected.

When nematodes carrying TRV feed on developing potato tubers, the virus causes a disease called corky ringspot in the United States and spraing in Europe. Tuber symptoms vary with the cultivar and the location but may include irregular tuber shape; various internal blotches, rings, and arcs; and corky tissue (Figs. 17.4 and 17.5). Symptoms of corky ringspot may or may not be visible on the outer surface of the tuber. Disease incidence can be very high and can result in the rejection of an entire crop.

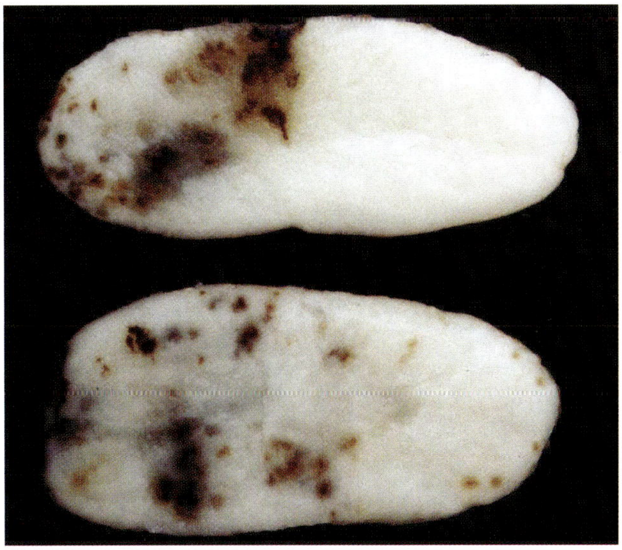

Fig. 17.4. Severe symptoms of corky ringspot, caused by *Tobacco rattle virus*, in Russet Burbank.

In rare cases, TRV will cause foliar symptoms in which leaves on one or more stems of a plant exhibit mottling. At first these symptoms resemble the yellowing caused by *Alfalfa mosaic virus*, but the yellowing is more ringlike in TRV infections (Fig. 17.6). This symptom is most often seen early in the season when an infected tuber produces a systemically infected plant. Generally the plant will grow out of this condition, and symptoms will not be apparent later in the season. The virus is infrequently transmitted to daughter tubers from an infected parent plant.

ELISA may detect TRV in shoots showing stem mottle early in the season, but later in the season RT-PCR may be necessary to detect TRV in foliage. The concentration of TRV in dormant tubers is frequently low, and therefore ELISA is generally unreliable for detection. Instead, RT-PCR is routinely used for detection of TRV in dormant tubers.

Potato virus X (PVX) is commonly found in potatoes but is usually of less economic importance than PLRV and PVY. Seed certification and testing efforts have greatly reduced the incidence of PVX in seed lots in the last two decades. Unlike PLRV and PVY, PVX is not insect-transmitted. It is transmitted only by infected seed tubers and mechanically by equipment and foliar contact. It has been reported that the virus can be spread by contaminated seed-cutting equipment

Fig. 17.5. Severe symptoms (rings and arcs) of corky ringspot, caused by *Tobacco rattle virus*, on Yukon Gold tubers.

Fig. 17.6. Mosaic and ringspots caused by *Tobacco rattle virus* on Russet Norkotah foliage.

and by tuber contact in storage. Foliar symptoms vary somewhat by cultivar but may include rough, leathery leaves and a mild to severe mosaic. Some infected plants show no symptoms, and for this reason the virus was once called potato latent virus. Even in the absence of symptoms, yield can be reduced to some degree. Mixed infections of PVX with either PVY or *Potato virus A*, however, result in very severe symptoms called rugose mosaic and can drastically reduce yields.

PVX is readily detectable in foliage by ELISA, since the concentration of the virus in infected plants is generally high. It can also be detected in dormant tubers and sprouts by ELISA unless the concentration of the virus is very low. In this case, RT-PCR can be used to detect PVX infection.

Potato virus A (PVA) is related to PVY and shares some of the same characteristics. This virus is tuber-perpetuated and aphid-transmitted in a nonpersistent manner, like PVY. The occurrence of PVA is sporadic, and infections seldom cause significant yield reductions in most cultivars and production areas. Mixed infections with PVY and PVA, however, can produce severe foliar symptoms and result in yield reductions. Tuber symptoms of PVA infection in the major North American cultivars have not been reported. PVA infection is generally identified by ELISA.

Potato virus S (PVS) is widespread in cultivated potatoes but is generally considered to have relatively minor effects on yield, although it may reduce yields by up to 20% in some cultivars. Early-generation materials often have a low incidence of PVS, because of elimination of the virus by heat treatment and meristem culture. After a few field generations, however, the virus can be found in nearly all commercial plantings. Like PVY and PVA, PVS is aphid-transmitted in a nonpersistent manner and readily spreads within a crop.

Potato virus M (PVM) is related to PVS, but its occurrence is very sporadic, and it is usually not widespread within a field. For this reason, the direct effect of PVM on the yield of most North American cultivars is unknown. PVM is aphid-transmitted in a nonpersistent manner.

Both PVS and PVM can be detected by ELISA. Recently, however, an isolate of PVM has been found that is serologically distinct from the more common strains. This could complicate serological detection of PVM by ELISA. In this case, RT-PCR tests may be necessary for positive identification.

Alfalfa mosaic virus (AMV), another aphid-transmitted virus, is occasionally found in potatoes. AMV infection causes a bright yellow chlorosis commonly called calico (Fig. 17.7). The foliar symptoms are obvious, and roguing of infected plants in seed fields reduces the incidence of the virus in seed lots. The virus is common in alfalfa but rarely moves into potato fields. Like the foliar symptoms, AMV symptoms in tubers are dramatic: large, dark brown to black rings and arcs develop throughout the tuber (Fig. 17.8). Management measures specifically to control AMV are generally not necessary.

The distinct foliar calico symptom caused by AMV generally makes specific tests for identification unnecessary. The virus can be detected by ELISA, if necessary. Tuber infections are usually identified by RT-PCR.

Tomato spotted wilt virus (TSWV) infects many vegetable and flower crops and has recently been recognized as a problem in potatoes in some areas. This virus is transmitted in a persistent manner by several species of thrips. Foliar symptoms include necrotic spots, rings, or irregular lesions on leaves and elongated brown lesions on stems. Some of the leaf lesions are similar to those caused by early blight, but the distinct rings are characteristic of virus infection. Tuber symptoms include discolored blotches and spots but vary with the cultivar and the location. A serological test, such as ELISA, can identify TSWV infection in symptomatic foliage. Tuber infections are identified by RT-PCR.

Potato mop-top virus (PMTV) has been reported in North and South America, Europe, and Asia. It has been confirmed in Maine and detected in tubers from several locations across the United States and Canada. This virus is transmitted by the soilborne powdery scab pathogen, *Spongospora subterranea,* and can remain viable in soil for many years (Chapter 19). Foliar symptoms include mosaics, chevrons, arcs, and calico-like symptoms (Fig. 17.9), somewhat resembling those of AMV or TRV. The foliar symptoms may be more pronounced in cool climates. Tuber symptoms include necrotic arcs, rings, and blotches (Fig. 17.10), which often resemble those caused by TRV. ELISA reagents are commercially available for detection of PMTV. RT-PCR has also been used for identification of the virus.

Potato spindle tuber viroid (PSTVd) has been reported around the world, but the incidence of this pathogen has declined sharply in North America in the last few decades, as a result of the adoption of molecular-based detection methods in certification programs. PSTVd can be spread mechanically and by foliar contact, and transmission of the viroid by aphids in mixed infections with PLRV has been reported. Unlike the major potato viruses, PSTVd can be transmitted in true potato seed. For this reason, avoiding infections in breeding programs is important. Various strains of PSTVd have been described that differ in their impact on potatoes.

Foliar symptoms of PSTVd infection are often very faint but may include stunting and stiff, upright leaves. These symptoms can also be caused by some viruses. PSTVd-infected tubers are elongated and spindle-shaped, and hence the name. Yield losses from mild strains of the pathogen can be negligible, but severe strains can cause significant losses. Since viroids do not have a protein coat, they cannot be detected by serological tests such as ELISA. PSTVd is usually detected and identified by a nucleic acid hybridization test. RT-PCR also can be used for detection.

Fig. 17.7. Calico symptoms on a potato plant infected with *Alfalfa mosaic virus.* (Courtesy D. A. Johnson)

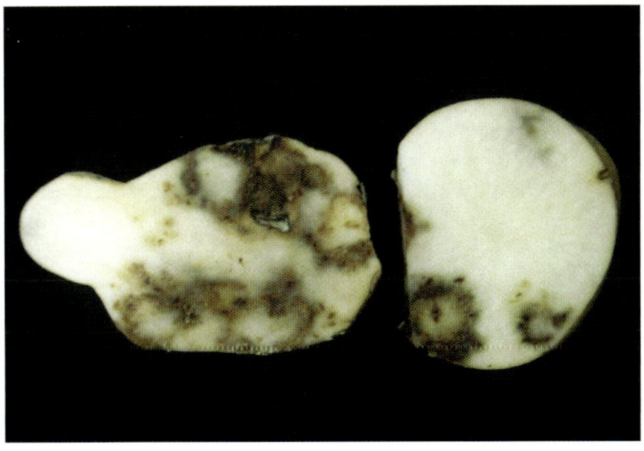

Fig. 17.8. Symptoms caused by *Alfalfa mosaic virus* in Russet Burbank tubers. (Reproduced from Diseases of Root and Tuber Crops, 2002, APS Digital Image Collections, American Phytopathological Society, St. Paul, MN)

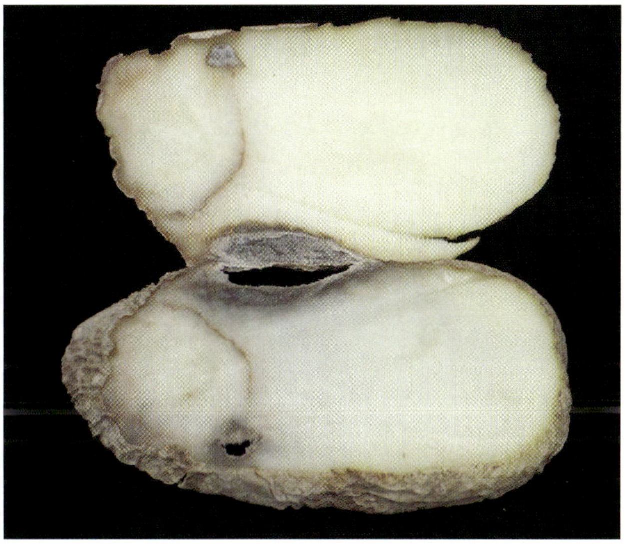

Fig. 17.9. Necrotic, dark brown arc in tuber flesh caused by *Potato mop-top virus.*

Fig. 17.10. Bright yellow blotches on potato foliage caused by *Potato mop-top virus.* (Reproduced, by permission, from R. A. C. Jones, C. Fribourg, and S. A. Slack, 1982, Potato Virus and Virus-Like Diseases, Plant Virus Slide Series, set 2, O. W. Barnett and S. A. Tolin, eds., Clemson University, Clemson, SC)

Diseases Caused by Phytoplasmas

Aster yellows (AY) is caused by a phytoplasma with a wide host range, including many vegetable crops. However, several strains or biological types of AY have been described that differ in the host plants they affect and the symptoms they cause. The aster leafhopper, *Macrosteles quadrilineatus,* is most often responsible for the transmission of the AY phytoplasma (Chapter 16). The most striking symptom of AY infection in potatoes is called **purple top**, because infected plants show leaf rolling, chlorosis, and a purple discoloration of shoot tips. Aerial tubers are frequently produced on infected plants, as a result of phloem tissue damage that prevents carbohydrates from moving to the developing underground tubers. The disease is a serious problem in some potato-growing regions of the United States and Mexico.

Another phytoplasma associated with purple top symptoms in potatoes in the Pacific Northwest is related to or synonymous with the beet leafhopper–transmitted virescence agent (BLTVA). This phytoplasma was originally described as a pathogen of periwinkles and numerous vegetable crops in the western United States. As the name implies, BLTVA is transmitted by the beet leafhopper, *Circulifer tenellus*. The symptoms of BLTVA infection are similar to those of AY. The symptoms in potatoes grown in the Columbia Basin of Washington and Oregon are similar to purple top (Fig. 17.11) and may have been attributed to AY in the past.

Another phytoplasma, distinct from both the AY phytoplasma and BLTVA, has been reported in Mexico, producing hair sprouts on tubers. The insect vectors of the different phytoplasmas described in Mexico have not yet been reported.

Phytoplasmas related to BLTVA have also been reported on potatoes in North Dakota, Alaska, and Canada. They cause a symptom called **witches'-broom**, because of the dense proliferation of shoots on infected plants.

There are few detailed reports on the effects of phytoplasma infections on tubers. The sugar balance in stored tubers can be altered by infection, leading to undesirable color development during processing. Yields may be reduced in severely infected fields, but the extent of the loss has been difficult to document. Reports from Mexico indicate that major yield losses have occurred in fields in which a high percentage of plants showed purple top symptoms.

Fig. 17.11. Symptoms of potato purple top, caused by a phytoplasma, in Ranger Russet.

Phytoplasmas are infrequently passed to daughter tubers, although infected seed tubers exhibit poor germination and produce fine hair sprouts, resulting in weak daughter plants. These weak plants produce few marketable tubers, and the pathogen is therefore considered self-eliminating in most cases.

In contrast to the viruses, ELISA reagents for phytoplasmas of potatoes are not available. Instead, detection and identification of these organisms rely on molecular tests such as hybridization assays and PCR. Differentiation between BLTVA and the AY phytoplasma requires a PCR test.

Management Strategies

The basic strategies for managing potato diseases caused by viruses, viroids, and phytoplasmas are planting certified seed tubers; managing vectors; sanitation practices, including cleaning and sanitizing equipment and storage areas and eliminating refuse tubers and volunteer potatoes; and planting disease-resistant cultivars when available (Box 17.1).

Management Prior to Planting

Certified Seed Tubers

Planting certified seed tubers is crucial for managing PLRV, PVY, PVX, PVA, PVS, PVM, AMV, and PSTVd (Chapter 6). Tuber-borne inoculum of TRV has traditionally been considered of minor importance for the spread of the virus, since TRV is rarely transmitted to daughter plants and even more rarely produces symptomatic daughter tubers. Nonetheless, most states have a "zero tolerance" for TRV-like symptoms in certified seed lots. Recent reports from Europe, however, suggest that symptomless tubers may harbor TRV and allow long-distance spread of the virus. Seed tubers infected with PMTV rarely give rise to systemically infected plants, and the daughter tubers are usually virus-free. For this reason, PMTV is considered self-eliminating if production fields are free of the virus and vector. However, planting seed tubers produced in areas free of powdery scab and PMTV prevents introduction of the virus into new areas.

Refuse Tubers and Volunteer Potatoes

Visual inspections of winter grow-outs and ELISA testing for PLRV have reduced, but not eliminated, the incidence of the virus in certified seed stocks. However, refuse tubers and volunteer potato plants can also serve as important sources of several viruses and need to be eliminated and managed, respectively.

Avoiding Infested Fields and Crop Rotation

A primary tactic for control of TRV is to avoid planting in areas where the stubby-root nematode occurs. Unfortunately, in the Pacific Northwest the nematode is widespread, and avoidance may not be practical. In addition, the nematode alone does not cause damage. Determining the number of nematodes in a soil sample is a common practice, but even fields that contain relatively few viruliferous stubby-root nematodes may suffer significant losses to corky ringspot. Soil samples can also be planted with bait plants, such as cucumber or tobacco, which are then observed or tested for the presence of TRV. Results of these soil sample tests assist growers in making a "treat or not treat" decision but may not be sensitive enough except when positive for TRV. Fumigation and foliar applications of registered chemicals that are nematicidal or nematistatic are used to suppress the population or reduce the feeding activity of stubby-root nematodes (Chapter 20). Biofumigant green manure crops, such as certain mustards, can reduce the numbers of stubby-root nematodes. These crops are not currently recommended as the sole method of control of stubby-root nematodes, but they can significantly reduce nematode numbers when used in conjunction with nematicides.

Growing alfalfa or certain species of mint for two or more years prior to potatoes may reduce the incidence of corky ringspot. These plants are good hosts of the nematode but not of the virus. Corn is an excellent host of the nematodes, and its roots can be infected by TRV. Therefore, growing a corn crop before a potato crop may increase corky ringspot. Some weeds, particularly nightshades, are hosts of the virus and nematodes, so maintaining weed-free potato fields or rotational crops may reduce disease incidence.

Management of PMTV is accomplished by avoiding planting potatoes in areas known to contain the powdery scab pathogen infected with the virus. The resting spores of *Spongospora* are long-lived, even in the absence of potato crops, and fumigation with currently available materials is of little use in controlling the vector and PMTV.

Host Resistance

Genetic resistance to PLRV has been identified in some cultivars, such as Abnaki. Most North American cultivars, however, are susceptible to PLRV infection, and the virus must be managed by other means. Genetically engineered resistance to PLRV has been developed in some potato cultivars, but they have not been widely accepted by the potato industry in the United States and thus are not commercially grown to any large extent.

There are sources of genetic resistance to some PVY strains in *Solanum tuberosum* subsp. *andigena* and *Solanum stoloniferum*. The Ry_{adg} and Ry_{sto} genes for PVY resistance from these plants have been incorporated into several European cultivars but have not yet been bred into major North American cultivars. There are other genetic factors that confer some resistance to PVY infection. For example, the cultivars Bannock Russet, Defender, and Ranger Russet show lower levels of PVY than Russet Burbank (Fig. 17.12). The cultivars Shepody, Gem Russet, and Russet Norkotah are very susceptible to PVY; however, they may not show dramatic foliar symptoms. These cultivars may be more attractive to aphid vectors than other cultivars. Potatoes with genetically engineered resistance to PVY have also been produced by a number of research groups, but this technology has not been accepted by the processing industry or the public, and consequently these breeding lines, like the PLRV-resistant transgenics, are not widely grown in the United States.

Some potato cultivars, such as Bintje, are highly resistant to TRV and rarely show symptoms of corky ringspot. Other

Box 17.1

Management Strategies for Diseases Caused by Viruses, Viroids, and Phytoplasmas

- **Plant certified seed tubers**
- **Manage aphid, leafhopper, and nematode vectors**
- **Clean and sanitize equipment and storage areas**
- **Eliminate cull piles and manage volunteer potatoes**
- **Plant disease-resistant cultivars**

Key Tactics for Specific Pathogens

***Potato leafroll virus* (PLRV)**

- Plant certified seed tubers
- Eliminate refuse tubers
- Manage volunteer potatoes
- Manage aphid vectors

***Potato virus Y* (PVY)**

- Plant certified seed tubers
- Eliminate refuse tubers
- Manage volunteer potatoes

***Tobacco rattle virus* (TRV)**

- Avoid planting in fields infested with stubby-root nematodes, which are vectors of TRV
- Reduce the stubby-root nematode population by soil fumigation or application of foliar nematicides
- Rotate with weed-free alfalfa

***Potato virus A* (PVA), *Potato virus S* (PVS), *Potato virus X* (PVX), *Potato virus M* (PVM), *Alfalfa mosaic virus* (AMV)**

- Plant certified seed tubers
- Eliminate refuse tubers
- Sanitize equipment and facilities used in cutting, handling, transport, planting, and storage

***Potato mop-top virus* (PMTV)**

- Plant in fields free of PMTV and its vector, the powdery scab fungus

***Potato spindle tuber viroid* (PSTVd)**

- Plant certified seed tubers
- Sanitize equipment and facilities used in cutting, handling, transport, planting, and storage

Aster yellows (AY) phytoplasma, beet leafhopper–transmitted virescence agent (BLTVA)

- Manage leafhopper vectors with foliar insecticides

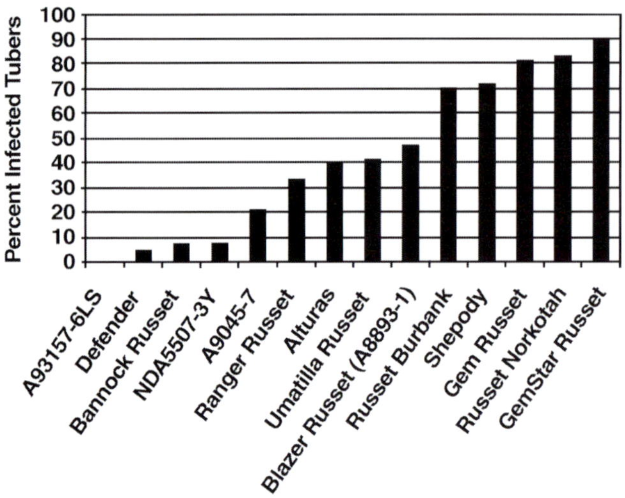

Fig. 17.12. Incidence of *Potato virus Y* infection in tubers of various cultivars grown in a virus disease nursery in Kimberly, Idaho, over periods of 2–9 years.

TRV-resistant cultivars are grown in Europe, but most North American cultivars, including Russet Burbank, Russet Norkotah, Yukon Gold, and Dark Red Norland, are susceptible. Efforts are under way to identify breeding lines that are resistant to the development of corky ringspot symptoms. Some breeding lines have been evaluated for several years in corky ringspot disease nurseries and show agronomically acceptable traits. Within a few years, TRV-resistant russet-type cultivars may be available.

Management During Planting

Since PVX, PVS, and PVM are easily transmitted mechanically, it is important to clean and chemically sanitize seed-cutting equipment, conveyor systems, trucks, bins, and other sources of possible tuber contamination. A picker-type planter can spread PVX during planting and should not be used on a seed lot known to contain this virus. Sanitation practices to eliminate contamination on cutting tools, bins, transportation equipment, and storage areas will reduce the spread of PSTVd, if it is present.

Management During Vegetative Growth

Vector Management

Foliar insecticides can reduce within-field spread of PLRV, since aphids need to feed for a relatively long period to acquire and transmit the virus (Chapter 16). One aphid management plan calls for application of a systemic insecticide at planting, to provide early-season aphid control by limiting colonization by winged aphids. This is followed by insect scouting and application of foliar insecticides if aphids are found in the crop. Aphid management during the growing season can greatly reduce secondary spread of the virus. It is especially important to reduce aphid populations from midseason to late in the season, when winged aphids are migrating, to reduce the proportion of infected tubers going into storage. It may be necessary to manage aphid populations in fields of cultivars not damaged by net necrosis, to reduce the spread of the virus to nearby fields of susceptible cultivars. The green peach aphid overwinters as eggs on *Prunus* trees, such as peach, cherry, and apricot, and it may be necessary to spray these trees early in the season, before aphids migrate.

Since **PVY** is nonpersistently transmitted, insecticide applications to manage its aphid vectors are not effective. The virus can be acquired and transmitted before an insecticide can kill the vector. Insecticide applications may increase the incidence of PVY by increasing the movement of aphids between plants. Sprays with a very rapid knockdown may be of some benefit in reducing secondary spread. Crop oils can reduce the spread of PVY by preventing the virus from adhering to aphid mouthparts, but these products must be applied at frequent intervals (Chapter 16).

Since PVY is nonpersistently transmitted, insecticide applications to manage its aphid vectors are not effective.

The application of insecticide specifically to reduce secondary infections of PVA, PVS, and PVM is usually not economically practical. Rather, measures directed at managing the vectors of PLRV and PVY reduce the spread of these other viruses as well.

Despite the widespread occurrence of the thrips vector, there are few reports of severe outbreaks of TSWV-like symptoms in potato. Since the occurrence of TSWV in the field is erratic, no specific management measures are recommended, other than reducing thrips if populations are high and a source of virus inoculum is present. However, in **minituber production** under greenhouse conditions, it may be necessary to manage thrips with insecticides, since the virus can spread rapidly and has a wide host range, including many greenhouse floral crops and weeds.

Reducing the incidence of diseases associated with **phytoplasmas** relies on management of the insect vectors (Chapter 16). Yellow sticky traps can be used to determine when the first flights of leafhoppers occur. The beet leafhopper overwinters as adults in potato-growing regions of the Pacific Northwest and becomes active when temperatures rise and spring rain leads to the germination of weedy hosts. These overwintered adults lay eggs on early-season plants, especially various mustards, and the next generation of insects will begin to migrate to other hosts when the spring hosts begin to dry. In potato-growing regions of the Pacific Northwest, this occurs around the middle of May, when the potatoes are still relatively small. Early applications of foliar insecticides to manage this migratory population by mid- to late May through early June have greatly reduced the incidence of phytoplasma-caused diseases in the Columbia Basin region of Washington and Oregon. Managing the insect in organic and low-input crops is more challenging, and losses in these crops can be significant. The rate of phytoplasma-bearing beet leafhoppers has consistently been around 15% in the Columbia Basin, and thus detection of any beet leafhoppers early in the season suggests that insect management measures should be considered.

In the upper Midwest, aster leafhopper vectors of the AY phytoplasma need to be monitored. Action thresholds based on the percentage of aster leafhoppers that test positive for the phytoplasma have been established in some states. Usually if only 1–2% of the collected leafhoppers are positive for the AY phytoplasma, then a spray program should be initiated, and the leafhopper population should be monitored.

CHAPTER 18

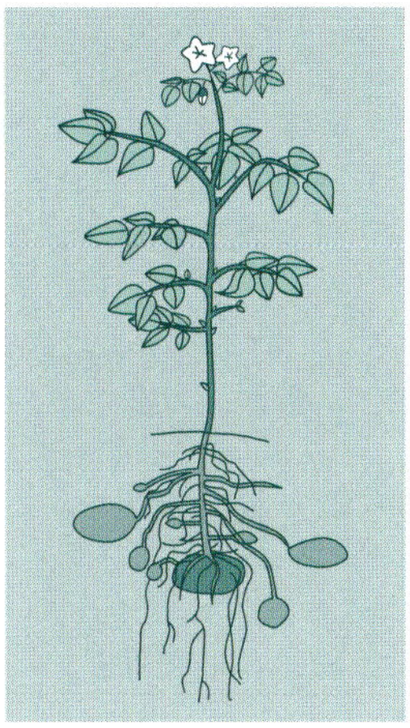

Solke H. De Boer
Canadian Food Inspection Agency
Charlottetown Laboratory, Charlottetown, Prince Edward Island

Managing Soft Rot and Ring Rot

Several potato diseases are caused by bacteria; the important ones in North America are soft rot and ring rot. The fact that these diseases are caused by bacteria has special significance for their management. Bacteria are quite different from fungi, which cause diseases such as early blight and silver scurf, and different from viruses, which cause diseases such as leaf roll and mosaic. Unlike fungi, bacteria cannot enter plants on their own but enter only through natural openings or wounds. Unlike viruses, they can live and multiply outside of plants.

Bacteria are single-celled organisms and extremely small. It would take about 25,000 bacteria placed end to end to cover a distance of 1 inch. They can also multiply rapidly. In fact, the damage done by the bacteria causing potato diseases is due, in part, to their capacity to generate large populations very quickly. Under optimum conditions, a population of bacteria can double in a period ranging from about 20 minutes to 4 hours. With a 20-minute generation time, a single cell can theoretically give rise to a population of an astounding 4,722,366,482,869,650 million cells in just 24 hours. With a 4-hour generation time, a single cell would give rise to a modest population of only 64 cells in 24 hours, but given a week it would theoretically yield 4,398,046,511,104 cells. Obviously, a useful management strategy for bacterial diseases will focus on avoiding optimum growing conditions for the bacteria.

The Soft Rots

Soft Rot Bacteria

The soft rot diseases of potato are caused by a group of bacteria often referred to as the pectolytic bacteria. They are called pectolytic because they produce enzymes that break down pectic molecules, which play an important role in holding plant cells together. When pectolytic enzymes break down pectic compounds in a diseased plant, the plant cells fall apart and leak out nutrients, on which large numbers of diverse bacteria thrive, causing the foul-smelling decay characteristic of soft rot. Pectolytic bacteria that cause disease in potato belong to two genera, *Pectobacterium* and *Dickeya.*

The damage done by the bacteria causing potato diseases is due, in part, to their capacity to generate large populations very quickly.

Dickeya spp. (formerly *Erwinia chrysanthemi*) are important pathogens of potato in Europe and several other countries, but they have seldom been associated with potato disease in the United States or Canada. In Europe, strains of *Dickeya* spp. cause a blackleg-like disease, called stem wet rot, during periods of high field temperature. *Dickeya* spp. also cause seed piece decay in Australia.

Two species of *Pectobacterium* cause diseases of potato: *P. atrosepticum* (formerly *E. carotovora* subsp. *atroseptica*) and *P. carotovorum* subsp. *carotovorum* (formerly *E. carotovora*

Fig. 18.1. Stem bases with inky black decay characteristic of blackleg (center and right), with a healthy stem and seed piece (left). (Courtesy W. G. Hoyman)

subsp. *carotovora*). *P. atrosepticum* is the causal agent of blackleg, the main foliar manifestation of the soft rot disease complex (Box 18.1 and Figs. 18.1–18.8). *P. carotovorum* subsp. *carotovorum* also causes aerial stem rot, a disease of potato stems. Two main features distinguish blackleg from aerial stem rot: blackleg is always initiated from the seed piece, whereas aerial stem rot infection starts in the aboveground portion of the stem where it has been damaged or injured by wind, hail, insect feeding, or crop management activities such as hilling and spraying; and blackleg turns affected tissue black, whereas aerial stem rot usually appears as various shades of brown or pale green. Aerial stem rot may cause some stems to turn the inky black associated with blackleg, but this is not common. All the pectolytic bacteria cause tuber rot.

Distinctions are made among the pectolytic bacteria because they have different characteristics that have implica-

Box 18.1

Symptoms of Diseases Caused by Soft Rot Bacteria

Blackleg

Symptoms Developing Early in the Growing Season

At emergence, plants are stunted, stiff, yellowish, and not vigorous. The belowground portion of the stem decays and turns black (Fig. 18.1).

Symptoms Developing Later in the Growing Season

Stem decay starts belowground and is very dark to black. Leaves are wilted. Sometimes decay breaks out at leaf axils. Foliage may or may not turn yellow. Diseased stems may decay completely and become lost in the potato canopy (Fig. 18.3).

Aerial Stem Rot

Stem decay starts at a broken point or the site of an injury in the stem (Figs. 18.4 and 18.5). The color of the decayed tissue varies from light to dark brown. The decayed region may become dry and brittle. Decay may occur where stems lie in wet furrows.

Stem Wet Rot

Symptoms similar to those of blackleg develop in the field at high temperature. The pith region of the stem decays more extensively than other stem tissue. Pith may be decayed from the belowground portion of the stem to the growing tip without external symptoms of decay.

Tuber Soft Rot

A wet, smelly decay develops in tubers (Fig. 18.6). Tissue does not cohere but becomes slimy and soft. The color of decayed tissue is variable, because of the extensive growth of secondary microorganisms.

Hard Rot

Dark, arrested decay lesions form on tubers, typically centered on lenticels (Fig. 18.8). Lesions may also be initiated at small wounds or injuries in the tuber skin.

Fig. 18.2. Dark black rot of the stem of a potato plant with blackleg. (Courtesy D. A. Johnson)

Fig. 18.3. Wilt and death of foliage of a potato plant with blackleg. (Courtesy D. A. Johnson)

Fig. 18.4. Aerial stem rot lesion developing from a wound (left). (Courtesy D. A. Johnson)

tions for disease development and management. For example, *Dickeya* spp. are high-temperature organisms, growing best at about 98°F, whereas *P. atrosepticum* prefers a lower temperature, growing best in the 70–75°F range. Furthermore, *P. atrosepticum* differs from the other types in that it occurs almost exclusively on potato and does not naturally infect other crops. It does not survive well in soil or in surface water. Bacteria classified as *P. carotovorum* subsp. *carotovorum* are quite different from the *P. atrosepticum* strains. *P. carotovorum* subsp. *carotovorum* is a most variable group of bacteria, containing members that differ from one another in various ways. The strains that are classified in this subspecies grow well over a wide range of temperature, even as high as 98°F. They can survive in soil and surface water. Strains of *P. carotovorum* subsp. *carotovorum* have been isolated from coastal salt water and various bodies of fresh water, including irrigation water. They have even been found in mountain snows and clouds, probably reaching such high elevations in aerosols created by ocean surf.

Disease Cycles

The main source of inoculum of the pectolytic bacteria causing blackleg, aerial stem rot, and tuber soft rot is the potato seed tuber. Potato tubers become infected during the growing season. Contaminated seed pieces often decay during the growing season and, as they decay, release large numbers of bacteria into the root zone, contaminating the surface of developing progeny tubers. Lenticels are a particularly favorable niche in which the bacteria survive during harvest, storage, and replanting.

The **blackleg** disease cycle illustrates how the soft rot bacteria survive and increase during their association with a potato crop (Fig. 18.9). When a potato tuber contaminated with *P. atrosepticum* is planted, either whole or as a cut seed piece, one of three things can happen. A **first scenario** is that the seed piece rots before a plant is established, a condition known as seed piece decay. A **second scenario** is that a plant is produced but becomes diseased during the growing season. If blackleg occurs early in the growing season, the plant will die before setting any new tubers. However, if blackleg develops later in the growing season, progeny tubers may become infected by bacteria from the stolon and may decay in the field. At harvest, such tubers will smear inoculum over harvesters, conveyers, and bin pilers, contaminating additional tubers. Further spread of bacterial inoculum takes place in storage and during subsequent handling of tubers. A **third scenario**, which is the most common, is that an asymptomatic plant is established and produces a crop of ostensibly healthy progeny tubers. However, the progeny tubers will have become contaminated from the decaying seed piece. Depending on harvest and storage conditions, most tubers will remain healthy during storage, but a small number will decay. When tuber decay starts, often at contaminated lenticels, and then ceases, a condition known as hard rot develops. Hard rot is characterized by dry, firm lesions of arrested

Fig. 18.5. Pale green rot of the stem (center) of a potato plant with aerial stem rot. (Courtesy D. A. Johnson)

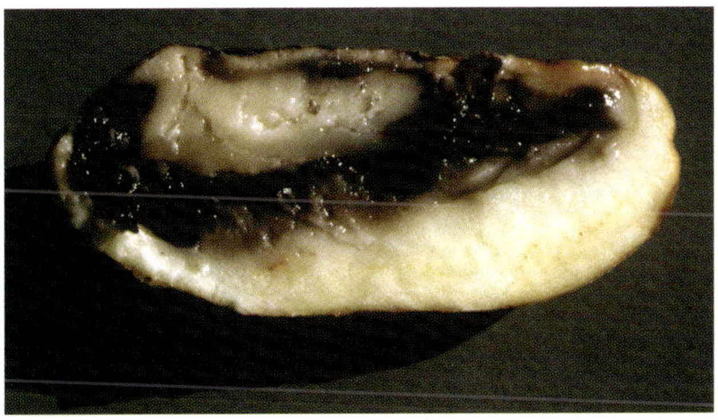

Fig. 18.6. Soft rot of internal tuber tissues. (Courtesy D. A. Johnson)

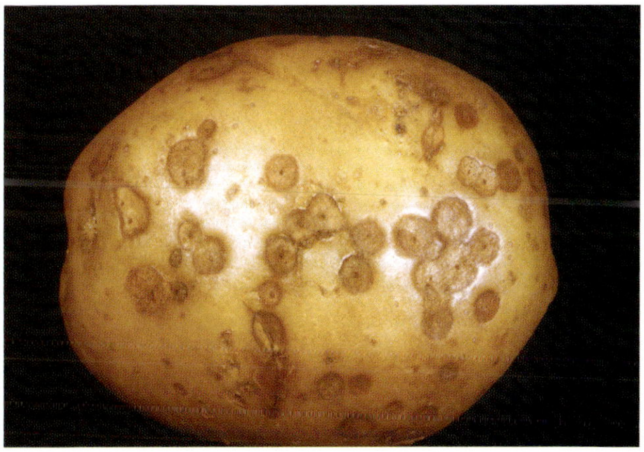

Fig. 18.7. Soft rot infection of tuber lenticels. (Courtesy W. G. Hoyman)

Fig. 18.8. Hard rot surrounding tuber lenticels. (Courtesy D. A. Johnson)

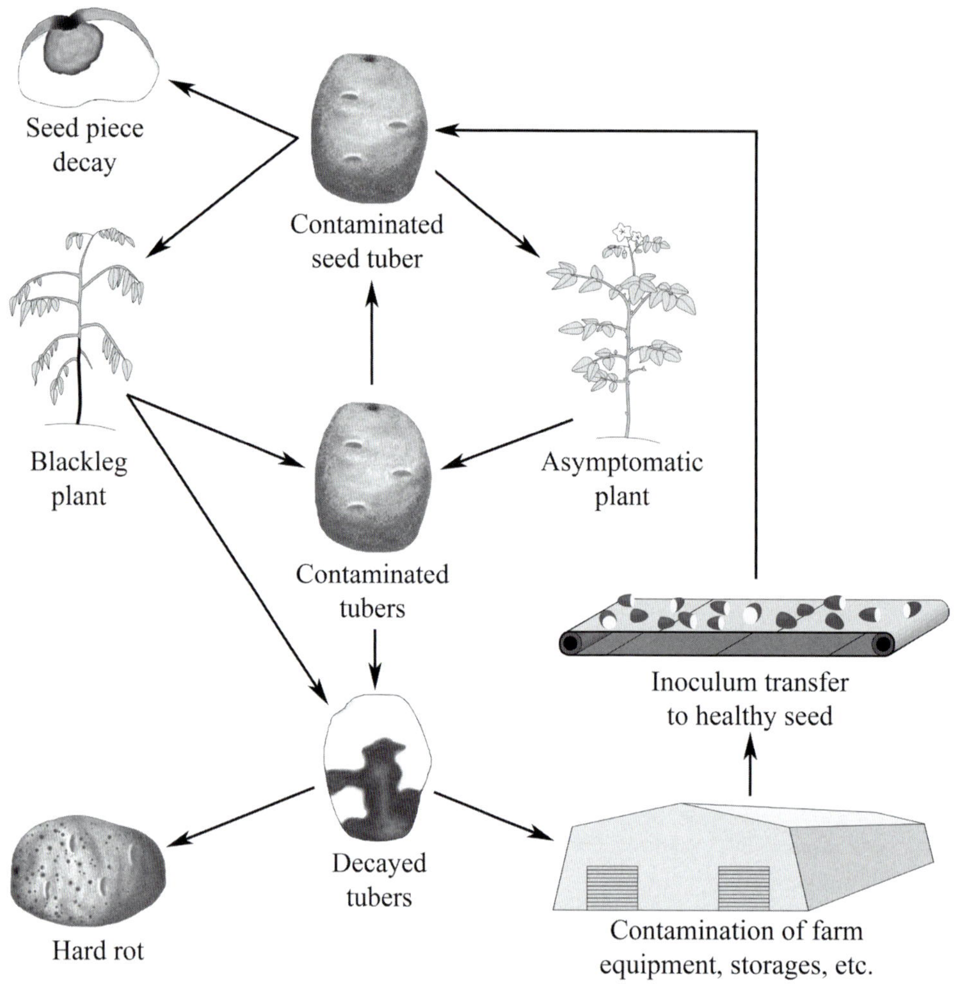

Fig. 18.9. Blackleg disease cycle on potato. (Drawing by Maegan Davis)

decay surrounding individual lenticels. If the rot continues to develop, the entire tuber may decay. When decaying tubers are present among tubers removed from storage, inoculum is spread in the same way as during harvest. Once a seed lot is contaminated with *P. atrosepticum,* the level of contamination increases during subsequent generations, although they may remain asymptomatic. When growing conditions favor blackleg, the disease may show up erratically throughout the field. However, asymptomatic tubers produced on infected plants, when used as seed, often produce plants without disease symptoms.

The disease cycle of **aerial stem rot** (bacterial stem rot) parallels that of blackleg in that *P. carotovorum* subsp. *carotovorum* also incites seed piece decay, and decaying seed pieces contaminate progeny tubers produced on asymptomatic plants. Unlike *P. atrosepticum, P. carotovorum* subsp. *carotovorum* does not usually infect potato stems from the seed piece directly through the vascular tissue. Some evidence suggests, however, that certain strains do infect the vascular tissue when field temperatures are high, causing symptoms resembling those of blackleg. Normally, aerial stem rot is initiated at sites of injury on potato stems. Injuries can be caused by hilling or spraying equipment, chewing insects, hail, wind damage, and fungal infections. The bacteria that infect such injuries may come from decaying seed pieces, irrigation water, rain, or neighboring infected plants. As in blackleg, stems infected with aerial stem rot often wither and die during the growing season and disappear from the foliage canopy. Tubers that become contaminated with *P. carotovorum* subsp. *carotovorum* during the growing season may decay if bacteria gain access to internal tissue through wounds or lenticels that have become swollen in waterlogged soil. *P. carotovorum* subsp. *carotovorum* is spread from decaying tubers during harvest, bin piling, and grading in the same way as *P. atrosepticum.*

The higher the temperature and the more free water available, the faster the bacteria multiply, and the greater the amount of soft rot.

Tuber soft rot is problematic when storage conditions are not ideal, because many of the tubers going into storage will be contaminated with pectolytic bacteria. The disease is caused by both *P. atrosepticum* and *P. carotovorum* subsp. *carotovorum* and by *Dickeya* spp. The surfaces of progeny tubers become contaminated with bacteria from decaying seed tubers. Progeny tubers are also sometimes infected by bacteria entering them through the vascular tissue of the stolon. Injuries to tubers during harvest, bin piling, and grading create entry points for the bacteria, in addition to the natural openings at the stolon attachment site and the lenticels. The initiation and rate of tuber decay depend on temperature and humidity. The higher the temperature and the more free water available, the faster the bacteria multiply, and the greater the amount of soft rot. In storage, soft rot bacteria spread from tuber to tuber to form pockets of decay within the pile. However, the spread of soft rot is limited if the tubers remain dry and cool. Ironically, contaminated seed tubers that do not decay during storage are the main culprit in keeping the disease cycle going. The soft rot bacteria are carried forward to the next generation when tubers contaminated with bacteria are used for seed. The disease cycle can be broken by the introduction of disease-tested minitubers, but contamination from external sources builds up during each field generation. External sources of inoculum include cull piles, lower-class tubers, irrigation water, and rain. The pathogen can be spread by equipment, insects, and other means that carry bacteria from a source of contamination to tubers.

Contaminated seed tubers that do not decay during storage are the main culprit in keeping the soft rot disease cycle going.

Bacterial Ring Rot

The Disease

Ring rot is a particularly severe disease of potato, because it can spread without being noticed and cause complete crop failure. The disease is caused by the bacterium *Clavibacter michiganensis* subsp. *sepedonicus.* It is a remarkably uniform subspecies, compared with the soft rot bacteria—all strains are pretty much identical. Ring rot bacteria invade the xylem and associated cells of the plant vascular tissue. The xylem functions chiefly in conducting water and dissolved minerals throughout the plant but also supports the transport of food resources to the tuber. The vascular tissue in potato tubers is located below the skin and is often visible as a ring of slightly different-colored tissue in a cross section—hence the name *ring rot,* from the appearance of decay in the tuber vascular ring.

The Ring Rot Pathogen

The ring rot bacterium is persistent and, when applied to almost any surface, survives for a long time, especially if the surface is dry and cool. Thus, once a farm becomes contaminated with the pathogen, extraordinary effort must be made to get rid of it. Bacteria from infected tubers can become smeared on farm equipment—planting machinery, harvesters, conveyers, or bin pilers—and can survive there for weeks or months if left alone (Box 18.2). They also survive on the walls and floors of storages and on interior and exterior surfaces of transport trucks.

Box 18.2

Sources and Means of Spread of Ring Rot Bacteria

- Seed tubers
- Cull potatoes
  - Volunteer plants
  - Potato debris
- Seed cutters
- Planters
- Harvesters
- Conveyers
- Bin pilers
- Storage bins
- Trucks
  - Wheels
  - Trailer compartment
- Insects
- Animals
- People
  - Boots
  - Gloves
  - Pocket knives

The ring rot bacterium does not survive well in soil or water. When experimentally added to soil, it cannot be recovered after only a few days unless the soil is very dry or very cold. There is a lot of microbial activity in a typical soil environment, and the ring rot bacterium does not compete well against other bacteria. Thus, where there is microbial competition, the ring rot bacterium quickly disappears. However, when microbial activity has essentially ceased because it is too dry or too cold, the ring rot bacterium survives better than most. When present on equipment or any other surface in a dried state, it can persist for many months.

Inside the plant, the ring rot bacterium meets with very little microbial competition and hence survives well. By nature, the bacterium grows slowly, but it increases and spreads in the plant. Consequently, infected volunteer potato tubers left in the field are another source of inoculum, as is infected potato plant debris. As long as infected plant tissue is present in a field, a source of inoculum is present, but once the potato tissues have been disintegrated by the action of soil microorganisms, ring rot inoculum also disappears. There is experimental evidence that the ring rot bacterium can survive in association with sugar beets and with certain solanaceous weeds. However, evidence is scant, and such inoculum is not thought to be significant in the ring rot disease cycle.

Disease Cycle

The ring rot bacterium is introduced into a farm when a seed tuber contaminated from an outside source is planted. The seed tuber could have become infected during the previous growing season, or it could have become contaminated after harvest, by contact with inoculum on equipment, in storage, or in a transport vehicle. Once infected, the seed tuber is very likely to produce an infected plant. Ring rot, however, develops slowly. The actual rate of development depends on the growing conditions and the potato cultivar. In general, conditions that do not favor potato growth are favorable for ring rot development. Ring rot symptoms (Box 18.3 and Figs. 18.10–18.16), even in a susceptible cultivar,

Box 18.3

Symptoms of Bacterial Ring Rot

Foliage

See Figs. 18.10–18.12.

- Wilting, starting at lower leaves and at first recovering at night
- Rolling of leaf margins
- Interveinal leaf chlorosis and necrosis
- Milky exudate expressed when the lower stem is cut and squeezed
- Dwarf rosette, characterized by short internodes (in some cultivars only)

Tubers: External Symptoms

See Fig. 18.13.

- Dark blotches under the skin
- Sharp, jagged surface cracks exposing decayed tissue

Tubers: Internal Symptoms

See Box 7.1 and Figs. 18.14–18.16.

- Slight discoloration of the vascular ring (this may be the only visible symptom)
- Rot in the vascular ring at the stem end, initially cream-colored but later light to dark brown
- Cheesy, cream-colored exudate expressed from vascular tissue when pressure is applied
- Corky tissue formed around hollows in the decayed vascular ring

usually do not become visible until near the end of the growing season. In a resistant cultivar, symptoms occur very rarely, if at all. The reaction of most cultivars is intermediate between highly susceptible and resistant. No cultivars are immune.

Progeny tubers become infected when they are entered by bacteria moving through the stolon from the stem. The

Fig. 18.11. Interveinal chlorosis and wilt of leaflets of a potato plant with ring rot. (Courtesy D. A. Johnson)

Fig. 18.12. Dwarf rosette and marginal leaflet burn in a potato plant with ring rot. (Courtesy D. A. Johnson)

Fig. 18.10. Interveinal chlorosis of leaflets of a potato plant with ring rot. (Courtesy D. A. Johnson)

Fig. 18.13. Tuber surface cracking characteristic of ring rot. (Courtesy D. A. Johnson)

incidence of tuber infections, however, does not consistently reflect the level of foliar symptoms. Sometimes many tubers show symptoms of ring rot while little is visible in the way of foliar symptoms; in other instances, foliar symptoms predominate over tuber symptoms.

Infected tubers sometimes become cracked or have dark surface blotches, but often they do not show any external symptoms (Box 18.3). Internal tuber symptoms vary from extensive decay of the vascular system to no visible symptoms. Early ring rot symptoms appear as slight discoloration in the vascular ring.

> **Latent infections can persist for one or more generations without the grower being aware of it, and in the meantime ring rot bacteria are spread to other lots.**

The ring rot disease cycle (Fig. 18.17) is similar to the blackleg disease cycle in many ways, but there are some important differences. Whereas contamination with the blackleg bacteria occurs largely on the tuber surface, ring rot may occur as internal latent infections. In a latent infection, the bacteria reside and multiply in the plant or tuber without causing visible disease symptoms. Latency is the curse of the ring rot disease. Latent infections can persist for one or more generations without the grower being aware of it, and in the meantime ring rot bacteria are spread to other lots. Furthermore, even when symptoms do develop, they are often not yet visible during field inspection of seed crops. If field inspection is done later in the growing season, symptoms are often obscured by senescence and other foliar diseases.

Brown Rot

Brown rot of potato, also known as southern bacterial wilt or simply bacterial wilt, is another serious potato disease but is not considered further in this chapter. Its symptoms are very similar to those of bacterial ring rot: rapid wilting of foliage, the presence of bacterial ooze in vascular tissue, and decay in the vascular ring of tubers. **Oozing of bacterial slime from the tuber eyes** is one distinguishing characteristic that differentiates brown rot from ring rot. The bacterium causing brown rot is *Ralstonia solanacearum*. One form of the bacterium, known as race 1, occurs in the southern United States and is controlled in much the same way as the ring rot and soft rot bacteria, by the use of disease-free seed and farm hygiene. Since the bacterium survives in soil, crop rotation is also important. Another form of the pathogen, race 3 biovar 2, is not endemic to the United States or Canada but is particularly well adapted to potato in temperate climates. It is listed as a quarantine pest, because of its potential for affecting potato production if it were to be introduced into potato ecosystems. The risk of introduction of this biovar is very real, since it has already been intercepted in geranium cuttings imported into the United States and Canada. Race 3 biovar 2 survives very well in association with the solanaceous weed *Solanum dulcamara*, commonly known as climbing nightshade, bittersweet nightshade, or deadly nightshade, which grows on banks of rivers and streams and, once infected, releases bacteria into the water. Irrigation of potato crops from such infested surface water is an important means by which the disease could be spread.

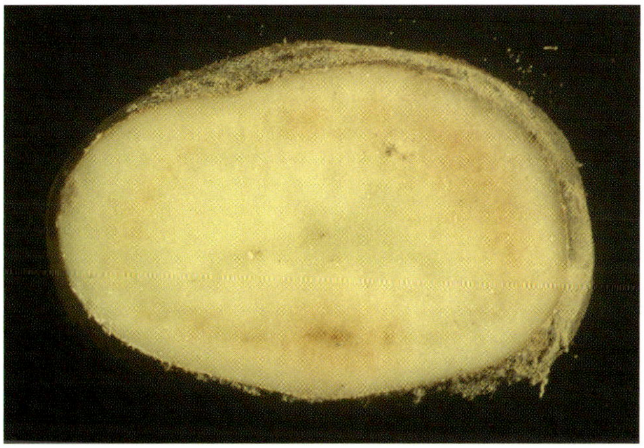

Fig. 18.14. Slight vascular discoloration in a tuber with ring rot. (Courtesy D. A. Johnson)

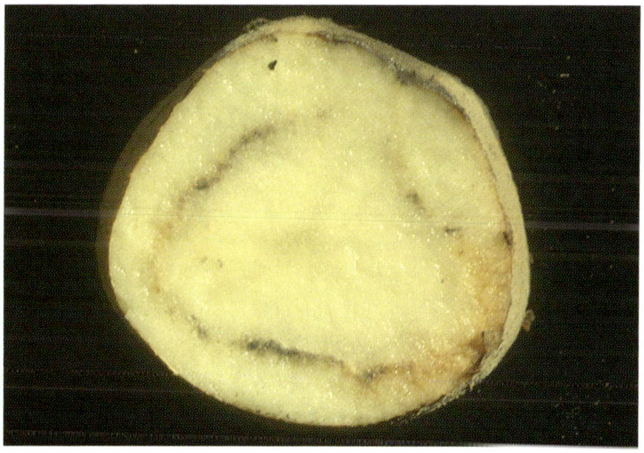

Fig. 18.15. Tuber vascular discoloration characteristic of ring rot. (Courtesy D. A. Johnson)

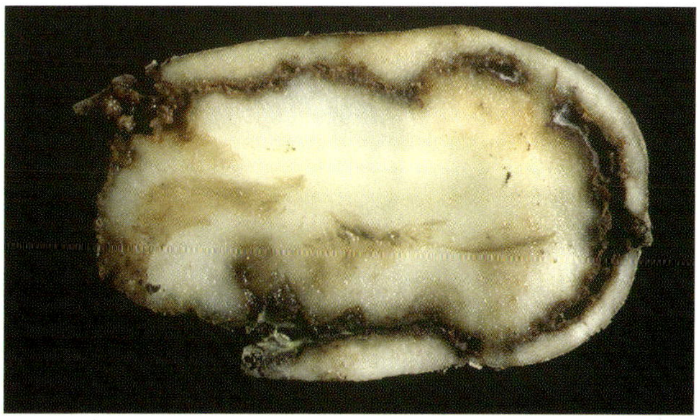

Fig. 18.16. Decay of tuber vascular tissue in a tuber with ring rot. (Courtesy D. A. Johnson)

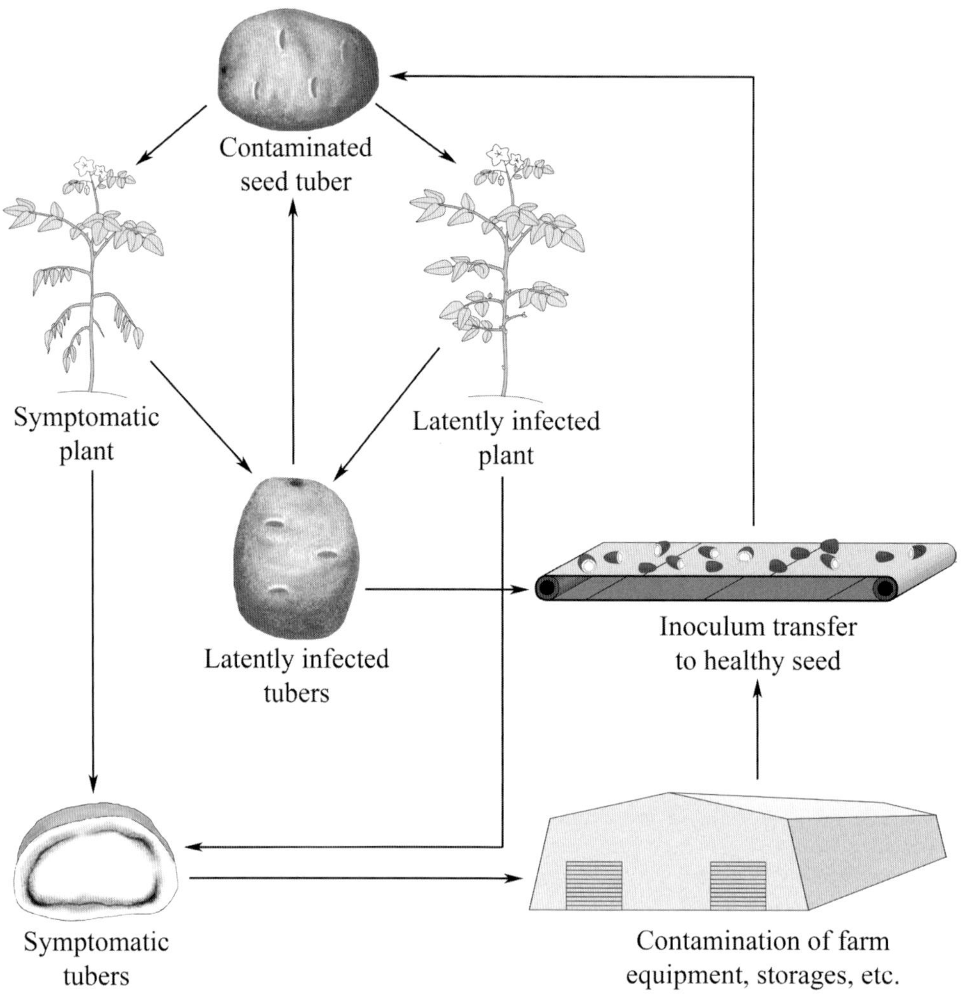

Fig. 18.17. Ring rot disease cycle on potato. (Drawing by Maegan Davis)

Management Strategies

Bacterial diseases of potato cannot be managed by chemical pesticides. The incidence and severity of these diseases can be reduced only by the implementation of appropriate tactics in a management strategy that includes careful selection of seed for planting, general farm hygiene, and good agronomic and storage practices (Box 18.4).

Management Prior to Planting

Seed Selection

Both soft rot and ring rot bacteria are chiefly tuber-borne, and thus planting disease-tested seed tubers is of paramount importance in any disease management strategy. The cleanest seed is directly derived from tissue culture. Minitubers, if produced with care in a proper facility, should also be pathogen-free. With every generation of increase, however, the load of soft rot bacteria increases, and the risk of contact with a ring rot source rises. Hence, the use of short-generation seed is an important tactic in management strategies for both soft rot and ring rot.

It is inevitable that field-grown tubers will become contaminated with the soft rot bacterium *P. carotovorum* subsp. *carotovorum*. In some regions this is also so for the blackleg bacterium, *P. atrosepticum*, although in other regions, perhaps because of better isolation, field-grown tubers remain relatively free of blackleg. Laboratory tests for determining the blackleg load in seed potato lots have been developed but generally have not been cost-effective for routine use.

Laboratory testing of seed potato lots for bacterial ring rot infection, on the other hand, is practical and routine in some certification programs (Chapter 6). Generally, seed lots are evaluated by random sampling of 400 tubers, which are tested in two composite lots of 200 by enzyme-linked immunosorbent assay (ELISA) or an immunofluorescence test. Molecular tests have also been developed but have largely been used for confirmation of the presence of bacteria in suspect lots rather than for routine indexing. The sensitivity of laboratory testing is greatly limited by sample size. The larger the sample, the greater the probability of detecting the ring rot bacterium, if it is present in a few tubers. It is not possible to obtain 100% certainty that a seed lot is free of ring rot by laboratory testing. The specifics of the entire seed production system must be evaluated to determine the likelihood that a seed lot is free of ring rot. Ring rot contamination from postharvest contact with contaminated surfaces cannot be

Box 18.4

Key Disease Management Tactics

Blackleg

- Use certified, short-generation seed tubers
- Treat seed tubers with fungicide
- Warm seed tubers prior to planting
- Plant in well-drained soil
- Maintain farm sanitation

Aerial Stem Rot

- Increase plant spacing
- Avoid overirrigation
- Avoid overfertilization
- Avoid damaging stems
- Maintain farm sanitation

Tuber Soft Rot

- Allow tuber maturation prior to harvest
- Harvest when soil is cool
- Promote wound healing in storage
- Maintain cool conditions for long-term storage
- Maintain airflow in storage
- Avoid condensation during storage
- Maintain farm sanitation

Bacterial Ring Rot

- Use certified, short-generation seed tubers
- Test seed for ring rot
- Disinfect equipment, storages, and transport trucks
- Maintain farm sanitation

detected by a laboratory test. Laboratory testing is only for detecting infections that have developed inside tubers and stems during the growing season.

Potato cultivars differ in susceptibility to soft rot and ring rot, and the choice of cultivar may be a factor in avoiding these diseases. Cultivars with a high level of resistance to ring rot, however, come with an increased risk of developing undetected latent infections, which can be spread to other cultivars in the farm operation. Some cultivars are more resistant to blackleg than others, but all cultivars can be infected. No potato cultivars are resistant to tuber soft rot.

Farm Hygiene

Bacterial pathogens can be spread by many means. Farm equipment and human activities are major means of dispersal, but bacterial pathogens can also be spread by insects, birds, rodents, and other animals and can be dispersed through the air in the form of aerosols. It is therefore of utmost importance to keep the level of inoculum in the farm environment at an absolute minimum. This is done by maintaining effective hygiene. Cull piles need to be removed or buried as soon as possible. Equipment should be cleaned and disinfected between each potato lot. Stray potatoes that fall from trucks, conveyers, and bin pilers should be discarded and not left lying around. Such potatoes are often bruised and very susceptible to decay, and once they are decayed, bacteria are easily dispersed from them.

It is particularly important to avoid **ring rot** by ensuring that all equipment and transport vehicles brought onto a farm are appropriately disinfected. Both the wheels and the interior of tractor trailers should be cleaned and disinfected before entering potato-handling areas, particularly on seed farms, where small amounts of inoculum can be passed on to several potato generations.

Potato producers that operate a feed lot and use cull or low-grade potatoes as animal feed must be particularly attentive to hygiene in their operations. The survival rate of ring rot and soft rot bacteria that pass through the intestinal tract of cattle or other animals is not known, but tubers that are not consumed or only partially consumed usually carry soft rot bacteria and may carry ring rot bacteria. Strict protocol must be observed to avoid any contact between feed potatoes and potatoes that may be used for seed. If there is a chance that potatoes from a feed lot end up as volunteers in fields destined for potato production, the risk of ring rot contamination must be considered.

Management During the Growing Season

Agronomic Practices

Agronomic practices such as planting, hilling, irrigating, spraying, and harvesting have a bearing on crop health. How these practices are carried out is particularly important in managing the bacterial soft rot diseases; they are less important in managing ring rot.

Undamaged seed tubers produce the healthiest crop. **Seed piece decay**, in particular, will increase when the seed tuber is damaged by bruising and wounding during handling. The distance that seed is dropped from the end of a conveyer should be minimized, to prevent bruising; it is preferably no more than 6 inches. Seed cutting also increases tuber susceptibility to decay and should be done only if required as a cost-effective crop management practice. All seed-cutting and handling equipment should be thoroughly cleaned between lots to prevent cross-contamination and carryover of inoculum.

Fungicide treatment of seed pieces is recommended to reduce seed piece decay, even though most fungicides do not have bactericidal activity. Fungi such as *Fusarium* create infection sites for soft rot bacteria, and management of such fungi therefore reduces the potential for seed piece decay.

Seed should be warmed to 50–55°F before planting and should be planted in soil in which the temperature is 50–55°F at the planting depth. Planting cold seed in warm soil should be avoided at all cost. The seed "sweats" under these conditions, creating a moist environment, which favors the multiplication of bacteria and increases susceptibility to seed piece decay. Irrigation immediately after planting should also be avoided if at all possible. Irrigation should begin only after the plants have emerged.

Waterlogged soil favors **seed piece decay** and **blackleg**. Therefore, potatoes should be planted in well-drained soil and, if necessary, irrigated only to provide adequate soil moisture. Seed piece decay, blackleg, and tuber soft rot all become worse in water-saturated soil. Generally, conditions that favor potato growth are less favorable for the soft rot diseases; more disease can be anticipated with sustained growing conditions that are either too cool or too hot for optimal potato growth. Wet soil also promotes lenticel infection in progeny tubers. Lenticel infection does not necessarily lead to decay, but tubers with infected and open lenticels are certainly more susceptible.

Waterlogged soil favors seed piece decay and blackleg.

Like the diseases of underground plant parts, **aerial stem rot** is exacerbated by wet conditions. Dense canopies of foliage and long periods of leaf wetness, as may occur with overhead irrigation, favor soft rot bacteria. Potato stems that are injured or broken by wind, hail, or insects or by spraying or hilling operations are particularly susceptible. The susceptibility of the aboveground portion of the plant can be mitigated by avoiding cultivars that produce long vines and dense foliage, by increasing the spacing between rows, and by avoiding overfertilization (especially with nitrogen) to prevent excessive growth of the foliage. Careful control of the timing, duration, and frequency of irrigation is also an important tool for managing the crop. Limiting irrigation to the morning, to permit drying of the foliage later in the day, is preferred. Also, to minimize the occurrence of aerial stem rot, less frequent irrigation for periods of longer duration is better than short irrigation periods. These practices are helpful in reducing the incidence of the bacterial soft rot diseases, but none of them will prevent these diseases altogether.

Harvesting conditions and practices have a significant impact on the storability of the crop. Tubers to be stored should be mature before they are harvested. They should be left in the soil for at least 7–10 days after vines are killed, to ensure better skin set, making them less susceptible to wounding during harvest. In the semiarid environment of the Columbia Basin, however, tubers can be successfully stored when harvested from green vines (Chapter 10). Tubers should be harvested, if possible, only when the soil temperature is a cool 50–65°F. They are much more susceptible to decay in storage if the pulp temperature is above 65°F during harvest. Even if the tuber skins are mature and have set well, bruising and mechanical damage must be avoided to minimize subsequent decay.

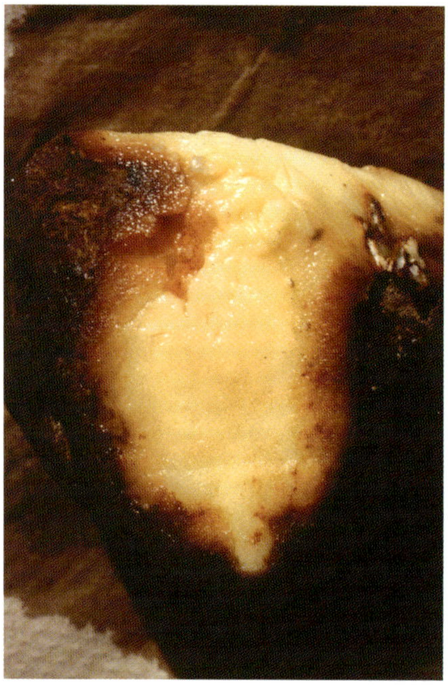

Fig. 18.18. Tuber soft rot symptoms following initial infection by the late blight organism. (Courtesy D. A. Johnson)

Ring rot is a zero-tolerance disease and thus must be managed largely during seed production and selection. Production of potatoes from seed known to be contaminated with ring rot is never recommended, because it increases the inoculum load in the farming district and thus jeopardizes other crops, particularly those grown for seed.

To prevent the spread of ring rot, it is important that machinery and equipment be cleaned and disinfected between lots. Most disinfectants are deactivated by interaction with organic material and soil. Therefore, cleaning all surfaces prior to disinfection is of utmost importance. An adequate period of exposure to disinfectants is required to kill bacteria. Usually a 10-minute exposure is recommended. Cleaning and disinfecting of transport vehicles and storages are also important, for they too can harbor the ring rot bacterium for several years.

Management During Storage and Packaging

Potato tubers store best under well-ventilated, cool conditions. If possible, harvest should not occur during rain. Excessive moisture should be avoided. Wet tubers do not store well. Decayed tubers should be graded out prior to storage. For the first 10–14 days in storage, the temperature should be modulated in the 50–55°F range to encourage wound healing. Even under the best harvesting conditions, some bruising and mechanical damage will occur, and sites of injury can serve as foci of soft rot infection during storage. After wounds have healed, the temperature should be lowered to minimize the growth of soft rot bacteria. Storage temperatures should be below 50°F. Potatoes destined for the fresh market and for processing should not be too cold, to avoid the conversion of starches to sugars. Seed potatoes can be stored

A film of water on tubers causes the tuber surface to become anaerobic, inhibiting natural resistance to infection while creating a favorable environment for the growth of soft rot bacteria.

at lower temperatures. Humidity should be kept high, about 90–95%, to minimize shrinkage. However, good ventilation should be supplied to prevent condensation of moisture on tuber surfaces and provide adequate levels of oxygen. A film of water on tubers causes the tuber surface to become anaerobic, inhibiting natural resistance to infection while creating a favorable environment for the growth of soft rot bacteria. Storage piles should be monitored for wet spots or a rise in temperature, since both are indicators that decay pockets are developing.

Tubers that are to be stored should not be washed. If they are to be washed before being marketed, they must be dried (usually by forced air) before being packaged. Placing wet tubers in plastic bags is a recipe for disaster.

Tubers infected with the late blight pathogen are often quickly invaded by soft rot bacteria (Fig. 18.18), creating large losses in storage. It is advisable to process tubers as soon as possible from fields in which a significant amount of late blight, blackleg, or tuber soft rot is present. If tuber rot is present when the crop goes into storage, increased ventilation to dry out any wetness that may be present will reduce the chance of meltdown in storage.

Although **ring rot** ought to be avoided at all cost, potato crops sometimes become infected with the disease. If the disease is severe, many tubers will become decayed prior to harvest, and the crop is best abandoned in the field. However, if the incidence of ring rot is low, the crop can often be salvaged for processing or even fresh-market sales. Visibly infected tubers should be graded out before tubers are put in storage, and the same storage principles should be applied as for soft rot–affected lots, to maintain as cool and dry conditions as practicable.

CHAPTER 19

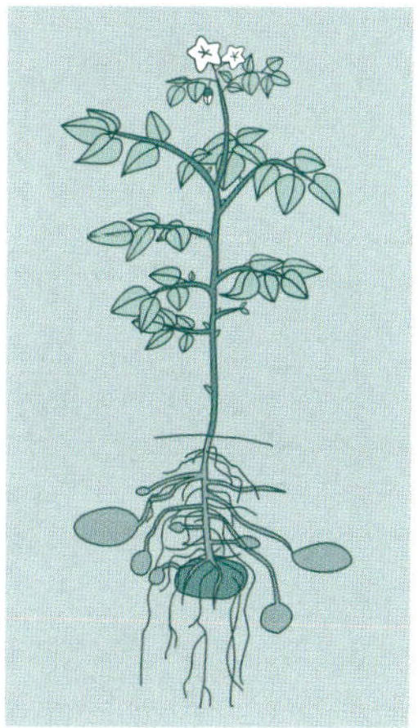

Mary L. Powelson
Department of Botany and Plant Pathology
Oregon State University, Corvallis

Randall C. Rowe
Department of Plant Pathology
Ohio State University, Wooster

Managing Diseases Caused by Seedborne and Soilborne Fungi and Fungus-Like Pathogens

The potato crop is faced with a broad diversity of fungal and fungus-like pathogens that persist in association with seed tubers or in soil. These pathogens cause diseases that affect the health of the seed piece, the growth of the crop, and tuber quality and yield. The most important of these diseases are seed piece decay; shoot, stolon, and stem canker; vascular wilt; scab; and rots and blemishes of the tuber. The pathogens that cause these diseases occur commonly throughout potato-producing areas of the United States and Canada. The significance of the diseases they cause varies in importance, depending on the geographic location, the cultivar, and the intended use of the crop.

Most, but not all, of these diseases have a simple disease cycle, in which only one cycle of infection, pathogen growth, and pathogen reproduction occurs during each production season. Because of this, effective management strategies focus on (1) reducing the amount of inoculum on the seed tuber or in the soil, (2) planting resistant cultivars, if they are available, (3) providing an environment that is unfavorable for infection, and (4) protecting the crop with fungicides when necessary.

Diseases and Pathogen Biology

Diseases at Planting

Fusarium seed piece decay starts as a brown dry rot on the edges of cut seed pieces. Brown to black flecks first appear on the cut surfaces and develop into depressions within 1 week and into pits within 2 weeks after cutting. Internal cavities usually develop, often containing white to pink mycelium (Fig. 19.1). Soft rot bacteria often enter decaying seed pieces, leading to a wet, slimy rot that usually results in a very poor stand.

Fusarium seed piece decay is caused by the fungi *Fusarium solani* var. *coeruleum* and *F. sambucinum*. The prevalence of these two pathogens varies among geographic areas. Both pathogens survive from season to season in diseased tubers and as chlamydospores or mycelium in decayed plant tissue in soil. Seed pieces become infected primarily from inoculum that is spread during seed handling and seed cutting. Infection occasionally results from contact with inoculum in the soil. The pathogens enter seed pieces through wounds, primarily the unhealed surfaces of cut seed. The rate of seed piece decay is accelerated in moist soils at temperatures above 55°F. The host range of these pathogens is narrow, potato being the only host.

Pythium seed piece decay can cause substantial problems, usually associated with excessive precipitation or early irrigation of poorly drained soils. If seed pieces are planted in soil that becomes saturated for several days in succession and soil temperatures rise above 70°F, infected seed pieces decay into a soft, watery mass, resulting in delayed emergence and poor stand establishment.

Pythium seed piece decay is caused primarily by the fungus-like oomycete *Pythium ultimum*. The pathogen persists in soil for years as dormant oospores. It enters seed pieces through wounds resulting from seed handling and cutting. As soil temperatures increase, the vulnerability of seed pieces to infection also increases. The pathogen has a wide host range, causing damping-off and root rot of many crops.

Seed piece decay may also be caused by soft rot bacteria (Chapter 18).

Fig. 19.1. Fusarium seed piece decay. Dry rotted tissue at cut edges, with cavities containing white to pink mycelium. (Courtesy G. Secor)

Fig. 19.2. A, Rhizoctonia canker on young sprouts. **B,** Rhizoctonia stolon canker on stolons. Both diseases cause sunken, dark brown to black lesions. **C,** White, powdery mold growth on the main stem just above the soil line, formed by the pathogen in its sexual stage.

Diseases Affecting Crop Growth

Rhizoctonia canker affects emerging sprouts, stems, and developing stolons. A canker is a brown to black, sunken lesion, which causes the death of infected tissue. When infection occurs before sprouts emerge, the result is a nonuniform stand. Infected sprouts that do emerge often remain stunted (Fig. 19.2A). When lesions completely girdle very young stolons (Fig. 19.2B), tuber set is reduced. A deep sunken canker may girdle the main stem, causing yellowing, purpling, and upward curling of leaves and the formation of aerial tubers. These three diseases—sprout canker, stolon canker, and stem canker—are caused by the fungus *Rhizoctonia solani*. During midseason, with the high humidity under a dense canopy, the fungus may develop into its sexual phase (*Thanatephorus cucumeris*), which appears as a white, powdery mold on stems, extending 1–2 inches above the soil line (Fig. 19.2C). Because it resembles the mold produced by *Sclerotinia* on potato stems, this growth can be mistaken for white mold (Chapter 21). This phase of Rhizoctonia canker has no known effect on plant vigor.

R. solani survives from season to season as thick-walled resting structures (sclerotia) on the surface of tubers or in soil or as mycelium associated with decomposing plant residues. Although sprout canker, stolon canker, and stem canker are all caused by the same fungus, the source of inoculum differs among these diseases. The seed piece is the principal source of inoculum for sprout canker. In contrast, the soil is the inoculum source for stolon and stem cankers. For all three diseases, cool (55–60°F), moist soils are optimal for infection. The pathogen directly penetrates the epidermis to enter host tissue.

There are several strains of *R. solani*, but the primary one that attacks potato is anastomosis group 3 (AG-3). This strain generally does not cause disease in other plant species, except for a few weeds and crop plants closely related to potato, such as black nightshade and eggplant. Thus the host range of this pathogen is narrow. Rotational crops may harbor other strains of *R. solani*, but these generally cause little damage to potato.

Potato early dying (also called Verticillium wilt or early die) is a disease of the vascular (water-conducting) tissues. The first symptom is early senescence of individual plants or groups of plants scattered among other plants that appear healthy. Foliar symptoms develop as uneven chlorosis and wilting of the leaves (Fig. 19.3A), which may mimic normal senescence. Areas between the leaflet veins often turn yellow and later brown. Leaf yellowing and death proceed upward from the base of infected stems, which often remain erect. A tan discoloration of the vascular system can often be seen in stems sliced open near the base (Fig. 19.3B). This symptom is not diagnostic, however, and may be due to other diseases or stresses. Vascular browning may develop in tubers of some cultivars.

Two fungi, *Verticillium dahliae* and *V. albo-atrum*, are the primary causes of potato early dying. *V. dahliae* is widespread in the United States and predominates in production areas where average daily summer temperatures commonly exceed 80°F. *V. albo-atrum* grows at lower temperatures and causes potato early dying in areas where average daily temperatures normally do not exceed 70°F during the growing season, such as Maine, the Red River Valley, southern Canada, and winter production areas in Florida. In some areas, a species of root-

Fig. 19.3. Potato early dying. **A,** Uneven foliar necrosis and chlorosis. **B,** Tan discoloration of the vascular tissues, evident in a cut stem base, associated with potato early dying. This symptom is not diagnostic, since it may be due to other diseases or stresses.

lesion nematode (*Pratylenchus penetrans*) may interact with *V. dahliae* as a critical component of potato early dying (Chapter 20). When both *V. dahliae* and *P. penetrans* infect plants, severe symptoms and significant yield losses can occur, even if the pathogen population density is low enough that either pathogen individually would have little or no effect. Several other species of *Pratylenchus* can also affect potato directly, but their importance in potato early dying is less clear.

Verticillium species survive from season to season in soil, with *V. dahliae* as small, black, thick-walled resting structures called microsclerotia and *V. albo-atrum* as darkened mycelia. Both pathogens can be carried on the surface and in vascular tissue of seed tubers. *Verticillium* has a very wide host range, comprising over 200 dicotyledonous species, including herbaceous annuals and both herbaceous and woody perennials. It can also persist on roots of some cereal crops without causing disease. Despite the wide host range of *Verticillium*, some strains of this pathogen are host-adapted and thus much more aggressively pathogenic on certain plant species. Four vegetative compatibility groups (VCG) of *V. dahliae* are well recognized, but only one (VCG 4) is a significant pathogen of potato. The VCG 4 strain has been further subdivided into VCG 4A and 4B, the former being more aggressive on potato and more interactive with *P. penetrans*.

Pratylenchus species survive from season to season as eggs, juveniles, or adults. *P. penetrans* also has a wide host range, including many monocot and dicot crop and weed species (Chapter 20).

Both *Verticillium* and *Pratylenchus* infect roots. *Verticillium* penetrates near the root tip and grows into the xylem tissue, where it multiplies and spreads throughout the plant via the vascular tissue. In contrast, *Pratylenchus* burrows into the root cortical tissue, but does not enter the vascular tissue. Infection by both *Verticillium* and *Pratylenchus* and disease development are favored by moist soils and high rates of evapotranspiration. These conditions result in rapid unrestricted flow of water carrying conidia of *Verticillium* through the vascular tissue. Temperature also plays a key role in pathogenesis. The optimum temperature for the growth of *V. dahliae* is 75°F; for *V. albo-atrum*, 68°F. The optimum soil temperature for *P. penetrans* is about 75°F. Disease severity tends to increase as the mean air temperature increases from 65 to 85°F. The growth of *Verticillium* is restricted above 85°F, and pathogenesis is halted.

Diseases Affecting Tuber Quality

Eight diseases can significantly affect tuber quality: common scab, powdery scab, pink rot, leak, black scurf, Fusarium dry rot, black dot, and silver scurf. Tuber quality can also be affected by certain nematode species (Chapter 20).

Common scab occurs only on the tuber. Initial symptoms are browning and abnormal cell expansion in and below the tuber periderm. As the lesions mature, they become circular and often coalesce. At harvest, tan to dark brown, rough-textured lesions cover portions of tuber surfaces (Fig. 19.4).

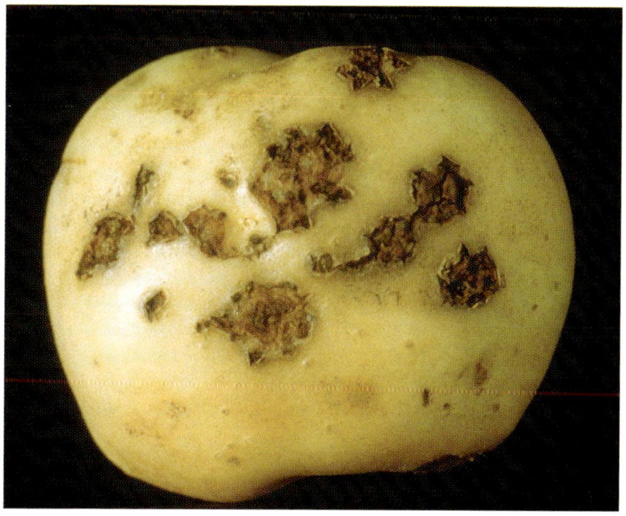

Fig. 19.4. Common scab. Tan to dark brown, rough-textured lesions, which may be superficial, slightly raised, or deeply sunken. (Courtesy B. Christ)

Scab lesions may be superficial, slightly raised, or deeply sunken. In North America, two species of filamentous bacteria cause common scab: *Streptomyces scabies,* which is found throughout all production areas, and *S. acidiscabies,* which has been found only in the northeastern United States. The latter is the cause of so-called acid scab. Both species survive as spores and can be both seedborne and soilborne. The bacteria enter tubers through lenticels. Infection is favored by warm (above 70°F), dry soils. The disease is most severe in soils of pH 5.5–7.5; however acid scab can occur in soils below pH 5.0. The host range of these bacterial pathogens is relatively narrow. In addition to potato, they cause common scab on the fleshy taproots of beet, radish, rutabaga, turnip, carrot, and parsnip.

Powdery scab, in contrast to common scab, is a disease of both roots and tubers. Small galls or wart-like growths form on infected roots (Fig. 19.5A). Root galls are not associated with any apparent aboveground plant symptoms, but yield may be affected in some cultivars. Lesions form on infected tubers, initially as small, purple to brown spots, which develop into tan, blister-like or pimple-like swellings. The pathogen gains entry into the tuber through the lenticels and eyes. Infection results in abnormal enlargement and division of cells in and below the tuber periderm. Eventually, the periderm ruptures, resulting in scab-like outgrowths (Fig. 19.5B). The tissue beneath the lesion eventually decays, leaving a shallow depression filled with olive brown to dark brown, powdery masses of persistent resting spores called cystosori. Tuber symptoms often appear similar to those of common scab. Observation of torn edges of the periderm around the edges of slightly sunken lesions can help differentiate powdery scab from common scab. However, confirmation of powdery scab requires microscopic examination.

Powdery scab is caused by a fungus-like protozoan pathogen, *Spongospora subterranea.* The pathogen survives from season to season as cystosori on tubers and in soil, where it can remain viable for 6 years or more. The source of inoculum for infection of roots and tubers is infected seed tubers or inoculum present in the soil. Cool (55–65°F), moist soils favor infection. *S. subterranea* can cause disease across a wide range of soil pH (4.7–7.6). The strain of the pathogen that attacks potatoes has a fairly narrow host range. It can complete its life cycle only by forming cystosori on roots of potato or a few closely related weeds and crop plants, such as black nightshade, jimsonweed, pepper, and tomato. Several cycles of infection and spore production can occur during a growing season.

Pink rot affects roots and tubers in the field and tubers in storage. In infected tubers, a dark line is often present at the margin of decay and may be visible through the skin. Three diagnostic symptoms are characteristic of this disease. When tubers in the initial stages of rot are cut open, the cut infected surface turns salmon pink after exposure to air for 20–30 minutes (Fig. 19.6). As the disease develops, the tuber acquires a "cooked potato" texture, in contrast to the crisp texture of a healthy tuber. When an infected tuber is cut and gently squeezed, a small amount of a clear liquid can be expelled from it. In storage, diseased tubers produce an ammonia odor that is very diagnostic for this disease.

Pink rot is most commonly caused by the fungus-like oomycete *Phytophthora erythroseptica,* which is endemic in many soils. The pathogen survives from season to season as oospores. Tuber infection in the field is favored by soils approaching saturation and a pulp temperature above 68°F. In this environment oospores germinate and produce sporangia, which then release small zoospores capable of moving in soil water. Zoospores that come in contact with tubers can enter directly through lenticels and eyes to cause infection. However, most tuber infections occur when the pathogen grows

Fig. 19.5. Powdery scab. **A,** Root galls, forming wart-like growths. **B,** Lesions on tubers are tan, blister-like or pimple-like swellings. A shallow depression beneath each lesion is filled with an olive to dark brown mass of powdery spores. Torn edges of periderm at the margins of powdery scab lesions help to differentiate this disease from common scab. (Courtesy D. A. Johnson)

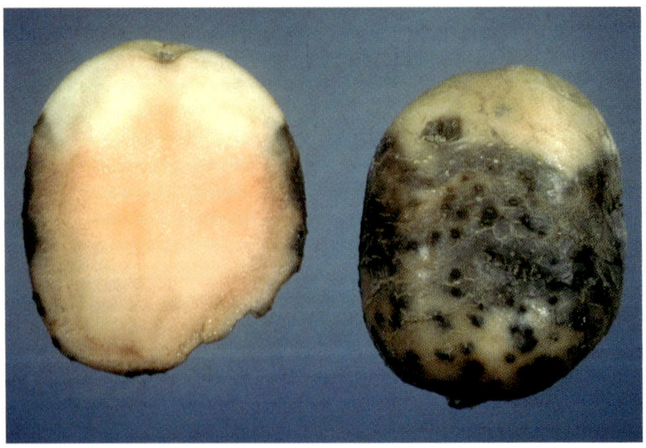

Fig. 19.6. Pink rot. Rot on the exterior of a tuber (right). The cut surface of an infected tuber turns pink with exposure to air (left).

into the tubers from infected stolons. Tubers can also become infected during the harvest, if they come in contact with pieces of diseased tissue dislodged from other infected tubers. Tubers that are excessively skinned are extremely susceptible. Conditions that favor tuber infection during harvest are excessive wounding and tuber pulp temperatures above 68°F. The host range of *P. erythroseptica* includes over 30 plant species, including crop and weed species closely related to potato, such as tomato, bell pepper, eggplant, and several species of nightshade. Potato strains of the pathogen have been isolated from roots of spinach, small grains, and some legumes.

Leak (also called shell rot) is initiated in the field, and symptoms become more severe during storage. The initial symptom, visible in a cut tuber, is a gray or brown, watery rot of internal tissues, surrounded by a sharply defined reddish brown to black border (Fig. 19.7). When squeezed, symptomatic tubers usually ooze a dark, watery liquid. As the disease progresses through the tuber, all that often remains is the thin, papery periderm—hence the name *shell rot*.

Leak is caused by several species of the fungus-like oomycete *Pythium*, primarily *P. ultimum*. These pathogens survive from season to season as oospores in soil. Conditions favoring germination and production of zoospores are the same as those for the pink rot pathogen. *Pythium* species gain entry into the tuber only through wounds. Infection is greatly favored in warm tubers (pulp temperatures above 70°F). *Pythium* species have wide host ranges, causing damping-off and root rot in diverse plant species.

Black scurf is the tuber disease that growers call "the dirt that won't wash off." The key diagnostic symptom is dark brown to black sclerotia on the tuber surface. The sclerotia are superficial and irregularly shaped. They vary in appearance from dark, netted or scurfy residues to individual black masses ranging from pinhead to pea size (Fig. 19.8). They do not penetrate or damage the tuber. The pathogen, *Rhizoctonia solani*, is the same fungus that causes cankers of shoots, stolons, and stems. Black scurf is caused by soilborne inoculum. Cool (55–60°F), moist soils are optimal for infection.

Fusarium dry rot is a postharvest disease that causes losses in storage and transit of both seed and commercial potatoes. The initial symptom is a shallow, brown lesion on the tuber surface. The lesion slowly expands and eventually becomes sunken and wrinkled. When a tuber is cut through an affected areas, a dry, crumbly, brown decay is evident (Fig. 19.9).

Dry rot of stored tubers is most commonly caused by the fungus *Fusarium sambucinum*, although other species of *Fusarium* may be involved. This pathogen survives from season to season in diseased tubers and as chlamydospores and mycelium in soil. Soilborne inoculum is spread during harvest, and infection occurs when the pathogen enters tubers through harvest wounds. Dry rot develops most rapidly at high relative humidity and temperatures of 60–70°F; however, the disease will continue to develop slowly even at much lower temperatures. This tuber-rotting pathogen does not infect other plant species.

Black dot has recently been recognized as a disease that can significantly impact potato production. The name comes from the abundant small, black sclerotia that develop on tubers and on senescing roots, stems, and stolons. Skin blemish lesions are commonly observed at the stem end of infected tubers. Lesions are silvery to brown and generally have poorly defined margins (Fig. 19.10), compared to silver scurf lesions, which are also silvery but generally have well-defined margins (Fig. 19.11). Basal portions of infected stems are commonly an amethyst color, revealed when a stem is sliced lengthwise. Dark brown to black lesions may occur on leaves, petioles, and aboveground stems. Foliar lesions may appear similar to those of early blight (Chapter 21) but lack the concentric rings typical of early blight lesions. Infected plants may display yellowing and dying foliage similar to symptoms of potato early

Fig. 19.8. Black scurf. Dark brown to black sclerotia on the tuber surface.

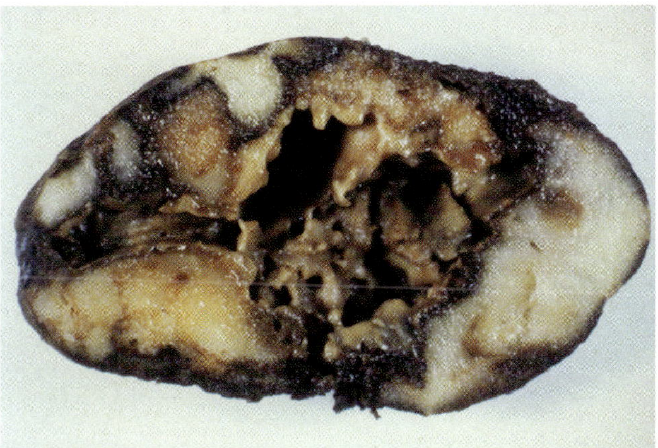

Fig. 19.7. Leak. Gray to brown, watery, rotted internal tissue surrounded by a sharply defined reddish brown to black border.

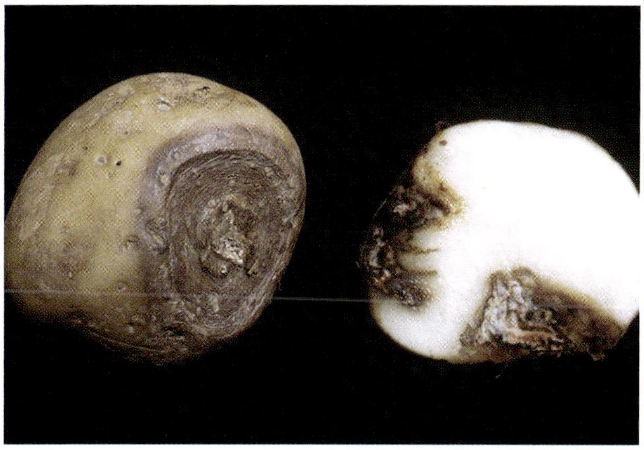

Fig. 19.9. Fusarium dry rot. Dry, crumbly, brown decay of internal tissue, often with white mold visible in decayed cavities.

dying. Both diseases are commonly present together, but none of these foliar symptoms are specific to black dot.

Black dot is caused by the fungus *Colletotrichum coccodes*. The pathogen survives as sclerotia in soil and on infected seed tubers. During the growing season, fruiting bodies called acervuli develop from sclerotia. The acervuli produce spores (conidia) that infect belowground tissues. In addition, windblown conidia can infect foliage by entering abrasion wounds caused by blowing sand. Most infections remain latent until tissues begin to senesce or the plant is stressed, at which time lesions bearing numerous sclerotia readily develop. The host range of this pathogen includes crops and weeds closely related to potato, such as tomato, pepper, and black nightshade, as well as some crops grown in rotation with potato, including mint and legumes such as yellow mustard, spring canola, soybean, and alfalfa. It can also survive on some grasses, such as oat, timothy, and giant foxtail, and several weed species, including velvetleaf (*Abutilon theophrasti*).

Silver scurf is a cosmetic skin blemish of the tuber that results in no internal damage. Tan to silvery gray, irregularly shaped patches form on the tuber periderm (Fig. 19.11). The lesions are circular, with a definite margin, and vary from spots the size of a pinhead to patches that cover most of the tuber surface. As the disease progresses, individual lesions may coalesce. The silvery appearance of older lesions, for which the disease is named, is most obvious when the tuber is wet. The lesions remain superficial, causing no damage to underlying tissues. After some time in storage, the surface of symptomatic tubers may become shriveled and wrinkled as a result of excessive moisture loss. Symptoms of silver scurf can be confused with those of black dot (Fig. 19.10 and Box 7.1).

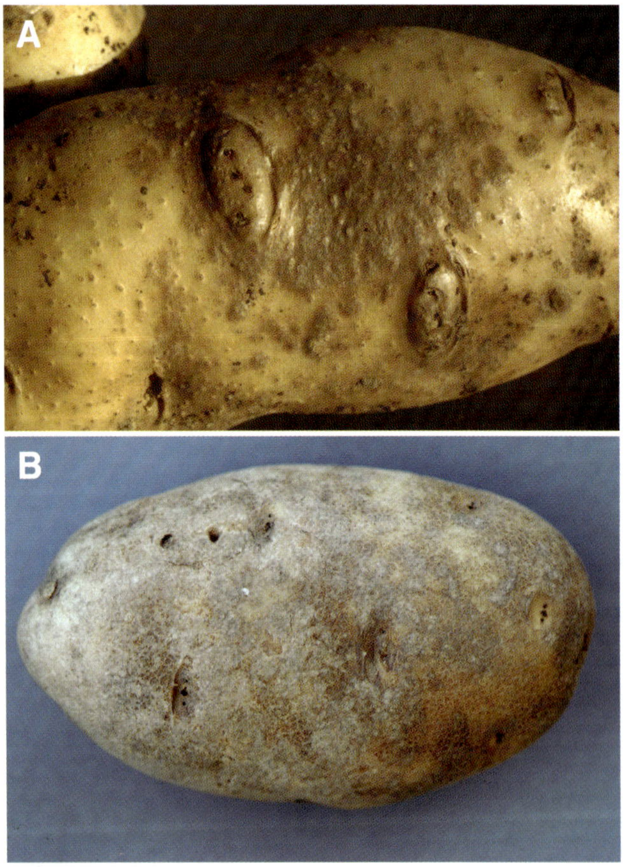

Fig. 19.10. Black dot. Silvery to brownish black lesions with poorly defined borders on Shepody, a light-skinned cultivar (**A**), and on Russet Burbank (**B**). (Courtesy D. A. Johnson)

In contrast to most of the diseases discussed in this chapter, silver scurf shares with powdery scab the distinction of having more than one cycle of pathogenesis (infection, pathogen growth, and reproduction) during the year. Although the pathogen can survive in soil for a limited time, infected seed tubers are the primary source of inoculum for the infection of daughter tubers—the first cycle of disease. Conidia are formed on the surface of diseased seed pieces and are washed off by rain and irrigation. The pathogen gains entry into developing tubers through lenticels and by direct penetration of the periderm. Symptoms are more severe on the stem end of tubers, and the severity increases over time; the longer tubers remain in the ground after vine kill, the more severe the symptoms.

The second cycle of disease, in which the fungus is spread from diseased to healthy tubers, occurs in the storage facility. Conidia form on the surface of diseased tubers at a relative humidity above 90% (and especially above 95%) and temperatures above 38°F. They are carried by air currents and through the ventilation system and are deposited on the surface of healthy tubers. When free moisture is present on tuber surfaces, a result of fluctuating temperatures and relative humidity in the storage facility, the conidia germinate and tuber infection occurs. Lesions due to secondary infection in storage generally become apparent after 4–5 months of storage. In contrast to the lesions formed in field infection, the lesions that develop in storage are not restricted to the stem end of the tuber but may cover a major portion of the tuber surface.

Silver scurf is caused by the fungus *Helminthosporium solani*. This pathogen survives from season to season on diseased seed tubers and as conidia in soil. Diseased seed tubers are regarded as the principal source of inoculum for infection of daughter tubers, because the survival of conidia in soil is thought to be poor. The pathogen has a very narrow host range; potato is its only known host.

Sources of inoculum, host ranges, and potato tissues attacked by the seedborne and soilborne pathogens discussed in this chapter vary considerably (Table 19.1). The seed tuber is the principal source of inoculum of the pathogens causing Fusarium seed piece decay, Rhizoctonia sprout canker, and silver scurf. Both the seed tuber and the soil are equally

Fig. 19.11. Silver scurf. Tan to silvery gray, irregularly shaped lesions with definite margins.

Table 19.1. Inoculum sources, infection sites, survival structures, and host ranges of seedborne and soilborne pathogens of potato

		Source of inoculum		Infection site					
Disease	**Pathogen**	**Seed-borne**	**Soil-borne**	**Sprout**	**Stolon**	**Root**	**Tuber**	**Survival structures**	**Host range**
Fusarium seed piece decay	*Fusarium solani* var. *coeruleum*, *F. sambucinum*	×	×	. . .	. . .	. . .	×	Chlamydospores	Narrow
Pythium seed piece decay	*Pythium ultimum*	. . .	×	. . .	. . .	×	×	Oospores	Broad
Rhizoctonia sprout canker	*Rhizoctonia solani*	×	. . .	×	. . .	. . .	. . .	Sclerotia	Narrow
Rhizoctonia stolon and stem canker	*Rhizoctonia solani*	. . .	×	×	×	. . .	. . .	Sclerotia	Narrow
Potato early dying	*Verticillium dahliae*, *V. albo-atrum*	×	×	. . .	. . .	×	×	Microsclerotia, darkened mycelia	Broad
	Pratylenchus penetrans	. . .	×	. . .	. . .	×	. . .	Eggs, juveniles, adults	Broad
Common scab	*Streptomyces scabies*, *S. acidiscabies*	×	×	. . .	. . .	. . .	×	Spores	Intermediate
Powdery scab	*Spongospora subterranea*	×	×	. . .	. . .	×	×	Cystosori	Narrow
Pink rot	*Phytophthora erythroseptica*	. . .	×	. . .	×	×	×	Oospores	Intermediate
Leak	*Pythium ultimum*	. . .	×	. . .	. . .	×	×	Oospores	Broad
Black scurf	*Rhizoctonia solani*	. . .	×	. . .	. . .	. . .	×	Sclerotia	Narrow
Fusarium dry rot	*Fusarium sambucinum*	. . .	×	. . .	. . .	. . .	×	Chlamydospores	Narrow
Black dot	*Colletotrichum coccodes*	×	×	. . .	×	×	×	Sclerotia	Intermediate
Silver scurf	*Helminthosporium solani*	×	×	. . .	. . .	. . .	×	Conidia	Narrow

important sources of inoculum of the common scab and powdery scab pathogens. Propagules capable of long-term survival in soil are central to the development of Pythium seed piece decay, Rhizoctonia stolon and stem canker, potato early dying, pink rot, leak, black scurf, Fusarium dry rot, and black dot. The pathogens that cause Fusarium seed piece decay and dry rot, Rhizoctonia canker, powdery scab, black scurf, and silver scurf all have narrow host ranges. In contrast, the pathogens that cause common scab, pink rot, and black dot have intermediate host ranges, and those that cause Pythium seed piece decay, potato early dying, and leak can infect a broad range of plant species (Table 19.1). A wound is essential for infection by the pathogens causing seed piece decay, leak, and dry rot, whereas the others can gain entry into potato plants by direct penetration or through natural openings. A full understanding of all these details of pathogen biology and disease development will lead to a better appreciation of the techniques used in managing these diseases and the necessity of combining several practices to develop an effective integrated management program.

Management of Seedborne and Soilborne Pathogens

Planting certified seed potatoes is an important tactic for the management of some **seedborne pathogens.** However, screening for the pathogens causing the diseases described in this chapter is not a component of any current certification programs in the United States or Canada. Therefore, visual inspection of seed tubers is an essential first step in determining the amount of disease (Chapter 7). This is accomplished through official shipping-point and arrival inspections and by careful observation of seed tuber quality prior to planting. This information will help the producer make informed decisions on the suitability of a seed lot for planting, how to handle the seed tubers prior to and at planting, and what disease management tactics to implement during crop growth.

The first step in successful management of diseases caused by **soilborne pathogens** is to be knowledgeable about the disease history of each field planned for potato production. This information comes from accurate records kept from previous potato crops grown in each field and the adjacent fields. In extreme cases, legal quarantine regulations may apply with respect to certain high-risk diseases that have been identified in localized areas. Recent examples include planting regulations arising from outbreaks of potato wart and potato cyst nematode (Chapter 20). Based on knowledge of the potential disease situation, appropriate management tactics can be implemented, applied one to several years prior to planting the crop, in the year before potatoes are planted, during crop production, and in storage.

Regardless of whether a disease originates from inoculum on seed tubers or inoculum present in the soil, the key to successfully managing these diseases is an integrated approach that combines cultural, genetic, and chemical tactics (Table 19.2).

Several Years Prior to Growing Potatoes

Crop Rotation

Healthy soil is the foundation of a healthy potato crop. Several of the soilborne pathogens discussed in this chapter have a broad host range and long-term resting structures. Therefore, it is prudent to condition soil used for potato production so that the survival of these pathogens is not favored and so that disease-suppressive microorganisms are favored (Chapters 3 and 5).

Table 19.2. Timing of management practices for suppression of diseases caused by seedborne and soilborne pathogens of potatoes[a]

	Prior years	The year prior to a potato crop			Planting				Tuber initiation		Tuber maturation	Harvest			Storage environment		
Disease	**Crop rotation**	**Soil fumigation**	**Organic amendments**	**Nonorganic amendments**	**Seed handling**	**Seed treatment**	**Planting depth**	**In-furrow fungicide**	**Foliar fungicide**	**Water management**	**Water management**	**Bruise management**	**Timing**	**Fungicide**	**Temperature**	**Humidity**	**Airflow**
Fusarium seed piece decay	. . .	. . .	. . .	. . .	++	+	++	+	. . .	. . .	. . .	. . .	. . .	. . .	. . .	. . .	. . .
Pythium seed piece decay	. . .	. . .	†	. . .	++	. . .	++	. . .	. . .	. . .	. . .	. . .	. . .	. . .	. . .	. . .	. . .
Rhizoctonia sprout canker	. . .	. . .	. . .	. . .	. . .	+	+++	+	. . .	. . .	. . .	. . .	. . .	. . .	. . .	. . .	. . .
Rhizoctonia stolon and stem canker	++	. . .	. . .	. . .	. . .	+	+	+	. . .	. . .	. . .	. . .	. . .	. . .	. . .	. . .	. . .
Potato early dying	+	+++	+	. . .	. . .	. . .	. . .	. . .	. . .	+	. . .	. . .	. . .	. . .	. . .	. . .	. . .
Common scab	+	. . .	†	+	. . .	+	. . .	. . .	. . .	+	. . .	. . .	. . .	. . .	. . .	. . .	. . .
Powdery scab	++	. . .	. . .	. . .	. . .	. . .	. . .	. . .	. . .	+	. . .	. . .	. . .	. . .	. . .	. . .	. . .
Pink rot	+	. . .	. . .	. . .	. . .	. . .	. . .	+	++	. . .	+++	++	+	+	++	. . .	++
Leak	. . .	. . .	. . .	. . .	. . .	. . .	. . .	+	++	. . .	+++	++	++	. . .	++	. . .	++
Black scurf	++	. . .	. . .	. . .	. . .	. . .	. . .	. . .	. . .	. . .	. . .	. . .	++	. . .	. . .	. . .	. . .
Fusarium dry rot	. . .	. . .	. . .	. . .	. . .	. . .	. . .	. . .	. . .	. . .	. . .	++	++	++	++	++	++
Black dot	++	. . .	. . .	. . .	. . .	. . .	. . .	. . .	. . .	. . .	. . .	. . .	++	. . .	. . .	. . .	. . .
Silver scurf	+++	. . .	. . .	. . .	. . .	+	. . .	+	. . .	. . .	. . .	. . .	++	. . .	+	+++	+

[a] +++ = Key management tactic.
++ = Moderately effective when used alone but best used in combination with other tactics.
+ = Effective only when used in combination with other tactics.
† = Organic amendments can increase the severity of the disease.

Crop rotation is a sound practice for multiple reasons, but generally it is effective as a disease management tool only as part of an integrated program. Crop rotation is most efficacious if it is implemented as a preventative rather than a curative disease management tactic. The objective of crop rotation should be to minimize populations of soilborne pathogens before they become large enough to impact tuber yield or quality. Production of rotational crops for 2 or 3 years between crops of potatoes is an effective tactic for the suppression of pathogens that have a limited host range or do not produce long-term survival structures. The pathogens causing Fusarium seed piece decay, Rhizoctonia canker, common scab, powdery scab, pink rot, black scurf, Fusarium dry rot, black dot, and silver scurf all have narrow to intermediate host ranges, and the pathogens causing common scab and silver scurf have no long-term survival structures.

Good rotational choices for the management of common scab are alfalfa, corn, rye, small grains, dry beans, and soybeans. Rotations with carrot, beet, red clover, spinach, turnip, or radish should be avoided. Growing sugar beets just prior to a crop of potatoes may increase problems with Rhizoctonia stem and stolon canker. Conversely, rotation out of potatoes for 2 or 3 years, preferably to grasses or cereals, can reduce the incidence and severity of the soilborne phase of Rhizoctonia canker and black dot. Tomatoes and peppers, being closely related to potatoes, may harbor potato pathogens or increase pathogen populations and should be avoided in rotation with potatoes. For the same reason, closely related weed species, such as black nightshade, should be well controlled in potato and rotation crops. *Verticillium* and *Pythium,* the pathogens causing potato early dying and leak, have broad host ranges among potential rotation crops, and both produce long-term survival structures. Short-term rotations of 2 or 3 years to corn, small grains, or alfalfa are useful as part of a total management program for potato early dying and leak, but this practice is generally successful only in fields where the pathogens have not established high populations. Recent studies of rotations with crops such as broccoli have demonstrated significant reductions in populations of microsclerotia of *V. dahliae* and suppression of wilt symptoms in susceptible crops subsequently grown in these fields. This effect is due, in part, to chemicals released from these crops having a specific effect on the survival of microsclerotia.

The Year Prior to Growing Potatoes

Soil Fumigation

Soilborne pathogens that produce long-term survival structures, such as *Verticillium,* have a very slow natural mortality rate. Thus, rotation out of potatoes or other susceptible species for many years may be required before the pathogen population declines below the threshold capable of causing economic loss. In heavily infested fields, other sanitation practices may be viable alternatives. Application of soil fumigants can be an effective but expensive tactic for management of potato early dying, though it is not recommended for other diseases caused by soilborne pathogens. Decisions about fumigation can be based on threshold soil populations of *Verticillium* and *Pratylenchus penetrans,* as determined by preseason soil sampling. Commercial sampling services may be available.

Several soil fumigants have been widely used to manage potato early dying. Metam-sodium products perform well when applied through sprinkler irrigation systems and are most effective in areas with coarse-textured soils. Where potatoes are grown in fine-textured soils and the crop is not sprinkler-irrigated, or in areas where sprinkler-applied fumigants are not acceptable, metam-sodium or other fumigants, such as 1,3-dichloropropene (1,3-D) combined with chloropicrin, can be shank-injected. The fumigant 1,3-D is an effective nematicide, but it has little effect on *Verticillium,* while chloropicrin is more effective against *Verticillium* than nematodes, and hence the two are applied in combination. Depending upon the chemical, the application rate, and soil environmental conditions at the time of application, fumigation has reduced soil populations of *V. dahliae* by 85–95%, with a corresponding suppression of potato early dying.

Despite this success, the performance of metam-sodium at some production sites has been inconsistent, possibly as a result of enhanced biodegradation of this chemical. Enhanced biodegradation is an accelerated decay of pesticides in soil through microbial activity. Populations of soil microbes that break down methyl isothiocyanate, the active ingredient of metam-sodium, have increased dramatically at some sites or: where metam-sodium has been applied annually for several years. Because most potato soils in the United States and Canada are fumigated no more than every 3 or 4 years, biodegradation of metam-sodium is probably not a major consideration. Nevertheless, it is prudent to be aware of the potential for reduced efficacy associated with frequent fumigation with this chemical.

Organic Amendments

One broad strategy being explored for the management of soilborne pathogens is the use of organic amendments, such as green manure crops, composts, and animal manures. In addition to suppression of some soilborne pathogens, other beneficial effects of increased soil organic matter include less surface crusting, increased water infiltration, and increased populations of soil microorganisms essential for the maintenance of healthy, highly productive soils.

Many **green manure crops** have been studied for suppression of soilborne pathogens of potatoes, including sudangrass, oat, wheat, barley, corn, rye, broccoli, canola, and winter pea. Green manure treatments can require a full season out of crop production, with the green manure crop grown in the season immediately preceding potato. In some locations, green manures have been effective in suppressing potato early dying and common scab (Chapter 3).

Green manures of rye, millet, and oat have been reported to reduce the incidence of common scab, whereas plowed-down legumes, particularly red clover, may aggravate the disease. Growers should use caution if applying animal manure or other animal wastes the year prior to potato production, as they may provide a food base for the scab pathogen. Heavy application of poultry manure, which tends to increase soil pH, can exacerbate common scab. Other types of organic amendments, including municipal wastes and pulp and paper wastes, have been shown to reduce the severity of potato early dying and common scab. When green manure crops or other organic amendments are used prior to planting potatoes, it is important to plow down the residue early enough in the previous season to allow microbial decomposition to occur. Undecomposed residues present during early potato growth may tie up nitrogen, which, when released later in the production cycle, may lead to deleterious effects on yield and quality.

Green manures and organic amendments have not been widely integrated into most potato production systems. Ongoing research continues to evaluate the use of these amendments in potato cropping systems as a tactic to manage a diversity of soilborne pathogens.

Nonorganic Amendments

In fields where common scab has been a problem, soils should be maintained at or slightly below pH 5.2. Application of gypsum, sulfur, or acid-forming fertilizers, such as ammonium sulfate, before or at planting helps maintain the soil pH between 5.0 and 5.2 and suppress the disease. Application of ammonium sulfate, rather than ammonium nitrate, may have a broader impact on soil microbial communities, which may indirectly affect pathogen activity. Consideration should be given to the potential for common scab before liming the soil, because raising the soil pH may aggravate this disease.

Planting

Disease-Resistant Cultivars

In 2000, approximately 75% of the potato acreage in the United States was cropped to only seven cultivars: Russet Burbank, Russet Norkotah, Superior, Shepody, Atlantic, Norland, and Ranger Russet. Among these, Ranger Russet is moderately resistant to potato early dying, Russet Burbank and Superior are moderately resistant to common scab, and Russet Burbank, Russet Norkotah, and Ranger Russet are resistant to powdery scab on tubers. All of these cultivars are susceptible to Fusarium dry rot, leak, pink rot, black dot, Rhizoctonia canker, and black scurf. In 2005, these same seven cultivars accounted for only 58 and 45% of U.S. and Canadian seed potato production, respectively, which indicates an expanding diversity among commercially grown cultivars. Cultivar resistance is not presently a tactic widely used to manage the diseases discussed in this chapter, but many new cultivars are being developed for the North American market, and some have varying levels of resistance to certain seedborne and soilborne pathogens. Because the North American potato industry is becoming more receptive to utilizing new cultivars that better meet the special needs of food processors and consumers, cultivars with enhanced disease resistance will become more widely available to growers as management tools and should be grown where appropriate (Table 6.2).

Handling Seed Tubers

Precutting and prehealing seed tubers several weeks prior to planting may expedite the planting operation and reduce losses to bacterial soft rot, but it puts the seed pieces at risk of seed piece decay caused by either *Fusarium* or *Pythium*. This is because the temperature range most favorable for the healing of cut surfaces, 55–60°F, also favors these pathogens. To reduce the risk of seed piece decay, seed tubers should be warmed before cutting to minimize injury and to promote rapid emergence. Several seed piece treatments are currently labeled for management of seed piece decay (Table 19.3) and should be applied immediately following cutting. If soil conditions are appropriate, seed pieces should be planted immediately. Alternatively, cut seed can be stored for up to several weeks at 55–60°F and 95–99% relative humidity with good air movement through the pile in order to hasten the healing process (Chapter 7). Cut seed should be planted in warm soil (50–55°F) and covered with as little soil as is practical. Planting in soil where the difference between the seed piece temperature and the soil temperature is greater than 10°F should be avoided. These three practices—application of a seed piece treatment, proper healing, and shallow planting in warm soil—favor rapid emergence, which allows the plant to become established before the seed piece decays and minimizes problems with Rhizoctonia sprout canker. A drawback to shallow planting is the production of daughter tubers near the soil surface, where they may be exposed to late blight infection later in the growing season (Chapter 21). To minimize this problem, a substantial hill should be formed over each row during postemergence cultivation.

Green manure or cover crops should not be plowed down immediately ahead of planting, because soil populations of the Pythium seed piece decay pathogen can increase dramatically on freshly incorporated vegetation. Delaying planting for at least 2 weeks after plow-down or killing vegetation with a herbicide a week or so before plow-down will minimize seed piece decay.

Green manure or cover crops should not be plowed down immediately ahead of planting, because soil populations of the Pythium seed piece decay pathogen can increase dramatically on freshly incorporated vegetation.

Seed tubers should be inspected upon receipt, and any seed lot in which more than 5% of the tuber surfaces is covered with sclerotia of *Rhizoctonia* should be rejected, because of the excessive risk of sprout canker. Seed treatments and in-furrow treatments with a registered fungicide provide some control of seedborne inoculum of *Rhizoctonia*. However, planting disease-free seed tubers is the most efficacious option.

The same practices recommended for the management of seed piece decay apply to the management of stem and stolon canker, the chronic phase of *Rhizoctonia* disease. Green tissue is much less susceptible to attack by *Rhizoctonia,* and therefore any practice that favors **rapid emergence** will help suppress the soilborne phase of the pathogen. Shallow planting, at a depth of 2–3 inches rather than 5–6 inches, allows for quicker emergence. Caution should be exercised with preirrigation in cool weather, since wet soil may remain cool longer and thus delay emergence.

Some tactics for protection against tuber diseases problematic at harvest and in storage are applied at planting or early in the growing season. **Seed treatments** with mancozeb and fludioxonil are effective in managing *H. solani,* the seedborne

Table 19.3. Fungicides for management of diseases caused by seedborne and soilborne pathogens of potato[a]

Fungicide and application type	Fusarium seed piece decay	Pythium seed piece decay	Rhizoctonia sprout canker	Rhizoctonia stolon and stem canker	Common scab	Pink rot	Leak	Fusarium dry rot	Black dot	Silver scurf
Seed piece treatments										
Maneb	×	...	×	...	...	...	...	...	...	...
Mancozeb	×	...	×	...	×	...	...	...	...	×
Fludioxonil + mancozeb	×	...	×	×	...	...	...	...	...	×
Flutolanil + mancozeb	×	...	×	...	...	...	...	...	...	×
Thiophanate-methyl + mancozeb	×	...	×	...	...	...	...	...	...	×
In-furrow treatments										
Azoxystrobin	×[b]	...	×	×	...	...	...	...	...	×
Flutolanil	...	...	...	×	...	...	...	...	...	...
Pentachloronitrobenzene (PCNB)	...	...	...	×	×[c]	...	...	...	...	...
Mefenoxam	...	...	...	...	...	×[d]	×[d]	...	...	...
Foliar fungicides										
Phosphorous acid	...	...	...	...	...	×	...	...	...	...
Mefenoxam	...	...	...	...	...	×[d]	...	...	...	...
Postharvest treatments										
Phosphorous acid	...	...	...	...	...	×	...	...	...	...
Thiabendazole	...	...	...	...	...	...	...	×[d]	...	...

[a] Trade names of fungicides are given in Table 14.5.
[b] Effective only on freshly cut seed.
[c] Results in suppression only.
[d] Strains of the pathogen may be resistant.

pathogen causing silver scurf, whereas virtually all isolates of the pathogen are resistant to thiabendazole. Fungicides with activity against pink rot and leak, when applied to the furrow at planting or to the foliage at tuber initiation (Growth Stage III), can reduce the amount of infection in daughter tubers (Table 19.3). An advantage of in-furrow application is that it provides more uniform control across the field than foliar application. Resistance of the pink rot and leak pathogens to metalaxyl and mefenoxam, two fungicides commonly applied for the management of these diseases, has been reported in several locations in the United States and Canada (Chapter 14). To determine if lack of control is a result of fungicide resistance or ineffective application, infected tubers can be tested by a reputable laboratory. Foliar application of phosphorous acid is effective against pink rot.

A large number of **fungicides** are currently labeled as seed piece, in-furrow, and postharvest treatments (Table 19.3). Seed treatments provide protection for both cut and uncut tuber surfaces, and soil-applied fungicides provide protection for growing roots and sprouts. Selecting a fungicide depends on which diseases are viewed as most important in the specific production situation and whether harvested tubers will be marketed fresh or for processing. *Selected trade names of fungicides are given in Table 14.5.*

Growth Stage III: Tuber Initiation

Water Management

Knowledge of the disease history of a field is paramount in the effective use of irrigation to manage potato early dying, powdery scab, and common scab. Both potato early dying and powdery scab are favored by moist soils from Growth Stage II to Growth Stage III. Manipulating the amount of irrigation water applied during this period can limit the severity of both diseases. Avoiding overirrigation or even keeping the soil a little dry (60–65% available soil moisture) may reduce the amount of root infection by *V. dahliae* and tuber infection by *S. subterranea*. However, this practice can exacerbate common scab in fields with a history of that disease and may also lead to translucent end, a physiological disorder, in susceptible cultivars, such as Russet Burbank (Chapter 23). The activity of the common scab pathogen is reduced in moist soils. Careful timing of irrigation to maintain 80–90% available soil water (near field capacity) during Growth Stage III and for the next 6–8 weeks can limit the severity of common scab. However, high soil moisture must be maintained through the entire period for this practice to be effective, and the practice is most effective with cultivars that are moderately resistant to common scab. The relative potential importance of these three diseases must be evaluated, and soil moisture should be managed accordingly.

Growth Stage V: Maturation

Water Management

Pink rot and leak are favored by wet soils, especially near the end of the season, when crop water use has declined. In irrigated fields with a history of either disease, the first line of defense is to manage irrigation water late in the growing season to avoid saturating the soil.

Harvest

Fusarium dry rot, pink rot, leak, and black scurf are best managed through implementation of harvest procedures that minimize tuber bruising. To ensure that tuber skins are mature and harvest injury is minimized, harvest should be delayed until the vines have been dead for 1–2 weeks and tuber pulp temperatures are between 50 and 65°F. However, tubers can be successfully stored when harvested from green vines in some areas, such as the semiarid environment of the Columbia Basin (Chapter 10). Application of the fungicide thiabendazole to tubers going into storage should be considered if Fusarium dry rot is anticipated. This is best accomplished during bin loading, by applying the fungicide as a fine mist at some point on the conveyor line where the tubers are tumbling. However, some strains of *Fusarium* are resistant to thiabendazole and thus are not controlled by this treatment. As with pink rot and leak, a reputable laboratory should be consulted to determine if the strains of the dry rot pathogen that are present are resistant or sensitive to this fungicide. Phosphorous acid can also be applied to tubers going into storage, to reduce the development of pink rot.

Generally, it is best not to allow healthy tubers to remain in soil too long after maturation, as the longer they stay in the soil, the greater the likelihood they will become diseased. Harvesting tubers soon after vine death limits the development of silver scurf lesions and the formation of sclerotia of *Rhizoctonia* on tuber surfaces. To minimize problems with leak, harvesting should be delayed until the tuber periderm is mature and tuber pulp temperatures are below 65°F. In fields known to be at risk of pink rot, it is best to harvest tubers when pulp temperatures are 45–50°F and the skin is set sufficiently to prevent skinning during the harvest operation.

Storage

If any significant amount of one or more tuber diseases is evident during the harvest operation, certain precautions should be taken in the management of the storage environment when the tubers are put into storage. The three basic tools of storage management are temperature, relative humidity, and airflow. The wound-healing or curing period is especially important with pathogens that gain entry through wounds, such as *Fusarium*, *Pythium*, and *Phytophthora*. Infection by these pathogens can be minimized by providing appropriate conditions for rapid suberization right after tubers are placed in storage: high humidity (95%), optimum tem-

The three basic tools of storage management are temperature, relative humidity, and airflow.

perature (50–55°F), and good ventilation (25 cubic feet per minute per ton of potatoes) to avoid condensation (Chapter 10). If there is a significant amount of pink rot or leak in tubers going into storage, fans should be operated continuously to increase air movement, and the storage temperature should be dropped below 50°F, with outside air used as needed, to cool the crop as quickly as possible. Tubers known to be infected with late blight (Chapter 21) may also need to be managed in this way. Many diseases are favored by high relative humidity, and bacterial soft rot is favored by the condensation of free moisture on tuber surfaces (Chapter 18). To reduce the potential for condensation, the tuber pulp temperature should be kept at 50°F, and the ventilating air temperature should be within a few degrees (2°F) of the tuber temperature (Chapter 10). To manage silver scurf, the relative humidity should be kept below 90% for at least the first month; this halts spore production and subsequent germination. However, a side effect of this practice is a potential increase in tuber shrinkage and pressure bruise. The spread of silver scurf can be limited by maintaining a temperature of 38°F and providing forced-air ventilation.

Summary

Key factors affecting the severity of diseases caused by seedborne pathogens are the health and condition of seed tubers, fungicide resistance in the pathogen population, and soil temperature and moisture. With the exception of seed tuber health, these same factors plus soil pH, percent organic matter, and pathogen population size are the key variables for diseases caused by soilborne pathogens.

Fortunately, there are a number of effective cultural and chemical tactics to manage diseases caused by seedborne and soilborne pathogens. Although no single tactic will manage any one disease, integrating several of the practices discussed will aid potato producers in maintaining crop quality while minimizing economic loss.

CHAPTER 20

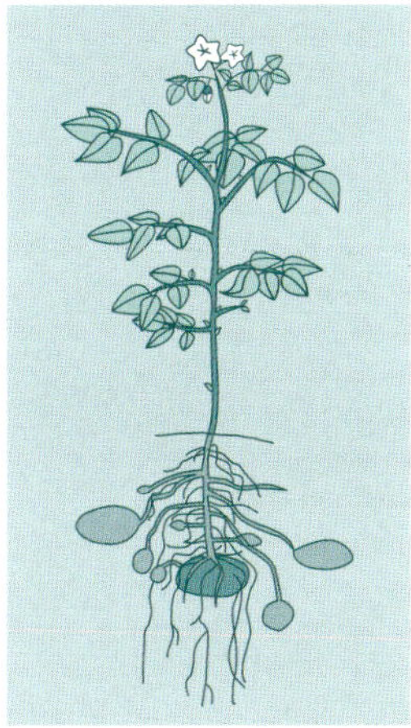

Ann E. MacGuidwin
Department of Plant Pathology
University of Wisconsin, Madison

Managing Diseases Caused by Nematodes

Nematodes are round, unsegmented worms (not true worms), living in every habitat on earth. Nematode species that live in the soil feed on bacteria, fungi, and soil-dwelling invertebrates. Productive soils contain many nematodes, and their abundance and diversity are good indicators of microbial activity. As nematodes feed on microbes, they release waste products rich in nitrogen that is immediately available to plants. Most nematode pests of agricultural crops feed exclusively on roots and belowground plant structures.

Plant-parasitic nematodes are microscopic, ranging from 0.004 to 0.04 inches in length, and live inside roots or in the soil close to roots. They feed on roots with a needle-like mouthpart called a stylet, which is inserted into plant tissues (Fig. 20.1). Some species affect root growth by destroying cells and interfering with the uptake of water and nutrients. Others stimulate physiological changes that upset the balance of hormones and other metabolic systems in the host plant. Some species do not affect plant growth but invade tubers and reduce tuber quality. Plant-parasitic nematodes complete their life cycles, from egg to juvenile to adult, in about 20–50 days, depending on temperature and soil moisture.

Nematode damage to potato ranges from a very subtle restriction in plant growth to severe stunting of plants or disfiguration of tubers. Symptoms caused by some nematode pests are easily confused with other problems, and a diagnosis should be confirmed with a soil test for the presence of particular nematode species. Infestations of nematodes build slowly over time and are usually patchy within a field, so field history is useful for making a diagnosis. Some nematode species are limited in geographic distribution, whereas others are found in nearly all production areas. Some species interact with other pathogens, mostly fungi, in disease complexes, such as potato early dying. Some nematode species are vectors of plant-pathogenic viruses, such as *Tobacco rattle virus,* which can be transmitted to potato by stubby-root nematodes.

Common Nematode Pests

Of the many types of nematodes that infect potatoes (Table 20.1), the four most widespread and economically significant in North America are root-lesion nematodes (*Pratylenchus* species), root-knot nematodes (*Meloidogyne* species), sting nematodes (*Belonolaimus* species), and stubby-root nematodes (*Paratrichodorus* species). Three others—two species of cyst nematodes (*Globodera rostochiensis* and *G. pallida*) and the potato rot nematode (*Ditylenchus destructor*)—occur in only a few localities in North America but are significant pests in other potato production areas of the world.

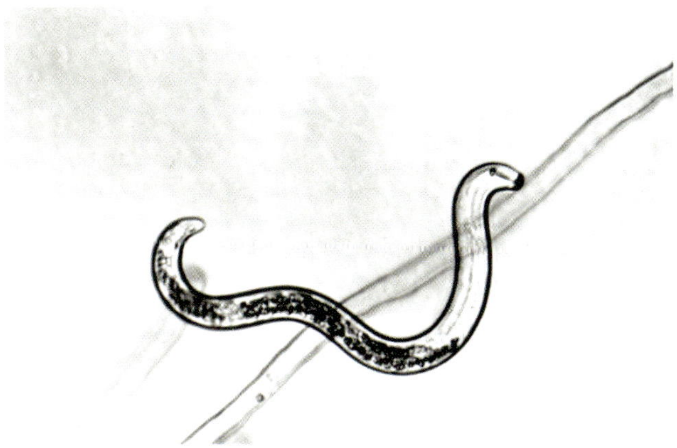

Fig. 20.1. Root-lesion nematode feeding on a root.

Root-Lesion Nematodes

Root-lesion nematodes, *Pratylenchus* spp., are one of the most common potato pests. There are many species, some of which are more harmful to potato than others. The species of most concern, *P. penetrans, P. neglectus, P. brachyurus,* and *P. scribneri,* have overlapping geographic distributions, but generally one species predominates within a region: *P. neglectus* in the Northwest, *P. penetrans* in the central and eastern regions, and *P. brachyurus* in the South. Species identification, however, is not offered by most diagnostic laboratories. Damage thresholds for root-lesion nematodes vary among geographic regions, because of the different species likely to be present and differences in cultural and environmental factors in different regions.

Table 20.1. Nematode pests of potato in the United States and Canada

Nematode	Pacific West[a]	Mountain West[b]	West North Central[c]	East North Central[d]	Middle Atlantic[e]	New England[f]	West South Central[g]	South Atlantic[h]	East South Central[i]	Western Canada[j]	Central Canada[k]	Eastern Canada[l]
Belonolaimus longi-caudatus	...	...	×	×	...	...	×	×	...	...	...	...
Ditylenchus destructor	×	×	...	×	×	...	×	×	...	×	...	×
Globodera pallida	...	×[m]	...	...	...	...	...	...	...	...	...	×
Globodera rostochiensis	...	...	...	...	×[n]	...	...	...	...	...	...	×
Helicotylenchus spp.	×	×	×	×	×	×	×	×	×	×	×	×
Hoplolaimus spp.	×	×	×	×	×	×	×	×	×	...	×	...
Meloidogyne chitwoodi	×	×	...	...	...	...	×	...	...	...	...	...
Meloidogyne hapla	×	×	×	×	×	×	×	×	×	×	×	×
Meloidogyne incognita	×	×	...	×	...	...	×	×	×	...	...	...
Meloidogyne javanica	×	×	...	...	...	...	×	×	×	...	...	...
Paratrichodorus spp.	×	×	×	×	×	×	×	×	×	...	×	...
Pratylenchus brachyurus	×	...	...	...	...	...	×	×	×	...		...
Pratylenchus crenatus	×	...	×	×	×	×	...	...	...	...	×	×
Pratylenchus neglectus	×	×	×	×	×	×	×	×	...	×	×	×
Pratylenchus penetrans	×	×	×	×	×	×	×	×	×	×	×	×
Pratylenchus scribneri	×	×	×	×	×	...	×	×	×	...	×	×
Tylenchorhynchus spp.	×	×	×	×	×	×	×	×	×	...	×	×
Xiphinema spp.	×	×	×	×	×	×	×	×	×	×	×	...

[a] California, Oregon, and Washington.
[b] Arizona, Colorado, Idaho, Montana, Nevada, New Mexico, Utah, and Wyoming.
[c] Iowa, Minnesota, Missouri, Nebraska, North Dakota, and South Dakota.
[d] Illinois, Indiana, Michigan, Ohio, and Wisconsin.
[e] New Jersey, New York, and Pennsylvania.
[f] Connecticut, Maine, Massachusetts, New Hampshire, Rhode Island, and Vermont.
[g] Arkansas, Kansas, Louisiana, Oklahoma, and Texas.
[h] Delaware, Florida, Georgia, Maryland, North Carolina, South Carolina, Virginia, and West Virginia.
[i] Alabama, Kentucky, Mississippi, and Tennessee.
[j] Alberta and British Columbia.
[k] Manitoba, Ontario, and Saskatchewan.
[l] New Brunswick, Newfoundland, Nova Scotia, Prince Edward Island, and Quebec.
[m] Identified in southern Idaho only.
[n] Restricted to New York.

Symptoms

At high population densities, root-lesion nematodes can reduce foliar growth and tuber yield and quality. Aboveground symptoms are stunting and delayed canopy closure. Root lesions, the hallmark symptom of this species, range from a darkened root system to subtle discoloration that is difficult to see without a microscope. The most useful data are the population density of nematodes as revealed by a soil and root assay. Even low numbers of some species are cause for concern, because of the interaction between some root-lesion species, predominantly *P. penetrans,* and *Verticillium dahliae* in potato early dying disease. Advisory services often use two thresholds for root-lesion nematodes to account for the impact of this pest with and without the additional damage potential of *Verticillium.*

Pathogen and Disease Development

Root-lesion nematodes are migratory endoparasites—they usually reside inside roots but move freely between root and soil habitats. Eggs deposited in plant tissue or soil hatch as second-stage juveniles. The juveniles molt (shed the cuticle) twice to develop into third-stage and then fourth-stage juveniles, which molt again to reach the adult stage. Only adults reproduce. Root-lesion nematodes in all stages are capable of infecting potato roots, stolons, and tubers. Juveniles and adults feed on the root surface and in the root cortex, with no preference for any particular root region (Fig. 20.1). When plant cells die in response to feeding, the nematodes move out of roots in search of new feeding sites. Some species produce both males and females, but some produce only females, which reproduce without males. Root-lesion nematodes complete their life cycle in 1–2 months, depending on soil moisture and temperature.

The damage potential of root-lesion nematodes depends on the amount of initial inoculum and weather conditions the first 2 months after planting. Their movement is impeded in waterlogged soils; thus, soil moisture conditions early in the growing season can have a significant impact on the rate of infection and population increase. Root-lesion nematodes can overwinter in all life stages. Reproduction begins soon after infection, and the nematode population density may increase to the point that plants are stunted. Infection often causes a lag in shoot growth that delays canopy closure by 3–10 days, and surveying a field for patches of open rows at the time of canopy closure can help identify the distribution of nematodes in the field.

Root-lesion nematodes have a wide host range, which enables them to persist at a high population density between potato crops. Some crops support higher rates of reproduction than others, so the crop immediately preceding potato,

Table 20.2. Efficacy of cover crops in the management of nematode pests of potato[a]

Cover crop	Latin name	Southern root-knot nematode (*Meloidogyne incognita*)	Columbia root-knot nematode (*Meloidogyne chitwoodi*)	Root-lesion nematodes (*Pratylenchus* spp.)	Sting nematode (*Belonolaimus longicaudatus*)
Forage pearl millet	*Pennisetum glaucum*	...	...	Good	...
Sorghum-sudangrass	*Sorghum bicolor* × *S. sudanense*	Good	Good	Good	...
Pangola digitgrass	*Digitaria decumbens*	Good	...	...	...
Transvaal digitgrass	*Digitaria decumbens*	...	...	...	Good
Perennial ryegrass	*Lolium perenne*	...	Good	...	...
Cereal rye	*Secale cereale*	Good	...	...	Poor
Black oat	*Avena strigosa*	...	...	Good	...
Sesame	*Sesamum indicum*	Good	...	...	...
Marigold	*Tagetes patula, T. erecta*	Good	...	Good	Poor
Showy crotalaria	*Crotalaria spectabilis*	Good	...	Poor	Good
Sunn hemp	*Crotalaria juncea*	Good	...	Poor	...
Cowpea	*Vigna unguiculata*	Good	...	...	...
Hairy indigo	*Indigofera hirsuta*	Good	...	Good	Good
Velvetbean	*Mucuna deeringiana*	Good	Good	...	...
Rapeseed	*Brassica napus*	...	Good	...	...
White mustard	*Sinapis alba*	...	Good	...	...
Oilseed radish	*Raphanus sativus*	...	Good	...	Good

[a] Evaluations compiled from reports published in scientific journals. The efficacy of some cover crops varies among cultivars. In some cases efficacy depends on incorporation of the cover crop as green manure.

including the fall cover crop, has a strong influence on the level of initial inoculum at the beginning of the potato growing season (Table 20.2).

Root-lesion nematodes have a wide host range, which enables them to persist at high population density between potato crops.

The interaction of root-lesion nematodes with *V. dahliae* in potato early dying is well documented. Co-infection by root-lesion nematodes (primarily *P. penetrans*) and *V. dahliae* may result in severe disease, including premature senescence, and low yields (Chapter 19). In some cultivars, certain root-lesion species have a synergistic relationship with the fungus: the damage caused by co-infection with both is greater than the sum of the damages that would be caused the two organisms acting separately.

Root-Knot Nematodes

There are more than 60 described species of root-knot nematodes. These pests damage a wide range of crops, but only a few species cause serious losses in potato. In the northeastern and north central United States and in eastern and central Canada, the northern root-knot nematode (*Meloidogyne hapla*) is the predominant species associated with potato. In the Pacific Northwest, both the northern root-knot nematode and the Columbia root-knot nematode (*M. chitwoodi*) attack potato. The Columbia root-knot nematode is widely distributed throughout the United States but is the predominant species in the Pacific Northwest. The southern root-knot nematode (*M. incognita*), the Javanese root-knot nematode (*M. javanica*), and the peanut root-knot nematode (*M. arenaria*) are associated with potato in the southern United States. The identification of species of root-knot nematodes present in a field can sometimes be deduced from the cropping history, geographic region, and plant symptoms. Species identification is provided by some labs and is important for planning crop rotations in regions where more than one species of root-knot nematode is present.

Symptoms

Root-knot nematodes are named for the common response of roots to infection. Invasion and feeding by these nematodes cause a swelling of cells at the feeding site and an increase in cell multiplication around the site, resulting in a visible swelling, or gall. The formation of root galls and the location, shape, and size of the galls vary among different nematode species and different host plants. On potato roots, moderately large galls form in response to infection by the southern root-knot nematode, but very small ones result from infection by the northern root-knot nematode. Gall formation on tubers also varies with the nematode species. Infection by the Columbia root-knot nematode and the southern root-knot nematode often causes bumpy tubers (Fig. 20.2), but infection by the northern root-knot nematode seldom does. Other symptoms associated with root-knot nematodes include internal browning of tubers (Fig. 20.3), formation of small tubers, poor shoot growth, and foliar chlorosis.

Pathogen and Disease Development

Root-knot nematodes are sedentary endoparasites—once they are in position and commence feeding, they remain permanently attached to tissues inside the root or tuber and never leave the organ. Like other plant-parasitic nematodes, they hatch from eggs as second-stage juveniles and migrate through the soil in search of host roots and tubers. After entering plant tissue, each juvenile establishes a feeding site composed of two to six cells. The bodies of juveniles enlarge as they feed. They soon lose the muscles necessary to leave the root or tuber, and after this point in their development only the stylet and head move during feeding. Feeding by these nematodes triggers physiological changes in the plant that cause nutrients to accumulate at the feeding site. The root or tuber cells at the feeding site, called giant cells, are highly active and become enlarged. These cells provide nutrients to the nematode for the remainder of its life. Second-stage juveniles mature through the third and fourth juvenile stages within host tissue before reaching adulthood. Most root-knot species are composed entirely of females. Adult females are swollen and round or oblong. Females deposit eggs in an egg mass, a collection of 200–1,000 eggs bound together by a sticky substance produced in specialized glands. Males, when they occur, remain wormlike in shape and do not feed.

One to five generations may be produced per year. Eggs and second-stage juveniles persisting through the winter in soil, root debris, and tubers act as inoculum in the subsequent crop. Root-knot nematodes may be located quite deep in the soil. The Columbia root-knot nematode, for example, has been recovered from depths of 5–6 feet below the soil surface. Root-knot nematodes are adept at surviving harsh environments. The eggs are particularly well suited for survival. Second-stage juveniles can avoid freezing if the soil cools slowly or is fairly dry, allowing them time to accumulate antifreeze compounds or rid their bodies of excess water. Both eggs and juveniles are susceptible to various natural enemies, such as egg-parasitic fungi and predators. The extent of mortality in root-knot nematode populations between seasons varies in different years and locations, ranging from 20 to 90%.

The several species of root-knot nematodes that attack potato differ in their preferences for other host plants. For example, the northern root-knot nematode attacks most vegetable and fruit crops and alfalfa but does not infect corn, grasses, and cereals. In contrast, the Columbia root-knot nematode infects corn, grasses, and cereals, and some populations (race 2) attack alfalfa, while some (race 1) do poorly on alfalfa. The southern and peanut root-knot nematodes are also divided into races that differ in host specificity.

In the southern United States, an association between the southern root-knot nematode and the wilt bacterium, *Ralstonia solanacearum,* has been demonstrated, in which the presence of the nematode increases the incidence and severity of bacterial wilt, even when only a small population of the bacterial pathogen is present.

Fig. 20.2. Bumps on a tuber, caused by the root-knot nematode *Meloidogyne chitwoodi.* (Courtesy D. A. Johnson)

Fig. 20.3. Internal browning caused by root-knot nematodes. (Courtesy G. S. Santo)

Potato Rot Nematode and Potato Cyst Nematodes

The potato rot nematode, *Ditylenchus destructor,* and the potato cyst nematodes, *Globodera rostochiensis* and *G. pallida,* damage potatoes only in highly localized areas of North America, but they are important pests in many other potato-growing regions of the world. Infestations of the potato rot nematode are scattered throughout the United States. Known infestations of the potato cyst nematode *G. rostochiensis* are restricted to New York. *G. pallida* has been identified in a field in southern Idaho. Both species have been identified in Canada (Table 20.1). In North America, potatoes harvested from fields known to be infested with potato rot or potato cyst nematodes are subject to quarantine and regulatory standards for sanitation. There is zero tolerance for international export of potatoes from fields infested with either pest. The economic impact of potato rot and potato cyst nematodes in the United States and Canada has not been extensive in recent years, because of enforcement of regulatory measures designed to limit their spread. Early detection is essential to ensure that the impact of these nematodes on U.S. and Canadian potato production remains low.

Symptoms

The potato rot nematode damages tubers without causing noticeable symptoms on roots or shoots. In the early stages of infection, small, white, chalky lesions are evident in peeled tubers. As the nematode reproduces, a dry rot develops inside the tubers, and the tuber surface becomes cracked (Figs. 20.4 and 20.5). Symptoms are often not detected until tubers are harvested or in storage. The population density of the potato rot nematode inside symptomatic tubers is usually very high, and all life stages are easily retrieved by soaking rotted tissue in water. Detection in soil is problematic, as it is not unusual for samples collected in the vicinity of infected tubers to be free of *D. destructor.*

Potato cyst nematodes affect plant growth and vigor (Fig. 20.6). The severity of symptoms is related to the number of nematodes present at planting. Infection reduces root biomass, which can lead to stunting of plants, yellowing and wilting of foliage, and formation of small tubers. Damage thresholds vary by location and are affected by soil type, potato cultivar, and environmental conditions. Potato cyst nematodes are not endemic in the United States and Canada, so infestations are most likely to begin with infected seed tubers. Once introduced, however, they can become widely distributed by the movement of eggs in water, windblown soil, and animal feces. Symptoms and yield loss may increase slowly over time, and the pathogen may not be detected until many years after it was introduced.

Pathogen and Disease Development

The epidemiology of the potato rot nematode is poorly understood. Unlike the chronic infestations typical of other nematode pests, soil infestations with this species appear suddenly one year and are undetectable the next. Detection of this nematode in soil samples is problematic, and the definitive diagnosis is based on symptomatic tubers. It is very rare to detect the nematode in consecutive potato crops, regardless of the length of rotation. Once infested, fields should remain suspect, because the nematode can reappear without warning. The potato rot nematode has a very wide host range and can even feed on fungi under laboratory conditions. There has been speculation that the nematode persists on alternate plant and fungal hosts in the field, but this possibility has not been confirmed. The life cycle of the potato rot nematode can be completed in as little as 14 days in the laboratory but has not been described under field conditions.

Potato cyst nematodes are endoparasites, with a life cycle similar to that of root-knot nematodes. White, round, pinhead-sized females start appearing on potato roots around the time of flowering (Fig. 20.7). Eggs are retained inside the body of the female, which then dies, while the body retains its shape as a cyst. The cyst protects the eggs, which lie dormant in the soil until host plants are present. Eggs can remain viable for many years. Potato, tomato, and some solanaceous weeds are hosts of the potato cyst nematode. The white females attached to roots can be seen with the naked eye, but the cysts, when loose in the soil, can be detected only by a soil test.

Sting Nematodes and Stubby-Root Nematodes

Sting nematodes (*Belonolaimus* spp.) and stubby-root nematodes (*Paratrichodorus* spp.) live exclusively in the soil

Fig. 20.4. Surface cracks and dry rot caused by the potato rot nematode. (Courtesy A. Phibbs)

Fig. 20.5. Internal dry rot caused by the potato rot nematode.

Fig. 20.6. Stunting of plants caused by the potato cyst nematode. (Photo by Christopher Hogger, used by permission of Mactode Publications)

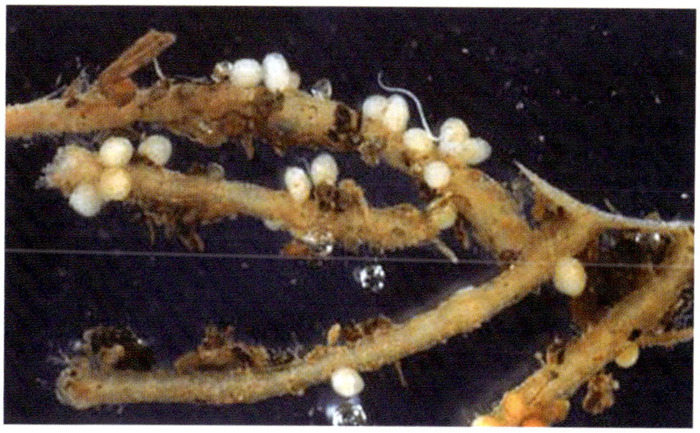

Fig. 20.7. Females (white) and cysts (dark) of the potato cyst nematode on roots. (Copyright Cornell University, used by permission)

and do not enter potato roots or tubers. These nematodes, like many other species with a similar ectoparasitic lifestyle, are fairly mobile in the soil profile and are easier to detect when temperature and moisture are not extreme for the region. Soil samples should be collected in the row in the vicinity of plant roots where nematode population densities are highest. Sting nematodes occur only in sandy soils and have been reported in the southeastern coastal plains, Kansas, and Nebraska. Stubby-root nematodes occur in every potato-producing region but are most damaging in sandy soils.

Symptoms

Sting nematodes have a long stylet and prefer to feed deep inside root tips, killing cells and halting root growth. The potential for damage caused by root feeding is high, and even low numbers of nematodes can impair root growth, particularly in young plants. Stunted plant growth, nutrient deficiency, water stress, and misshapen tubers are symptoms caused by sting nematodes (Figs. 20.8 and 20.9). The deformation of club-shaped roots damaged by nematode feeding can be mistaken for injury caused by dinitroaniline herbicides. Sting nematode infestations in the field can be quite patchy, and it is not unusual for nematode damage to occur in small circular to elliptical areas that remain fairly constant over time.

Fig. 20.8. Plants stunted as a result of sting nematode injury (foreground). Plants in the background were treated with a nematicide. (Courtesy W. T. Crow)

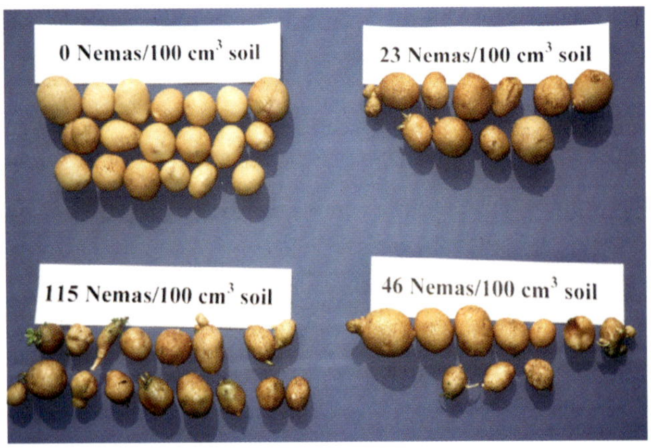

Fig. 20.9. Effects of sting nematodes on tuber yield and quality. (Courtesy W. T. Crow)

Stubby-root nematodes feed on the root surface and cause short, abbreviated roots. Plants are able to tolerate some feeding damage, and action thresholds vary among localities according to soil type and temperature conditions. At high nematode population densities, symptoms similar to those described for sting nematodes occur. Stubby-root nematodes are vectors of *Tobacco rattle virus,* the causal agent of corky ringspot (Fig. 20.10), a tuber disease characterized by a distinctive pattern of internal necrosis (Chapter 17). Corky ringspot occurs in many production regions but is particularly a problem in Florida and the Columbia Basin of Washington and Oregon.

Pathogen and Disease Development

Sting nematodes are among the largest of the plant-parasitic species (up to ⅛ inch long), so they are able to move only through the large pores typical of sandy soils. Irrigation and rain provide water films that facilitate the movement of sting nematodes. They seek out growing root tips and use a long, protrusible stylet to reach deep inside root tissues. The first generation of nematodes causes the most damage. Sting nematodes have a wide host range, but crops vary widely in the amount of nematode reproduction that is sustained.

Stubby-root nematodes are much smaller than sting nematodes and have a unique curved, toothlike stylet. A stubby-root nematode scrapes its stylet over the root surface and produces saliva that hardens into a strawlike structure, through which the nematode withdraws the contents of plant cells. When the cells die, the nematode moves to a new feeding location. Males are rare in *Paratrichodorus* spp., and populations composed almost entirely of females build up rapidly. The life cycle can be completed in as little as 16 days in warm soils, and longer in cool soils. Stubby-root nematodes have a wide host range, including corn, turf, cabbage, and soybean.

Diagnosing Nematode Problems

To obtain an accurate diagnosis of a suspected nematode problem, soil and root samples should be submitted to ex-

Fig. 20.10. External symptoms of corky ringspot. (Reprinted from D. P. Weingartner, J. R. Schumaker, and G. C. Smart, Jr., 1983, Why soil fumigation fails to control potato corky ringspot disease in Florida, Plant Disease 67:130–134)

perienced professionals who offer nematode assay services. On-site diagnosis based on symptoms should be confirmed by assay. Management of most nematode pests of potato is based on damage thresholds.

Sampling

Collecting samples that will accurately represent the nematode population in a field is the most important step in obtaining a reliable diagnosis. Local crop advisors or other experienced professionals should be asked to recommend proper sampling procedures. Appropriate methods depend on the nematode species present in the field, taking into account the unique biology and life history of each species. Different sampling schemes may be required at different times of year or for different crops. Soil samples can be collected with soil core sampling tubes, trowels, or shovels. Separate samples should be collected from areas of the field that differ in soil type, topography, or cropping history. When the crop is well established, the samples should be collected from the root zone, including areas between rows. Some labs extract root pieces from soil samples for analysis. Others prefer separate root samples from intact plants. A sampling depth of 6–9 inches is adequate for the detection of most plant-parasitic nematodes, but some, including the sting and Columbia root-knot nematodes, may be situated much deeper in the soil, and greater sampling depths are requiring for detection.

Nematodes are generally not uniformly distributed across a field but rather are clumped in areas of higher concentration interspersed with areas of lower concentration.

Fig. 20.11. Clusters of poorly emerged, stunted, weak plants (dark areas) damaged by the root-lesion nematode *Pratylenchus penetrans* in an aerial infrared photograph of a 125-acre potato field. Healthy plants appear red. (Courtesy P. B. Hamm)

Nematodes are generally not uniformly distributed across a field but rather are clumped in areas of higher concentration interspersed with areas of lower concentration (Figs. 20.11 and 20.12). Because of their aggregated pattern, intensive sampling is generally needed to accurately estimate nematode populations in a field. Population estimates can be improved by either collecting a large number of samples or by combining a large number of soil cores in each sample. These two alternatives each require about the same amount of effort, and since processing may be costly, it is generally more economical to bulk a large number of soil cores into a single sample. A common recommendation is to collect 20 soil cores for every 5 acres. Representative 5-acre sections can be sampled in large fields. The cores can be combined, mixed well, and then subsampled before submission to an assay lab. Most laboratories process only 1 pint of soil or less per sample.

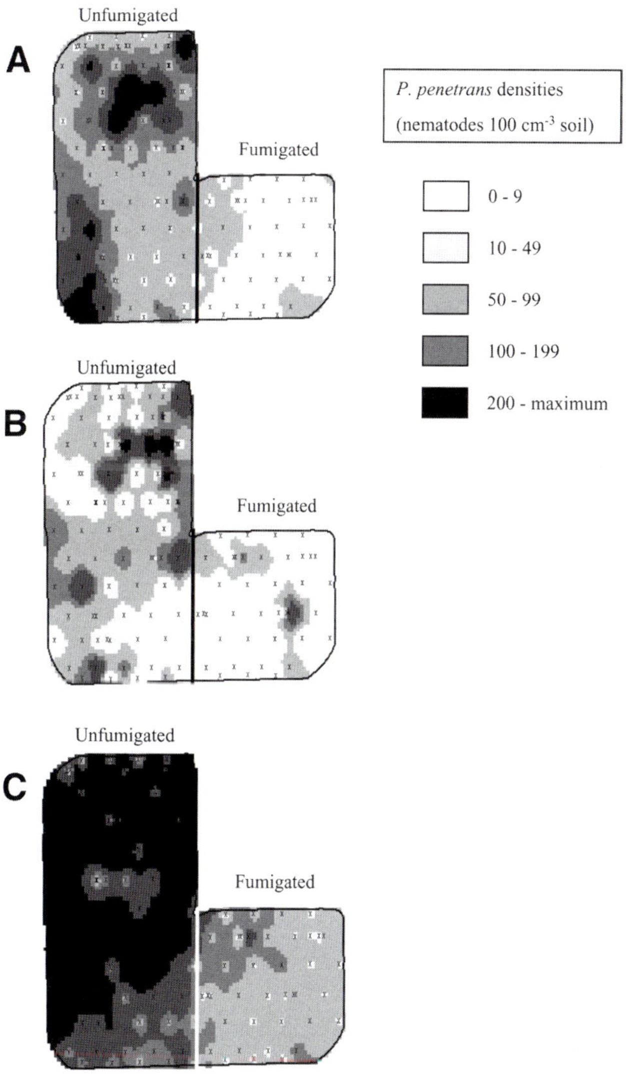

Fig. 20.12. Field map showing the pattern of root-lesion nematodes (*Pratylenchus penetrans*) based on 119 soil samples collected during a rotation of potato in year 1 (**A**), corn in year 2 (**B**), and pepper in year 3 (**C**). (Reprinted, by permission, from G. D. Morgan, A. E. MacGuidwin, J. Zhu, and L. K. Binning, 2002, Population dynamics and distribution of root lesion nematode [*Pratylenchus penetrans*] over a three-year potato crop rotation, Agronomy Journal 94:1146–1155)

The timing of sampling depends on the objectives. To detect and monitor nematode populations, sampling is best done at midseason, when their numbers are usually at their highest. For decisions regarding postseason treatment or fall cover crops, samples should be collected at least 1 month before harvest. Decisions regarding treatments administered at the time of planting or after planting usually require sampling in the spring or the previous fall. In northern areas, samples should be collected before the ground is frozen or after the ground has thawed.

Proper handling of soil samples after collection is critical to obtaining reliable results. Nematodes are very sensitive to rapid desiccation and extreme heat, and they die if soil samples are exposed to sun and air. In the field, samples should be placed in sealed plastic bags and kept out of direct sunlight. If samples must be stored prior to shipping to the diagnostic lab, they should be refrigerated at about 40°F.

Assay Procedures and Interpretation

Several methods have been devised for extracting nematodes from soil and roots. Some methods require specialized equipment, but in all cases the end result is a suspension of nematodes in water. The water sample is examined under a microscope to identify and count the nematodes. Since no extraction technique can recover all of the nematodes in a sample, the number counted in a sample is usually adjusted to correct for the recovery efficiency of the procedure. This varies tremendously with the extraction technique, the soil texture, and the nematode species and life stages present.

Some diagnostic labs vary their procedures, depending on the objective of the assay and the nematodes they expect to find. For example, samples submitted in the fall or spring following soil fumigation usually contain many dead nematodes. Since it is often difficult to distinguish between dead and living nematodes, the lab may use a technique that recovers only nematodes capable of movement. If samples were collected to evaluate pesticide efficacy, the lab should be alerted to ensure that proper methods are used. For root-inhabiting nematodes (root-lesion and root-knot species), both soil and root assays should be requested, even for samples collected after harvest or before planting. Many nematodes overwinter in roots and must be included in population estimates.

Interpreting Assay Results

Nematode management decisions are based on the anticipated risk of damage due to an estimated population of a particular nematode species. There is a close relationship between crop performance and nematode population when the crop is planted. Data from samples collected shortly before planting a potato crop can be interpreted by comparison with locally adapted action thresholds established for particular nematode species.

Predicting the effect of nematode population size on the performance of a subsequent potato crop is more complicated. Samples may be collected during the growing season of the rotation crop to predict the preseason nematode population in the following year. To do so, it is necessary to estimate changes in the population size during the current season and during the interval between harvesting the current crop and planting the potato crop. Nematode overwintering is poorly understood, so estimates of future populations may be inaccurate.

The extent to which laboratories offer recommendations varies and should be understood before samples are submitted. Local authorities, including Cooperative Extension specialists, can often aid in the interpretation of assay results. Information submitted with a sample (such as the purpose of sampling, soil characteristics, the size of the sampling site, the sampling methods used, and cropping history) can simplify and aid in the interpretation of results.

Information on methods of estimating nematode populations can be obtained from the laboratory performing the assays. The laboratory report should indicate whether an estimate is based on raw or adjusted counts of nematodes and whether the extraction technique excluded dead nematodes. The units (e.g., number of nematodes per pint of soil) and habitat (e.g., soil or roots) represented by the counts should also be clearly stated.

Damage Thresholds

Damage thresholds for **root-lesion nematodes** vary by species but are generally in the range of 50–200 nematodes per 100 cm^3 (cubic centimeters) of soil for the nematodes alone and 25–100 nematodes in combination with *V. dahliae.* Population densities can build quickly, so cultural practices aimed at reducing root-lesion nematodes should be deployed when the nematode is first detected.

Root-knot nematodes may reduce plant growth, yield, and tuber quality. Damage thresholds for tuber quality are very low, and management procedures should begin when species inciting tuber symptoms, such as the Columbia root-knot nematode, are first detected. Management of species that limit plant growth but do not disfigure tubers should be based on damage thresholds.

The **potato rot nematode** and **potato cyst nematodes** are very serious pests, and steps to reduce population densities and prevent further dissemination should begin as soon as these nematodes are detected. Efforts must be made to prevent the introduction of these nematodes into new areas.

Sting nematodes are very damaging, and most labs set thresholds below 10 nematodes per 100 cm^3 of soil at planting. Detection of any sting nematodes should signal consideration of a nematode management plan, if not immediate action.

Damage thresholds for **stubby-root nematodes** vary by production area, ranging from 50 to 200 nematodes per 100 cm^3 of soil at planting. If corky ringspot disease is of concern, any population of stubby-root nematodes can be damaging (Chapter 17).

Management Strategies

Management principles are similar for diverse types of nematodes, and the information presented here is intended as a general guide for all nematode pests. The main management strategies are prevention of the introduction of nematodes into uninfested areas, planting resistant cultivars, cultural practices, and chemical control. Management practices for nematodes that attack potatoes are mostly implemented either prior to the growing season or prior to planting potatoes.

Preventing Introduction

The best way to avoid nematode problems is to not bring them onto the production site from infested locations. Seed certification programs do not test for **root-lesion nematodes**, and they are so common that most fields in potato production areas are already infested. Root-lesion nematodes can infect seed potatoes or survive in soil that adheres to tubers, but seed tubers have not been shown to be a major means of spread.

Planting seed tubers in soil infested with **root-knot nematodes** carries a risk of spreading the nematodes. Not all states include root-knot nematodes in their certification standards, so it may be necessary to examine the seed source directly. Tubers infected with root-knot nematodes may be bumpy or have small internal discolorations. Suspect tubers should be examined with a hand lens for the presence of females or submitted to an assay lab for diagnosis.

Root-knot nematodes can be carried in seed tubers and transported in soil that adheres to tubers, machinery, and animals and in surface irrigation water. The presence or absence of nematodes should be considered when equipment is moved from one field to another and when cultural activities are planned.

The **potato rot nematode** and **potato cyst nematodes** are very limited in their geographic distribution, and diligence in planting nematode-free seed is important in preventing their spread. All potato production regions in the United States and Canada restrict the sale and movement of tubers infected with the potato cyst nematode. The transport of soil and plant material from areas known to be infested with the potato cyst nematode is restricted by federal quarantine. Purchasing certified seed tubers offers protection against the potato cyst nematode, because nematodes are disseminated in soil adhering to tubers. In some areas, restrictions are also imposed on the potato rot nematode in potatoes grown for seed. Seed infected with the potato rot nematode is often symptomatic, and tubers with dry rot should be submitted for nematode assay.

Sting nematodes are restricted to certain geographic regions and are limited to sandy soils. Sting, stubby-root, and other plant-parasitic nematodes can be transported by flowing water and windblown soil. Activities that prevent the movement of soil by natural forces, animals, or machinery can reduce the movement of nematodes from field to field. **Stubby-root nematodes** are very common in potato fields across North America, so the focus should be on preventing the buildup of nematode populations to damaging levels.

Host Resistance

Potato cultivars with resistance to root-lesion, root-knot, sting, or stubby-root nematodes are not commercially available for U.S. or Canadian potato production. Suitable potato germplasm conferring resistance to root-knot nematodes has been identified, and efforts are under way to breed commercially acceptable resistant cultivars. Efforts are also under way to develop commercial cultivars with resistance to Verticillium wilt. The use of *Verticillium*-resistant cultivars should decrease losses in fields infested with root-lesion nematodes, because of the synergistic disease interaction between *P. penetrans* and *V. dahliae* (Chapter 19). Rotation with crops that are not hosts of root-knot nematodes can reduce damage in a subsequent potato crop. For example, some soybean cultivars have good resistance to the southern root-knot nematode and may be a suitable choice for crop rotation.

Potato cultivars resistant to both species of **potato cyst nematodes** are available and may be mandated by regulatory agencies for infested fields. Resistant cultivars stimulate the hatching of nematode eggs and are readily infected by juvenile nematodes but do not support their development to the adult stage, in which they are capable of reproduction. Resistant cultivars sustain less yield reduction than susceptible cultivars and decrease the rate of nematode reproduction.

Cultural Practices

Various cultural practices are useful for reducing nematode population densities. The magnitude of the reduction depends on the nematode species and cultural practice.

Cultural practices can help reduce tuber damage caused by the Columbia root-knot nematode, but cultural practices alone are not likely to achieve a sufficient reduction in damage. They should be viewed as a means of reducing the amount or frequency of soil fumigation or allowing the use nonfumigant nematicides instead of soil fumigants for processing potatoes grown in soil infested with the Columbia root-knot nematode. The level of reduction achieved by cultural practices may be sufficient to manage other root-knot species, such as the northern root-knot nematode.

Some crops produce decomposition products that are toxic to nematodes when incorporated as **green manure**. Although some green manure crops support nematode reproduction as they grow, they release lethal gases as they decompose. Sorghum-sudangrass hybrids produce hydrogen cyanide as they decompose. Mustard, rapeseed, and other brassica crops contain glucosinolates that degrade into lethal gases similar to the active ingredient in some soil fumigants. The efficacy of the plant volatiles in killing nematodes depends on the dose and exposure time. Cover crops should be plowed down when temperature and conditions are conducive to decomposition (Chapter 5). Cover crops vary in their susceptibility to root-lesion and root-knot nematodes, and it is important to time planting and incorporation appropriately for the nematode pest of concern (Table 20.2).

Root-lesion nematodes have a wide host range, including cereals, legumes, vegetables, fruit, and many weed species. Short-season crops that can be followed by cultivation or crops planted late in the season, such as fall cover crops, tend to support lower root-lesion nematode population densities than long-season crops that remain undisturbed in the fall. Root-lesion nematodes overwinter in all life stages, and intact root systems offer them protection as well as nourishment.

Root-lesion nematodes are mobile in soil throughout their life cycle. Plant products that physiologically impair nematode sensory and motor systems can effectively reduce the impact of root-lesion parasitism on plants as well as the buildup of nematode populations. Forage pearl millet (*Pennisetum glaucum*), marigold, and black oat (cultivar Saia) planted as cover crops reduce the population density of *Pratylenchus penetrans* as they grow.

Cover crops can be used to decrease population densities of **root-knot nematodes**, particularly crops that release products toxic to nematodes as they decompose. Sudangrass, sorghum-sudangrass hybrids, brassicas (mustard, oil radish, rapeseed, and broccoli), and marigolds are effective against root-knot nematodes. These crops can be planted the year before potato for a full season or a partial season following a short-season vegetable crop or cereal. The residue should be incorporated when temperatures are high enough to facilitate decomposition. Brassica residue incorporated in the spring can be harmful to plants as well as nematodes, so sufficient time should be allowed between incorporation and planting. Plant matter buried in soil stimulates microbial activity, so the timing, amount, and carbon-to-nitrogen ratio of the biomass added to the soil may influence nutrient immobilization (Chapter 8).

Soil solarization, heating soil by covering the surface with clear plastic sheeting, can improve yields in fields with a high potential for damage due to root-lesion and other nematodes. The exact mechanism is not understood, but the effect is presumed to be due to an elevation of temperature that is stressful to nematodes but beneficial to their natural enemies. Solar energy penetrating the plastic is directed back into the soil by a film of water that forms on the surface of the plastic closest to the soil. The duration of solarization required for a reduction in nematode population ranges from 30 to 90 days and is a function of soil texture (sandy soils warm faster), soil temperature, soil moisture, and the amount of cloud cover.

Cover crops and solarization can be combined in a practice called **biofumigation**. A cover crop is managed as a green manure, and the soil surface is covered with plastic sheeting immediately after incorporation. Covering the residue with plastic sheeting warms the soil, accelerates decomposition, and traps gases in the soil. As toxic gases accumulate, they replace oxygen in the soil, so that nematodes are exposed to high concentrations of the toxic gases and a decreased level of oxygen. Adapting biofumigation and solarization to large-scale potato production systems remains a challenge, but there is considerable interest in developing this strategy as an alternative to soil fumigation.

Crop rotation is effective in managing root-knot nematodes, because most species have a limited host range. The cropping sequence and the number of years between potato crops needed to reduce populations below thresholds depend on the root-knot species present and its population density. One year of corn, wheat, or rye is usually sufficient to reduce a population of northern root-knot nematodes. However, these crops are good hosts of the Columbia root-knot and root-lesion nematodes and favor the reproduction of these pests. Thus, cropping decisions in fields infested with several types of nematodes must be based on management of the key nematode pest—the species that is economically most limiting.

The value of crop rotation has been demonstrated for the **potato cyst nematode**, a pest with a host range restricted to the family Solanaceae. Planting nonhost crops has a small direct effect on the nematode population, because very few eggs hatch unless host root exudates are present, but it does provide time for natural enemies to destroy the nematode eggs. Small invertebrates, such as mites, springtails, and rotifers, and egg-parasitic fungi are known to attack eggs of the potato cyst nematode. Building soil organic matter by adding amendments or cover crops may create conditions conducive to natural enemies of nematodes.

Sting nematodes, **stubby-root nematodes**, and other nematode pests that live in soil rather than roots have wide host ranges. It is difficult to break their life cycle and curtail the buildup of population densities through crop rotation. There are differences in the extent of reproduction supported by different crops and, in some cases, different cultivars of the same crop. Disturbing soil and exposing root systems to air and sun immediately after harvest can reduce nematode populations following a rotation crop. Many weeds are hosts of sting and stubby-root nematodes, so nematode populations may **rebound** (Chapter 3) between growing seasons. Turning in green manures of crop residue, cover crops, or weeds stimulates biological activity in the soil. Although much work remains to be done, research indicates that the increase of nematode pest species is buffered by a diverse soil community.

Weed management between potato crops is also important, since many species of weeds can serve as reservoirs for root-knot and other nematodes. Root systems may remain viable after shoots die, so it is not always safe to assume nematodes have been denied a food source on the basis of reduced aboveground plant biomass. Adult female nematodes spend considerable time feeding in preparation for egg production. Nutrients stored during this time are sufficient to sustain egg deposition for weeks, even if the plants harboring the females die. Cultivation to disrupt root systems and expose them to the elements interferes with nematode reproduction.

Biological amendments that stimulate activity in soil or bolster plant responses to root-knot and other nematodes are available. Some products are composed of live organisms; others contain chemicals that stimulate responses of soil organisms. In some cases, the target is the nematode egg or juvenile. In other cases, the target is natural enemies of nematodes. For example, chitin-based products can stimulate fungi capable of parasitizing nematode eggs. The efficacy of many biological products varies according to the nematode population and local conditions, so on-farm testing of products is necessary.

Soil flooding is feasible as a root-knot nematode management tactic in some potato production areas, mostly in the southern United States. Flooding kills nematodes by reducing the amount of oxygen available for respiration and increasing concentrations of substances toxic to nematodes, such as organic acids, methane, and hydrogen sulfide. A period of 7–9 months of continuous flooding or multiple flooding and drying cycles is required to kill eggs of root-knot nematodes. Flooding is most effective when the soil and air remain warm. Several cycles of flooding (for a minimum of 2 weeks) alternating with drying and disking may be as effective as continuous flooding. Allowing flood water to flow onto uninfested fields or areas may prove to be harmful, however, since water is an efficient means of dispersing nematodes.

Chemical Tactics

All **soil fumigants** can be injected into the soil, and some must be applied in this manner. Chisel equipment is used to deliver a fumigant to a depth of 4–12 inches, either in the row or broadcast. Prior to fumigation, the soil should be in seedbed condition and free of clods and undecomposed plant material. The best results are obtained when the soil is moist and between 50 and 70°F at a depth of 6 inches. Methyl bromide and chloropicrin are highly volatile materials, and when they are applied the soil must be covered with tarps to prevent premature escape of the gas. The fumigants 1,3-dichloropropene, metam-sodium, and metam-potassium are less volatile, and the soil may be sealed with a cultipacker or a similar implement after treatment. All fumigants can damage plants, but they dissipate from the soil over a period ranging from about 2 days (methyl bromide) to 2 weeks (1,3-dichloropropene). Dissipation occurs more quickly under warm and dry conditions.

Fumigants, such as 1,3-dichloropropene, metam-sodium, and metam-potassium, are likely to kill or disable nematodes at depths of about 2–10 inches. Nematodes residing deep in the soil profile at the time of application but migrating to the soil surface later, when a crop is planted, may escape the effect of the fumigant. For this reason, some portion of most nematode populations survives after fumigation. The percentage of the population surviving may be high, and increasing the rate of application does not necessarily increase the depth to which a fumigant penetrates the soil.

Soil fumigation effectively reduces **root-lesion nematode** populations (Fig. 20.12) and is cost-effective when the population density exceeds the damage threshold. Metam-sodium is the fumigant most widely used for joint control of root-lesion nematodes and *Verticillium*. Soil fumigation is also effective against **root-knot nematodes**. Soil fumigants are commonly used to control the Columbia root-knot nematode, because of the low tolerance for nematode-infected tubers in the processed crop. 1,3-Dichloropropene can maintain cull percentages at acceptable levels in most infestations, but in some cases a combination of 1,3-dichloropropene and metam-sodium is needed. Careful attention to product labels is important, because the allowed rates of application of a combination of chemicals may be different from the rates for those chemicals used alone.

Fumigation of fields infested with **potato cyst nematodes** or the **potato rot nematode** may be required by state or provincial statute, depending on the location and the intended use of the potato crop. Methyl bromide, metam-sodium, and 1,3-dichloropropene are effective in reducing the population density of potato cyst and potato rot nematodes but do not eradicate these pests. Follow-up assays of fumigated fields and monitoring potato crops for symptoms are essential when these regulated pests are present.

Nonfumigant nematicides, such as ethoprop, oxamyl, and aldicarb, are registered for use on potato. The product label should be consulted for restrictions specific to certain conditions or locales. Granular or liquid formulations of nonfumigants are applied in the spring to the soil surface and incorporated or banded in the row or furrow. Thorough mechanical incorporation and adequate moisture are needed to promote solubilization and movement of the nematicide and thus ensure that it comes in contact with root systems and nematodes. Some nonfumigants affect nematodes directly, by contact action, and others are taken up systemically by the plant and affect nematodes after feeding or infection has occurred. Some nonfumigants are also insecticidal and protect against both insects and nematodes at rates used for nematode management. However, the reverse may not hold true; materials used at rates recommended for insect management may not protect against nematodes.

Nematicides and soil fumigants will reduce but not eradicate nematode populations. Nematode population densities, other crops in the rotation, and soil texture and conditions influence the efficacy of nematicides and dictate the frequency of their use. Depending on the chemical, exposure rate, and timing, nematodes may be immobilized rather than killed. The schedule for applying nematicides or soil fumigants should be determined on the basis of nematode population densities as indicated by soil assays. Pesticides may have unintended impacts on natural enemies of nematodes, so they should be used only when needed. Nematicides have high mammalian toxicity, and thus only certified applicators are authorized to apply them.

Box 20.1

Key Nematode Management Tactics

Root-Lesion Nematodes

- Growing cover crops
- Rotation with short-season crops that can be followed by cultivation
- Weed management between potato crops
- Soil solarization
- Soil fumigation
- Irrigation and fertilizer management to maintain plant health

Root-Knot Nematodes

- Crop rotation
- Planting only certified seed tubers
- Precautions to prevent movement of infested soil on equipment and in water
- Weed management between potato crops
- Soil fumigation

Potato Rot Nematode and Potato Cyst Nematodes

- Effective crop rotation
- Observation of quarantine restrictions to prevent introduction into new fields and regions
- Selection of resistant cultivars
- Planting only certified seed tubers
- Soil fumigation

Sting Nematodes and Stubby-Root Nematodes

- Precautions to prevent movement of infested soil and water from field to field
- Nonfumigant nematicide treatment

Oxamyl offers some protection against **root-lesion nematodes**, but it is economical only if nematode population densities are high and Verticillium wilt is not a problem. Populations of root-lesion nematodes exposed to nonfumigant nematicides often rebound in midseason, and there may be a net gain in nematode numbers, although potato yields are increased as a result of protection against nematodes early in the season.

Nematicides and soil fumigants will reduce but not eradicate nematode populations.

Ethoprop and oxamyl are less expensive than fumigants and are applied as granular or liquid formulations just before or at the time of planting. Nonfumigants used in combination with cover crops can be effective for managing **root-knot nematodes** once population densities are reduced to low levels. Spot treatment with nonfumigant nematicides is feasible in fields where nematodes occur in an aggregated pattern.

Nonfumigant nematicides are generally more economical than soil fumigants for reducing population densities of **sting nematodes** and **stubby-root nematodes**. Chemical tactics may be the best option when population densities of sting or stubby-root nematodes are above the damage threshold or when treatment is indicated by the production history of the field. Chemical management of stubby-root nematodes is also indicated when corky ringspot is a problem. Nematicides such as terbufos or carbofuran used on rotation crops may reduce sting and stubby-root nematodes in a subsequent potato crop.

New classes of nematicides (such as avermectins) and chemicals aimed at disrupting nematode life cycles (such as hatch inhibitors and pheromones) are being developed to augment currently available pesticides. However, existing nematicides are being withdrawn from registration or are being placed under use restrictions faster than new products are being made available. Because of this trend, it is particularly important that integrated nematode management programs be adopted, including several management strategies, instead of relying exclusively on nematicides to protect potatoes or rotational crops (Box 20.1). It is also essential to take all necessary steps to prevent environmental contamination with nematicides and other pesticides and exposure of people to hazardous levels of these materials.

CHAPTER 21

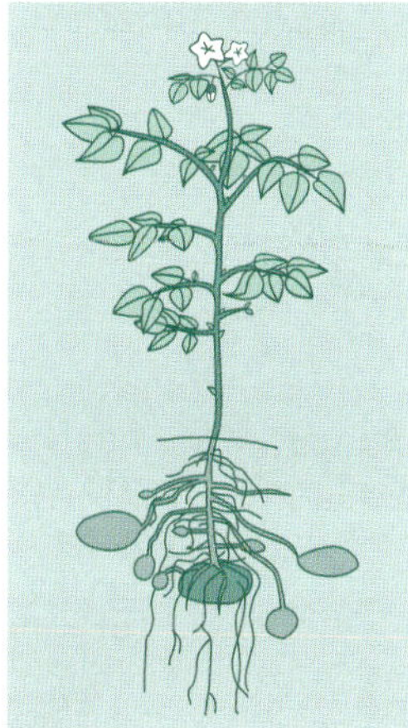

Walter R. Stevenson
Department of Plant Pathology
University of Wisconsin, Madison

William W. Kirk
Department of Plant Pathology
Michigan State University, East Lansing

Zahi K. Atallah
Department of Plant Pathology
University of Wisconsin, Madison

Managing Foliar Diseases: Early Blight, Late Blight, and White Mold

Managing foliar diseases of potato is a season-long focus of growers around the world. Three diseases—early blight, late blight, and white mold—account for the majority of fungicide use in potato production. In areas where environmental and cultural conditions favor disease development, the cost of management with a conventional fungicide program can easily exceed 10% of total production costs. Without control these diseases affect leaf and stem health and, in the case of late blight, tuber storability. Early blight and late blight alone or in combination can limit yield, tuber size, quality of fresh-market and processing tubers, storability of tubers, dry matter content of tubers, and marketability of the crop. The timing of infection and associated crop losses vary regionally with yearly weather patterns and with the effectiveness of management tactics.

Early Blight

Early blight occurs annually to some degree in most production areas. However, the onset of disease symptoms and rate of disease progress determine the impact of the disease on the crop. Disease intensity varies among production areas and is a function of three variables: the frequency of wetting of foliage by rainfall, fog, dew, or irrigation; the nutritional status of the foliage; and the susceptibility of the cultivar. While losses rarely exceed 20%, intensive fungicide treatment helps contain losses to less than 5%. In eastern North America, foliar infection is the most critical phase of the disease, whereas in the West, particularly the Rocky Mountain production areas, tuber infection can be more damaging than foliar infection.

Symptoms

Foliar symptoms of early blight first appear as small (⅛ inch in diameter or less), circular, dark brown spots on the lower (older) leaves. The spots enlarge to ⅜ inch in diameter and, as they increase in size, become restricted by leaf veins and take on an angular shape (Fig. 21.1). Early in the growing season, lesions on young, fully expanded succulent leaves may be up to ½ inch in diameter, and because of their size they may be confused with late blight lesions. Leaf lesions are relatively easy to diagnose in the field: the lesion surface is characterized by dark concentric rings alternating with bands of light tan tissue (Fig. 21.2). In some parts of the world, early blight is called target spot, because of the distinctive appearance of leaf lesions. A narrow band of chlorotic tissue often surrounds each lesion, and extensive chlorosis of infected foliage develops over time. Elongated, superficial, brown or black lesions may also form on stems and petioles. By the end of the season, the upper leaves may be peppered with numerous small lesions (Fig. 21.3), and lesions may coalesce to cover a large area of the leaf. Severely infected leaves eventually wither and die but usually remain attached to the plant. Severe foliar infection by the early to middle part of the tuber bulking stage can result in the formation of small tubers, low yield, and low dry matter content in tubers.

> **Early blight intensity varies among production areas and is a function of three variables: the frequency of wetting of foliage by rainfall, fog, dew, or irrigation; the nutritional status of the foliage; and the susceptibility of the cultivar.**

Tuber symptoms of early blight include circular to irregular lesions that are slightly sunken and often surrounded by a raised purple to dark brown border. The underlying tissues are leathery to corky in texture, dry, and usually dark brown (Fig. 21.4). Early blight lesions reduce the quality and marketability of fresh-market tubers. Tuber infection also presents a challenge to processors, as tubers often require additional peeling to remove the darkened lesions and discolored underlying tissues.

Pathogen and Disease Development

Early blight is caused by the fungus *Alternaria solani*. Spores and mycelium of the pathogen survive between cropping seasons in soil, on potato tubers, and in crop debris from potato and other susceptible solanaceous crops and weeds, including hairy nightshade (*Solanum sarrachoides*). The overwintering spores and mycelium of *A. solani* can withstand a wide range of environmental conditions, including prolonged exposure to sunlight and repeated cycles of drying, freezing, and thawing. In spring, spores serve as the primary inoculum, initiating the disease. Plants grown in fields or adjacent to fields where potatoes were infected with early blight during the previous season are most prone to infection, since large quantities of overwintering inoculum are likely to be present from the previous crop. Inoculum is readily moved within and between fields, as the spores are easily spread by air currents, windblown soil particles, splashing rain, and irrigation water (Fig. 21.5).

Spores of *A. solani* (Fig. 21.6) are produced on potato foliage and plant debris over a temperature range of 41–86°F (optimum 68°F). Alternating wet and dry periods with temperatures in this range favor spore production. Few spores are produced on plant tissue that is continuously wet or dry. The dissemination of inoculum follows a diurnal pattern in which the number of airborne spores increases as leaves and plant debris that are wet with dew or other sources of nighttime moisture dry off, relative humidity decreases, and wind speed increases. The number of airborne spores generally peaks in midmorning and declines in late afternoon and at night.

Spores landing on leaves of susceptible plants germinate and penetrate tissues directly through the epidermis, through stomata, and through wounds, such as those caused by sand abrasion, mechanical injury, or insect feeding. Free moisture from rain, irrigation, fog, or dew and favorable temperatures are required for infection, but at higher temperatures fewer

Fig. 21.1. Early blight lesions, caused by *Alternaria solani*.

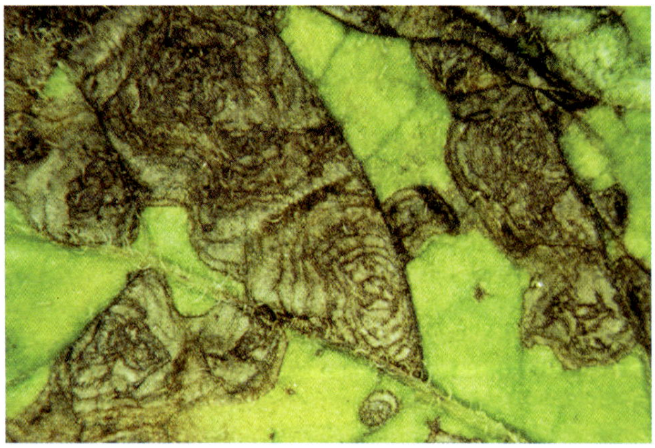

Fig. 21.2. Early blight lesion with a typical angular shape and distinctive alternating dark brown and tan concentric rings.

Fig. 21.3. Leaves peppered with numerous small early blight lesions, remaining attached to the plant. (Reprinted from W. R. Stevenson, R. Loria, G. D. Franc, and D. P. Weingartner, eds., 2001, Compendium of Potato Diseases, 2nd ed., American Phytopathological Society, St. Paul, MN)

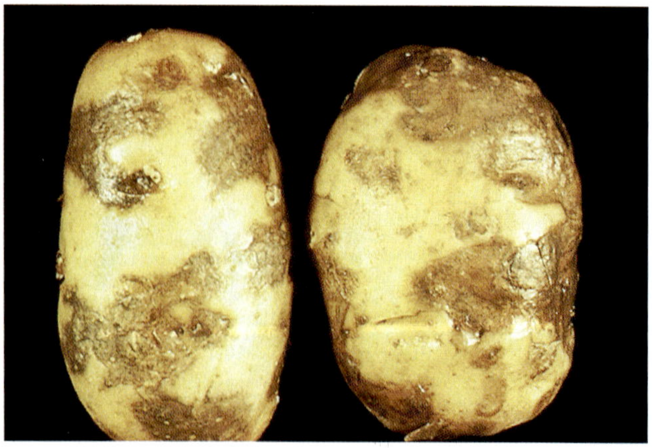

Fig. 21.4. Sunken, dark lesions on tubers with early blight.

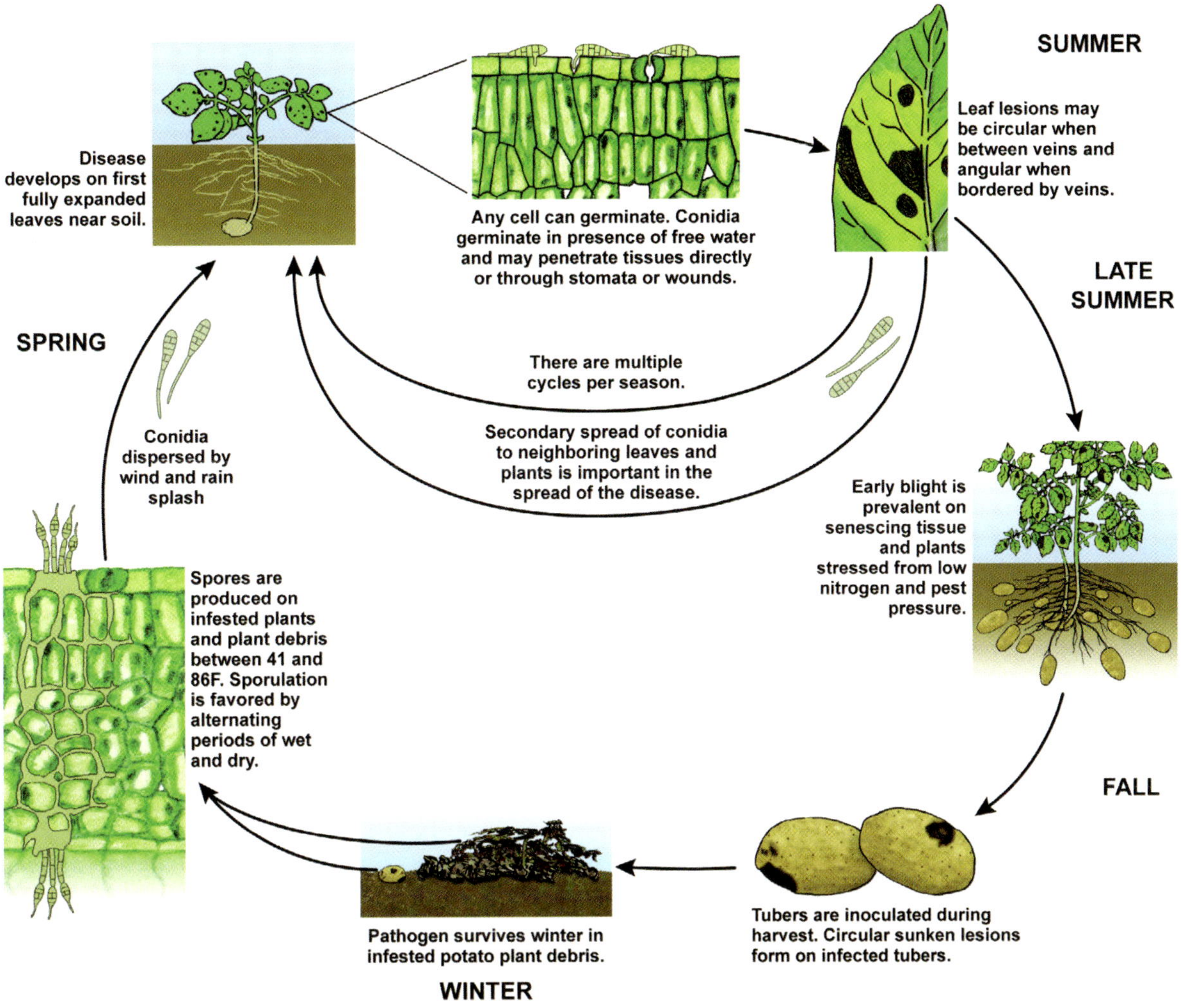

Fig. 21.5. Life cycle of the early blight pathogen, *Alternaria solani.* (Reprinted from P. Wharton and W. Kirk, 2007, Early blight, Extension Bulletin E-2991, Michigan State University)

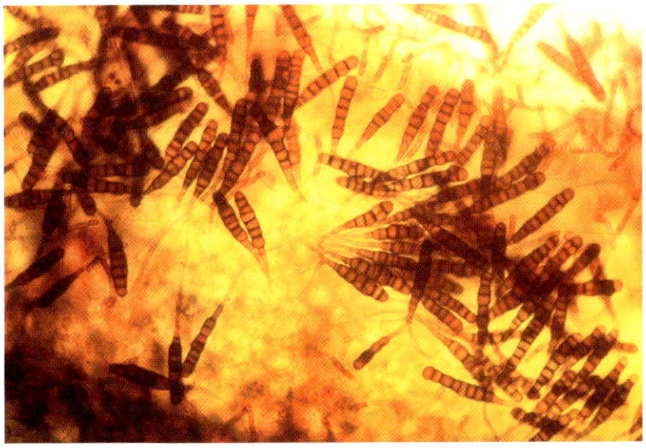

Fig. 21.6. Conidia of *Alternaria solani.* (Reprinted from Stevenson et al., 2001)

hours of leaf wetness are required (e.g., infection occurs after 12 hours at 50°F but after only 8 hours at 59°F).

Several cycles of disease occur within a single growing season. Secondary spread of the pathogen begins when spores are produced on lesions and carried to neighboring leaves and plants. This process is favored by alternating wet and dry conditions with long periods of high relative humidity and leaf wetness. Early blight is largely a disease of older plant tissue or plants subjected to stress induced by adverse weather, nitrogen deficiency, or other pathogens and pests affecting plant health. Early in the growing season the disease develops first on fully expanded leaves near the soil surface and progresses slowly on juvenile tissues near the growing point. The rate of disease spread increases after flowering and can be quite rapid later in the season, during tuber bulking and periods of plant stress. Early blight lesions are usually present on most leaves of unprotected plants late in the growing season.

During harvest and tuber handling, spores of *A. solani* come in contact with tubers. Spores that have accumulated on the soil surface or are dislodged from desiccated vines at harvest may become distributed over the tuber periderm. Infection requires free moisture. The pathogen penetrates tubers through lenticels and sites of mechanical injury to the skin. Immature tubers and white- and red-skinned cultivars are very susceptible to infection, as they are prone to abrasion and skinning during harvest. Coarse-textured soil tends to enhance tuber abrasion, and wet conditions at harvest favor infection. During storage, individual lesions may continue to develop, but secondary spread does not occur. Symptomatic tubers may shrivel through excessive water loss; the amount of water loss depends on storage conditions and disease severity. Early blight lesions, unlike late blight lesions, are usually not sites of secondary infection by other decay organisms. Seed tubers may be infected with *A. solani* and exhibit symptoms, but these infections appear to play only a minor role in the disease cycle.

Late Blight

Late blight is one of the most destructive diseases of potato throughout the world. The pathogen, *Phytophthora infestans,* is favored by wet, cool weather (50–75°F), high humidity, and frequent rainfall. Under such conditions, the disease can spread rapidly and has the potential to completely defoliate a field within 2–3 weeks of the first visible symptoms if no control measures are taken. *P. infestans* can infect tubers at any growth stage and after harvest. Furthermore, blighted tubers are susceptible to bacterial soft rot, which further compounds the problem in storage.

Symptoms

The first symptoms of late blight on foliage are small, light to dark green, circular to irregularly shaped, water-soaked lesions (Fig. 21.7). They usually appear first on the lower leaves, where the microclimate is more humid. Lesions may occur on upper leaves, succulent stems, or flower pedicels if weather conditions are favorable and the pathogen is carried into the field on air currents. Lesions often begin to develop on the compound leaf near the point of attachment to the petiole (which is often cupped) or leaf edges, where dew is retained longest. During cool, moist weather, lesions expand rapidly into large, dark brown or black spots, which often appear greasy (Fig. 21.8). The lesions are not limited by leaf veins, and a lesion formed at leaf tip or edge can cause the expanding leaf to be misshapen (Fig. 21.9). As new infections occur and existing lesions coalesce, entire leaves may become blighted and are killed within a few days.

The late blight pathogen is favored by wet, cool weather (50–75°F), high humidity, and frequent rainfall.

Stem lesions (Fig. 21.10) are often initiated at the point of leaf attachment, and the leaf adjacent to a stem lesion be-

Fig. 21.7. Young late blight lesions are brown and irregularly shaped. (Reprinted from W. Kirk, P. Wharton, R. Hammerschmidt, F. Abu-El Samen, and D. Douches, 2004, Late blight, Extension Bulletin E-2945, Michigan State University)

Fig. 21.8. Late blight lesions expand into large, dark brown spots surrounded by a light green border. (Reprinted from Kirk et al., 2004)

Fig. 21.9. A, Late blight lesions are not limited by leaf veins and expand rapidly. **B,** Entire leaves die within days of infection. (Reprinted from Kirk et al., 2004)

Fig. 21.10. A, Late blight stem lesions often start where leaves attach to the stem and then spread along the stem (arrows). **B,** The leaf adjacent to a stem lesion may become detached shortly after infection. (Reprinted from Kirk et al., 2004)

comes detached shortly after infection. The lesion continues to develop along the length of the stem and can remain active even in hot, dry weather.

In the early morning or during cool, damp weather, a white velvety growth may be seen on the underside of symptomatic leaves and on stems at lesion borders (Fig. 21.11). This white velvety growth is a diagnostic sign of late blight, distinguishing it from other foliar diseases of potato, such as gray mold. However, it disappears quickly as the leaf and stem surfaces dry, and thus it is often not visible. A pale green to yellow border is often present around leaf lesions. Fields severely affected by late blight have a distinctive odor resulting from the rapid breakdown of potato tissue. The odor is similar to that produced by chemical vine kill or after severe frost.

Tuber infection is characterized by irregularly shaped, slightly depressed, brown to purplish areas on the skin. These symptoms are most obvious on white, smooth-skinned tubers. A tan to reddish brown (copper-colored), dry, granular rot develops beneath the skin in the discolored area, extending into the tuber usually less than ½ inch (Fig. 21.12). The extent of tuber rotting depends on the susceptibility of the cultivar, temperature, and length of time after initial infection. The margin of diseased tissue is not distinctly delimited from healthy tissue and is marked by brown fingerlike extensions into the healthy tissue of the tuber. In time, the entire tuber becomes blighted and discolored. Late blight infection is often accompanied by bacterial soft rot.

Identification of late blight can be made by microscopic examination of the pathogen collected from infected leaves, stems, or tubers (Fig. 21.13). The pathogen can be quickly identified by the distinctive size and shape of the spores and spore-bearing stalks.

Pathogen and Disease Development

Phytophthora infestans, the causal agent of late blight, is not a true fungus but a "brown algae–like" water mold (Fig. 21.14), in the class Oomycetes of the Kingdom Stramenopila. *P. infestans* forms large, clear, lemon-shaped spores called sporangia on stalks called sporangiophores. Sporangia germinate in two distinct ways. At temperatures between 44–55°F and in the presence of free water, sporangia germinate by forming and releasing zoospores. Each sporangium produces eight to 12 of these motile spores (Fig. 21.15), which can swim freely in a film of water, attach themselves to a leaf surface (encyst), and infect plant tissues. An encysted zoospore produces a germ tube, which penetrates a stoma (pore) in the leaf surface or directly penetrates the leaf epidermis or tuber periderm.

Fig. 21.11. White, velvety growth of the late blight pathogen on the surface of an infected stem, seen early in the morning in damp weather.

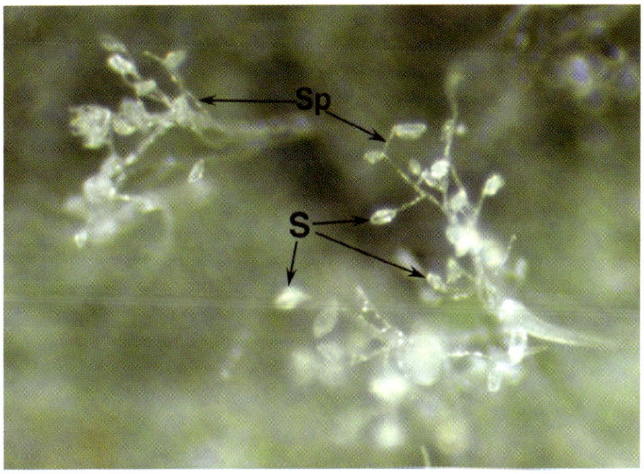

Fig. 21.13. Sporangia (S) and sporangiophores (Sp) of the late blight pathogen, *Phytophthora infestans.* (Reprinted from Kirk et al., 2004)

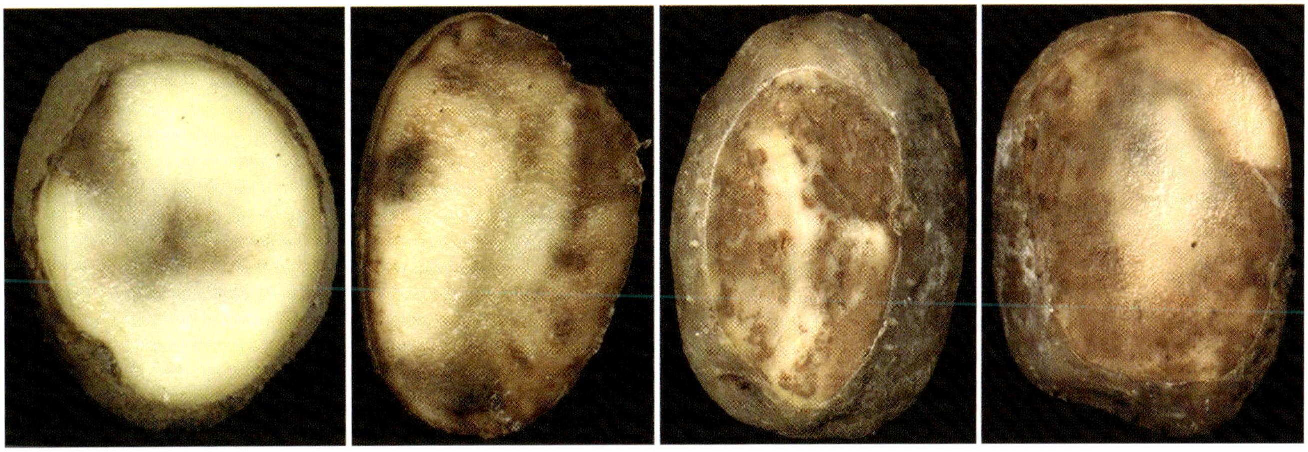

Fig. 21.12. Tuber symptoms of late blight, characterized by discoloration of the skin and a tan to reddish brown, dry, granular rot beneath the discolored skin. (Reprinted from Kirk et al., 2004)

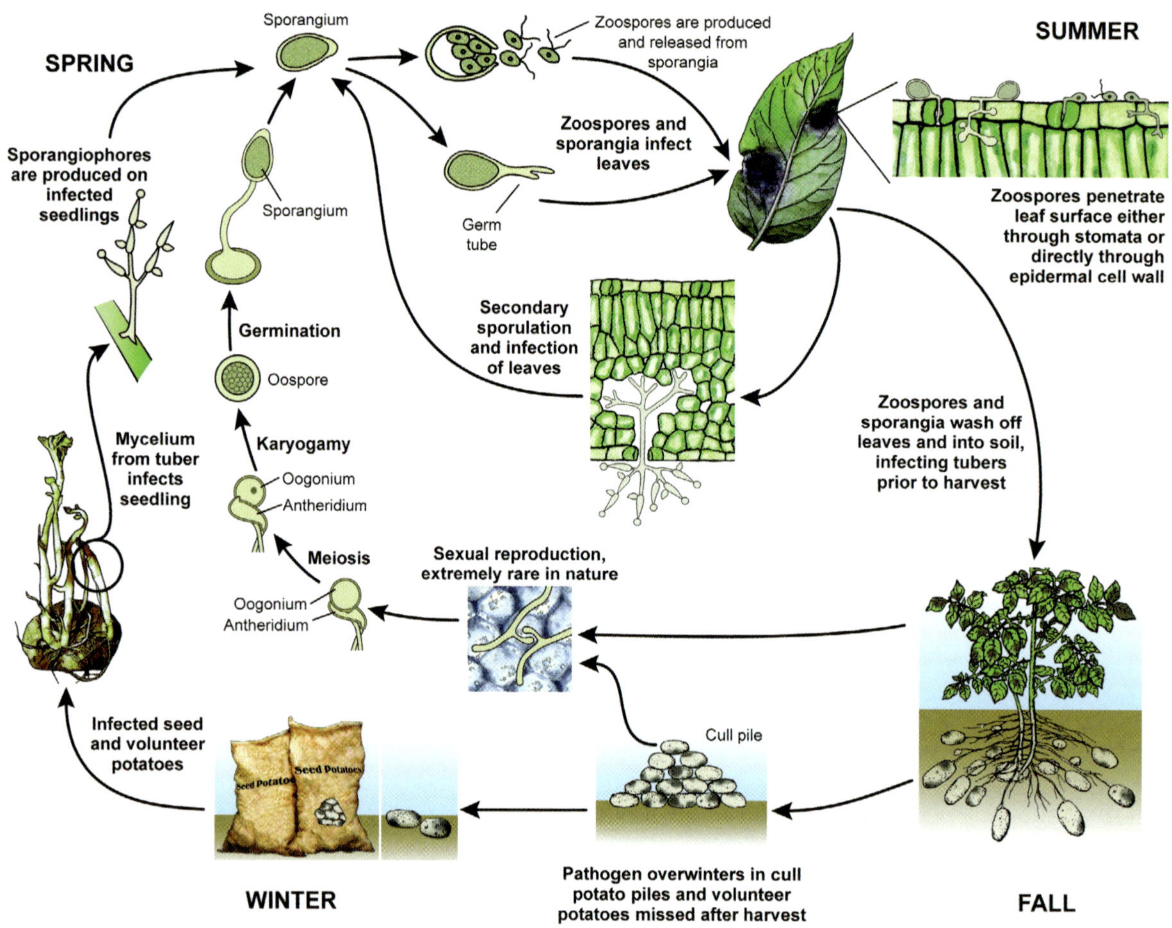

Fig. 21.14. Life cycle of the late blight pathogen, *Phytophthora infestans*. (Adapted from Kirk et al., 2004)

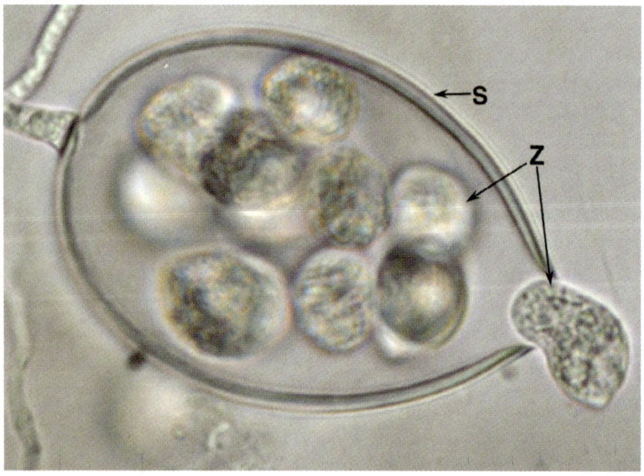

Fig. 21.15. A sporangium (S) of the late blight pathogen may produce eight to 12 zoospores (Z), which are released through a pore at the tip of the sporangium. (Reprinted from Kirk et al., 2004)

At temperatures of 55–70°F, sporangia germinate directly by forming a single germ tube, which can penetrate plant tissues. Night temperatures of 50–60°F accompanied by light rain, fog, or heavy dew, followed by days of 60–75°F with high relative humidity, are ideal conditions for late blight development. The dissemination of inoculum, like that of the early blight pathogen, follows a diurnal pattern. The number of airborne sporangia increases as wet foliage dries, ambient relative humidity decreases, and wind speed increases. The number of airborne sporangia peaks in the morning and declines in late afternoon and at night.

Tubers may become infected if sporangia produced on the foliage are washed down into the soil by rain or irrigation water. Waterborne spores appear to follow stems and stolons in a water film into the soil, come in contact with tubers, and initiate infection. Tubers near the soil surface are thus more likely to become infected than tubers deep in the hill. Tubers can be exposed to sporangia and zoospores transported by rain and irrigation water through cracks in the soil.

Isolates of the late blight pathogen are distinguished into two mating types, A1 and A2. Isolates of the pathogen in North America are currently classified primarily as the A2 mating type. In other production areas (e.g., Mexico and Europe), however, both mating types are present, and the pathogen is much more genetically diverse, making disease management in the field even more difficult. In production areas where the life cycle may include both mating types and both asexual and sexual stages are present, the pathogen survives between seasons as oospores (sexually derived, thick-walled, resistant structures) in the soil and mycelium in tubers. In most areas of North America where *P. infestans* is currently asexual, it survives from season to season as mycelium in tubers in storage or cull piles and in tubers that were missed during harvest and remain unfrozen over the winter (volunteer potatoes). In the spring, airborne sporangia of the pathogen are released from infected tubers in cull piles or volunteers and are transported by air currents to potato foliage. Infected seed potatoes are also an important source of inoculum. Tubers with easily seen symptoms usually rot and do not contribute to epidemics when planted, whereas tubers with asymptomatic infections of sprouts or eyes will commonly produce infections. The pathogen may spread from tuber to tuber during seed cutting and in the handling, storage, and planting of cut potato seed. Many infected seed pieces rot in the soil before emergence. Potato sprouts that emerge from an infected tuber may die after emergence, but some may survive and exhibit late blight symptoms. Sporangia of *P. infestans* are spread from infected to healthy plants by wind, splashed rain, mechanical transport, and animals. Genotypes of the pathogen currently in North America are capable of completing a single life cycle (from the production of sporangia to infection and then to spore production) in as little as 3–4 days in favorable weather. Many reproductive cycles can be completed in one growing season, and thus an explosive increase in disease occurs once the pathogen becomes established in a field.

Several genotypes of *P. infestans* have been reported in North America (US-1, US-6, US-7, US-8, US-11, US-14, and US-17). Prior to the mid-1990s, only genotype US-1 was present in North America. Currently, the predominant genotype in North America is US-8. Most isolates of this genotype are more aggressive than isolates of US-1 and are more likely to cause stem lesions. Genotypes US-1, US-6, US-11, and US-17 are mating type A1; genotypes US-7, US-8, and US-14 are mating type A2. To complete sexual recombination, mating type A1 and mating type A2 must come into contact with each other, to produce oospores. Oospores are resistant to freezing and other environmental extremes and can survive in diseased leaves and stems or free in soil for at least the period between cropping seasons. Sexual recombination and the production of overwintering oospores offer the potential for the development of new, more aggressive pathogen genotypes. In areas of the world where overwintering oospores are observed, late blight has become an important factor in virtually every growing season when conditions favor disease development. When only one of the mating types is present, the pathogen can survive from season to season only as mycelium in tubers. If infected tubers freeze and die over the winter or if they are buried deeply enough in soil to prevent emergence, the disease cycle is broken.

White Mold

White mold (Sclerotinia stem rot) is a recurrent problem in potato fields with dense canopies, high rates of nitrogen fertilization, and overhead irrigation that exceeds crop needs. The disease is prevalent in the Pacific Northwest, but in production areas in the Midwest it is of secondary importance. White mold incidence is most notable when agricultural practices promote extensive canopy growth, which tends to maintain high relative humidity within the canopy and long periods of canopy wetness. These conditions are essential for disease development.

Symptoms

Water-soaked lesions appear on stems. The lesions are tan to beige at first and subsequently appear bleached. In humid environments, they may become covered with white fungal growth, and hard, black sclerotia are sometimes found on the lesion surface. More commonly, sclerotia are formed within the stem. Lesions girdle the stem and hollow the internal pith, leaving a papery shell covering the sclerotia. Symptoms are not usually observed on belowground tissues (roots, stolons, and tubers).

Pathogen and Disease Development

White mold is caused by the fungus *Sclerotinia sclerotiorum* (Fig. 21.16). The pathogen causes disease in more than 400 plant species. *S. sclerotiorum* produces ascospores (sexual spores), which are dispersed by wind over short to medium distances. It is estimated that around 90% of all ascospores are deposited within 325 feet of the source, although distances of 2–2.5 miles have been reported in the scientific literature. The pathogen survives from season to season and can be spread in the form of sclerotia, which are melanized hyphae of various shapes and sizes. Sclerotia are able to withstand adverse environmental conditions, pesticides, and even fumigants. They form mainly inside hollowed potato stems after infection and colonization of the stem tissue. Sclerotia incorporated into the soil during tillage are capable of surviving for several years in the absence of a susceptible host and weather favoring disease. Only sclerotia in the top 2 inches of the soil are biologically active; those at greater depths remain dormant. Seed lots of numerous crops grown in rotation with potato (e.g., sunflower, soybean, bean, and pea) may contain loose sclerotia, which are introduced into the field when the seed is planted.

White mold is a recurrent problem in fields with dense canopies, high rates of nitrogen fertilization, and overhead irrigation that exceeds crop needs.

Sexual recombination occurs in sclerotia and leads to the carpogenic germination of apothecia in shaded and moist soil, mainly around the crowns of plants and the sides of irrigation furrows. Apothecia are tan, cup-shaped, mushroom-like structures (⅛–⅜ inch in diameter), lined with thousands of saclike structures called asci, each containing eight ascospores. The apothecia are commonly confused with the gray or brown reproductive structures produced by the bird's-nest fungus (*Cyathus* spp.).

Under agricultural conditions, *S. sclerotiorum* is primarily homothallic, which means that all ascospores from the same apothecium are genetically identical. Recent studies have noted that significant levels of outcrossing occur in the Pacific Northwest and in Australia. Differences in aggressiveness among isolates of *S. sclerotiorum* collected in Canada have been reported, but these differences were not observed among isolates collected in the Pacific Northwest.

Sclerotia germinate to produce either apothecia or mycelium. Mycelium growing directly from a sclerotium may infect adjacent plant tissue, if it is within ½ inch. The crown of the plant is then affected, and the entire plant may die prior to the closure of foliage between rows (Fig. 21.17). Although mycelium produced from sclerotia can infect green potato stems, ascospores are unable to infect healthy, unwounded potato foliage or stems.

Ascospores, carried within fields or from neighboring fields by wind currents, are the primary inoculum infecting potato. Potato flowers play a critical role in the disease cycle. The production of apothecia and release of ascospores occur over an extended period coinciding with the blooming of potato plants. Potato blossoms become infected while on the plant

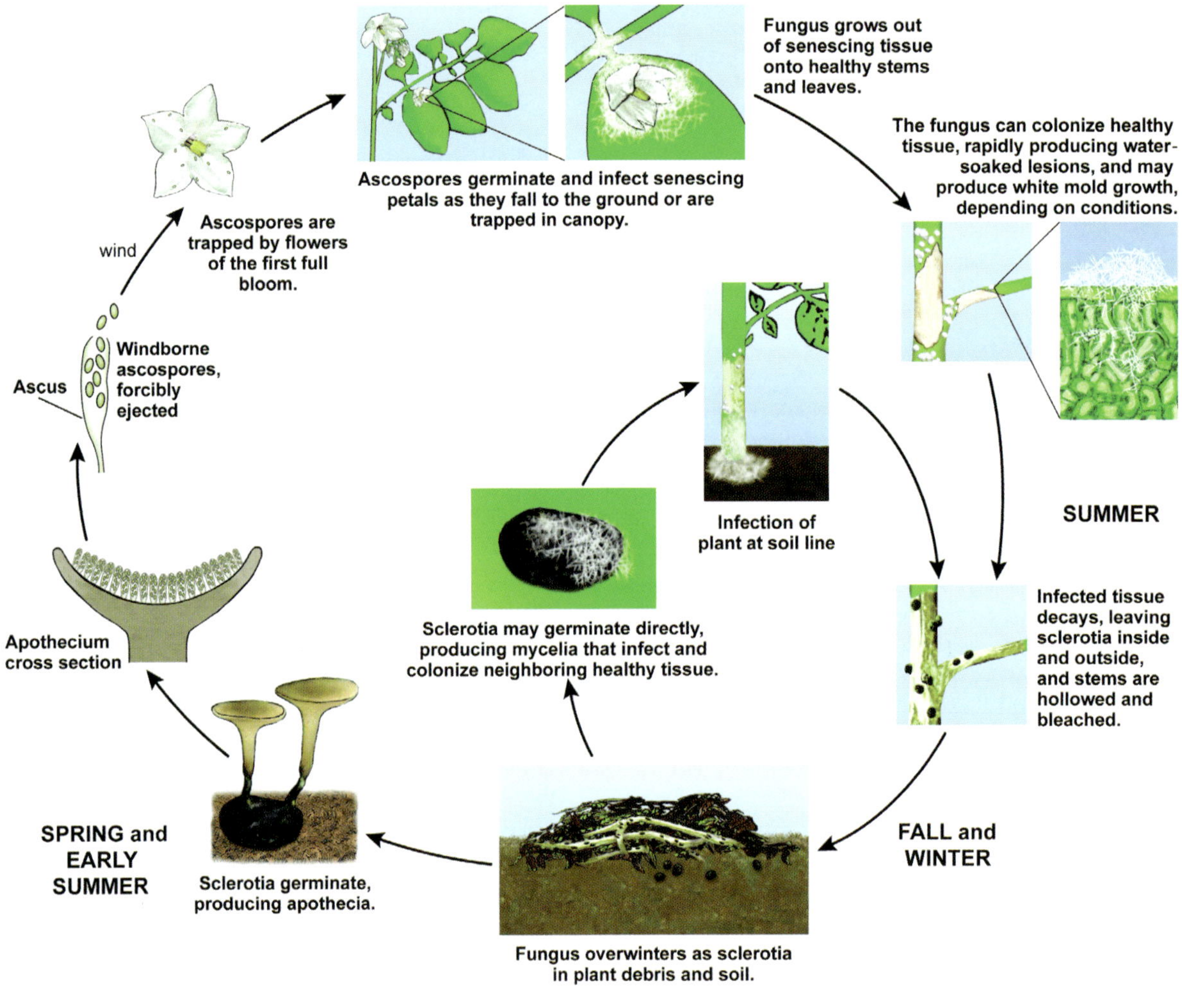

Fig. 21.16. Life cycle of the white mold pathogen, *Sclerotinia sclerotiorum.* (Reprinted from P. Wharton and W. Kirk, 2007, White mold, Extension Bulletin E-2989, Michigan State University)

Fig. 21.17. Direct infection of the crown of a potato plant by *Sclerotinia sclerotiorum.* (Courtesy D. A. Johnson)

Fig. 21.18. Potato blossom, on the ground, colonized by mycelium of *Sclerotinia sclerotiorum.* (Courtesy D. A. Johnson)

and later drop to the ground or onto leaves and stems in the canopy. In high humidity, the white mold pathogen colonizes contaminated flowers on the ground (Fig. 21.18) or lodged in the canopy (Fig. 21.19), mainly on leaf axils. Ascospores are also deposited on senescent and dead plant material on the ground, where they germinate and produce mycelia. Potato stems or leaves in contact with a colonized blossom or mycelium on the ground then become infected (Fig. 21.20).

Management Strategies

An effective management program for foliar diseases is an essential part of holistic potato health management in most production locations. Management of early blight, late blight, and white mold requires an integrated program of cultural and chemical practices designed to both minimize sources of inoculum and reduce disease development (Box 21.1). This includes preplant decisions, management strategies during crop growth, and harvest preparations.

Management Prior to Planting

Cultivar Selection

Some early-maturing cultivars can be harvested when the weather generally does not favor the development of early blight or late blight or before these diseases have progressed to the point of affecting yield or quality. For most growers, however, selecting a cultivar with the highest level of available disease resistance that is compatible with marketing needs is an important base component of a multilayered integrated pest management program. Growing disease-resistant cultivars helps reduce production risks associated with these diseases, decrease the number of fungicide sprays needed for management, and lower the cost of production. In some locations where environmental conditions are generally unfavorable for disease development or where disease pressure is low, it may be possible to grow disease-resistant cultivars without fungicide inputs. Breeders are now placing greater emphasis on developing cultivars with improved resistance to many diseases, including early blight and late blight.

Historically, early blight has generally progressed faster on early-maturing cultivars than on late-maturing ones. However, with the release of cultivars having resistance to multiple diseases, there are now greater differences in early blight susceptibility among cultivars of similar maturity (Fig. 21.21), and having a choice of cultivars allows growers to reduce fungicide inputs. Early blight management may be more successful in cultivars with resistance to other diseases (e.g., Verticillium wilt) or disease complexes (e.g., early dying), which are thereby subject to less stress later in the growing season, since stress affects cultivar susceptibility to early blight.

Most cultivars planted in North America are susceptible to **late blight**, but some cultivars, such as Jacqueline Lee (table stock) and Defender (lightly russeted long type), are highly resistant to genotype US-8 (the most widespread genotype in North America) of the late blight pathogen. They are currently being used as parents for the production of new chip and processing cultivars. Several moderately susceptible cultivars (e.g., Pike) are also available for commercial use.

There has been little progress in breeding for resistance to **white mold**, and at present the most widely cultivated commercial cultivars of potato are equally susceptible to this disease.

Field Selection

Select fields with good drainage and fertility that will promote overall crop health. Avoid narrow fields bordered by trees and other natural obstructions to air movement, as they tend to prolong periods of leaf wetness, making management of foliar diseases more difficult. Small, irregularly shaped fields in which it is difficult to uniformly apply fungicides should not be planted with potato. If aerial application of fungicides is planned, avoid planting in fields with trees and power lines, since these potential obstructions may impede uniform fungicide application.

In fields irrigated with center pivots and where **late blight** commonly occurs, avoid planting the area within 80 feet of the field center, since this area commonly remains wet for long periods when irrigating.

Fields with volunteer potatoes from previous crops should be avoided, since carryover plants are potential sources of inoculum of the **early blight** and **late blight** pathogens,

Fig. 21.19. Potato stem colonized by *Sclerotinia sclerotiorum,* in an infection that started from a colonized flower lodged on the stem. (Courtesy D. A. Johnson)

Fig. 21.20. White mold lesion on a stem, with white mycelium and developing sclerotia (right). (Courtesy D. A. Johnson)

Box 21.1

Key Disease Management Tactics

Early Blight

- Select fields that have been out of potato production for at least 2–3 years.
- Select cultivars with useful levels of field resistance.
- Provide adequate levels of nitrogen and phosphorus to meet crop needs.
- Match irrigation with crop needs. Do not overirrigate.
- Manage pests that stress potato plants and thus increase susceptibility to early blight.
- Use a disease-forecasting program and scout fields to determine when to initiate fungicide application and determine a schedule of applications.
- Apply fungicides carefully, using an effective material in the recommended amount at the optimum time.
- Harvest tubers when skins are well set, and avoid unnecessary abrasion.

Late Blight

- Eliminate cull piles before planting the next crop.
- Control volunteer potatoes and other potential sources of overwintering inoculum.
- Select cultivars with useful levels of field resistance.
- Plant certified seed tubers.
- Consider fungicide treatment of seed, especially if the seed is precut.
- Manage irrigation carefully to avoid extended periods of leaf wetness.
- Carefully monitor the crop at least weekly, especially during extended periods of wet, cool weather, to detect early symptoms of late blight.
- Monitor weather conditions and apply fungicides carefully, using a disease-forecasting program as a guide in scheduling.
- In most production areas, allow the vines to remain completely dead for 2–3 weeks before harvest.
- Continue fungicide applications until vines are dead.
- Carefully monitor tubers in storage for early symptoms of late blight, including tuber decay, and adjust storage conditions accordingly.

White Mold

- Select fields that have been out of potato production for at least 2–3 years.
- Avoid overfertilizing the crop and promoting excessive vine growth.
- Carefully irrigate to meet the needs of the crop, especially during and after the bloom period. Do not overirrigate.
- Time fungicide applications to coincide with the bloom period, to achieve maximum benefits.
- Consider treatment with a biocontrol agent to reduce the survival of sclerotia of the pathogen.

which could initiate early-season infection of a newly emerged crop.

The **white mold** fungus can survive for 5 years or more, and therefore growers should keep detailed field records on disease occurrence in susceptible rotational crops.

Effective crop rotation, planting, fertilization, irrigation, and fungicide decisions depend on accurate field records.

Rotation

Select fields that have been out of potato production for at least 2–3 years. Long rotations with small grains, corn, or legumes such as alfalfa or clovers are preferable to short rotations with row crops such as bean, soybean, sunflower, or canola, which are hosts of the white mold fungus. Inoculum of the early blight fungus can successfully overwinter to the following year in soil and on plant debris. The late blight pathogen can survive between cropping seasons on volunteer and cull potatoes. Sclerotia of the white mold fungus can survive in the soil for extended periods. If possible, rotational sequences should be planned on a whole-farm or area-wide basis, to avoid planting potato in fields immediately adjacent to fields used for potato production in the previous year. Field selection requires long-term planning to take advantage of the benefits provided by the separation of potato plantings in both time and distance.

Elimination of Overwintering Inoculum

Potential sources of overwintering inoculum of the **late blight** pathogen should be identified and destroyed or controlled before a new crop is planted, and definitely before the crop emerges. These sources include cull piles, infected volunteer potatoes that have survived the winter, and weed hosts, such as hairy nightshade. Refuse from the previous potato crop can be a source of inoculum of the **early blight** pathogen and should be incorporated into the soil prior to planting the next potato crop. Sclerotia of the **white mold** fungus survive in the soil for several years, and the inoculum bank may increase or decrease, with the amount of inoculum depending on the susceptibility of rotational crops, disease development in these crops, and biological control promoting sclerotial decay. Soil fumigation is not an effective management tactic for white mold, early blight, or late blight.

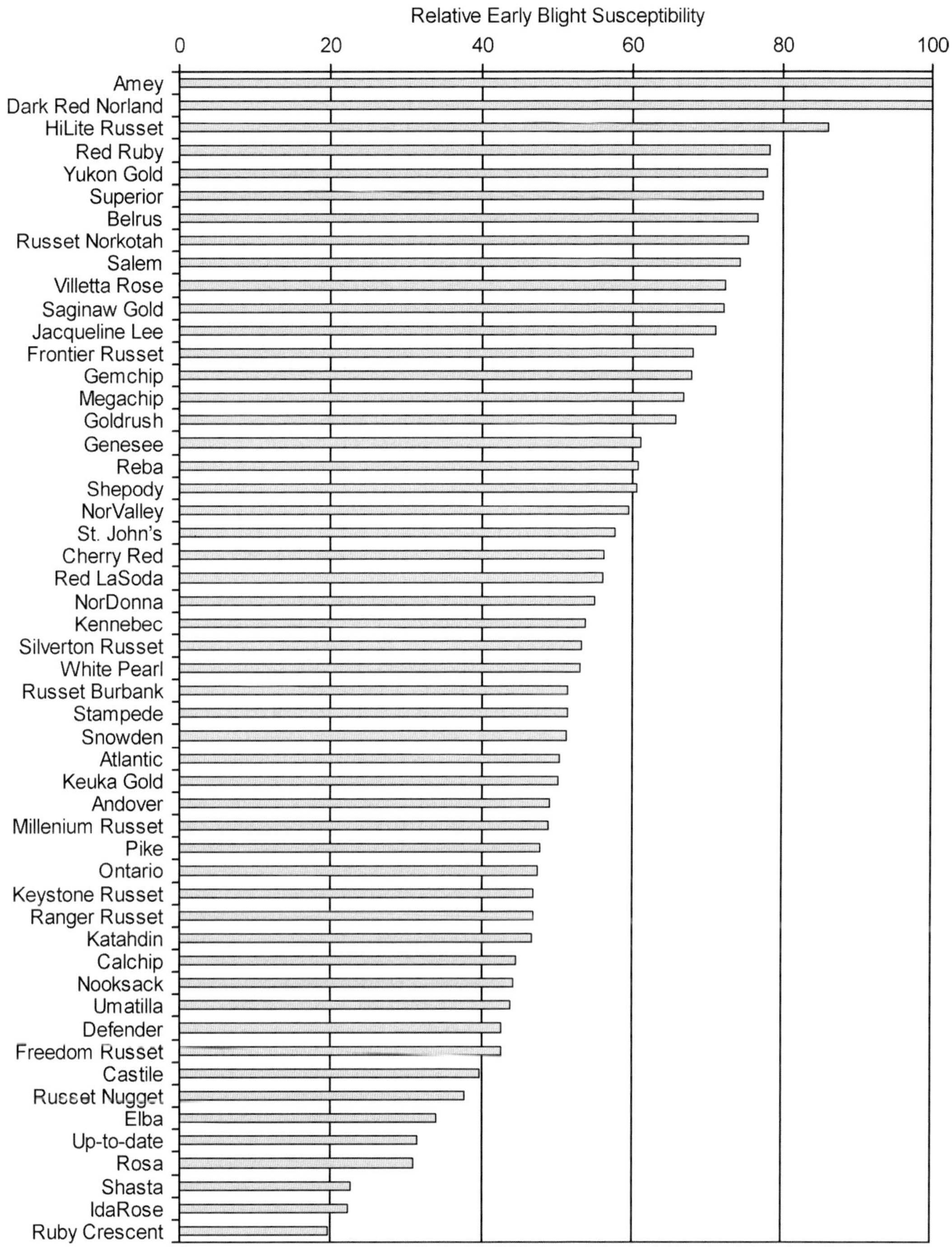

Fig. 21.21. Susceptibility of potato cultivars to early blight, relative to that of Dark Red Norland, a highly susceptible early-season cultivar.

Only **certified seed tubers** should be planted. While infection of up to 1% of tubers by the late blight pathogen is allowed in certified seed lots, the use of certified seed helps to reduce the potential of introducing the late blight pathogen into the production area and may help reduce the need for fungicides. The importance of pathogen-free seed in the overall disease management plan cannot be overemphasized.

Improper **disposal of cull potatoes** is often a major factor in initiating late blight epidemics. Blighted tubers that survive the winter and sprout the following spring are an important source of spores. Blighted potatoes should not be dumped in cull piles. Piles larger than about 500 cwt may not freeze throughout, even in production areas with harsh winter weather. The safest method of disposal is to spread cull potatoes sparingly (about 400 cwt per acre) during the winter on a field not intended for potato culture the following spring, where they are exposed to several cycles of freezing and thawing. Grinding or chopping cull potatoes before spreading will promote rapid decomposition. Alternatively, culls can be hauled to a commercial landfill for burial, they can be buried on-site at least 3 feet deep, or they can be composted with other organic materials. Burial or composting sites should be visited often to check for tuber sprouting or emergence that must be eliminated. Feeding culls to livestock may be appropriate after the next crop has been planted and is emerging, as long as all tubers given to livestock are completely consumed within 24 hours. Extra care is required to prevent escapes. Existing cull piles should be destroyed by burning or burying or at least should be treated with a general herbicide to kill existing foliage and prevent further sprout formation. Tubers remaining on the surface of the field after harvest should be left to freeze and not disked or plowed under, since tubers that are disked or plowed under often survive the winter. Destruction of volunteer potatoes in the spring, by means of cultural practices and application of an herbicide registered for use in the rotational crop, is also important in minimizing sources of primary inoculum.

Seed Cutting and Handling

Seed tubers infected with the late blight pathogen are sources of inoculum that can be spread from tuber to tuber during seed cutting, handling, and planting. Before purchase and seed delivery, growers should carefully review all records pertaining to field, bin, winter grow-out, and shipping-point inspections. When there is potential seed tuber infection, growers should avoid holding cut seed in a warm, humid environment for a prolonged period, since these conditions favor the sporulation of the pathogen and continued infection of tubers. Growers should also apply an effective broad-spectrum seed treatment fungicide at cutting, to minimize the spread of the pathogen from seed piece to seed piece prior to planting.

Management During the Growing Season

Plants that are stressed by other diseases, such as early dying, Rhizoctonia stem canker, root-lesion nematodes, and seed piece decay, are more susceptible to **early blight** than healthy plants. Likewise, stress associated with insect feeding, especially by Colorado potato beetles and leafhoppers, promotes the development of early blight. Weeds increase crop stress by competing for space, light, nutrients, and water. Excessive weed growth may also reduce air movement over the crop canopy, thus prolonging the duration of leaf wetness and favoring the formation of spores, spore germination, and infection by the early blight and late blight pathogens. Hilling helps to reduce weeds early in the season and covers the developing tubers with sufficient soil to minimize infection later in the season by spores of the late blight pathogen, which can be washed off the foliage and carried into the soil.

Fertilization

Host nutrition has a great effect on foliar diseases. Deficiencies of nitrogen and phosphorus increase susceptibility to early blight. Overfertilization with nitrogen, however, stimulates excessively lush vegetative growth, creating a microenvironment favorable for spore germination and plant infection. High crop density and excessive nitrogen fertilization may increase the frequency of disease occurrence by creating a dense plant canopy, which traps moisture, increases relative humidity, and prevents air movement. Additionally, excessive foliar nitrogen can delay maturity and reduce yield. Thus a balance must be reached, in which sufficient nitrogen is provided to satisfy the demands of the crop over the duration of the growing season without delaying maturity, enhancing susceptibility to **early blight**, or affecting the management of other diseases. Growers must carefully adjust nitrogen application according to the requirement of each cultivar, to avoid increasing the risk of disease. Soil and petiole analysis for nitrate content may be useful in monitoring the crop's needs and making adjustments in fertilization rates (Chapter 8).

Irrigation

Irrigation plays a key role in the development of **early blight**, **late blight**, and **white mold**. Sprinkler irrigation periodically wets foliage and, along with rainfall and frequent dews, creates the alternating wet and dry periods favorable for disease development. Overhead irrigation increases the likelihood of foliar diseases and thus the need for fungicides. Excessive irrigation can erode hills, exposing shallow-set tubers to foliar pathogens. Excessive irrigation can also wash sporangia and zoospores of the late blight pathogen into the soil, where they can infect tubers. In addition, nitrates and other leachable nutrients are removed from the root zone of the crop by excessive irrigation or rainfall providing more water than is lost by evapotranspiration. Nitrogen deficiency severe enough to cause premature senescence of potato plants increases their susceptibility to early blight and increases the difficulty of reducing disease spread. The depletion of nitrate in the root zone depends to a large extent on soil characteristics. In sandy soils with low organic matter content, several inches of rainfall in a short time can cause enough nitrate leaching to affect early blight development.

The duration of periods of uninterrupted plant wetness is a critical factor in infection by pathogens causing foliar diseases. When irrigating, try to avoid prolonging the duration of wetness of the foliage (Chapter 9). Careful irrigation is especially critical to the management of white mold and aerial (bacterial) stem rot during and immediately after the blooming of potato plants, when extended periods of wetness are highly conducive to stem infection.

Monitoring Disease Development

Crops must be closely monitored during the entire growing season to detect the appearance and spread of disease. Routine scouting of potato fields for foliar diseases increases the likelihood of identifying problems before they are beyond control and provides information on the effectiveness of current management practices. Scouting should begin just before crop emergence, to check for potatoes growing where tubers may have been discarded, such as tree lines, and where overwintering inoculum might be present. After emergence, the field should be monitored at least weekly until harvest, to evaluate crop and disease development at representative sites throughout the field. Special attention should be given to areas most prone to disease development. Such areas are usually near windbreaks, woodlots, or other obstructions to air movement, where plants tend to dry out more slowly; near the center of pivot irrigation systems or along wheel tracks; and near bodies of water, such as farm ponds and streams. Areas where fungicide application is difficult (for example, beneath power lines, around utility poles, along highways, or near residential property) should also receive special attention. Care should be taken not to spread late blight inoculum from field to field when scouting. Disposable pants and rubber boots should be worn when scouting fields and then washed and disinfected after leaving each field. Late blight inoculum can be spread from field to field on sprayers and other farm equipment that comes in contact with infected plant foliage, and such equipment should be carefully pressure-washed and disinfected before entering a field. Areas where plants are exhibiting symptoms of late blight should be treated last, to avoid redistributing inoculum within the field.

Disease Prediction Programs

Several computer programs have been developed to forecast the development of early blight and late blight. These programs track weather conditions to predict when environmental conditions are favorable for disease development and recommend when fungicide applications should be made. **Late blight** forecasting systems express the effects of temperature, relative humidity, and rainfall or irrigation on disease development as "severity values." Local late blight forecasts for many potato production areas in North America are available in published newsletters and from e-mail lists, Web sites, and toll-free telephone lines providing recorded information. Accurate weather data are required for accurate disease prediction, and currently the most reliable data are obtained by in-field, site-specific weather monitoring. Disease forecasts generalized for production regions are helpful in alerting growers over a broad area, but local management decisions are more sound if based on site-specific information.

Fungicides

While cultural practices help to delay or minimize the impact of foliar diseases, the application of fungicides is usually a necessary part of season-long disease management programs. There is currently a long list of fungicides, representing groups of chemicals having different modes of action, for the management of **early blight** and **late blight**, including protectant and broad-spectrum materials such as maneb, metiram, mancozeb, chlorothalonil, and fixed coppers, which have been used in the potato industry for decades to manage both diseases. In recent years, several new fungicides with different chemical modes of action have been labeled for the management of these diseases. The newer products typically have specific modes of action, some products controlling only early blight and some controlling only late blight. These products, having a high degree of specificity, must be used as part of a season-long program that includes mixtures with broad-spectrum materials and alternation with materials having different modes of action (Chapter 14). Fungicides must not be applied at rates exceeding the amount stated on the product label, to help prevent the development of pathogen resistance to chemicals having the same mode of action. In the last decade, resistance management has become a determining factor in the selection and use of fungicide products. Resistance of *P. infestans* to metalaxyl and mefenoxam is widespread, and the performance of strobilurin fungicides for early blight control is declining in some production areas as a result of resistance in *A. solani*. The number of fungicide applications required for management varies regionally. Growers in the West often make only one to three applications per year in the absence of late blight, but it is not uncommon for growers in the central, eastern, and southeastern production areas to make 10 or more applications on late-maturing cultivars. *Selected trade names of fungicides are given in Table 14.5.*

The timing of fungicide application is critical for effective management of white mold. Application at full bloom of the primary flower inflorescence is more effective than application at or before row closure or after blossom drop.

In the absence of cultivars with resistance to **white mold**, fungicide treatment remains the most widely used tactic for managing this disease. Several fungicides, with diverse modes of action, are labeled for control of white mold. The timing of fungicide application is critical for effective management. Application at full bloom of the primary flower inflorescence is more effective than application at or before row closure or after blossom drop. In field trials, applications made at the time of closure between plant rows were too early for effective control, and applications made when 20% of the blossoms had already dropped were too late. Fungicide application at initial full bloom most effectively reduced the incidence of stem infection.

Biological Control

The development of biological controls for foliar diseases has lagged behind the development of conventional chemical measures. Recently, the commercialization of the biocontrol agent *Coniothyrium minitans,* a parasite of sclerotia the white mold fungus, has shown promise in the management of the pathogen. Results have varied among regions, but there is an indication that when incorporated into the soil in fields with a history of white mold, this biocontrol agent can help to reduce the survival of soilborne sclerotia of *S. sclerotiorum.* Repeated incorporation at the beginning of each cropping season and immediately after the harvest of rotation crops susceptible to white mold can further deplete the soil bank of sclerotia of *S. sclerotiorum.*

The use of biocontrol materials is not compatible with soil fumigation. Soil fumigation can also reduce populations of naturally occurring microbes that attack soilborne plant pathogens. In fields where a soil fumigant is used to manage the pathogens involved in the early dying complex and where there is a history of white mold, biocontrol materials should be applied at least 2–3 weeks after the fumigant, when the fumigant is no longer effective.

Soil applications of *C. minitans,* while helping to reduce soilborne sclerotia, are not effective in protecting potato blossoms or senescent tissues on the ground against ascospore showers originating from nearby untreated fields.

Management During Harvest and Storage

The **late blight** pathogen does not normally remain alive for more than 2 weeks after the foliage dies, and thus in most production areas it is recommend that the vines be *completely dead* for 2–3 weeks before harvest, to reduce the risk of contaminating tubers with spores of the pathogen. Growers in the semiarid environment of the Columbia Basin have harvested potatoes from plants with green foliage infected with late blight without incurring an increase in tuber blight. In all situations involving late blight, foliar fungicide applications should continue until harvest or until all vines and vine regrowth are dead. Tubers should be harvested only after the skins are well set and should be handled gently to avoid abrasions, which increase the risk of infection by the early blight fungus and other pathogens of stored tubers. Tubers from areas of fields known to have been affected by late blight or from wet areas should not be stored or should be stored separately for a short time. Infected tubers expose the entire storage to the risk of late blight and associated diseases, such as soft rot. Whenever possible, avoid harvesting during wet weather. The crop should not be overirrigated close to harvest. If wet tubers are harvested, precautions must be taken to quickly remove free moisture from their surfaces.

Condensation of moisture on tubers in storage, from circulating air that is warmer than the tubers, will favor sporulation by the late blight pathogen, and late blight may spread in the pile. During bin loading, reduce the loading rate to allow for adequate inspection of tubers on the conveyor and removal of suspect tubers. If a significant level of late blight (affecting more than 1% of the tubers) is found at harvest, carefully consider prompt processing or marketing, short-term storage, or storage at reduced temperature and relative humidity.

Chip stock and processing potatoes are normally stored at 45–55°F and 90–95% relative humidity, but these conditions are ideal for the development of late blight tuber rot. Storage bins should be monitored constantly for increases in temperature, wet spots, depressions in the pile or a reduction in the height of the pile, increase in humidity in the storage area, pungent ammoniac gases, and runoff at the base of piles. These are telltale signs of disease, and such storages should be managed and processed immediately.

Contaminated fresh-market tubers should be cooled to as low a temperature as is practical, preferably 38–40°F, as quickly as possible after harvest. Most pathogens do not grow much at temperatures of 38°F or lower, but some development will occur at higher temperatures. Good circulation of air through the pile, to keep the relative humidity below 85% and thus dry out infections, must be maintained in storages with rot associated with late blight and other soft rot diseases. Blighted potatoes should not be held any longer than is absolutely necessary.

CHAPTER 22

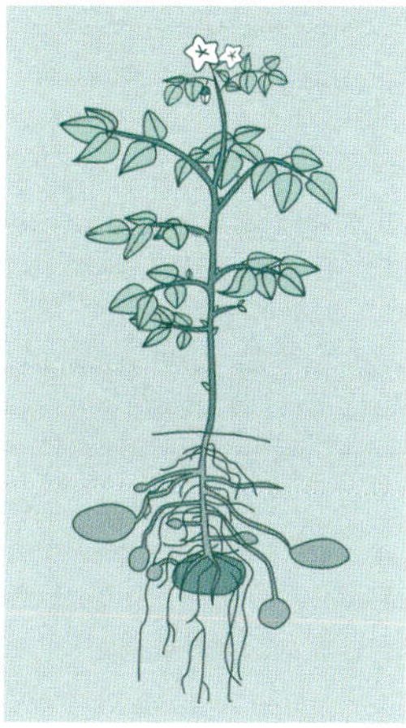

Rick A. Boydston
U.S. Department of Agriculture, Agricultural Research Service
Prosser, Washington

Pamela J. S. Hutchinson
Aberdeen Research and Extension Center
University of Idaho, Aberdeen

Robin R. Bellinder
Department of Horticulture
Cornell University, Ithaca, New York

Managing Weeds

Weeds must be managed throughout the entire crop rotation to optimize potato tuber yield and quality. Weeds of most arable croplands were not originally present in those areas but have been introduced there, mainly by human activities. Understanding factors that favor weeds and allow them to persist in crop rotations is key to designing cropping systems that discourage and reduce the impact of weeds. Soil in all potato production fields contains billions of weed seeds. Weed seeds in the soil, called the **soil seed bank**, are the primary source of weeds with which the potato grower must contend year after year. Because of the great quantity of seeds produced by most weed species and the longevity of the seeds in the soil, appropriate weed management strategies must be integrated into all potato health management programs.

Weeds directly limit crop yields by shading crop plants and by competing for other resources, such as space, water, and nutrients. Competition from weeds at any stage of potato growth reduces yield, but it is most detrimental during the first month or two after potatoes emerge. Dense weed infestations restrict the growth of potato plants, which produce undersized tubers as a result. Potato yield can be reduced by 20–50% in fields with inadequate weed management. Some weeds also affect crop quality by direct injury (e.g., by piercing tubers with sharp rhizomes). Water stress due to weed competition can result in physiological disorders in tubers and can lower their dry matter content.

Weeds also indirectly affect potato production in various ways. Insects such as psyllids, aphids, and leafhoppers may multiply on weeds before entering potato fields. Some viruses, nematodes, fungi, and bacteria that infect potato survive or increase on weeds. Excessive weed growth can impose stress on the crop, leading to early senescence and increased susceptibility to diseases, such as early blight. Weed growth can also interfere with the drying of foliage during the day, thus lengthening periods of leaf wetness and increasing the chances of infection by pathogens causing foliar blights. Cultivating a crop to manage weeds, by causing soil compaction and root pruning, can lower yields by more than 5% and may cause some loss of tuber quality. Misuse of herbicides can result in injury to a potato crop during the current season, tubers grown for seed, or subsequent crops grown in rotation.

Heavy weed growth can interfere with harvest operations. Weed species that do not naturally senesce to a dry state by harvest or do not dry after vine-killing operations or freezing weather are especially difficult in this regard, but even species that dry by harvest can cause problems in harvesting, because of their root masses. Grasses and other weeds with fibrous root systems are particularly problematic, as soil clings to their roots and must be separated from potato tubers and removed, slowing the harvest operation. Vine beaters, cutters, or other mechanical devices may be needed to prevent weeds from impeding the operation of potato harvesters, or hand picking may be necessary.

Weed management is only one aspect of overall potato production, and it should be designed to integrate with other production practices and avoid interfering with them.

Weed management is only one aspect of potato production, and it should be designed to integrate with other production practices and avoid interfering with them. Even with an adequate weed management program, weed escapes and adverse effects sometimes associated with normal control practices generally limit yields by as much as 7%. Potatoes as volunteers are also weeds and serve as hosts of insect pests and pathogens, and they affect cultivar purity in subsequent crops. Improved control practices can minimize, but most likely can never completely eliminate, losses due to weeds.

Weed Types

Weeds are generally classed as annual, biennial, and perennial species. Within these groups, weeds are also distinguished into broadleaves, grasses, and sedges.

Annual species complete their life cycle in a single year and reproduce by seed. They die naturally at the end of the season—after producing seed. **Summer annuals** germinate in the spring or summer and then enter a period of vegetative growth and seed production. Examples of summer annual weeds are hairy nightshade and redroot pigweed. **Winter annuals** germinate in the fall, overwinter in a dormant, vegetative state, and then begin growing again in the spring, producing seed during the summer. Examples of winter annual weeds are hairy vetch, downy brome, shepherd's purse, and field pennycress.

Since annual weed species depend solely on seed to survive from year to year, reducing the soil seed bank can help in the management of these species. There are three management strategies for reducing the soil seed bank: (1) prevent emerged weeds from producing seed, (2) remove seed that is produced, and (3) kill or promote the destruction of seed in the soil. Viable weed seed is lost to predation by soil microorganisms, insects, rodents, and birds. Weed seed viability slowly declines over time, and seed decay is often accelerated in warm soil or soil with fluctuating temperature and moisture. Seed longevity increases with the depth of burial.

> **There are three management strategies for reducing the soil seed bank: prevent emerged weeds from producing seed, remove seed that is produced, and kill or promote the destruction of seed in the soil.**

Weed seeds germinate only under specific conditions of temperature, moisture, and oxygen. Exposure to light stimulates some annual weeds to germinate, while it inhibits the germination of other species. A certain portion of the seed bank is dormant and will not germinate even when conditions are favorable. However, germination of a substantial portion of the seed bank can be stimulated through ground preparation and irrigation. Emerged seedlings can be killed by timely tillage or appropriate herbicides prior to planting or emergence of the crop.

Biennial species require two growing seasons to complete their life cycle. The first year's growth is entirely vegetative, with top growth usually confined to a rosette of leaves. The taproot is often fleshy, serving as a food storage organ. During the second year, the flower stalk rises from the crown. After producing seed in the second year, the plant dies. Examples of biennial weeds are wild carrot and western salsify.

Few biennial or winter annual weed species cause serious problems in potatoes, because they are disrupted by tillage and harvest operations associated with potato production.

Perennial species live three or more years and reproduce by various types of vegetative structures (stolons, creeping roots, rhizomes, tubers, and bulbs) in addition to seed. They can regenerate shoots each year, using food reserves stored in vegetative structures in the soil, so they are not dependent on seed germination for their survival. They can also resprout when top growth has been removed mechanically or by other means, as long as the underground storage organ is intact. Examples of perennial weeds are quackgrass and Canada thistle.

Weed Identification and General Management Strategies

Proper weed identification is the first step in designing an effective weed management program. Many weed identification manuals, Cooperative Extension publications, and Internet sources are available. Fields should be individually surveyed, and written records of the weed species present and their distribution should be retained. Keeping accurate records of the weeds present from year to year is important in determining changes in weed species in response to cropping practices and will aid in the selection of management tactics targeting those weeds.

Annual Weeds

Most annual broadleaf weeds detrimental to potato production are species common to any arable land (Table 22.1). Examples include various species of **pigweed** (Fig. 22.1), **kochia** (Fig. 22.2), **common lambsquarters** (Fig. 22.3), **Russian thistle** (Fig. 22.4), and **nightshades** (Fig. 22.5). In a Michigan study, when one redroot pigweed and one barnyardgrass plant per 39 inches of potato row were established at potato planting, marketable tuber yield was reduced

Table 22.1. Key weeds in potatoes

Common name	Scientific name
Annual broadleaf weeds	
Catchweed bedstraw	*Galium aparine*
Cocklebur, common	*Xanthium strumarium*
Knotweed, prostrate	*Polygonum aviculare*
Kochia	*Kochia scoparia*
Ladysthumb	*Polygonum persicaria*
Lambsquarters, common	*Chenopodium album*
Mallow, common	*Malva neglecta*
Nightshade, black	*Solanum nigrum*
Nightshade, cutleaf	*Solanum triflorum*
Nightshade, eastern black	*Solanum ptycanthum*
Nightshade, hairy	*Solanum sarrachoides*
Pigweed, prostrate	*Amaranthus blitoides*
Pigweed, redroot	*Amaranthus retroflexus*
Pigweed, tumble	*Amaranthus albus*
Powell amaranth	*Amaranthus powellii*
Purslane, common	*Portulaca oleracea*
Russian thistle	*Salsola tragus*
Smartweed, Pennsylvania	*Polygonum pensylvanicum*
Sunflower, common	*Helianthus annuus*
Annual grass weeds	
Barnyardgrass	*Echinochloa crus-galli*
Crabgrass, large	*Digitaria sanguinalis*
Foxtail, green	*Setaria viridis*
Foxtail, yellow	*Setaria glauca*
Sandbur, long-spine	*Cenchrus longispinus*
Wild oat	*Avena fatua*
Wild proso millet	*Panicum miliaceum*
Witchgrass	*Panicum capillare*
Perennial weeds	
Bermudagrass	*Cynodon dactylon*
Canada thistle	*Cirsium arvense*
Dandelion	*Taraxacum officinale*
Field bindweed	*Convolvulus arvensis*
Quackgrass	*Elymus repens*
Russian knapweed	*Centaurea repens*
Yellow nutsedge	*Cyperus esculentus*

by 19–33%. Annual broadleaf species can generally be controlled by timely cultivation together with the application of appropriate herbicides. The diversity of weed species typical of potato production fields, however, is sufficient to require a well-planned weed management program.

Annual weeds in the "white" or "thread" stage, shortly after germination and just prior to emergence, are extremely susceptible to shallow cultivation. Harrowing or hilling during this period is very effective in eliminating many annual weed seedlings. Most small-seeded annual broadleaf seedlings are also very susceptible to uprooting and burial by cultivation when less than 2 inches tall. They also are susceptible to minimal doses of burn-down herbicides applied during the period from their cotyledon (seed leaf) stage through their second-true-leaf stage (Fig. 22.6), but before potato emergence. This period generally spans 7–14 days after weed emergence. By the time the plants have formed six true leaves, they are well established and can survive a day or more after being uprooted in soil, increasing the likelihood of rerooting and recovery following cultivation. Although the seedling vigor of these species is vastly inferior to that of potato sprouts, broadleaf weeds usually germinate in the warmer surface soil above planted potato seed pieces and have the potential to compete effectively with potatoes after establishment.

Sunflower and **common cocklebur** are large-seeded, warm-season broadleaf species that can appear after potato emergence in some production areas. These weeds are difficult to control effectively by either cultivation or herbicides, although, when small, they are susceptible to a postemergence application of metribuzin plus rimsulfuron. A crop rotation including small grains allows the use of herbicides effective against these species in a rotation-wide management strategy.

Annual nightshades (Fig. 22.5) are especially troublesome in potato crops, since they survive below the canopy of potato plants and often develop at a similar growth rate. In addition, since nightshades and potatoes are both in the plant family Solanaceae, many herbicides that control nightshade are harmful to potatoes. Most nightshade species commonly found in potatoes can be suppressed by cultivation together with application of selective herbicides active against nightshade species, such as dimethenamid-P (chloroacetamide), rimsulfuron (sulfonylurea), flumioxazin (protoporphyrinogen oxidase inhibitor), EPTC (thiocarbamate), and linuron (urea) (linuron is labeled for use in potato production east of the Rockies only). Nightshade management should involve a

Fig. 22.1. Redroot pigweed (*Amaranthus retroflexus*), with an indentation at the tip of the true leaf.

Fig. 22.2. Kochia (*Kochia scoparia*).

Fig. 22.3. Common lambsquarters (*Chenopodium album*).

Fig. 22.4. Russian thistle (*Salsola tragus*). (Courtesy R. Parker)

rotation-wide strategy that allows the use of effective herbicides and cultivation in rotation crops.

Dodders (*Cuscuta* spp.) are yellow, vining annual weeds that can cause severe losses in potato crops. They germinate on the soil surface, twine around a host plant, and form haustoria, which attach to the host and enter its cells to rob it of water and nutrients. They thus survive parasitically on the host plant, without any attachment to the soil. Fields known to be thoroughly infested with dodder should be avoided for potato production. If an infestation is confined to a small area, all dodder plant parts and all affected potatoes should be removed in bags and burned, or the area should be treated with an herbicide that completely destroys both dodder and potato vegetation.

Fig. 22.5. **A,** Black nightshade (*Solanum nigrum*). **B,** Hairy nightshade (*S. sarrachoides*). **C,** Cutleaf nightshade (*S. triflorum*), with long, narrow cotyledons. (Photo **A** by D. Wilson, courtesy S. Nissen)

Volunteer grain sometimes causes problems in potato crops. Seed of winter or spring cereals, including **wheat**, **barley**, **rye**, and **corn**, often remains from a previous crop. Grain seedlings may be effectively controlled by allowing time for or promoting germination and growth prior to normal tillage for potato seedbed preparation. Irrigation following harvest of small grains can promote germination if soil temperatures are high enough and if a long postharvest period is not required before seed dormancy has passed. Potato seedbed preparation, drag-off, and hilling normally dispose of volunteer small grains, which also are susceptible to most preemergence herbicides used for weed control in potatoes. However, since corn usually germinates after potato plants emerge, these tillage operations are seldom adequate to con-

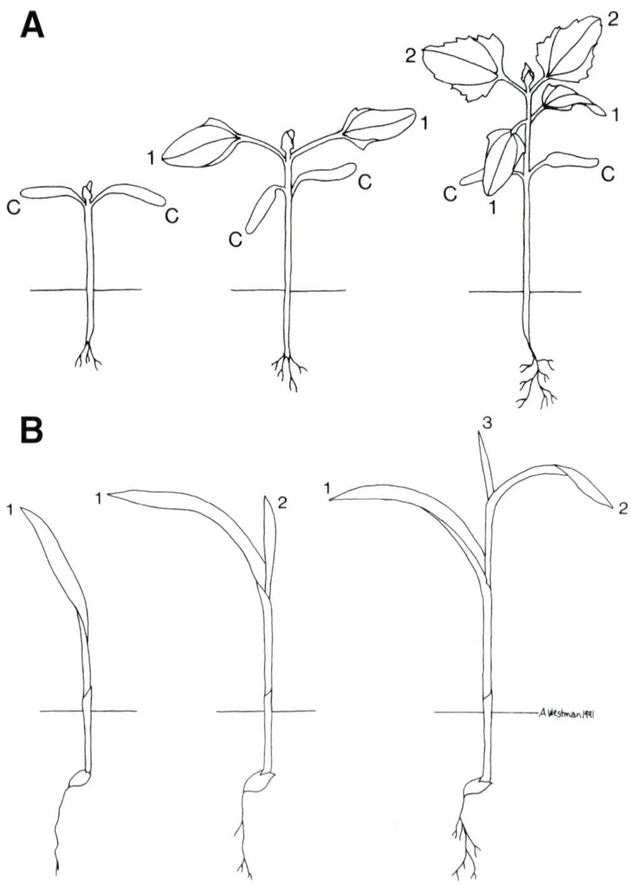

Fig. 22.6. Annual weed seedlings in early growth stages, when management tactics should be implemented. C = cotyledon; 1 = first true leaf; 2 = second true leaf; 3 = third true leaf. **A,** Common lambsquarters (*Chenopodium album*), an annual broadleaf weed, in the cotyledon stage, first-true-leaf stage, and second-true-leaf stage. **B,** Giant foxtail (*Setaria faberii*), an annual grass weed, in the one-, two-, and three-leaf stages. (Drawing by Anne Westman)

trol volunteer corn. Postemergence herbicides are available for controlling corn in potato crops.

Most **annual grass weeds** that are present in potato fields (Table 22.1), with the exception of wild oat (*Avena fatua*), are warm-season grasses, which germinate relatively late, after potatoes have emerged. Annual grasses such as **barnyardgrass** (Fig. 22.7), **crabgrass** (Fig. 22.8), **foxtails**, **wild proso millet**, **witchgrass** (*Panicum capillare*) (Fig. 22.9), and **sandbur** can be managed by cultivation together with the application of preemergence herbicides. Postemergence herbicides such as clethodim, sethoxydim, and rimsulfuron can also be used to manage most annual grass weeds in potatoes.

Many annual weeds can germinate throughout the growing season, with peak emergence occurring at specific periods, depending on optimum soil temperature and moisture for each species, the timing of cultivation, and rainfall or irrigation, which can stimulate emergence. Significant annual weed germination can occur in the late summer, following an early-harvested potato crop. Late-emerging weeds often flower and produce seed quickly, in response to the shortening day length. Annual weeds emerging late in the summer should be controlled with cultivation or herbicides if they

Fig. 22.7. Barnyardgrass (*Echinochloa crus-galli*).

Fig. 22.8. Large crabgrass (*Digitaria sanguinalis*).

begin flowering before a killing frost, to prevent a potentially large increase in the seed bank.

Perennial Weeds

Perennial weeds (Table 22.1) often reduce yields more than annual weeds, because of their more vigorous growth and fewer options for management by herbicides and cultivation. Potato yields have been reduced by as much as 70% by **Canada thistle**, 16% by **field bindweed**, and 32% by **Russian knapweed**. Season-long interference by 14 quackgrass shoots per square foot can reduce potato yield by as much as 78%. Quackgrass and yellow nutsedge (Fig. 22.10) not only greatly limit yield but also lower quality, because their rhizomes penetrate and cause deformation of tubers.

Fig. 22.9. Witchgrass (*Panicum capillare*).

Fig. 22.10. Yellow nutsedge (*Cyperus esculentus*).

Cultivation suppresses many perennial weeds, but since the buds from their creeping roots and belowground stems (rhizomes) continue to emerge until most of their shoots are flowering, cultivation simply delays their growth and sometimes spreads them to areas of the field that were previously uninfested. Covering emerging shoots with soil does not kill perennial weeds. Most perennials tolerate herbicides labeled for use in potatoes. Fields heavily infested with perennials should be avoided for potato production, because management of these species is generally not satisfactory in potato crops.

Serious perennial weed problems require **persistent management programs implemented over several rotational cropping cycles** targeting individual species. Most perennial weeds move food reserves from leaves and stems downward to storage organs as the days shorten and the plants prepare for overwintering and subsequent growth the next spring. As a result, many perennial weeds are susceptible to herbicide treatment during late summer and early fall, when the herbicide is more likely to be translocated to the roots or overwintering storage organs. Therefore, an effective management program for perennial weeds includes rotations and planting times that allow infested areas to be crop-free during late summer and early fall. Herbicides such as glyphosate or 2,4-D applied to perennial weeds late in the growing season are moved into underground buds and thus limit the reproductive ability of the weeds. For this strategy to be successful, green leaves must be present at the time of application. If possible, fields should be disked 7–12 days after the initial herbicide application. Perennial weed infestations often require two or more seasons of properly timed treatment to achieve adequate control.

On fallow ground and in many rotation crops, 2,4-D, dicamba, and clopyralid (synthetic auxins) and glyphosate (EPSP synthase inhibitor) are particularly effective in controlling broadleaf perennials. Combinations of these herbicides are often more effective than the individual herbicides alone, and sometimes they can be applied at a lower rate when used in combination than when used alone. Consult product labels for specific information regarding the use of these herbicides.

Herbicide application together with timely cultivation offers the most efficient control of quackgrass and yellow nutsedge. **Quackgrass** can be suppressed early in the season with EPTC, but potato production will still be hampered by surviving or later-emerging plants. Quackgrass can also be suppressed in a potato crop by postemergence application of rimsulfuron, sethoxydim, or clethodim, but it is most effectively managed by the nonselective herbicide glyphosate applied in the fall, after harvest. Fall-applied glyphosate is translocated with carbohydrates into underground vegetative rhizome buds, effectively reducing the number of buds that emerge the following season. Drying quackgrass roots by spring-tooth tillage of dry soil two or more times in midsummer or late summer following harvest of a small grain crop can also be effective. This practice, however, is not recommended in areas with coarse-textured, highly erodible soils, where glyphosate treatment is preferable, to minimize soil erosion.

Yellow nutsedge is often misidentified as a grass, but it can be easily distinguished from grasses by looking at a cross section of the stem, which is triangular, whereas grass stems are round or flat. Yellow nutsedge reproduces predominately by means of underground tubers, which sprout after the soil has warmed. In many areas, it emerges 2–3 weeks later than annual weeds such as common lambsquarters, kochia, and mustards. Nutsedge grows vegetatively until midsummer and then begins tuber formation as the days begin to shorten. Early-season management with cultivation or herbicides reduces competition from the vegetative nutsedge plant and the number of these plants that live to produce new tubers. Since yellow nutsedge forms tubers and possibly even small plants late in the growing season, management at this time is crucial. Early planting of vigorously growing potato cultivars is advantageous in fields known to be heavily infested, since nutsedge does not compete well in heavy shade. Glyphosate can be used to manage nutsedge outside the crop season, and EPTC, S-metolachlor, dimethenamid-P, and rimsulfuron selectively suppress this weed in potatoes.

Integrated Weed Management

Cultural Practices

Appropriate cultural practices are the foundation of any sound weed management program. Cultural practices that prevent new weed species from being introduced and established can greatly decrease weed management efforts and inputs required over time (Box 22.1). The overall health of the potato crop and the growth habit of the cultivar can have an impact on the need for weed management. Vigorous indeterminate cultivars (e.g., Russet Burbank) will suppress weeds more than less vigorous determinate cultivars (e.g., Russet Norkotah). A healthy potato crop with few or no planter skips will suppress weed development once vines close the rows. Proper planting time, seeding depth, seed spacing, fertility, and irrigation all contribute to producing a more competitive crop, which will tolerate and suppress weeds, so that there is less need for direct management. Potato row spacing varies between production areas and may be difficult to change, because of equipment limitations. Potatoes planted on narrow rows close the canopy quicker and suppress weeds more than potatoes planted on wider rows. However, potatoes planted on wider rows allow slightly later cultivation.

Potatoes should be planted in soil with adequate moisture and not irrigated prior to emergence unless the soil becomes excessively dry at the seed piece depth. Irrigation prior to potato emergence promotes the germination of shallow weed seed, which could benefit or hinder weed management, depending on cultivation and herbicide application timing. Frequent application of small amounts of irrigation water promotes the germination and survival of annual weeds, whereas less frequent, longer irrigation sets promote drying of the surface soil between irrigations and may kill germinated weed seedlings in the white or thread stage.

A well-planned **crop rotation** is a cornerstone of effective weed management. Certain weeds naturally become associated with particular crops, because of similar life cycles, growth requirements, or herbicide tolerances. If a crop is grown continuously, weeds associated with it (such as annual nightshades in potatoes) tend to dominate and proliferate year after year. A diverse crop rotation discourages domination by any one set of weed species and provides the opportunity to control troublesome species during various portions of the rotation. A winter annual crop in the rotation, such as winter wheat, can reduce populations of common summer annual weeds associated with potato. Rotation to alfalfa often suppresses weeds that are difficult to manage, because of the competitiveness of alfalfa and the frequent mowing of this crop, and it allows the use of herbicides that are not registered for use in potato.

A well-planned crop rotation is a cornerstone of effective weed management.

Rotation-wide management plans often lead to the best overall weed management, and such plans are the only effective strategy against some species, such as field bindweed and Canada thistle. Effective herbicides registered for use in small grains or corn can be used to manage field bindweed and Canada thistle, which are difficult to control in potato. When herbicides are selected for use in rotational crops, compatibility with potato and drift to potato fields in the vicinity must always be considered (Figs. 13.10–13.20). Some imidazolinone and sulfonylurea herbicides used in small grains and uncropped areas are effective, long-lasting herbicides, but they can persist and injure potatoes, which can be highly susceptible to even a trace amount in the soil (Figs. 13.12, 13.16, 13.19, and 13.20). Herbicide labels contain explicit restrictions on crop rotation and should be followed closely.

Irrigation in late summer or early fall following the harvest of a rotational crop or an early-harvested potato crop encourages the germination of weed seeds, and the resulting seedlings can be destroyed by tillage, herbicide application, or killing frosts in late fall or spring. Full-season potato growth is necessary for late-season competition and weed management. Implementation of optimal management strategies to maintain potato health will produce an early-emerging, vigorous, competitive potato crop, with rapid row closure and continued health and competitive ability until vine kill, leaving less opportunity and resources for developing weeds. **Late-season weed management** is accomplished mainly by maintaining sufficient crop cover. If possible, late-season weeds should be killed before they produce seeds. Chemical or mechanical vine-killing procedures should target weeds with immature seeds along with potato vines, to reduce the buildup of weed seeds in the soil. However, mature weed seeds will not be killed by vine killing.

Late-season weeds should be killed before they produce seeds.

In areas with an extended growing season after the potato harvest, **fall-planted cover crops** can suppress weeds and pathogenic nematodes, reclaim nutrients, and reduce soil erosion. Sorghum-sudangrass and mustards emerge quickly when planted in mid- to late August in most potato production regions and develop canopies that can smother annual weeds germinating in the fall. However, if a cover crop emerges slowly or fails to emerge, weeds should be controlled with herbicides or tillage, to prevent the production of mature weed seeds before a killing frost.

Box 22.1

Preventative Weed Management

Various management strategies can be implemented to minimize the spread of weeds from field to field and the introduction of troublesome weed species to fields where they were not previously present.

- Plant only certified seed of all crops in the rotation. Inspect seed lots before purchase, and check the tag for weed seed tolerance. By law, certified seed must be labeled for weed seed content.
- Practice good field sanitation. Vehicles, equipment, and livestock can spread weed seeds. Mud generally contains weed seeds and should be washed from vehicles, equipment, and livestock before they are permitted access to potato production areas.
- Install screens at headgates and pumps to exclude weed seeds from river or canal irrigation water.
- Compost manure properly before spreading it on fields, to destroy much of the weed seed that might be present.
- Manage weeds in field borders, uncultivated areas, fencerows, roadsides, and ditch banks. Weeds in field borders can harbor insect pests, nematodes, and viruses, and the weeds themselves may spread into the field. Rather than relying on soil sterilants for weed management in these areas, plant a good grass cover for biological management, and then keep it weed-free and healthy.
- Do not ignore new weed species. If an unknown or new weed species is noticed, take samples to an expert for identification. If practical, remove and destroy new weeds before they form seed.

Cultivation

Cultivation is an essential part of an integrated weed management program in potato. It is an economical alternative to hand weeding when herbicides cannot be used, as in organic farming or in areas where herbicide use is prohibited. Cultivation is effective against annual weed species if performed while weed seedlings are still young and small (Box 22.2). It is only partially effective against perennial weeds.

The timing and type of cultivation should be determined by and coordinated with other weed management and cropping practices (e.g., hilling, irrigation, reservoir tillage, etc.). Weeds that emerge earlier than potatoes can often be controlled with shallow cultivation and hilling followed by preemergence or early postemergence herbicide application. In situations where no herbicide input is desired, additional cultivations may be needed to reduce the impact of weeds. Cultivation to a shallow depth, when weeds are small (or just

Box 22.2

Cultivation Principles and Techniques

Principles of Cultivation

- **Very small weeds are killed more easily than larger weeds.** Many germinated but not yet emerged seedlings, in the "white" or "thread" stage, are killed by shallow cultivation. Cultivate before annual broadleaf species grow beyond the two-true-leaf stage or before annual grasses have more than three leaves (Fig. 22.6). Larger weeds can reroot, and those on hills can reemerge through soil thrown up to cover them.
- The longer annual weeds are allowed to compete with potato plants, the more effect they will have on potato yield.
- Most annual weeds germinate in the top 1–2 inches of soil, so cultivating too deeply will only expose new weed seeds to more favorable germination depth and conditions.
- Weeds in dry soil are easier to kill with cultivation than weeds in moist soil.

Cultivation Techniques

- Planting techniques vary in different production areas and have a direct effect on cultivation timing and frequency. When cultivation is performed after planting, the number of tillage and cultivation operations can vary from one or two to as many as four or five, depending on the planting depth and hill height at planting.
- When seed pieces are planted at or near the soil surface and a 10- to 12-inch hill is formed, fewer cultivations and hilling operations are generally performed, and they are done primarily to prevent frost damage and tuber greening.
- When potatoes are planted 4–5 inches deep and covered with only a slight hill, more frequent cultivation may be necessary. Hill building can begin slowly, and tillage is performed to reduce soil crusting, increase soil aeration, and manage weeds.
- In a similar situation or if the field is left flat after planting, hill formation and cultivation can be combined in a one-pass operation shortly after weeds have emerged. This allows maximum opportunity to cover and subsequently kill weeds in the row while minimizing field traffic.
- When a single hilling-cultivation operation is performed at or just prior to potato emergence, preemergence herbicides can be applied to the clean bed created by hilling for effective weed management, and further cultivation is usually not necessary.
- Test, adjust, and monitor cultivation equipment to ensure maximum weed management and minimum crop damage.
- If annual weeds reach the proper stage for cultivation before the potato sprouts are within 2 inches of the soil surface, shallow cultivation with a flex-tine or spike-tooth harrow (often referred to as drag-off) will kill weeds without injuring potato sprouts. Check the depth of harrowing to avoid injuring potato sprouts, and make sure that all weeds are being uprooted.
- Hilling just as potato plants are emerging will not reduce yield as long as no more than 10% of the total plants in the field have emerged. Hilling after more than 10% of the potatoes have emerged should be done with equipment adjusted to throw soil around the base of the plants, covering small weeds without covering many potato leaves.
- If annual weeds are emerging when the potato plants are still small and would be covered by soil thrown into the row by a cultivator, then a flex-tine harrow could be used to kill weeds instead of a cultivator.

prior to weed emergence) and when the soil is not too moist, is especially important in systems where herbicides cannot be used. Once weeds become too large and well established, cultivation becomes less effective. In organic and other cropping systems that restrict herbicide use, weeds emerging late in the growing season may cause crop and harvest losses. Hand weeding can supplement cultivation, but it is seldom economical. Vine chopping prior to harvest can eliminate the top growth of many late-season weeds, but the soil mass associated with their fibrous roots may still slow harvest. Selecting potato cultivars that grow a full canopy and can be harvested early may be advantageous if late-season weeds become a problem.

Cultivation has several **disadvantages**. A major disadvantage is direct damage to potato roots and tops. In an Idaho study, U.S. No. 1 tuber yield and total tuber yield were reduced by 17 and 10%, respectively, in a weed-free study when hilling at 10% potato emergence was followed by two cultivations—the first when the plants were 4–8 inches tall and the second when they were 12–18 inches tall—compared with the yield of plants grown with hilling only or hilling plus one cultivation when they were 4–8 inches tall.

Another major disadvantage of cultivation is indirect damage due to soil compaction. Compaction reduces soil aeration, and consequently plant growth and tuber health are adversely affected. Compaction also leads to the formation of clods, which can cause tuber bruising at harvest. Soils containing the amount of moisture required for potato production are usually wet enough to become easily compacted by cultivation. When a four-row cultivator is used, 25% of

- Newly emerged potatoes are normally strong enough to survive flex-tine harrowing with no more than about 5% shoot breakage. The forward speed of the tractor is critical during this operation—too high a speed results in excessive damage to the potato plants. Plants with broken shoots are set back as much as they would be from a freeze. However, a small amount of shoot breakage is sometimes preferable to covering all the potato plants with a typical cultivation-hilling operation or allowing weeds to develop.
- If more action is needed, add a second gang of harrows behind the first. It is helpful to pull a ripper shank behind the tractor wheels while harrowing. This reduces compaction and helps guide the tractor more precisely for the next operation, which should not be needed until the potatoes are large enough that they will not be covered by cultivation.
- It is important to plan the hill-building process. When multiple cultivations are to be performed, throw only enough soil with the cultivator to cover the entire hill with 1–2 inches of soil. Small, emerged weeds within the row, which the cultivator cannot uproot, often die when covered with this amount of soil, whereas less soil does not uniformly cover and kill all weeds throughout the field. In contrast, coverage with more than 2 inches of soil builds the hill too quickly, so that later cultivations cannot add enough to cover and kill young weeds emerging after the previous cultivation. In addition, potato plants could suffer delayed growth if covered with too much soil.
- Begin cultivation in the driest part of the field first, so weeds will be killed by drying prior to scheduled irrigation or expected rainfall. Most uprooted annual weeds in the two-true-leaf stage or smaller lying in warm, dry soil for at least 1 day will die. Cultivating dry soil also minimizes soil compaction and its effects on yield and clod formation.
- Irrigation should be scheduled considering the need for relatively dry soil to optimize response to cultivation, although irrigated fields should never be allowed to fall below 75% available soil water in the root zone during Growth Stages I and II of the potato crop.
- The last hilling cultivation, often called lay-by, must form the final hill while still throwing enough soil around the base of the potato plants to kill any small weeds beginning to grow in the row. The lay-by cultivation should be performed well before row closure, since cultivation performed too late can cause crop damage and midseason stress, especially when performed at the time of tuber initiation. Stress at this stage of development can lead to early maturity of the crop, malformed tubers, and other quality losses. Where row closure is expected 60–80 days after planting, potato plants usually grow rapidly enough that lay-by cultivation should be done when there is still 12 inches of space between plants across the rows. At that time, an 18-inch tire can pass without significant crop damage, and root pruning is minimized. Within 5 days, the rows will grow too close to avoid damage. Within 7–10 days, the rows will be completely closed, and a vigorous crop canopy will then suppress most weed development. **Avoid pruning potato roots during cultivation.**
- In some areas, the lay-by cultivation includes reservoir tillage (for example, by means of the Dammer Diker tillage system), which leaves pitted furrows to help reduce water runoff on sloping ground or enhance water infiltration. When reservoir tillage is performed after herbicide application, the treated soil may be disturbed, diluted by untreated soil, and redistributed unevenly, and consequently weed control will be reduced. Some growers have added a spray boom to their reservoir tillage implements in order to apply herbicides after the pitted furrows have been formed, thus avoiding these problems. The label instructions for some herbicides allow sprinkler application.

the field is covered with wheel tracks, and every row of potatoes is adjacent to and affected by compacted soil. With a six-row cultivator, four out of every six rows are adjacent to and affected by compacted soil. In comparison, herbicide sprayer swaths are usually two to five times wider than most cultivators, so they cause proportionally less soil compaction. Application of herbicides by a sprinkler irrigation system (chemigation) or by aircraft totally eliminates compaction due to ground equipment moving across a field.

Cultivation has several other disadvantages. It promotes new flushes of weed growth by exposing more seeds to conditions favorable for germination. It is less effective in wet soils, in which weeds can readily reroot. It provides less control of weeds growing within crop rows than of weeds between rows. In addition, weeds germinating after potato row closure cannot be managed by cultivation, although many late-emerging weeds do not survive under the potato canopy. By comparison, herbicides generally provide more uniform management of perennial weed species than cultivation and in some situations may be applied in a band over the crop row to target weeds within the row. Moreover, herbicide application usually takes less time than cultivation, and cultivation requires more labor and has higher fuel costs if a repeat operation is necessary.

Furthermore, cultivation can promote soil erosion on sloping land and stir oxygen into the soil, increasing microbial activity and potentially leading to a loss of soil organic matter. Cultivation can spread certain soilborne pathogens and weed seeds throughout a field. Some potato viruses (e.g., *Potato virus X*, *Potato virus S*, and *Potato virus Y*) and *Potato spindle tuber viroid* can be transmitted by cultivators brushing against potato foliage. The increased risk of transmission of these disease agents during cultivation is especially undesirable in seed potato production.

Despite these disadvantages, cultivation can play a key role in weed management programs, as long as it is done properly and in a timely manner (Box 22.2). Potential losses due to cultivation must be balanced against the costs of and potential losses due to alternative weed management practices and the potential losses due to not managing the weeds. Some weed seedlings may emerge after row closure, but the growth of these weeds is generally suppressed by the crop. Exceptions are perennial weeds, such as Canada thistle, and shade-tolerant annual weeds, such as nightshades. Nightshades survive under the crop canopy but are not released from competition until the potato vines begin to die from frost or late-season disease. If frost or hail in late spring or early summer following row closure reopens the vines, weeds are released from competition with the potato plants and can develop into a severe problem. In these cases, postemergence selective herbicides may be needed.

Herbicides

Herbicide application can be a very effective component of a weed management program. When herbicide application is effectively integrated with timely cultivation, the two can complement one another and overcome some of the disadvantages of either used alone. The first step in devising an appropriate herbicide plan is to properly identify the weed species requiring management in each field. It is also necessary to understand how the growth habits of these species and their susceptibility to available management methods impact the management tools available. Answers to the following questions will influence the management strategies utilized:

- Are the weeds annuals, which develop from seeds, or perennials, which grow from rhizomes or other underground structures surviving from year to year?
- When do they emerge?
- Are they susceptible to timely cultivation or herbicides registered for use in potatoes?

Herbicide applications should be planned carefully and well in advance, to avoid possible complications (Chapter 13). Choose an herbicide or combination of herbicides that, together with cultivation and other practices, will manage all weed species that are likely cause problems. Special management programs may be needed for troublesome species, especially some perennial weeds. A large population of a single surviving weed species can pose as much competition for a potato crop as multiple weed species and may be difficult to manage. Most fields contain numerous weed species, and it is highly unlikely that a single herbicide can manage all of them. Therefore, herbicide combinations (applied sequentially or as tank mixtures) should be selected, to provide broad-spectrum management of most or all weed species present. These combinations should be comprised of herbicides with varying modes of action and overlapping weed spectra, especially if herbicide-resistant weed populations are present. The use of herbicides with multiple modes of action also can provide a means of combating the development of herbicide-resistant weed populations. *Be sure that a combination is permitted by the labels of all of the herbicides included!* Weed biotypes with a natural tolerance for an herbicide may be controlled when it is applied in combination with another herbicide having a different mode of action (Chapter 14).

Make sure the herbicides under consideration are currently registered for use on potatoes in your region. Pay special attention to sections of the herbicide label relating to potatoes, including restrictions that may apply to certain cultivars and cultural practices. *Make sure any plant-back restrictions on the label fit into your overall crop rotation plan.* If necessary, seek additional information from qualified specialists.

To achieve the best weed management possible, it is important to understand in detail how herbicide utilization will fit into the overall potato health management program. Determine exactly how and when herbicides are to be applied and whether multiple applications are necessary and are according to label instructions. Most herbicides used in potatoes are applied before the weeds and potatoes have emerged. Therefore, accurate records of major weed species present in each field in previous seasons are critical to selecting the proper herbicide. When postemergence herbicides are used to manage annual weeds, application should occur when the weeds are less than 3 inches tall (Figs. 22.1–22.10). Herbicides are more effective against annual weeds when applied in their early growth stages, and less herbicide may be required. Choose an effective application rate, but do not exceed the labeled dosage for existing soil and weed conditions. Excessive doses could injure the potato crop or even the next crop in the rotation (Figs. 13.10, 13.12–13.16, 13.19, and 13.20).

Check the herbicide label to determine whether cultivation or operation of other equipment in the field before or after application will jeopardize successful weed management. Final hilling and other surface conditioning often must be done before herbicide application. Determine whether soil moisture is critical for herbicide effectiveness. Some herbicides require incorporation and activation by rainfall or irrigation within 7 days after application, and if this does not or cannot occur, then shallow mechanical incorporation may be necessary.

Herbicide Selectivity

The ability of herbicides to kill some plants without appreciably injuring others is the result of complex interactions among physiological, chemical, environmental, and physical factors. Herbicide selectivity is almost always relative and dose-related. The degree to which a plant is capable of responding to an applied herbicide is a measure of its susceptibility to that chemical under specific environmental conditions. Tolerance of certain plant species to an herbicide at high doses is termed **physiological selectivity**. In contrast, **placement selectivity** is the ability of an herbicide to be used safely on a crop species only when direct contact with the crop is prevented or minimized. For example, if the roots of the crop develop below the depth at which weeds germinate, certain herbicides that affect the crop only through root uptake and that are immobile in the soil can be placed in the shallow weed germination zone without injuring the crop.

Herbicide selectivity is dependent on the chemical mode of action, placement in the soil, and application timing. Rapid-acting herbicides, which kill on contact, are useful for annual weed control. Slow-acting herbicides, which are usually moved systemically within the plant, have a chronic effect and are generally more useful in controlling perennial weeds.

Application Timing

Herbicides can be applied as **preplant-incorporated**, **preemergence**, or **postemergence** treatments. These terms can cause confusion about whether they refer to emergence of the crop or emergence of the weeds. *Preplant-incorporated* herbicides are applied and incorporated into the soil before the crop is planted. *Preemergence* usually refers to application prior to weed emergence, but it can also refer to application after planting but prior to crop emergence. *Postemergence* refers to application after weed emergence, but it also can refer to application after crop emergence. Herbicides that must be applied prior to weed emergence usually do not affect emerged weeds. Herbicides that must be applied to potatoes as preemergence treatments typically cause injury if applied to emerged potatoes, although the application of some herbicides by chemigation is allowed shortly after potato emergence, since the large volume of irrigation water washes the herbicide off potato foliage, reducing the risk of injury (Chapter 13).

Herbicide Incorporation

Herbicides are incorporated into the soil for three reasons: (1) reduction of herbicide loss due to exposure to sunlight and volatilization, thus avoiding reduced weed control and injury to susceptible nontarget plants in adjacent areas, (2) herbicide activation by placement in the moist subsurface soil, where the herbicide will be available for absorption and uptake from the soil water solution, and (3) herbicide placement in proximity to germinating weed seeds or plant parts where absorption occurs. EPTC, for example, is volatile and can be rapidly lost if not incorporated within several hours of application. Herbicides that are not volatile or degraded by sunlight are also frequently incorporated when they are applied after potato planting but prior to potato or weed emergence. Weeds are thus killed as they grow through the layer of treated soil.

Soil-applied herbicides usually work best when there is adequate rainfall or irrigation to move them into the weed seed germination zone, activate them, and make them available in solution for absorption by weed seedlings. Herbicides requiring incorporation must be mechanically incorporated if adequate rainfall has not occurred or if irrigation water is unavailable within a short period after application.

Depending on their chemical and physical characteristics, soil-applied herbicides can become bound to soil organic matter and certain clays, thus becoming less available for uptake by weeds and crop plants. Heavy-textured soils (clay, clay loam, and silty clay loam) and soils with high amounts of organic matter therefore generally require higher rates of herbicide for adequate weed management than lighter-textured soils (sand, sandy loam, loamy sand, and loam). Most herbicide labels list the proper application rates for soils of different textures and with different organic matter contents. The bioavailability and mobility of some herbicides and, ultimately, weed and crop responses to them are affected by soil pH, and recommended rates must be adjusted accordingly.

Preemergence herbicides are generally absorbed through the early-developing roots and shoots of weed seedlings, halting basic plant processes to cause death below or just above the soil surface. Herbicides currently registered in the United States for preemergence (both preplant-incorporated and surface-applied) weed management in potatoes are dimethenamid-P, EPTC, flumioxazin, linuron (east of the Rockies only), metribuzin, metolachlor, pendimethalin, rimsulfuron, S-metolachlor, and trifluralin. Most of these herbicides are applied and incorporated 7–21 days after planting, or as late as after the final hilling operation. Dimethenamid-P, flumioxazin, linuron, and trifluralin must be applied before potato emergence. The other herbicides mentioned can be safely applied after potato emergence, but only metribuzin and rimsulfuron are active against emerged weeds. *Selected trade names of herbicides used in potato production are given in Table 14.1.*

Postemergence herbicides are applied at various times following weed emergence and allow management of weeds in potatoes as needed. Nonselective herbicides, such as paraquat and glyphosate, may be applied to emerged weeds prior to crop emergence. Following potato emergence, metribuzin and rimsulfuron can be used for management of broadleaf and grass weed species, and clethodim and sethoxydim can be used to manage grass species. Adjuvants, such as crop oil concentrate, methylated seed oil, nonionic surfactant, and fertilizer solution, can markedly improve the performance of some postemergence herbicides. Herbicide label recommendations should be followed for optimum adjuvant selection, as improper adjuvant use can result in crop damage or poor weed control.

Reservoir tillage is the practice of making pitted furrows (for example, by means of the Dammer Diker tillage system) to improve water infiltration and reduce runoff on slopes. It is practiced in many potato-producing regions in the western United States. Soil disturbance (hilling operations, deep cultivation, or reservoir tillage) following preemergence herbicide application can dilute the herbicide by mixing it with untreated soil and may leave untreated soil on the surface, usually resulting in decreased weed management. It is important, therefore, to carefully plan the timing of tillage and herbicide application to obtain the greatest benefit from both operations. Timely cultivation or hilling during the first 35–40 days after planting, coupled with an herbicide application after the final soil disturbance, or a single combined cultivation and hilling operation followed soon after by a preemergence herbicide application to the clean bed can provide season-long weed management in many regions. Sequential applications of a preemergence herbicide after planting followed by one or more postemergence herbicides after the final soil disturbance may also provide more effective weed control than single applications.

Herbicide Carryover and Potato Injury

Herbicide degradation in soil is usually accomplished by microbial or chemical breakdown. Degradation processes require adequate soil moisture and temperatures within an optimal range. Herbicide carryover, or persistence from one growing season to the next, can be longer than normal following reduced rainfall or irrigation during the application year, a shortened growing season, a dry fall and winter, lower than normal temperatures, or other conditions favoring persistence of the herbicide. To avoid problems with herbicide carryover, determine if the herbicides being considered for use in potatoes are compatible with the other rotational crops to be planted in the field during subsequent years, and determine if the herbicides that have been used allow potatoes to be planted next in the rotation sequence.

Some herbicides persist in the soil in the year following application at concentrations high enough to limit potato yield and affect tuber quality (Figs. 13.10, 13.12, 13.16, 13.19, and 13.20). Potatoes cannot be planted the year following application of clopyralid, imazethapyr, imazamox, and atrazine. Read the labels and talk to area experts to find out if special efforts, such as deep plowing and postharvest irrigation, are needed to increase herbicide degradation and reduce the risk of injury to the following crop. In some areas where a supplemental label exists, potatoes may be planted the year following the use of imazamethabenz-methyl in

Multiple tactics, including cultural practices, cultivation, and herbicide application, should be utilized in an integrated management approach to prevent the buildup of a few dominant weed species.

small grains if the rate is reduced, recommended irrigation amounts are supplied during the grain's growing season and after grain harvest, and tillage is performed after the grain has been harvested. Know the field history and obtain pesticide records for each field prior to planting potatoes. If herbicide residues are suspected, sample the soil in the field and conduct a bioassay or laboratory analysis for herbicide residues. Simple bioassay methods and choices of bioassay crops are widely available from Cooperative Extension specialists and publications and from chemical manufacturers. To minimize the impact of herbicide carryover, follow the label directions for the time interval needed between herbicide application and planting of a sensitive crop, keep complete and accurate records of herbicides used on each field, avoid application overlaps, and apply herbicides only at labeled rates.

Summary

An effective weed management strategy is an essential component of holistic potato health management. Many planned steps should be taken, the first of which should be weed species identification. Understanding cropping production practices that allow weeds to persist or become economically limiting is essential in designing and implementing a plan for reducing the impact of weeds on the potato crop. Multiple tactics, including cultural practices, cultivation, and herbicide application, should be utilized in an integrated management approach to prevent the buildup of a few dominant weed species. Appropriately timed cultivation and herbicide applications can keep losses due to weeds below economic levels in potatoes. Weeds that are hard to manage in potatoes may be more effectively and economically managed in other crops grown in rotation with potatoes. Care should be taken to plan appropriate management tactics in all crops rotated with potatoes to minimize risk of crop injury. Effective weed management allows the potato crop maximum use of all the resources made available by the crop management plan and will result in a vigorous crop better able to respond fully to other management strategies aimed at disease and insect pests.

CHAPTER 23

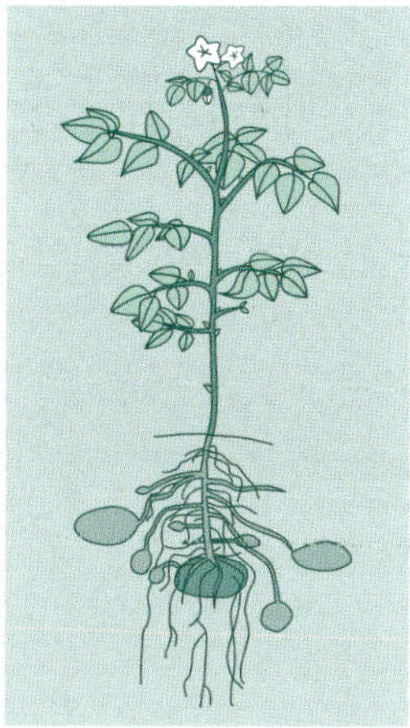

Larry K. Hiller
Department of Horticulture and Landscape Architecture
Washington State University, Pullman

Robert E. Thornton
Department of Horticulture and Landscape Architecture
Washington State University, Pullman

Managing Physiological Disorders

Potato plants, and particularly tubers, are susceptible to numerous noninfectious physiological disorders, which cause pronounced and often detrimental changes in their shape, function, and appearance. Tuber defects include external and internal blemishes, changes in shape and size, and alterations in the color of the skin and flesh. Most physiological disorders do not cause significant changes in the nutritional value of affected tubers. However, these disorders can cause large economic losses, as a result of their detrimental effect on appearance and culinary uses, which can lower the grade and yield of marketable tubers.

Physiological disorders are important whether the potato crop is intended for the fresh market or for processing, but economic losses are usually greater when tubers do not meet fresh-market grade standards. Tubers not acceptable for the fresh market are often acceptable for processing, although the use of low-quality tubers results in an increase in sorting and trimming waste and an overall increase in processing costs. Disposal of processing wastes is also a concern. Severely affected tubers are usually used to make dehydrated products or starch or are used for livestock feed—alternatives that provide minimal returns—and in extreme cases tubers must be discarded. Because of the economic losses associated with physiological disorders, cultural practices that minimize the occurrence of these disorders play an important role in any holistic potato health management plan.

Causes of Physiological Disorders

Physiological disorders are often referred to as **abiotic**, **nonpathogenic**, **nonparasitic**, or **noninfectious** diseases. The significant characteristic of all physiological disorders is that they are *not* caused by pathogenic organisms and are not infectious—that is, they are not transmitted from affected plants to healthy plants. Physiological disorders result from internal imbalances that alter or disrupt the normal growth and development of plants and tubers. Some of these disorders may be brought on or accentuated by environmental factors or disease.

Most tuber physiological disorders develop slowly, and symptoms may be observable only late in the crop growth cycle. In many cases, the damage has occurred by the time symptoms are observed. This often makes it difficult to correctly identify the cause of a disorder and to determine when the damage-causing event occurred. Disorders that affect plants directly, such as those resulting from hail, lightning, and freezing, are associated with easily identified causal factors. Other disorders, however, such as hollow heart, are the result of a complex interaction of environmental and cultural factors. To complicate matters even more, many of these disorders occur erratically from year to year and from location to location.

Tuber Physiological Disorders

Tuber physiological disorders are classed as external disorders, internal disorders, and physical damage. **External disorders** are visible on the exterior of affected tubers and are easily observed. **Internal disorders** can be detected only by cutting into tubers and examining internal tissues. Bruises are **physical damage** to tuber tissue resulting from impact. Some bruises are not associated with readily observable abrasions on the tuber surface and are visible only after cutting or peeling.

External Tuber Disorders

Tuber Malformation

Tuber malformation includes deformities known as bottleneck, dumbbell, pointed-end, and knobby tuber (Fig. 23.1). Affected tubers are also referred to as misshapen tubers.

Bottleneck, dumbbell, and pointed-end tubers result from irregular longitudinal growth, characterized by a constriction along the long axis of the tuber. **Bottleneck tubers** are smaller at the stem end than at the bud end, but many variations exist. **Dumbbell tubers** have a constriction in the middle. **Pointed-end tubers** have a definite constriction at either the stem end or (more commonly) the bud end. The reducing sugar content is often higher in the pointed end, and translucent end may develop.

Knobby tubers result from secondary lateral growth in one or more eyes after a cessation or inhibition of normal tuber growth. Knobs of various sizes and shapes are formed, depending on the stage of tuber development in which the disorder begins. The plant and the interior of the tuber usually have no symptoms.

The type and severity of deformity are influenced by the stage of tuber development during which growth is interrupted, the size of the tuber, and the severity of the associated stress factors. An interruption in growth early in tuber development may result in bottleneck tubers or tubers pointed at the stem end. An interruption during midseason results in dumbbell-shaped tubers. An interruption later in tuber development results in tubers pointed at the bud end.

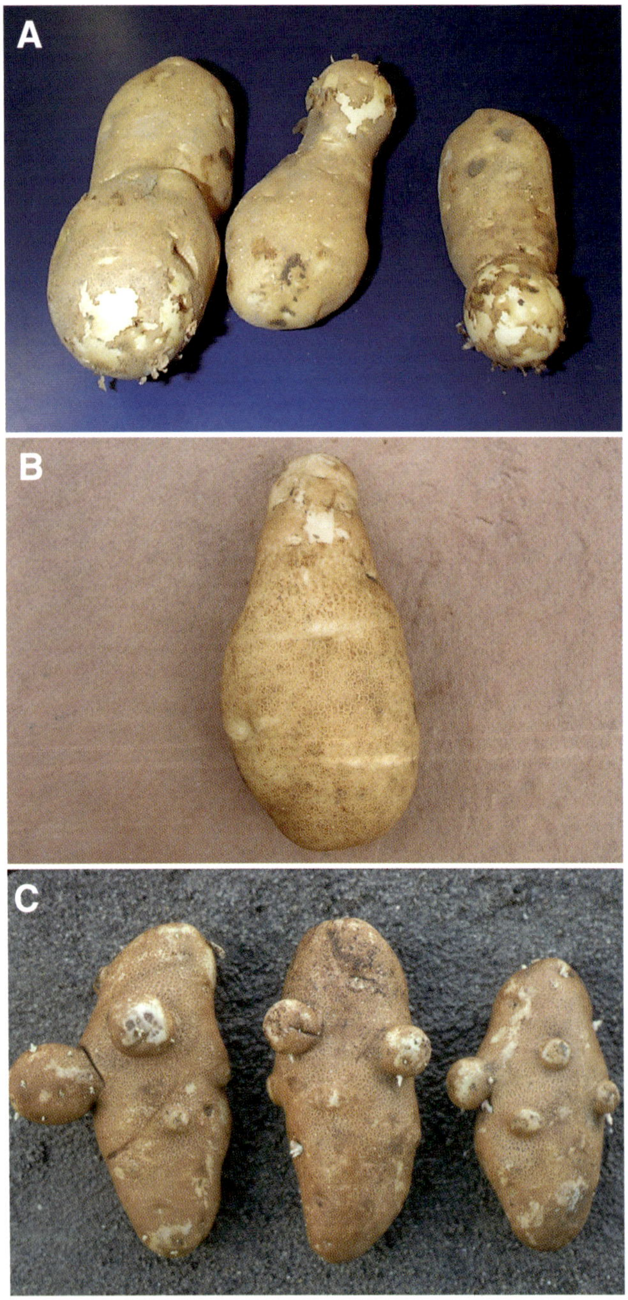

Fig. 23.1. A, Dumbbell malformation. **B,** Pointed-end malformation. **C,** Knobby tuber. (**A, C** courtesy N. Olsen; **B** courtesy S. B. Johnson)

Tuber malformation is often associated with periods of moisture or temperature stress that retard or stop tuber growth. Stress periods are usually followed by more favorable growing conditions and consequently a resumption of rapid tuber growth. Sudden or severe stress generally leads to a higher incidence of malformation than gradually increasing or continuous stress. Stress caused by inadequate soil moisture is apparently not the only factor resulting in malformation, since the disorder has been reported to occur under optimal irrigation regimes. High temperatures induce malformation by causing a temporary reduction in the tuber growth rate, an increase in plant and tuber respiration rates, a decrease in cell division, a deviation from normal plant hormone levels, and a temporary reduction in the supply of carbohydrates to the tuber. With the return of more favorable conditions, tuber growth and starch synthesis resume only in certain areas of the tuber, commonly next to buds, where cell division is more active.

Other factors contributing to the occurrence and severity of tuber malformation include the number of stems per plant (single-stem plants have fewer tubers per hill), poor plant stand, low tuber set per hill as a result of *Rhizoctonia* infection, the size and physiological age of seed tubers and pieces, improper soil fertility, and excessive vine growth. Certain management practices that result in rapid growth of foliage and tubers, such as excessive irrigation and nitrogen fertilization, after a stress period, increase tuber malformation.

Cultivars differ in susceptibility to tuber malformation, but no cultivar is completely resistant. Cultivars that form long tubers, such as Russet Burbank, tend to be more susceptible than those with round or blocky tubers.

To limit the incidence and severity of tuber malformation, establish a uniform stand of multiple-stem plants, and implement management practices to provide appropriate irrigation and fertility for uniform growth.

Second Growth

Three distinct types of second growth occur in potatoes: heat sprouts (also called aerial sprouts or heat runners), tuber chaining (also called chain tuberization or gemmation), and secondary tuber (also called little tuber, little potato, kinder tuber, or sprout tuber). **Heat sprouts** are produced from stolons or tubers and grow as new sprouts or develop into leafy aboveground stems. **Tuber chaining** is the formation of a series of small tubers on a single stolon. **Secondary tuber** is the simultaneous development of small daughter tubers and sprouts. The daughter tubers are formed on extremely short sprouts or directly on the seed piece (Fig. 23.2).

Heat sprouts and **tuber chaining** result from renewed growth after a period of interrupted or irregular development during normal tuber growth. These disorders can be initiated by soil temperatures above 75°F. Additional second growth can occur at temperatures below 75°F, depending on the length of previous exposure to temperatures above 75°F. Inadequate soil moisture alone does not result in the initiation of these disorders. During the growing season, however, high temperatures are often accompanied by moisture stress, and these two factors interact to limit the tuber growth rate, thus causing second growth.

Secondary tuber is commonly associated with physiologically old seed tubers. This disorder can develop in storage, in the field before plant emergence (generally when seed has been held in warm storage and then planted in low-temperature soil), or when very limited aboveground growth has occurred. The incidence of second growth is related to an alteration in the apical dominance of the primary tuber, a temporary deviation in the tuberization process, changes in the amounts and proportions of various growth hormones, and elevated concentrations of carbon dioxide and ethylene in the tuber zone.

Cultural practices that **promote uniform growth** of plants and tubers throughout the season help minimize second growth. Seed potatoes should be stored at 38–40°F until they are warmed for handling and planting. If physiologically old tubers, either from long-term or high-temperature storage, must be used, they should not be planted in cold, dry soil. To minimize the occurrence of these disorders, ensure uniform spacing for the desired plant population, apply fertilizer in the correct amount and with the correct placement, and maintain uniform soil moisture sufficient to meet the water requirements of the crop.

Tuber Cracking

There are three distinct kinds of physiological tuber cracking: growth cracks, thumbnail cracks, and air checks.

Growth cracks are shallow to moderately deep fissures in the surface tissues of tubers (Fig. 23.3). They occur more often on the bud end and usually extend lengthwise in long tubers. Growth cracks are usually associated with uneven availability of soil moisture and rapid water uptake resulting in large changes in tuber turgidity and growth rate. These conditions occur following rain or irrigation after a period of dryness or high-temperature stress. Other contributing factors include plant spacing and fertilizer rate, placement, and application timing. An inadequate amount of boron in the soil is also a contributing factor.

Growth cracks may also develop when plants are infected with *Potato mop-top virus* or *Potato spindle tuber viroid* (Chapter 17) or have been damaged by exposure to certain herbicides (Figs. 13.19 and 13.20). Growth cracks that result from disease or herbicide damage are difficult to distinguish from those due to physiological factors, and identification of their causes requires observation of foliar symptoms and sometimes laboratory diagnosis.

Fig. 23.2. Secondary tuber, a form of second growth, with small daughter tubers formed directly from the seed piece. (Courtesy D. A. Johnson)

Cultural practices that maintain a constant growth rate of plants and tubers during the entire growing season are required to minimize physiological growth cracking. These practices include obtaining the desired plant population and the appropriate number of stems per hill, providing uniform irrigation to achieve even distribution of soil moisture, and applying fertilizer at the correct rate and with the correct placement and timing.

Thumbnail cracking and **air checks** are shallow, random breaks in the skin of extremely turgid tubers that have been exposed to drying conditions. Similar shallow, irregular splits in the tuber surface may occur from physical impact during harvest and are often related to high tuber hydration and low temperatures. This surface splitting should not be confused with shatter bruise, which is an impact injury. These surface cracks usually heal slowly, especially in a low-humidity atmosphere, and thus affected tubers are subject to dehydration and tuber rot. Similar-appearing cracks that are not related to physical impact may develop in storage.

Greening

Exposure of tubers to sunlight, indirect daylight, or artificial light enhances chlorophyll development and results in a green color beneath the skin. Even light of relatively low intensity induces greening. The higher the intensity of the light and the longer the exposure, the more intense the color. Greening occurs more rapidly at temperatures above 55°F. All cultivars are prone to greening, but white-skinned cultivars and tubers with poorly developed skins are the most susceptible.

Greening occurs in the field in tubers that are located near the soil surface and are thus exposed to sunlight. Parts of tubers may become exposed if seed pieces have been planted too shallow or if hills have been improperly formed or eroded by wind, irrigation, or heavy rain. Some cultivars naturally set tubers near the soil surface, which increases the possibility of exposure to sunlight. Greening may develop late in the season prior to harvest, when the vines die and no longer shade the soil surface. It can also occur as a result of exposure of tubers to light after harvest.

Greening is reduced by planting seed tubers or pieces at a sufficient depth and forming hills of the proper shape and size to keep the tubers covered with soil.

Fig. 23.3. Severe growth crack. (Courtesy D. A. Johnson)

Sunscald

Sunscald is caused by frequent or prolonged exposure of tubers to intense sunlight and high temperatures. The tuber surface takes on a blistered, metallic appearance. The underlying tissues may become watery and turn brown. Affected areas may dry out and may be invaded by tuber-rotting organisms. Sunscald can be reduced or prevented by the practices recommended for preventing greening. In addition, tubers should be covered during transport, to avoid exposure to sunlight and wind.

Skinning

Immature tubers have a thin, delicate skin, which can easily be torn or frayed during harvest and handling. Skinning—often referred to as feathering—can also occur in mature tubers if they are roughly handled. Early-harvested potatoes are especially susceptible, particularly if the vines have not died or have not been killed long enough prior to harvest for the skins to have set properly. The bud end of tubers is more prone to skinning than the stem end. Wound healing of skinned areas proceeds quickly under conditions favorable for suberization, and the new tissue is similar to the original skin in its resistance to water loss and invasion by pathogens. However, if tubers are subjected to high temperatures, low humidity, and drying air (conditions imposed during transport after harvest), tissue exposed by skinning dehydrates easily and is subject to sunscald.

Unless tubers will be marketed immediately after harvest, they should not be harvested and handled until the skin has set, if external appearance is a significant market factor. Fertilizer rate and application timing should be managed according to the growing season and the anticipated harvest date to achieve skin set by harvest time. If vine desiccation or removal is necessary, a period of 1–3 weeks is needed between treatment and harvest, to allow the skin to toughen. During harvest and handling, equipment must be operated with chain speeds adjusted to minimize mechanical impact (Chapter 10). Cover harvested tubers in the field and during transport to protect them from heat, sunlight, and air movement.

Enlarged Lenticels

Lenticels are small, inconspicuous pores on the tuber surface. If tubers have an insufficient supply of oxygen, as in waterlogged or very compacted dry soil, lenticels become enlarged and prominent (Fig. 23.4). Enlarged lenticels may resemble scab lesions but are smaller and lighter in color. They detract from the appearance of tubers and provide entrance sites for rot-causing pathogens. Freshly harvested tubers that remain wet for any length of time can be sufficiently deprived of oxygen that their lenticels become enlarged, so that the risk of tuber soft rot increases (Chapter 18).

Enlarged lenticels detract from the appearance of tubers and provide entrance sites for rot-causing pathogens.

Practices that minimize the development of enlarged lenticels include providing good soil drainage, especially in low areas of fields where soil remains wet for extended periods; avoiding excess irrigation during the growing season; removing free surface moisture from tubers placed in storage by providing adequate ventilation; and preventing condensation of free moisture on surfaces of stored tubers (Chapter 10). Areas of fields where tubers have a significant amount of enlarged lenticels should not be harvested, but if they are harvested, tubers from these areas should be kept separate and marketed immediately.

Freezing and Chilling Injury

Chilling damage to tubers occurs at temperatures below 38°F. The symptoms are somewhat variable, depending on the temperature, exposure time, and cultivar. The surface of damaged tubers may appear wrinkled and flabby. Internal tissue of chilled tubers may turn bluish gray to black, or diffuse smoky gray areas may develop. Blackening of the vascular ring may also occur, and this may closely resemble net necrosis, caused by *Potato leafroll virus* (Fig. 17.2). Tubers that have been exposed to temperatures below 29°F become soft and watery. Frozen tissues tend to disintegrate and eventually dry out.

Tubers for commercial and seed use should be protected from temperatures below 38°F both in storage and during shipping. Harvesting operations should not proceed if tuber pulp temperatures are below 45°F (Chapter 10).

Internal Tuber Disorders

Brown Center and Hollow Heart

Brown center (also called incipient hollow heart, brown heart, or sugar center) and hollow heart are considered two different phases of the same internal disorder. They are thought to be initiated by the same conditions. Even though they are related, the two disorders often occur separately. Tubers that have brown center do not always develop hollow heart, and hollow heart may occur without being preceded by

Fig. 23.4. Enlarged lenticels on a tuber, resulting from exposure to excessively wet soil. (Courtesy D. A. Johnson)

brown center. Brown center is initiated when tubers are very small, the most susceptible period being from tuber initiation (late Growth Stage III) to the time when tubers weigh about 2 ounces (early Growth Stage IV). Hollow heart also occurs in tubers this small.

The first symptom of **brown center** is light brown pith tissue near the center of the tuber (Fig. 23.5). The brown area can vary in size and in intensity of color. Severe changes occur in tissues that turn brown, eventually resulting in rupture and death of affected cells. If the tuber grows slowly and uniformly after the discoloration develops, the dead cells may be spread apart by the growth of living cells interspersed among them, and the brown color may dissipate. However, if the tuber grows rapidly, the affected tissue may split apart, so that an internal cavity forms, and thus the disorder develops into hollow heart.

Hollow heart is characterized by a longitudinal or transverse, irregularly shaped cavity in the flesh of the tuber (Fig. 23.6). Such cavities can range from tiny openings to pockets nearly the size of the entire pith tissue. They form in the center or near the stem end or bud end of the tuber, depending on when the disorder is initiated. Cavities occurring closer to the stem end are invariably preceded by brown center and develop early in the season. This form of the disorder is called stem-end hollow heart. It is common in the cultivar Russet Burbank. Cavities occurring closer to the apical end form later in the growing season and usually are not preceded by brown center. This late-season disorder is called bud-end hollow heart. The cultivars Norgold Russet, Lemhi Russet, and Atlantic are noted for this form of the disorder. Bud-end hollow heart can also develop in Russet Burbank but is not as common as the stem-end disorder in that cultivar.

The exact causes of brown center and hollow heart are not known, but important factors in the initiation and development of these disorders are air and soil temperatures, soil moisture, tuber size, and growth rate. Other factors that may influence the incidence of hollow heart are soil fertility, fertilizer application method and timing, plant density and uniformity, planting date, the number of stems per seed piece, seed piece size, and seed quality, including physiological age.

Fig. 23.5. Brown center. (Courtesy N. Olsen)

One theory to explain the development of hollow heart is that plant stresses during or shortly after tuber initiation cause potato plants to take water, minerals, and carbohydrates from the cells of young, growing tubers, so that some tuber cells die. When the stresses are relieved and favorable growing conditions return, the rapid enlargement of the tubers causes the injured cells to separate, and a cavity is formed. A second theory suggests that hollow heart results from rapid tuber enlargement without any associated stresses. In either case, the disorder is more likely to develop in rapidly growing tubers.

Temperature is a factor in the initiation of brown center in Russet Burbank. A period of cool weather resulting in soil temperatures below 55°F for 5–7 days during tuber initiation and early bulking is sufficient to incite this disorder.

A second factor is the amount of **soil moisture** during tuber initiation. Maintaining more than 80–85% available soil water may result in an increase in both brown center and stem-end hollow heart (Chapter 9). The increase may be due to the slow warming of wet soils, which increases the likelihood of exposure of newly developed tubers to temperatures below 55°F. Maintaining 70–80% available soil water during cool periods when tuber initiation is occurring may decrease the incidence of brown center and hollow heart, but it does not entirely eliminate these disorders. Allowing fields to dry down to less than 65% available soil water should be avoided, as this may cause sufficient stress to induce second growth or tuber cracking. Excessively dry soil during tuber initiation limits total yield, increases the risk of common scab (Chapter 19), and results in lower tuber specific gravity. In the fall, when the crop is nearing maturity, very wet soil may increase the incidence of bud-end hollow heart. If possible, uniform soil moisture should be maintained throughout the growing season. If a soil water deficit does occur, a gradual return to a more favorable percentage of available soil water is preferable to a large, sudden addition of water.

A **late planting date** generally results in a lower incidence of brown center, since soils tend to warm as the season progresses. Late planting may also result in lower yield, but this can be offset by increased tuber quality. Any effect of the planting date, however, is closely associated with weather patterns in a given year. Planting seed pieces with brown center or hollow heart does not affect the incidence of these disorders in the subsequent crop. Uniform placement of seed tubers and pieces at the desired spacing, with few skips, and planting large or properly

Fig. 23.6. Hollow heart in a cut tuber. (Courtesy G. A. Secor)

aged seed pieces to obtain multiple-stem plants have been associated with lower incidence of hollow heart.

Fertility also can influence the incidence of brown center and hollow heart. Application of a small amount of nitrogen before planting and additional small amounts periodically throughout the growing season usually results in a lower incidence of hollow heart than a single preplant application supplying enough nitrogen for the entire season. Nitrogen applied at or shortly after tuber initiation may increase the tuber growth rate and thus favor the development of brown center and hollow heart (Chapter 8). The role of other nutrients in brown center and hollow heart is not well understood; however, maintaining a balance of nutrients to avoid plant stresses is important in managing these disorders.

A management program to limit brown center and hollow heart starts by selecting cultivars that are less susceptible to these disorders. Potato **cultivars** differ widely in their susceptibility to brown center and hollow heart. Norgold Russet, Lemhi Russet, Atlantic, and NorKing Russet are more susceptible than Russet Burbank. There are advanced breeding lines that are more resistant to hollow heart than Russet Burbank. Cultural practices that ensure **evenly spaced plants** with several stems provide adequate plant competition and are also helpful. If possible, choose a planting date to ensure that tuber initiation does not occur during a period of low soil temperature. Maintain uniform soil moisture and fertility, and strive to achieve uniform plant and tuber growth rates throughout the season.

Internal Brown Spot and Heat Necrosis

Symptoms of internal necrosis in tubers have been described by various names, of which the most common are *internal brown spot* and *heat necrosis*. The physiology of these two disorders is not completely known, but they differ in being associated with different tissues and with symptoms occurring predominantly at different locations within the tuber. Symptoms of both disorders can be confused with those of the virus disease corky ringspot (Figs. 17.4 and 17.5), which can be distinguished from the physiological disorders by a laboratory diagnostic test.

Internal brown spot is characterized by small, round or irregularly shaped, light tan, reddish brown, or rust-colored spots or blotches scattered throughout the tuber flesh (Fig. 23.7). The lesions can occur at any place within the tuber but are generally more common in tissues inside the vascular ring, predominantly those toward the bud end. Tubers affected with internal brown spot normally do not show external symptoms, although some reports have associated this disorder with malformed or large tubers or a thickening or abnormal russeting of the skin.

Fig. 23.7. Internal brown spot. In the tuber on the far left and the one at the top, the brown tissue appears to be more closely associated with the vascular ring; symptoms in this location are often confused with those of heat necrosis.

Internal brown spot may occur at various times throughout the growing season. It may develop soon after tuber initiation or during bulking in periods of rapid or uneven tuber growth. The symptoms tend to increase in severity during the growing season and can continue to intensify in storage. Tubers stored at temperatures above 60°F generally have a higher incidence of the disorder.

The cause or causes of internal brown spot are poorly understood. It has been suggested that the primary cause may be reabsorption of water or other tuber constituents to support vegetative development. The disorder has been associated with coarse-textured soils and with several climatic conditions, including hot, dry weather; high soil temperatures; and low or fluctuating soil moisture.

Calcium deficiency has also been suspected as a cause of internal brown spot, because calcium concentrations have been reported to be lower in tubers with this disorder than in unaffected tubers. However, internal brown spot lesions do not always develop in tubers with low levels of calcium. Deficiencies of other nutrients can also result in necrosis of cells in tuber flesh, but this may be due to their influence on the availability of calcium to affected tissues. Internal brown spot has not been consistently controlled by field applications of calcium, but the lack of success may be due to poor uptake and movement of calcium to the tubers, especially during the stress periods thought to induce the disorder.

It is difficult to develop a management program to minimize the incidence of internal brown spot, since the disorder and the factors associated with it are not well understood. Delayed planting has resulted in decreased incidence of internal brown spot, perhaps by reducing the proportion of large tubers (as a result of obtaining a uniform population of multiple-stem plants) and by shifting the critical developmental stage to a later part of the growing season. Irrigation and fertilization practices that promote even growth without periods of stress on the plants and tubers are helpful in lowering the incidence of the disorder.

Heat necrosis lesions are principally confined to the vascular ring tissue near the bud end, but they may form anywhere in the tuber. This disorder has been associated with external tuber deformations and with depressions in the skin and related cracking or rough areas. It is most common in tubers growing near the soil surface, where they are exposed to higher temperatures, especially after the vines begin to die. To lessen the chances that these conditions will occur, harvest should be timed so as to prevent exposure of tubers to high soil temperatures, harvested tubers should be protected from exposure to high temperatures, and tubers should be held under optimal storage conditions.

Translucent End

Translucent end (also called sugar end, incipient jelly end, or jelly end rot) affects the stem end of tubers (Fig. 23.8). The affected tissue has a lower dry matter content and higher reducing sugar content than unaffected parts of the same tuber (Fig. 23.9 and Chapter 10). Cultivars with long tubers, such as Russet Burbank, are prone to this disorder. It is often associated with tubers having pointed stem ends.

The stem end of Russet Burbank tubers normally has a higher starch content than the bud end, but seasonal variations and growing conditions can influence the relative amounts of both starch and reducing sugars. Different theories have been proposed to explain the lower dry matter content in the stem end of tubers with translucent end: (1) carbohydrates move out of the stem-end tissues and into the plant for renewed vegetative growth after a period of stress, (2) carbohydrates move from the stem end to the bud end for development and maturation of the bud end after a period of stress, (3) starch in the tissues that become translucent is converted to sugars as a result of stress, (4) carbohydrates produced during photosynthesis do not move to the stem end of affected tubers, and (5) sugars produced by photosynthesis, when moved to the tubers, are not converted to starch.

Translucent end is most often associated with a water deficit in the plant (Chapter 9). This may result from low soil moisture, high air temperature, drying winds, or a high ratio of tops to tubers. It was once thought that translucent end occurred only near harvest. More recently, however, it has been shown that water stress during tuber initiation can result in an increase in the incidence of this disorder.

Cultural practices to minimize the occurrence of translucent end should promote good stands and maintain constant, uniform soil moisture and fertility to ensure uniform growing conditions and minimize moisture stress during periods of high temperature.

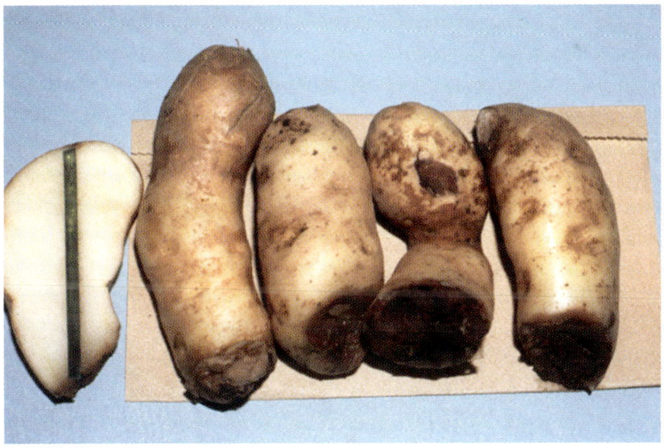

Fig. 23.8. Jelly end rot, a disorder usually associated with second growth and pointed stem end.

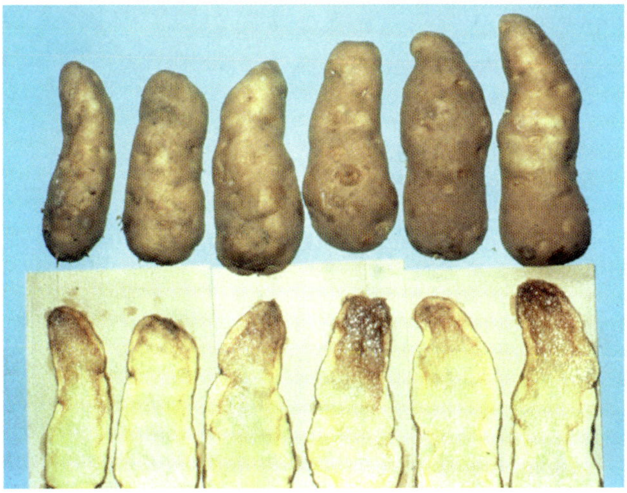

Fig. 23.9. Pointed tubers with translucent end. The darker fry color at the stem end (bottom row) results from an accumulation of reducing sugars in affected areas.

Vascular Discoloration

Discoloration of varying intensity may develop in the vascular system at the stem end of tubers. This discoloration is generally confined to the area within ½ inch of the point of stolon attachment, but in very severe cases it may extend throughout the entire vascular ring.

Vascular discoloration appears as speckled, light tan to reddish brown or even dark chocolate-colored streaks. Immature tubers are susceptible, but the disorder occurs more often in tubers that are nearing maturity. The discoloration can be found in tubers in the field. It can also be found, during the first 1–2 months in storage, in tubers that did not exhibit field symptoms but were exposed to the causal factors in the field. This physiological vascular discoloration is sometimes confused with vascular net necrosis, caused by *Potato leafroll virus* (Fig. 17.2), or with vascular discoloration related to *Verticillium* infection.

Physiological vascular discoloration is generally associated with low soil moisture and rapid death of green vines due to vine desiccants or other chemicals, frost, or mechanical removal, when accompanied by moisture stress. High-temperature stress at the time of vine killing generally increases the severity of the discoloration. Irrigation prior to vine killing decreases the amount of stem-end discoloration, regardless of the method of vine killing.

Blackheart

Internal blackheart occurs as the result of a deficiency of oxygen available to the tuber for respiration. It is relatively easy to diagnose: tissue in the center of the tuber is black to blue black, and the discolored area is irregular in shape and usually has a distinct border (Fig. 23.10). Sometimes a cavity

Fig. 23.10. Blackheart, with a distinct border at the edge of the discolored area. (Courtesy M. J. Pavek)

forms in the center of the tuber. The darkened areas are usually fairly firm, in contrast to those affected by Pythium leak, which are spongy and watery (Chapter 19). There is no odor associated with blackheart.

Blackheart develops when tubers are held in a low-oxygen environment, such as a poorly aerated storage, truck, or railcar; when gas diffusion through the tubers is slowed because of very low temperatures; or when respiration is elevated because of very high temperatures. This condition can also develop in the field when the soil is flooded. Proper soil drainage and good aeration in storage and transit prevent this disorder.

Bruise Damage

Three different types of bruise damage to potato tubers are distinguishable: blackspot, shatter bruise, and pressure bruise. Blackspot and shatter bruise both result from a bruising force or impact. Pressure bruise is often associated with tuber dehydration and occurs during storage as a result of the pressure of tubers being forced against each other or against the storage structure.

Blackspot is an internal discoloration that begins to appear just under the skin 6–8 hours after bruising has occurred. The discoloration intensifies over time to a bluish gray to black. The discolored tissue is variable in size and shape and occurs at varying depths in the tuber flesh (Fig. 23.11). Bruised areas are visible only if the outer layer of skin is removed. The stem end of the tuber is usually more susceptible to blackspot than the bud end. A bruising force sufficient to rupture cells is required, and thus blackspot can occur whenever tubers are handled.

Shatter bruise is characterized by cracking of the outer surface of the tuber. The cracks often extend into the tuber flesh, appearing as small multiple fissures or splits. Cells of damaged tissue around the margins of the fissures may discolor and appear similar to blackspot bruises. Unlike blackspot, shatter bruises may penetrate deeply into the tuber flesh (Fig. 23.11). They may not be readily observed at the time of damage without special diagnostic techniques, but they usually become evident after the affected tissue has become slightly dehydrated and the border tissue has shrunk.

Fig. 23.11. Blackspot in tissue below the skin of bruised areas. Shattered tissue, damaged by a crushing impact, is shown extending deep into the tuber on the left. (Courtesy R. C. Rowe)

Pressure bruises are flattened or slightly sunken areas that develop in storage at points of contact between tubers or at points where tubers are pressed against the storage structure or equipment. The bruise results from dehydration of the affected area when external pressure causes a shift in the distribution of water within the tuber. Pressure bruises vary in size and shape, depending on the extent of the contact and the amount of pressure. The problem increases if the relative humidity in the storage is allowed to drop below 90%. Slight bruises tend to disappear, but severe ones persist, and severely bruised tubers may be more susceptible to blackspot when handled. Certain lots of tubers may be more susceptible to pressure bruise than others, because of preharvest and storage factors that are not well understood.

Tubers must be in a physiologically susceptible condition and a bruising force sufficient to rupture cells must be applied for blackspot or shatter bruise to occur. Tuber turgidity (amount of hydration) and pulp temperature are two very important factors influencing the susceptibility of tubers to blackspot and shatter bruise. In general, susceptibility to shatter bruise increases as tuber hydration increases, and susceptibility to blackspot increases as tuber hydration decreases (Fig. 10.6). If tubers are extremely dehydrated, however, they become less susceptible to blackspot. As the tuber pulp temperature decreases, especially below 45°F, total bruise damage (blackspot plus shatter bruise) increases (Fig. 10.6). Shatter bruise is especially common in cold, crisp tubers. The optimal tuber hydration for minimal total bruise damage shifts somewhat with pulp temperature (Fig. 23.12). Other factors that influence susceptibility to blackspot and shatter bruise include tuber maturity, planting date, soil moisture, potassium fertilization, and vine condition just prior to harvest. All of these factors are related in some way to tuber maturity and condition. In addition, cultivars differ in susceptibility to these disorders.

Reduction of total bruise damage is a critical step in maintaining high tuber quality. Bruise reduction requires an integrated management program beginning at planting and continuing until the potatoes reach the consumer or proces-

Fig. 23.12. Effect of tuber pulp temperature and hydration on susceptibility to blackspot and shatter bruise. (Reprinted, by permission from The Potato Association of America, from D. A. Smittle, R. E. Thornton, C. L. Peterson, and B. B. Dean, 1974, Harvesting potatoes with minimum damage, American Potato Journal 51:152–164)

sor. Since cultivars vary in susceptibility to bruise, selection of a less susceptible cultivar is desirable, provided it is acceptable to the intended market. It is also important to provide adequate amounts of nutrients, especially nitrogen and potassium, and provide soil moisture and pest management favoring the development of healthy vines and good root growth until harvest. Close attention must be paid to soil moisture, tuber hydration, and pulp temperatures at harvest. Careful management of all handling procedures during harvest, storage, and post-storage operations is critical to minimizing bruising (Chapter 10). Since pressure bruising is a response to external pressure influenced by tuber hydration, the pile height may need to be reduced when tubers have been exposed to field conditions that result in tuber dehydration.

Plant Physiological Disorders

Various physiological disorders affect the appearance and function of potato foliage. These include damage from hail, lightning, and air pollution; exposure to chemicals, water, wind, and sun; and nutritional imbalances. Damage to foliage influences photosynthetic and respiratory activity, water relations, and other physiological processes vital to healthy plant growth and development. Because the growth of foliage is inherently related to the growth of tubers, foliar disorders and tuber disorders are often interconnected. Competition between foliage and tubers for water and nutrients can influence the occurrence and severity of many tuber disorders.

Hail Damage

Hail damage to potato plants during summer storms can range from leaf and stem scarring to partial defoliation to complete mutilation of leaves and stems. Losses are related to the amount of damage sustained, the growth stage of the crop when the damage occurs, the cultivar, cultural practices, and weather after the damage.

Slight leaf damage at any stage of growth will not affect yield, but the loss of more than 25% of the foliage can result in lower total yield and marketable yield. Increasing amounts of foliar damage will result in lower yields of marketable tubers, higher numbers of small and malformed tubers, and lower tuber specific gravity. Moderate hail damage sustained during Growth Stage II generally has little effect on yield or tuber grade. Stem damage deep enough to cut vascular tissues can limit yield, especially if it occurs before, during, or just after bloom (Growth Stage III). The greatest losses are inflicted when hail damage occurs early in Growth Stage IV, the 2- to 4-week period after bloom. Besides limiting yield, hail damage during this period results in a decrease in tuber starch content, proportional to the amount of foliage removed. Potato plants have a remarkable capacity for recovery from both leaf and stem damage, but partial loss of foliage at any stage can delay the maturity of the crop, a factor that must be considered in planning for harvest and storage.

Weather and soil fertility influence the degree to which plants will recover from hail damage. If the damage is moderate, favorable growing conditions after a hailstorm and management to provide appropriate soil moisture and fertility promote rapid recovery, without any major effect on normal growth. Damaged tissues normally heal quickly, and rapid healing can minimize the development of bacterial soft rot, which is often associated with hail damage.

Lightning Damage

Symptoms resulting from a lightning strike in a potato field normally appear within 2–24 hours as a well-defined circular or elliptical area of injured or dead plants (Fig. 23.13). The damage is relatively uniform in the interior of the affected area and diminishes near the margins. Occasionally, scattered individual plants or small groups of plants with injury symptoms are intermingled with less severely affected or unaffected plants. Areas of damage often follow water drainage patterns in the field and often occur in the lowest spots. Variations in the amount of soil moisture and differences in the intensity of the electrical discharge influence the injury pattern and distribution within a field. Because the affected area is highly visible and localized, inexperienced observers may assume that a disease is spreading from a point source.

Stems of lightning-damaged plants are soft, water-soaked, and discolored brown to black. They often collapse at the top and generally wilt irreversibly. Damaged tissues dry quickly and turn brown to tan, with surface layers appearing light tan to almost white. A characteristic sign of lightning damage is that the stem pith tissues collapse and form horizontal plates, appearing crosshatched or ladderlike. This can be observed by longitudinally slicing the stems of affected plants.

Tubers may initially appear unaffected, but dark brown to black necrotic spots and some skin cracking soon develop. Severely injured tubers appear cooked, with extensive collapse of the central pith and less damage to the outer cortical tissues. Breakdown and decay of the pith generally continue, often leaving a hole in the center of the tuber. Damaged tubers are highly susceptible to secondary pathogens and generally decay prior to harvest.

Tipburn

Tipburn is a physiological disorder in which the tips and margins of potato leaves and leaflets turn yellow and then gradually die, roll upward, turn brown or black, and become

Fig. 23.13. Dead foliage at the site of a lightning strike. (Courtesy D. A. Johnson)

Fig. 23.14. Tipburn, a climate-induced disorder, on actively growing leaves. (Courtesy R. C. Rowe)

brittle (Fig. 23.14). This is a weather-induced disorder, resulting from excessive moisture loss from leaves under hot, dry, windy conditions following a period of cool, moist weather. Under these conditions, the amount of water transpired by the leaves exceeds that taken up by the roots.

Tipburn is often mistakenly identified as late blight (Chapter 21), but tipburn lesions are uniformly brown and have very abrupt margins, rather than grading from brownish black to yellow to green, which is typical of late blight lesions. Furthermore, late blight lesions are located on the sides or centers of leaves as well as the tips. Sucking insects, such as leafhoppers and aphids, may accentuate tipburn. Root pruning during cultivation can be a contributing factor. Tipburn can be reduced by measures that conserve soil moisture; irrigation during hot, dry periods; and care in cultivation, to avoid root damage.

Air Pollution Damage

Potato foliage can be damaged by exposure to one or more atmospheric pollutants, such as the photochemical oxidants ozone and peroxyacetyl nitrate (PAN). The damage often occurs during temperature inversions, when warm, stagnant air is held close to the ground for several days. Oxides of sulfur may contribute to the problem in areas downwind from a power plant or smelter. The symptoms vary, depending on the concentration of pollutants in the air and the duration of exposure. The reported tolerance and sensitivity of cultivars vary from one geographic area to another.

Ozone damage usually results in a condition known as speckle leaf, a chlorotic or necrotic speckling of the upper leaf surface. Severe symptoms include more pronounced chlorosis, bronzing, and necrosis of leaves, resulting in early maturity of affected plants. PAN produces a silvering or glazing of the lower surface of affected leaves, with the most severe symptoms on young, fully expanded leaves. Damage to tubers has not been reported to result from exposure to air pollutants; however, lower yields are associated with premature senescence caused by pollution.

The only control for air pollution damage is to avoid growing potatoes in areas heavily polluted with ozone, PAN, or sulfur oxides. In marginally polluted areas where some damage regularly occurs, only tolerant cultivars should be planted.

Chemical Damage

Considerable damage can result from accidental or improper application of a wide range of agricultural chemicals. The symptoms and the severity of damage vary considerably, depending on the type of chemical involved, dosage, application method, growth stage, cultivar, and environmental factors. Various aberrations, distortions, chlorosis, and necrosis of foliage and external and internal symptoms, deformations, and stunting of tubers may result. The symptoms are usually more severe during a plant water deficit.

Growth-regulating chemicals, such as herbicides, can cause leaf distortions and chlorosis, which may be confused with plant symptoms resulting from infection by viruses. Tuber deformations, changes in skin color, and surface disfigurations may also occur. Chemical damage can result from herbicides applied directly to the potato crop or soilborne residues of herbicides used on preceding rotational crops (Chapters 13 and 22). Stem-end vascular discoloration of tubers may develop after application of preharvest vine defoliants, particularly if the plants are stressed (see Internal Tuber Disorders). Improper fertilizer application to foliage can result in leaf burn. Fertilizer damage to seed pieces can cause shoot and root burn followed by seed piece decay, resulting in reduced plant population and low plant vigor.

Proper diagnosis of chemically induced disorders is difficult and generally requires experience and laboratory analysis. Considerable caution must be exercised in diagnosing suspected chemical damage. Observe and record the condition of not only the potato plants but also nearby weeds or other crops that may have been exposed. Gather all pertinent information on agrochemicals applied to the current crop and previous crops, including the rate and method of application and prevailing weather conditions. Consultation with an experienced diagnostician is usually necessary to determine the causes of chemical damage.

Water Damage

Potato plants and tubers may be damaged by excessively wet conditions or flooding that lasts for more than a few days. The damage occurs primarily as a consequence of oxygen deprivation in the root system. Enlarged lenticels normally develop in tubers under excessively wet conditions (Fig. 23.4). Relations between potato health and water levels in the soil and in the plant are discussed in Chapter 9.

Management Strategies

The development of physiological disorders in potato plants and tubers is influenced by unfavorable environmental factors (air and soil temperatures, rainfall, humidity, and wind), improper cultural practices, and inadequate or excessive soil moisture and fertility. Because climatic conditions cannot be manipulated, management of physiological disorders depends on the careful management of factors over which the grower does have control. These include cultivar selection; seed quality, size, and placement; planting date; fertility; cultivation; irrigation; and pest management. Although the exact causes of many physiological disorders are not known, the basic management objective for minimizing these disorders is to provide the best possible conditions for uniform growth of the crop throughout the growing season (Box 23.1).

Cultivars differ greatly in susceptibility to physiological disorders. Be aware of the limitations of the cultivars being grown, and implement appropriate management plans. If possible, cultivars highly susceptible to certain physiological disorders should be avoided. Planting should be timed to allow **rapid emergence** and avoid periods of low soil temperature during tuberization. To obtain rapid emergence, it is important to keep seed potatoes cool in storage, avoid excessively long storage periods, and not plant physiologically old whole tubers or cut seed pieces in cold, dry soil. The size and physiological age of seed tubers and seed pieces influence the number of stems and tubers per plant. The goal is to **establish a uniform stand of plants** with the desired number of stems per plant throughout the field. Uniform seed spacing and depth are important to ensure adequate plant competition and to control tuber size and growth rate. Different cultivars require different plant spacing, and local recommendations should be obtained. The desired planting depth varies with the cultivar, soil texture, and climate. Seed must be planted shallow enough to allow rapid emergence, but it must be planted deep enough that tubers can develop within the hill, to avoid greening and sunscald and protect them from infection by the early blight and late blight pathogens.

The basic management objective for minimizing physiological disorders is to provide the best possible conditions for uniform growth of the crop throughout the growing season.

During the growing season, management practices should **promote *uniform* and continuous plant and tuber growth**. As much as possible, soil moisture should be managed so that the soil is not excessively wet or dry at any time. If irrigation is available, a strict irrigation management plan should be followed to accomplish this goal (Chapter 9). In rain-fed production, proper drainage should be provided, and planting should be timed to minimize problems associated with excess soil moisture. Careful management of fertility is critical and is closely tied to water management. The goals are to **avoid nutrient deficiencies** at any growth stage, by providing for the continuous nutritional needs of the crop, and to avoid growth spurts resulting from the application of excess fertilizer, especially during tuberization and after periods of environmental stress (Chapter 8).

Tubers must be **protected from environmental stresses** as much as possible as they develop. Providing sufficient soil covering over developing tubers before row closure is important in minimizing tuber exposure and thus subsequent greening and sunscald. Vine-killing procedures and timing should be chosen so as to minimize the development of vascular discoloration and allow tuber skins to set properly prior to harvest. The reduction of harvest damage requires careful management to maintain tuber quality and marketability (Chapters 7 and 10). To **minimize bruising**, harvest and handling equipment must be used properly, and harvested tubers must be protected from exposure to high or low temperature, sunlight, and excess moisture. Timely movement of the crop to properly designed and managed storages is essential to maintain quality of the harvested tubers.

Minimizing the development of physiological disorders is an essential part of an overall potato health management plan. Although there is still much to be learned about the causes of physiological disorders, the effects of environmental and cultural variables, and the management of these disorders, they can be minimized by providing the best possible growing conditions for the crop.

Box 23.1

Management Tactics for Reducing the Incidence of Physiological Disorders

- Store seed tubers at 34–40°F until they are warmed for handling and planting.
- Select cultivars that are less susceptible to brown center and hollow heart.
- Warm seed tubers to 50–55°F before handling.
- Plant in warm soil to promote rapid plant emergence; 50–55°F is optimal.
- Plant seed pieces at a sufficient depth, and form hills to keep developing tubers covered with soil.
- Establish a uniform stand of multiple-stem plants.
- Provide irrigation and fertility regimes for uniform plant and tuber growth.
- Avoid plant stress due to either excessive or insufficient irrigation.
- Provide good soil drainage.
- Plan the harvest to prevent exposure of tubers to high soil temperatures.
- Do not harvest if tuber pulp temperatures are below 45°F.
- Reduce bruising during planting, harvest, handling, and storage.
- Cover tubers during transport to avoid exposure to sunlight and wind.
- Provide adequate ventilation in storage to remove free moisture from the surface of tubers.
- Provide good aeration throughout the storage period.

The basic management objective for minimizing physiological disorders is to provide the best possible conditions for uniform growth of the crop throughout the growing season.

Glossary

A acre
a.i. active ingredient
cm centimeter (1 cm = 0.01 m = 0.3937 in.)
F Fahrenheit
g gram (1 g = 0.03527 oz)
gal gallon (1 gal liquid [U.S.] = 3.785 L)
ha hectare (1 ha = 2.471 acres)
in. inch (1 in. = 2.540 cm)
kg kilogram (1 kg = 2.205 lb)
L liter (1 L = 1.057 quarts liquid [U.S.])
lb pound (1 lb = 453.59 g)
m meter (1 m = 39.37 in.)
mg milligram (1 mg = 0.001 g)
ml milliliter (1 ml = 0.001 L)
oz ounce (1 oz = 28.35 g); fluid ounce (1 fl oz [U.S.] = 29.57 ml)
ppm parts per million

abiotic characterized by the absence of life, as in diseases not caused by living organisms
acid having a pH less than 7
acute characterized by symptoms that develop suddenly (*see* **chronic**)
aerial occurring in the air
aerobic living only in the presence of oxygen
agar a jellylike material derived from algae and used to solidify liquid culture media; a medium containing this material
agroecosystem the interacting, functional system comprising all living organisms in an agricultural area and their nonliving environment
alkaline having basic (nonacidic) properties; having a pH greater than 7
alkaloid any of various nitrogen-containing ring compounds produced by plants and causing physiological effects in animals
anaerobic living in the absence of oxygen
annual a plant that completes its life cycle and dies within 1 year (*see* **biennial**; **perennial**)
antagonist an organism or substance that counteracts or limits the action of another
anterior situated toward the front or head
antibiotic a chemical compound produced by a microorganism and killing or inhibiting the growth of other living organisms
apex (pl. **apices**; adj. **apical**) the tip of a root or shoot, containing the apical meristem
aphid any of various species of small, sucking insects of the family Aphididae (order Homoptera), producing honeydew and injuring plants when present in large populations
apothecium (pl. **apothecia**) the open, cuplike or saucerlike, ascus-bearing fruiting body (ascocarp) of certain fungi, often supported on a stalk
arthropod any member of the phylum Arthropoda, which consists of animals with articulated bodies and limbs, including insects, arachnids, and crustaceans
ascospore a sexual spore borne in an ascus
ascus (pl. **asci**) the saclike structure containing ascospores (typically eight), usually borne in a fungal fruiting body
asexual vegetative; lacking sex organs, gametes, or sexual spores; pertaining to the imperfect or anamorphic stage of a fungus
asexual reproduction reproduction not involving the union of gametes and meiosis
axil the angle formed by a leaf petiole and the stem
axillary pertaining to or placed within an axil

bactericide a chemical or physical agent that kills bacteria
bacterium (pl. **bacteria**) any of various prokaryotic, microscopic, single-celled organisms having a cell wall and increasing by binary fission
biennial a plant that produces seed and dies at the end of its second year of growth (*see* **annual**; **perennial**)
bioassay any test (assay) using a living organism
biological control exploitation of the natural competition, parasitism, or antagonism of organisms for the management of pests and pathogens (syn. **biocontrol**)
biotype a subdivision of a species, subspecies, or race characterized by some identifiable physiological trait, such as a specific virulence pattern
blight sudden, severe, and extensive spotting, discoloration, wilting, or destruction of leaves, flowers, stems, or entire plants
breeding line a plant strain used in a breeding program, usually having one or more desirable agronomic or breeding characteristics
broadcast application application of fertilizer by spreading or scattering on the soil surface

calcareous rich in calcium carbonate (lime)
callus specialized plant tissues that form over a wound or cut (cork cambium may form, and the cells produced will gradually seal the wound)
canopy the expanded leafy top of a plant or plants
carbohydrate any of various chemical compounds composed of carbon, hydrogen, and oxygen, such as sugars, starches, and cellulose
causal agent the organism or agent that produces a given disease

chlamydospore a thick-walled or double-walled, asexual resting spore formed from hyphal cells (terminal or intercalary) or by transformation of conidial cells and functioning as an overwintering stage
chlorophyll (adj. **chlorophyllous**) any of a group of green pigments found in chloroplasts and important in photosynthesis
chlorosis (adj. **chlorotic**) failure of chlorophyll development, caused by disease or a nutritional disturbance; fading of green plant color to light green, yellow, or white
chronic characterized by slow-developing, persistent, or recurring symptoms (*see* **acute**)
circulative pertaining to viruses that pass through the gut and circulate in the body of an insect vector before being transmitted to a host
circulative transmission virus transmission in which the virus must accumulate within or pass through the lymphatic system of an insect vector before it can be transmitted to a plant
coalesce to grow together into one body or spot
colonization establishment and ramification of a pathogen within a host plant
colonize to infect and ramify through plant tissue in the course of the growth of a pathogen
colony growth of a microorganism in mass, especially as a pure culture
conidiophore a simple or branched hypha on which conidia are produced
conidium (pl. **conidia**) an asexual, nonmotile fungal spore that develops externally or is liberated from the cell that formed it
contact pesticide a pesticide that kills on contact
cortex (adj. **cortical**) the region of parenchyma tissue between the epidermis and the phloem in stems and roots; the region beneath the rind of a sclerotium
cotyledon the seed leaf, or primary embryonic leaf within the seed, in which nutrients for the new plant are stored (monocots have one cotyledon, dicots have two)
cultivar (abbr. **cv.**) a cultivated variety of a plant species, resulting from deliberate genetic manipulation and having recognizable characteristics such as color; shape of flowers, fruit, or seeds; height; or form (syn. **variety**)
cultural practices horticultural management practices, such as application of nutrients, irrigation, and cultivation, sometimes used for disease management
cuticle (adj. **cuticular**) the noncellular outer layer of an insect or a nematode; the water-repellent, waxy layer of epidermal cells of plant parts, such as leaves, stems, and fruit
cv. *see* **cultivar**
cyst in fungi, a resting structure in a protective membrane or shell-like enclosure; in nematodes, the egg-laden carcass of a female; in bacteria, a specialized cell enclosed in a thick wall, often dormant and resistant to environmental conditions
cystosorus (pl. **cystosori**) a group of resting spores
cytoplasm the living protoplasm in a cell, exclusive of the nucleus

damping-off death of a seedling before or shortly after emergence due to decomposition of the root or lower stem, commonly distinguished into preemergence and postemergence damping-off
defoliation loss of leaves from a plant, whether normal or premature
desiccate to dry out
diagnostic (n. **diagnosis**) pertaining to a distinguishing characteristic important for the identification of a disease or other disorder
diapause a period of delayed development or growth accompanied by reduced metabolism and inactivity, especially in certain insects
dieback (v. **die back**) progressive death of shoots, leaves, or roots, beginning at the tips
disease abnormal functioning of an organism
disease cycle the succession of all events and interactions among a host, a parasite, and the environment in the process of disease, from initial infection of the host by a causal agent, through pathogenesis, to overseasoning, until another infection occurs
disinfect to kill harmful microbes, such as bacteria and viruses
disinfectant a chemical or physical agent that kills pathogenic microbes and viruses on the surface of plants, plant parts, and physical objects
disinfest to kill pathogens that have not yet initiated disease or other contaminating microorganisms that occur in or on inanimate objects, such as soil or tools, or on the surface of plant parts, such as seed
dispersal *see* **dissemination**
dissemination the spread of infectious material (inoculum) from diseased to healthy plants (syn. **dispersal**)
DNA deoxyribonucleic acid, the double-stranded helical molecule that contains genetic code information, consisting of repeating units, or nucleotides, each of which is composed of deoxyribose (a sugar), a phosphate group, and a purine (adenine or guanine) or a pyrimidine (thymine or cytosine) base
dormancy (adj. **dormant**) a condition of suspended growth and reduced metabolism of an organism, generally induced by internal factors or environmental conditions as a mechanism of survival
dorsal situated toward the back or top (as opposed to ventral)
dwarfing underdevelopment of a plant or plant organs, which may be caused by disease, inadequate nutrition, or unfavorable environmental conditions

ecology the study of the interactions between individual organisms, between groups of organisms, and between organisms and their environment
ecosystem a community of organisms and their environment; the interacting, functional system comprising all living organisms in an area and their nonliving environment
ectoparasite a parasite that feeds from the exterior of its host
egg mass a group of eggs held together by a gelatinous matrix
ELISA *see* **enzyme-linked immunosorbent assay (ELISA)**
emergence the growth of a seedling shoot through the surface of the soil
endemic native to a particular place; characterized by a low and steady level of natural disease occurrence
enzyme a protein that catalyzes a specific biochemical reaction
enzyme-linked immunosorbent assay (ELISA) a serological test in which the sensitivity of the reaction is increased by attaching an enzyme that produces a colored product to one of the reactants
epidemic an increase of disease in a population; a general and serious outbreak of disease
epidermis (adj. **epidermal**) the surface layer of cells of leaves and other plant parts
eradicant a chemical used to eliminate a pathogen from a host or an environment
eradication control of plant disease by elimination of the pathogen after it has become established or elimination of the plants that carry the pathogen
exudate a liquid excreted or discharged from diseased tissues, from roots and leaves, or by fungi

fallow cultivated land kept free from a crop or weeds during the normal growing season
fecund (n. **fecundity**) fruitful in offspring or vegetation; prolific
flaccid wilted, lacking turgor
flagellum (pl. **flagella**) a hairlike, whiplike, or tinsel-like appendage of a motile cell, bacterium, or zoospore, providing locomotion
focus (pl. **foci**) a small area of diseased plants within a population
foliar pertaining to leaves
friable easily crumbled into small aggregates or powder, as of some soils
fumigant (v. **fumigate**) a gas or volatile substance used to kill or inhibit the growth of microorganisms or other pests

fungicide (adj. **fungicidal**) a chemical or physical agent that kills or inhibits the growth of fungi

gall abnormal swelling or localized outgrowth, often roughly spherical, produced by a plant as a result of attack by a fungus, bacterium, nematode, insect, or other organism (syn. **tumor**)
genetic relating to heredity; pertaining to heritable characteristics
genus (pl. **genera**) a taxonomic category comprising a group of closely related (structurally or phylogenetically) species (the genus name, or generic name, is the first name in a Latin binomial)
germinate (n. **germination**) to begin growth of a seed, spore, sclerotium, or other reproductive body
girdle to circle and cut through a stem or the bark and outer few rings of wood, disrupting the phloem and xylem

haulm stems or stalks, collectively
herbaceous pertaining to primary, soft, nonwoody tissue of a plant or plant part; having the characteristics of an herb
herbicide a chemical that kills or inhibits the growth of plants, such as a weed or grass killer
host a living plant attacked by or harboring a living parasite and providing the invader with part or all of its nourishment
host range the range of plants on which an organism, particularly a parasite, feeds
hyaline transparent or nearly so; translucent; colorless
hybrid (v. **hybridize**) the offspring of two individuals of different genotypes
hypha (pl. **hyphae**; adj. **hyphal**) a tubular filament of a fungal thallus or mycelium, the basic structural unit of a fungus
hypocotyl the portion of the stem below the cotyledons and above the root

immune not capable of being infected by a given pathogen
in vitro in glass, on artificial media, or in an artificial environment; outside a host
in vivo within a living organism
infection the process in which an organism enters, invades, or penetrates and establishes a parasitic relationship with a host plant
infection court a site in or on a host plant where infection can occur
infectious pertaining to disease capable of spreading from plant to plant
infective with reference to a pathogen, able to attack a host and cause infection; with reference to a vector, carrying or containing a pathogen and able to transfer it to a host plant
infest (n. **infestation**) to attack as a pest (used especially of insects and nematodes); to contaminate, as with microorganisms; to be present in large numbers
inflorescence a flower or flower cluster
initial inoculum inoculum, usually from an overwintering source, that initiates disease in the field, as opposed to inoculum that spreads disease during the season (syn. **primary inoculum**)
injury damage caused by transitory interaction with an agent such as an insect, chemical, or unfavorable environmental condition
inoculate (n. **inoculation**) to place inoculum in an infection court; to insert a pathogen into healthy tissue
inoculum (pl. **inocula**) a pathogen or its parts capable of causing infection when transferred to a favorable location
intercrop to grow two or more crops simultaneously on the same area of land
internode (adj. **internodal**) the portion of a stem between two successive nodes
interveinal between veins of a leaf
isolate (n.) a culture or subpopulation of a microorganism separated from its parent population and maintained under controlled conditions; (v.) to remove from soil or host material and grow in pure culture

juvenile an immature form that appears similar to but is usually smaller than the adult and is not sexually mature, as in insect species with gradual metamorphosis or nematodes

knot a gall; a localized abnormal swelling

larva (pl. **larvae**) an immature stage of certain animals (especially insects) that undergo complete metamorphosis; the wormlike or caterpillar stage of the life cycle of such insects
latent present but not manifested or visible, as in a symptomless infection
latent infection infection unaccompanied by visible symptoms
leaf spot a plant disease lesion typically restricted in development in the leaf after reaching a characteristic size
leafhopper any of various species of mobile insects with sucking mouthparts in the order Homoptera
leaflet one of the separate blades or divisions of a compound leaf
legume a simple, dry, dehiscent fruit developed from a simple pistil and splitting at maturity along two seams; any member of the plant family Fabaceae (formerly Leguminosae)
lenticel a natural opening in the surface of a stem, tuber, fruit, or root, permitting gas exchange
lesion a localized diseased area or wound
life cycle the series of cyclical stages in the growth and development of an organism

mating types in heterothallic fungi, compatible strains necessary for sexual reproduction, usually designated + and – or A and B
mechanical injury injury of a plant part by abrasion, mutilation, or wounding
mechanical transmission spread or introduction of inoculum to an infection court (wounding) by human manipulation, accompanied by physical disruption of host tissues
meristem culture aseptic culture of a plant or plant part from a portion of the meristem
microclimate weather conditions on a small scale, as at the surface of a plant or within a planting
microorganism an organism small enough that as an individual it can be seen only with the aid of microscope (syn. **microbe**)
microsclerotium a microscopic, dense aggregate of darkly pigmented, thick-walled hyphal cells
midrib the central, thickened vein of a leaf
migratory moving from place to place on a plant or from plant to plant when feeding
molt to shed a cuticle or body encasement during a phase of growth
monoculture the cultivation of plants of the same species in close proximity, with few or no other plant species present
mosaic a disease symptom characterized by nonuniform coloration, with intermingled normal and light green or yellowish patches, usually caused by a virus (often used interchangeably with **mottle**)
motile capable of self-propulsion by means of flagella, cilia, or amoeboid movement
mottle a disease symptom characterized by an irregular pattern of light and dark areas in plant tissue, usually caused by a virus (often used interchangeably with **mosaic**)
mulch a layer of material, such as organic matter or plastic sheeting, applied to the surface of the soil for purposes such as retention of water and inhibition of weeds
mutation (n. **mutant**) an abrupt heritable or genetic change in a gene or an individual as a result of an alteration in a gene or chromosome or an increase in chromosome number
mycelium (pl. **mycelia**; adj. **mycelial**) the mass of hyphae constituting the body (thallus) of a fungus
mycoparasite a fungus that attacks another fungus

necrosis (adj. **necrotic**) death of cells or tissue, usually accompanied by blackening or browning

nematicide an agent, usually a chemical, that kills nematodes
nematode any of various unsegmented roundworms parasitic on plants or animals or free-living in soil or water
noninfectious disease any disease caused by an abiotic agent and not transmissible from one plant to another
nonpersistent transmission a type of virus transmission in which the virus is acquired and transmitted by a vector after a short feeding period and is retained by the vector for only a short time (syn. **stylet-borne transmission**)
nymph the juvenile stage in the life cycle of an insect with incomplete metamorphosis, superficially resembling the adult

Oomycetes (adj. **oomycetous**) a class of fungus-like organisms, typically with nonseptate mycelium, asexual sporangia and zoospores, and sexual oospores
oospore thick-walled, sexually-derived resting spore of organisms in the class Oomycetes
ooze a mass of bacterial cells mixed with host fluids
overwinter to survive or persist through the winter
oviposit to deposit or lay eggs with an ovipositor
ozone a highly reactive form of oxygen (O_3), capable of injuring plants

parasite (adj. **parasitic**) an organism that lives in intimate association with another organism on which it depends for its nutrition, but not necessarily as a pathogen
pathogen (adj. **pathogenic**) any disease-producing organism or agent
pathovar (abbr. **pv.**) a subdivision of a plant-pathogenic bacterial species, defined by host range (a pathovar of a bacterial species is analogous to a *forma specialis* of a fungal species)
penetration initial invasion of a host by a pathogen
perennial a plant that survives for several to many years (*see* **annual; biennial**)
persistent transmission a type of virus transmission in which the virus is acquired and transmitted by a vector after a relatively long feeding period and remains transmissible for a prolonged period while in association with the vector (syn. **circulative transmission**)
pest any organism that damages plants or plant products
pesticide a chemical used to control pests
petiole the stalk portion of a leaf
pH the negative logarithm of the effective hydrogen ion concentration, a measure of acidity (pH 7 is neutral; values less than pH 7 are acidic; values greater than pH 7 are alkaline)
phloem food-conducting, food-storing tissue in the vascular system of roots, stems, and leaves
photosynthate a product of photosynthesis
photosynthesis the manufacture of carbohydrates from carbon dioxide and water in the presence of chlorophyll, using light energy and releasing oxygen
predispose (n. **predisposition**) to make prone to infection and disease
primary inoculum inoculum, usually from an overwintering source, that initiates disease in the field, as opposed to inoculum that spreads disease during the season (syn. **initial inoculum**)
protectant an agent, usually a chemical, applied to a plant surface in advance of exposure to a pathogen, to prevent infection
protein any of numerous nitrogen-containing organic compounds composed of amino acids
protoplasm the living contents of a cell
protoplast the living contents of a cell, exclusive of the cell wall
proximal situated toward or near the point of attachment or main body
pupa (pl. **pupae**; v. **pupate**) the quiescent stage between the larva and the adult of certain insects
pv. *see* **pathovar**

quarantine legislative control of the transport of plants or plant parts to prevent the spread of pests or pathogens

race a subgroup or biotype within a species or variety, distinguished from other races by virulence, symptom expression, or host range, but not by morphology
resistant (n. **resistance**) possessing properties that prevent or impede disease development (*see* **susceptible**)
respiration a series of chemical reactions that make energy available through oxidation of carbohydrates and fat
resting spore a spore, often thick-walled, that can remain alive in a dormant state for some time, later germinating and capable of initiating infection
rhizome a mostly horizontal, jointed, fleshy, often elongated, usually underground stem
rhizosphere the microenvironment in the soil immediately surrounding plant roots
ringspot a disease symptom characterized by yellowish or necrotic rings enclosing green tissue, as in some plant diseases caused by viruses
RNA ribonucleic acid, any of several nucleic acids composed of repeating units of ribose (a sugar), a phosphate group, and a purine (adenine or guanine) or a pyrimidine (uracil or cytosine) base, transcribed from DNA, and involved in translation to proteins
rogue to remove and destroy individual plants that are diseased, infested by insects, or otherwise undesirable
rosette a disease symptom characterized by short, bunchy growth as a result of a shortening of the internodes with no comparable reduction in leaf size
rot softening, discoloration, and often disintegration of plant tissue as a result of fungal or bacterial infection
rotation cultivation of different kinds of crops in succession in the same field
russet a brownish, roughened area on the skin of fruit or tubers, resulting from cork formation

sanitation destruction or removal of infected and infested plants or plant parts; decontamination of tools, equipment, containers, work space, hands, etc.
saprophyte (adj. **saprophytic**) an organism that obtains nourishment from nonliving organic matter (syn. **saprobe**)
scab a roughened, crustlike diseased area on the surface of a plant organ
sclerotium (pl. **sclerotia**) a vegetative resting body of a fungus, composed of a compact mass of hyphae with or without included host tissue, usually with a darkened rind
scorch any symptom that suggests the action of flame or fire on the affected tissue, often at the margins of leaves
secondary infection infection resulting from the spread of infectious material from a primary infection or from a secondary infection without an intervening inactive period
secondary inoculum inoculum produced by an infection established previously in the same growing season
secondary organism an organism that multiplies in already diseased tissue but is not the primary pathogen
seed treatment application of a biological agent, chemical substance, or physical treatment to seed, to protect the seed or the resulting plant from pathogens or to stimulate germination or plant growth
seedborne carried on or in a seed
selective medium a culture medium containing substances that inhibit or prevent the growth of certain species of microorganisms
senesce (adj. **senescent**, n. **senescence**) to decline, as with maturation, age, or disease stress
serology (adj. **serologic, serological**) a method using the specificity of the antigen-antibody reaction for the detection and identification of antigenic substances and the organisms that carry them

sign an indication of disease from direct observation of a pathogen or its parts on a host plant (*see* **symptom**)
soft rot a mushy disintegration of fleshy fruit or vegetables as a result of a bacterial or fungal infection
soil drench application of a solution or suspension of a chemical to the soil, especially a pesticide to control soilborne pathogens
soilborne carried on or beneath the soil surface
solarization a disease control practice in which soil is covered with polyethylene sheeting to increase the heating of the soil by sunlight and thus control soilborne plant pathogens
sp. (pl. **spp.**) species (the abbreviations are used only after a genus name; the singular form is used to refer to an undetermined species; the plural form is used to refer to two or more species without naming them individually)
sporangiophore the sporangium-bearing body of certain fungi
sporangiospore a nonmotile, asexual spore borne in a sporangium
sporangium (pl. **sporangia**) a saclike fungal structure, the entire contents of which are converted into an indefinite number of asexual spores
spore a reproductive structure of fungi and some other organisms, containing one or more cells; a bacterial cell modified to survive an adverse environment
sporulate to produce spores
spot a limited necrotic area occurring as a disease symptom on leaves, flowers, or stems
sterilization (adj. **sterilized**) the total destruction of living organisms by various means, such as heat, chemicals, or irradiation
stolon a slender, horizontal stem that grows close to the soil surface, also called a runner; in fungi, a hypha that grows horizontally along the surface
stoma (pl. **stomata**; adj. **stomatal**) a structure composed of two guard cells and the opening between them in the epidermis of a leaf or stem, functioning in gas exchange, also called a stomate
strain a distinct form of a species or virus, differing biologically, physically, or chemically from other forms of the species or virus
stunting reduction in height of a vertical axis resulting from a progressive reduction in the length of successive internodes or a decrease in their number
stylet the stiff, slender, hollow feeding organ of plant-parasitic nematodes or sap-sucking insects, such as aphids and leafhoppers
stylet-borne transmission a type of virus transmission in which the virus is acquired and transmitted by a vector after a short feeding period and is retained by the vector for only a short time (syn. **nonpersistent transmission**)
substrate the substance on which an organism lives or from which it obtains nutrients; a chemical substance acted upon, often by an enzyme
sunscald injury of plant tissues burned or scorched by direct sun
susceptible (n. **susceptibility**) prone to develop disease when infected by a pathogen (see **resistant**)
symptom the visible reactions or alterations of a plant as a result of a disease
symptomless carrier a plant that, although infected with a pathogen (usually a virus), has no obvious symptoms
systemic pertaining to a disease in which the pathogen (or a single infection) spreads generally throughout the plant; pertaining to chemicals that spread internally through the plant

tissue a group of cells, usually of similar structure, that perform the same or related functions
tissue culture an *in vitro* method of propagating healthy cells from plant tissues
toxin a poisonous substance of biological origin
translocation movement of water, nutrients, chemicals, or food materials within a plant
translucent clear enough to allow the transmission of light
transmit (n. **transmission**) to spread or transfer, as in spreading an infectious pathogen from plant to plant or from one plant generation to another
transpiration water loss by evaporation from leaf surfaces and through stomata
tuber an underground stem adapted for storage, typically produced at the end of a stolon
turgid swollen; inflated; plump or swollen as a result of internal water pressure

vascular pertaining to the fluid-conducting tissues (xylem and phloem) in plants
vascular wilt disease a xylem disease that disrupts normal uptake of water and minerals, resulting in wilting and yellowing of foliage
vector a living organism (e.g., insect, mite, bird, higher animal, nematode, parasitic plant, human) able to carry and transmit a pathogen; in genetic engineering, a cloning vehicle, or self-replicating DNA molecule, such as a plasmid or virus, used to introduce a fragment of foreign DNA into a host cell
vegetative pertaining to the somatic or asexual parts of a plant, which are not involved in sexual reproduction
vegetative propagation asexual reproduction; in plants, the use of cuttings, bulbs, tubers, and other vegetative plant parts to grow new plants
vein a small vascular bundle in a leaf
vein banding a virus disease symptom in which regions along veins are darker green than tissue between veins
vein clearing disappearance of green color in or around leaf veins
vermiform worm-shaped
viable (n. **viability**) alive; capable of germination (of seeds, fungus spores, sclerotia, etc.); capable of growth
viroid an infectious, unencapsidated (naked), circular, single-stranded RNA
viruliferous virus-laden (usually used with reference to insects or nematodes as vectors)
virus a submicroscopic, intracellular, obligate parasite consisting of a core of infectious nucleic acid (either RNA or DNA) usually surrounded by a protein coat
volunteer a self-set plant; a plant seeded by chance

water-soaked characterized by water soaking, a disease symptom in which plant tissues or lesions appear wet, dark, and usually sunken and translucent
whorl a circular arrangement of like parts
wild type the phenotype characteristic of the majority of individuals of a species under natural conditions
wilt drooping of leaves and stems from lack of water due to an inadequate water supply or excessive transpiration; a vascular disease that interrupts normal water uptake
witches'-broom a disease symptom in which many weak shoots arise at or close to the same point, producing abnormal massed, brushlike growth

xylem the water- and mineral-conducting, food-storing, supporting tissue of a plant

zoospore a fungal spore with flagella, capable of locomotion in water

Sources of Additional Information

General References

Stark, J. C., and Love, S. L., eds. 2003. Potato Production Systems. University of Idaho Agricultural Communications, Moscow.

Stevenson, W. R., Loria, R., Franc, G. D., and Weingartner, D. P., eds. 2001. Compendium of Potato Diseases. 2nd ed. American Phytopathological Society, St. Paul, MN.

Western Regional IPM Project. 2006. Integrated Pest Management for Potatoes in the Western United States. 2nd ed. Publ. 3316. University of California, Division of Agriculture and Natural Resources, Oakland.

Aphids

Bacon, O. G., Burton, V. E., McLean, D. L., James, R. H., Riley, W. D., Baghott, K. G., and Kinsey, M. G. 1976. Control of the green peach aphid and its effect on the incidence of potato leaf roll virus. Journal of Economic Entomology 69:410–414.

Byrne, D. N., and Bishop, G. W. 1979. Relationship of green peach aphid numbers to spread of potato leaf roll virus in southern Idaho. Journal of Economic Entomology 72:809–811.

DiFonzo, C. D., Ragsdale, D. W., and Radcliffe, E. B. 1995. Potato leafroll virus spread in differentially resistant potato cultivars under varying aphid densities. American Potato Journal 72:119–132.

Flanders, K. L., Radcliffe, E. B., and Ragsdale, D. W. 1991. Potato leafroll virus spread in relation to densities of green peach aphid (Homoptera: Aphididae): Implications for management thresholds for Minnesota seed potatoes. Journal of Economic Entomology 84:1028–1036.

Mowry, T. M. 2001. Green peach aphid (Homoptera: Aphididae) action thresholds for controlling the spread of potato leafroll virus in Idaho. Journal of Economic Entomology 94:1332–1339.

Sexson, D. L., Wyman, J. A., Radcliffe, E. B., Hoy, C. W., Ragsdale, D. W., and Dively, G. 2005. Potato. Pages 92–108 in: Vegetable Insect Management. R. Foster and B. R. Flood, eds. 2nd ed. Meister Media Worldwide, Willoughby, OH.

Bruise Prevention

Potato Bruise Prevention Video Series. Tape 1, The Harvester. 1985. Tape 2, Harvester Chain Adjustment. 1989. Tape 3, Handling. 1990. Tape 4, Reducing Bruising in Fresh-Pack Warehouses. 1996. University of Idaho Agricultural Communications, Moscow.

Economics

Casavant, K. L., Infanger, C. L., and Bridges, D. E. 1999. Agricultural Economics and Management. Prentice-Hall, Saddle River, NJ.

Organic Production

Daniels, C. H., and Fults, J. 2002. Fact sheet for vinegar/acetic acid recommendations. Oregon Department of Agriculture Fact Sheet PIC-01002. http://oregonstate.edu/dept/nursery-weeds/weedspeciespage/acetic_acid_factsheet.pdf

Frazier, M. J., Olsen, N., and Kleinkopf, G. 2004. Organic and alternative methods for potato sprout control in storage. University of Idaho Current Information Series No. 1120. http://info.ag.uidaho.edu/pdf/CIS/CIS1120.pdf

Hunter, C. D. 1997. Suppliers of Beneficial Organisms in North America. California Environmental Protection Agency, Sacramento. Online, California Department of Pesticide Regulation, Pest Management Analysis and Planning Program, http://www.cdpr.ca.gov/docs/ipminov/bensuppl.htm

Organic Materials Review Institute. 2007. OMRI standards updates. http://www.omri.org

Stoltz, R. L., Baird, C. R., and Alvarez, J. M. 2002. Beneficial organisms associated with Pacific Northwest crops. PNW 343. Pacific Northwest Cooperative Extension Publications: Washington State University Cooperative Extension, Pullman; Oregon State University Extension Service, Corvallis; University of Idaho Cooperative Extension System, Moscow.

Weeds

Eberlein, C. V., Patterson, P. E., Guttieri, M. J., and Stark, J. C. 1997. Efficacy and economics of cultivation for weed control in potato (*Solanum tuberosum*). Weed Technology 11:257–264.

Hutchinson, P. J. S. 2005. Potatoes. Pages 170–187 in: 2005 Pacific Northwest Weed Control Handbook. R. D. William et al., eds. Pacific Northwest Cooperative Extension Publications: Washington

State University Cooperative Extension, Pullman; Oregon State University Extension Service, Corvallis; University of Idaho Cooperative Extension System, Moscow.

Hutchinson, P. J. S., and Eberlein, C. V. 2003. Weed management. Pages 240–283 in: Potato Production Systems. J. C. Stark and S. L. Love, eds. University of Idaho Agricultural Communications, Moscow.

Mallory-Smith, C., Hyslop, G. R., Thill, D., Colquhoun, J., and Morishita, D. 2002. Herbicide-resistant weeds and their management. PNW 437. Pacific Northwest Cooperative Extension Publications: Washington State University Cooperative Extension, Pullman; Oregon State University Extension Service, Corvallis; University of Idaho Cooperative Extension System, Moscow.

Mallory-Smith, C. A., and Retzinger, R. J. 2003. Revised classification of herbicides by site of action for weed resistance management strategies. Weed Technology 17:605–619.

Whitson, T. D., ed. 2000. Weeds of the West. Western Society of Weed Science, Newark, CA.

VanGessel, M. J., and Renner, K. A. 1990. Redroot pigweed (*Amaranthus retroflexus*) and barnyardgrass (*Echinochloa crus-galli*) interference in potatoes (*Solanum tuberosum*). Weed Science 38:338–343.

Potato Production Web Sites

The following list presents a sample of Web sites containing information on potatoes and potato production. Some of them are generalist sites, and others contain cultural and environmental information for specific regions. The longevity of these sites is not guaranteed, but they are all from solid sources. To obtain specific information, follow the links provided at each site.

Cultivars

Canadian Food Inspection Agency, http://www.inspection.gc.ca/ (search "potatoes")

European Cultivated Potato Database, http://www.europotato.org/

Potato Association of America, Variety Images and Descriptions, http://www.umaine.edu/paa/var.htm

Nutrient Content of Tubers

U.S. Department of Agriculture, National Agricultural Library, Food and Nutrition Information Center, http://fnic.nal.usda.gov/

Organic Production

U.S. Department of Agriculture, Agricultural Marketing Service, National Organic Program, http://www.ams.usda.gov/nop/

Pesticide Labels and Material Safety Data Sheets

Crop Data Management Systems, http://www.cdms.net/LabelsMsds/LMDefault.aspx

Pests and Diseases

Cornell University, Department of Plant Pathology, Vegetable MD Online, http://vegetablemdonline.ppath.cornell.edu/

University of Idaho Pest Management Center, http://www.ag.uidaho.edu/pmc/programs/ipm_links.htm

Michigan State University, Michigan Potato Diseases, http://www.potatodiseases.org/

Washington State University, Department of Plant Pathology, Potato Diseases, http://classes.plantpath.wsu.edu/dajohn/

University of Wisconsin, Department of Plant Pathology, Vegetable Pathology Web Site, http://www.plantpath.wisc.edu/wivegdis/

Regional Sites

Canadian Food Inspection Agency, http://www.inspection.gc.ca/ (search "potatoes")

Colorado State University, VegNet, http://www.colostate.edu/orgs/VegNet/vegnet/spuds.html

University of Florida, Institute of Food and Agricultural Sciences, Electronic Data Information Source, http://edis.ifas.ufl.edu/

University of Maine Cooperative Extension, Potato Program, http://www.umaine.edu/umext/potatoprogram/

University of Nebraska, Potato Education Guide, http://www.panhandle.unl.edu/potato/

North Dakota State University Extension Service, http://www.ag.ndsu.nodak.edu/

Oregon State University, Crop and Soil Science Department, Potato Information Exchange, http://oregonstate.edu/potatoes/

Washington State University, Department of Horticulture and Landscape Architecture, Potato Information and Exchange, http://potatoes.wsu.edu/

University of Wisconsin, Integrated Pest and Crop Management, http://ipcm.wisc.edu/

Index